Stochastic Processes with Applications

Stochastic Processes with Applications

Special Issue Editors

Antonio Di Crescenzo
Claudio Macci
Barbara Martinucci

MDPI • Basel • Beijing • Wuhan • Barcelona • Belgrade

Special Issue Editors
Antonio Di Crescenzo
University of Salerno
Italy

Claudio Macci
Università di Roma Tor Vergata
Italy

Barbara Martinucci
University of Salerno
Italy

Editorial Office
MDPI
St. Alban-Anlage 66
4052 Basel, Switzerland

This is a reprint of articles from the Special Issue published online in the open access journal *Mathematics* (ISSN 2227-7390) from 2017 to 2018 (available at: https://www.mdpi.com/journal/mathematics/special_issues/Stochastic_Processes_Applications).

For citation purposes, cite each article independently as indicated on the article page online and as indicated below:

LastName, A.A.; LastName, B.B.; LastName, C.C. Article Title. *Journal Name* **Year**, *Article Number*, Page Range.

ISBN 978-3-03921-728-1 (Pbk)
ISBN 978-3-03921-729-8 (PDF)

Cover image courtesy of Antonio Di Crescenzo.

Contents

About the Special Issue Editors . . . . **vii**

Preface to "Stochastic Processes with Applications" . . . . **ix**

Mario Abundo
The Randomized First-Hitting Problem of Continuously Time-Changed Brownian Motion
Reprinted from: *Mathematics* **2018**, *6*, 91, doi:10.3390/math6060091 . . . . **1**

Ravi Agarwal, Snezhana Hristova, Donal O'Regan and Peter Kopanov
Stability Analysis of Cohen–Grossberg Neural Networks with Random Impulses
Reprinted from: *Mathematics* **2018**, *6*, 144, doi:10.3390/math6090144 . . . . **11**

Giuseppina Albano and Virginia Giorno
On Short-Term Loan Interest Rate Models: A First Passage Time Approach
Reprinted from: *Mathematics* **2018**, *6*, 70, doi:10.3390/math6050070 . . . . **23**

Giacomo Ascione, Nikolai Leonenko and Enrica Pirozzi
Fractional Queues with Catastrophes and Their Transient Behaviour
Reprinted from: *Mathematics* **2018**, *6*, 159, doi:10.3390/math6090159 . . . . **35**

G. Ayyappan and S. Karpagam
An $M^{[X]}/G(a,b)/1$ Queueing System with Breakdown and Repair, Stand-By Server, Multiple Vacation and Control Policy on Request for Re-Service
Reprinted from: *Mathematics* **2018**, *6*, 101, doi:10.3390/math6060101 . . . . **61**

Raquel Caballero-Águila, Aurora Hermoso-Carazo and Josefa Linares-Pérez
Fusion Estimation from Multisensor Observations with Multiplicative Noises and Correlated Random Delays in Transmission
Reprinted from: *Mathematics* **2017**, *5*, 45, doi:10.3390/math5030045 . . . . **79**

Davide De Gaetano
Forecast Combinations in the Presence of Structural Breaks: Evidence from U.S. Equity Markets
Reprinted from: *Mathematics* **2018**, *6*, 34, doi:10.3390/math6030034 . . . . **99**

Antonio Di Crescenzo and Patrizia Di Gironimo
Stochastic Comparisons and Dynamic Information of Random Lifetimes in a Replacement Model
Reprinted from: *Mathematics* **2018**, *6*, 204, doi:10.3390/math6100204 . . . . **118**

Antonio Di Crescenzo, Virginia Giorno, Balasubramanian Krishna Kumar and Amelia G. Nobile
A Time-Non-Homogeneous Double-Ended Queue with Failures and Repairs and Its Continuous Approximation
Reprinted from: *Mathematics* **2018**, *6*, 81, doi:10.3390/math6050081 . . . . **131**

Maria Gamboa and Maria Jesus Lopez-Herrero
On the Number of Periodic Inspections During Outbreaks of Discrete-Time Stochastic SIS Epidemic Models
Reprinted from: *Mathematics* **2018**, *6*, 128, doi:10.3390/math6080128 . . . . **154**

Rita Giuliano and Claudio Macci
Large Deviation Results and Applications to the Generalized Cramér Model
Reprinted from: *Mathematics* **2018**, *6*, 49, doi:10.3390/math6040049 . . . . **167**

Antonio Gómez-Corral, Martín López-García
A Within-Host Stochastic Model for Nematode Infection
Reprinted from: *Mathematics* **2018**, *6*, 143, doi:10.3390/math6090143 . . . . **179**

Alexander I. Nazarov and Yakov Yu. Nikitin
On Small Deviation Asymptotics In L_2 of Some Mixed Gaussian Processes
Reprinted from: *Mathematics* **2018**, *6*, 55, doi:10.3390/math6040055 . . . . **203**

Luca Pratelli and Pietro Rigo
Convergence in Total Variation to a Mixture of Gaussian Laws
Reprinted from: *Mathematics* **2018**, *6*, 99, doi:10.3390/math6060099 . . . . **212**

Patricia Román-Román, Juan José Serrano-Pérez and Francisco Torres-Ruiz
Some Notes about Inference for the Lognormal Diffusion Process with Exogenous Factors
Reprinted from: *Mathematics* **2018**, *6*, 85, doi:10.3390/math6050085 . . . . **226**

Anna Sinitcina, Yacov Satin, Alexander Zeifman, Galina Shilova, Alexander Sipin, Ksenia Kiseleva, Tatyana Panfilova, Anastasia Kryukova, Irina Gudkova and Elena Fokicheva
On the Bounds for a Two-Dimensional Birth-Death Processwith Catastrophes
Reprinted from: *Mathematics* **2018**, *6*, 80, doi:10.3390/math6050080 . . . . **239**

Somayeh Zarezadeh, Somayeh Ashrafi and Majid Asadi
Network Reliability Modeling Based on a Geometric Counting Process
Reprinted from: *Mathematics* **2018**, *6*, 197, doi:10.3390/math6100197 . . . . **256**

About the Special Issue Editors

Antonio Di Crescenzo, Full Professor of Probability and Mathematical Statistics at the University of Salerno, received his degree in Information Science in 1988 from University of Salerno, and Ph.D. degree in Applied Mathematics and Informatics in 1996 from University of Naples Federico II. He has served as Research Associate at the University of Naples Federico II (1991–1998) and Associate Professor firstly at the University of Basilicata, Potenza (1998–2001) followed by the University of Salerno (2001–2018). He has published more than 90 research papers in probability and stochastic models. He co-organized several international meetings and participated in more than 60 international scientific meetings, especially regarding applications of probability. He is member of the Editorial Board of various journals, including *Mathematics* (MDPI) and *Methodology and Computing in Applied Probability* (Springer). His research interests include stochastic modeling, theory and simulation of stochastic processes with applications to biomathematics, queueing systems, reliability theory, aging notions, and information measures.

Claudio Macci, Associate Professor, received his M.S. degree in Mathematics in 1993 from Sapienza University of Rome, and Ph.D. degree in Mathematics from Tor Vergata University of Rome in 1998. He served as Assistant Professor at University of Turin from 1999 to 2003 before moving to Tor Vergata University of Rome, where has served as Associate Professor since 2013. He has published more than 80 papers in probability and statistics. His main research interest is in large deviations theory. Other research interests include Bayesian statistics, ruin probabilities in insurance, and fractional processes.

Barbara Martinucci Assistant Professor, received her M.S. degree in Mathematics in 2000 and earned her Ph.D. in Applied Mathematics and Informatics in 2005 from University of Naples Federico II. She has been serving as Assistant Professor of Probability and Mathematical Statistics at the Department of Mathematics of the University of Salerno since her appointment in November 2008. Her research interests lie in the general area of stochastic processes with a focus on random evolutions with finite velocity.

Preface to "Stochastic Processes with Applications"

The aim of the Special Issue "Stochastic Processes with Applications" is to present a collection of original papers covering recent advances in the theory and application of stochastic processes. The focus is especially on applications of stochastic processes as models of dynamic phenomena in various research areas, such as economics, statistical physics, queuing theory, biology, theoretical neurobiology, and reliability theory.

The volume contains 17 articles collected from June 2017 to September 2018. We appreciate that the contributions originate from three continents. Indeed, the geographical distribution of the 46 authors is as follows: Italy (13), Russia (12), Spain (9), India (3), Iran (3), Bulgaria (2), UK (2), Ireland (1), and USA (1). It is a pleasure to note that some of the authors that contributed to this volume subsequently served as Editors of other Special Issues of *Mathematics* on similar topics.

The stochastic processes treated in this book range within quite wide areas, such as diffusion and Gaussian processes, stochastic volatility models, epidemic models, neural networks, counting processes, fractional processes, Markov chains, and birth–death processes. Some investigations also involve stochastic processes describing phenomena subject to catastrophes and repairs, and related first-passage-time problems. Asymptotic results are also treated, as related to large and small deviations and convergence problems for mixed Gaussian laws. Various contributions are based on simulation tools or on statistical methods, such as maximum likelihood estimation and hypothesis testing. Computational methods are also adopted and are mainly based on recursive algorithms for estimation or forecasting. Some articles of the volume deal with results concerning applied fields, with special reference to mathematical finance, biomathematics, queueing, and reliability theory.

We are confident that the research presented herein will attract the interest of *Mathematics* readers and of the large community of scholars active in the realm of stochastic processes and applications.

Finally, we would like to express our deep gratitude to the researchers who contributed to this volume. Moreover, our warm thanks go to those who reviewed the manuscripts. The refereeing activity was essential to the realization of this Special Issue.

Antonio Di Crescenzo, Claudio Macci, Barbara Martinucci
Special Issue Editors

mathematics

MDPI

Article

The Randomized First-Hitting Problem of Continuously Time-Changed Brownian Motion

Mario Abundo

Dipartimento di Matematica, Università Tor Vergata, 00133 Rome, Italy; abundo@mat.uniroma2.it; Tel.: +390672594627

Received: 4 April 2018; Accepted: 25 May 2018; Published: 28 May 2018

Abstract: Let $X(t)$ be a continuously time-changed Brownian motion starting from a random position η, $S(t)$ a given continuous, increasing boundary, with $S(0) \geq 0$, $P(\eta \geq S(0)) = 1$, and F an assigned distribution function. We study the inverse first-passage time problem for $X(t)$, which consists in finding the distribution of η such that the first-passage time of $X(t)$ below $S(t)$ has distribution F, generalizing the results, valid in the case when $S(t)$ is a straight line. Some explicit examples are reported.

Keywords: first-passage time; inverse first-passage problem; diffusion

---

1. Introduction

This brief note is a continuation of [1,2]. Let $\sigma(t)$ be a regular enough non random function, and let $X(t) = \eta + \int_0^t \sigma(s)dB_s$, where B_t is standard Brownian motion (BM) and the initial position η is a random variable, independent of B_t. Suppose that the quadratic variation $\rho(t) = \int_0^t \sigma^2(s)ds$ is increasing and $\rho(+\infty) = \infty$, then there exists a standard BM $\widetilde{B}$ such that $X(t) = \eta + \widetilde{B}(\rho(t))$, namely $X(t)$ is a continuously time-changed BM (see e.g., [3]). For a continuous, increasing boundary $S(t)$, such that $P(\eta \geq S(0)) = 1$, let

$$\tau = \tau_S = \inf\{t > 0 : X(t) \leq S(t)\} \tag{1}$$

be the first-passage time (FPT) of $X(t)$ below S. We assume that τ is finite with probability one and that it possesses a density $f(t) = \frac{dF(t)}{dt}$, where $F(t) = P(\tau \leq t)$. Actually, the FPT of continuously time-changed BM is a well studied problem for constant or linear boundary and a non-random initial value (see e.g., [4–6]).

Assuming that $S(t)$ is increasing, and $F(t)$ is a continuous distribution function, we study the following inverse first-passage-time (IFPT) problem:

given a distribution F, find the density g of η (if it exists) for which it results $P(\tau \leq t) = F(t)$.

The function g is called a solution to the IFPT problem. This problem, also known as the generalized Shiryaev problem, was studied in [1,2,7,8], essentially in the case when $X(t)$ is BM and $S(t)$ is a straight line; note that the question of the existence of the solution is not a trivial matter (see e.g., [2,7]). In this paper, by using the properties of the exponential martingale, we extend the results to more general boundaries S.

The IFPT problem has interesting applications in mathematical finance , in particular in credit risk modeling, where the FPT represents a default event of an obligor (see [7]) and in diffusion models for neural activity ([9]).

Notice, however, that another type of inverse first-passage problem can be considered: it consists in determining the boundary shape S, when the FPT distribution F and the starting point η are assigned (see e.g., [10–13]).

The paper is organized as follows: Section 2 contains the main results, in Section 3 some explicit examples are reported; Section 4 is devoted to conclusions and final remarks.

2. Main Results

The following holds:

Theorem 1. *Let be $S(t)$ a continuous, increasing boundary with $S(0) \geq 0$, $\sigma(t)$ a bounded, non random continuous function of $t > 0$, and let $X(t) = \eta + \int_0^t \sigma(s)dB_s$ be the integral process starting from the random position $\eta \geq S(0)$; we assume that $\rho(t) = \int_0^t \sigma^2(s)ds$ is increasing and satisfies $\rho(+\infty) = +\infty$. Let F be the probability distribution of the FPT τ_S of X below the boundary S (τ_S is a.s. finite by virtue of Remark 3). We suppose that the r.v. η admits a density $g(x)$; for $\theta > 0$, we denote by $\widehat{g}(\theta) = E(e^{-\theta\eta})$ the Laplace transform of g.*

Then, if there exists a solution to the IFPT problem for X, the following relation holds:

$$\widehat{g}(\theta) = \int_0^{+\infty} e^{-\theta S(t) - \frac{\theta^2}{2}\rho(t)} dF(t). \tag{2}$$

Proof. The process $X(t)$ is a martingale, we denote by $\mathcal{F}_t$ its natural filtration. Thanking to the hypothesis, by using the Dambis, Dubins–Schwarz theorem (see e.g., [3]), it follows that the process $\widetilde{B}(t) = X(\rho^{-1}(t))$ is a Brownian motion with respect to the filtration $\mathcal{F}_{\rho^{-1}(t)}$; so the process $X(t)$ can be written as $X(t) = \eta + \widetilde{B}(\rho(t))$ and the FPT τ can be written as $\tau = \inf\{t > 0 : \eta + \widetilde{B}(\rho(t)) \leq S(t)\}$. For $\theta > 0$, let us consider the process $Z_t = e^{-\theta X(t) - \frac{1}{2}\theta^2\rho(t)}$; as easily seen, Z_t is a positive martingale; indeed, it can be represented as $Z_t = e^{-\theta X(0)} - \theta \int_0^t Z_s \sigma(s) dB_s$ (see e.g., Theorem 5.2 of [14]). We observe that, for $t \leq \tau$ the martingale Z_t is bounded, because $X(t)$ is non negative and therefore $0 < Z_t \leq e^{-\theta X(t)} \leq 1$. Then, by using the fact that, for any finite stopping time τ one has $E[Z_0] = E[Z_{\tau \wedge t}]$ (see e.g., Formula (7.7) in [14]), and the dominated convergence theorem, we obtain that

$$\begin{aligned} E[Z_0] = E[e^{-\theta X(0)}] = E[e^{-\theta\eta}] &= \lim_{t\to\infty} E[e^{-\theta X(\tau\wedge t) - \frac{1}{2}\theta^2\rho(\tau\wedge t)}] \\ &= E[\lim_{t\to\infty} e^{-\theta X(\tau\wedge t) - \frac{1}{2}\theta^2\rho(\tau\wedge t)}] = E[e^{-\theta S(\tau) - \frac{1}{2}\theta^2\rho(\tau)}]. \end{aligned} \tag{3}$$

Thus, if $\widehat{g}(\theta) = E(e^{-\theta\eta})$ is the Laplace transform of the density of the initial position η, we finally get

$$\widehat{g}(\theta) = E\left[e^{-\theta S(\tau) - \frac{\theta^2}{2}\rho(\tau)}\right], \tag{4}$$

that is Equation (2). □

Remark 1. *If one takes in place of $X(t)$ a process of the form $\widetilde{X}(t) = \eta S(t) + S(t)B(\rho(t))$, with $\eta \geq 1$, that is, a special case of continuous Gauss-Markov process ([15]) with mean $\eta S(t)$, then $\widetilde{X}(t)/S(t)$ is still a continuously time-changed BM, and so the IFPT problem for $\widetilde{X}(t)$ and $S(t)$ is reduced to that of continuously time-changed BM and a constant barrier, for which results are available (see e.g., [4–6]).*

Remark 2. *By using Laplace transform inversion (when it is possible), Equation (4) allows to find the solution g to the IFPT problem for X, the continuous increasing boundary S, and the distribution F of the FPT τ. Indeed, some care has to be used to exclude that the found distribution of η has atoms together with a density. However, as already noted in [2,7], the function $\widehat{g}$ may not be the Laplace transform of some probability density function, so in that case the IFPT problem has no solution; really, it may admit more than one solution, since the right-hand member of Equation (4) essentially furnishes the moments of η of any order n, but this is not always sufficient to uniquely determine the density g of η. In line of principle, the right-hand member of Equation (4) can be expressed in terms of the Laplace transform of $f(t) = F'(t)$, though it is not always possible to do this explicitly.*

A simple case is when $S(t) = a + bt$, *with* $a, b \geq 0$, *and* $\rho(t) = t$, *that is,* $X(t) = B_t$ $(\sigma(t) = 1)$; *in fact, one obtains*

$$\widehat{g}(\theta) = E\left[e^{-\theta(a+b\tau)-\frac{\theta^2}{2}\tau}\right] = e^{-\theta a} E\left[e^{-\theta(b+\frac{\theta}{2})\tau}\right] = e^{-\theta a} \widehat{f}\left(\frac{\theta(\theta+2b)}{2}\right), \tag{5}$$

which coincides with Equation (2.2) of [2], and it provides a relation between the Laplace transform of the density of the initial position η *and the Laplace transform of the density of the FPT* τ.

Remark 3. *Let* $S(t)$ *be increasing and* $S(0) \geq 0$, *then* τ *is a.s. finite; in fact* $\widetilde{\tau} = \rho(\tau) = \inf\{t > 0 : \eta + \widetilde{B}_t \leq \widetilde{S}(t)\} \leq \widetilde{\tau}_1$, *where* $\widetilde{S}(t) = S(\rho^{-1}(t))$ *is increasing and* $\widetilde{\tau}_1$ *is the first hitting time to* $S(0)$ *of BM* $\widetilde{B}$ *starting at* η; *since* $\widetilde{\tau}_1$ *is a.s. finite, also* $\widetilde{\tau}$ *is so. Next, from the finiteness of* $\widetilde{\tau}$ *it follows that* $\tau = \rho^{-1}(\widetilde{\tau})$ *is finite, too. Moreover, if one seeks that* $E(\tau) < \infty$, *a sufficient condition for this is that* $\rho(t)$ *and* $\widetilde{S}(t)$ *are both convex functions; indeed,* $\widetilde{\tau} \leq \widetilde{\tau}_2$, *where* $\widetilde{\tau}_2$ *is the FPT of BM* $\widetilde{B}$ *starting from* η *below the straight line* $a + bt$ $(a = S(0) \geq 0,\ b = \widetilde{S}'(0) \geq 0)$ *which is tangent to the graph of* $\widetilde{S}(t)$ *at* $t = 0$. *Thus, since* $E(\widetilde{\tau}_2) < \infty$, *it follows that* $E(\widetilde{\tau})$ *is finite, too; finally, being* ρ^{-1} *concave, Jensen's inequality for concave functions implies that* $E(\tau) = E(\rho^{-1}(\widetilde{\tau})) \leq \rho^{-1}(E(\widetilde{\tau}))$ *and therefore* $E(\tau) < \infty$.

Remark 4. *Theorem 1 allows to solve also the so called Skorokhod embedding (SE) problem:*

Given a distribution H, find an integrable stopping time τ^*, *such that the distribution of* $X(\tau^*)$ *is H, namely* $P(X(\tau^*) \leq x) = H(x)$.

In fact, let be $S(t)$ *increasing, with* $S(0) = 0$; *first suppose that the support of H is* $[0, +\infty)$; *then, from Equation (4) it follows that*

$$\widehat{g}(\theta) = E[e^{-\theta X(\tau) - \frac{\theta^2}{2}\rho(S^{-1}(X(\tau)))}], \tag{6}$$

and this solves the SE problem with $\tau^* = \tau$; *it suffices to take the random initial point* $X(0) = \eta > 0$ *in such a way that its Laplace transform* $\widehat{g}$ *satisfies*

$$\widehat{g}(\theta) = \int_0^{S(+\infty)} e^{-\theta x - \frac{\theta^2}{2}\rho(S^{-1}(x))} dH(x). \tag{7}$$

In the special case when $S(t) = a + bt$ $(a, b > 0)$ *and* $\rho(t) = t$, *Equation (7) becomes (cf. the result in [8] for* $a = 0$) :

$$\widehat{g}(\theta) = e^{\frac{a\theta^2}{2b}} \widehat{h}\left(\frac{\theta(\theta+2b)}{2b}\right), \tag{8}$$

where $h(x) = H'(x)$ *and* $\widehat{h}$ *denotes the Laplace transform of h.*

In analogous way, the SE problem can be solved if the support of H is $(-\infty, 0]$; *now, the FPT is understood as* $\tau^- = \inf\{t > 0 : \eta + B(\rho(t)) > -S(t)\}$ $(\eta < 0)$, *that is, the first hitting time to the boundary* $S^-(t) = -S(t)$ *from below.*

Therefore, the solution to the general SE problem, namely without restrictions on the support of the distribution H, can be obtained as follows (see [8], for the case when $S(t)$ *is a straight line).*

The r.v. $X(\tau)$ *can be represented as a mixture of the r.v.* $X^+ > 0$ *and* $X^- < 0$:

$$X(\tau) = \begin{cases} X^+ & \text{with probability } p^+ = P(X(\tau) \geq 0) \\ X^- & \text{with probability } p^- = 1 - p^+. \end{cases} \tag{9}$$

Suppose that the SE problem for the r.v. X^+ *and* X^- *can be solved by* $S^+(t) = S(t)$ *and* $\eta^+ = \eta > 0$, *and* $S^-(t) = -S(t)$ *and* $\eta^- = -\eta < 0$, *respectively. Then, we get that the r.v.*

$$\eta^{\pm} = \begin{cases} \eta^+ & \text{with probability } p^+ \\ \eta^- & \text{with probability } p^- \end{cases} \tag{10}$$

and the boundary $S^{\pm}(t) = S^+(t) \cup S^-(t)$ *solve the SE problem for the r.v.* $X(\tau)$.

If $\widehat{g}$ is analytic in a neighbor of $\theta = 0$, then the moments of order n of η, $E(\eta^n)$, exist finite, and they are given by $E(\eta^n) = (-1)^n \frac{d^n}{d\theta^n}\widehat{g}|_{\theta=0}$. By taking the first derivative in Equation (4) and calculating it at $\theta = 0$, we obtain

$$E(\eta) = -\widehat{g}'(0) = E(S(\tau)). \tag{11}$$

By calculating the second derivative of $\widehat{g}$ at $\theta = 0$, we get

$$E(\eta^2) = \widehat{g}''(0) = E(S^2(\tau) - \rho(\tau))), \tag{12}$$

and so

$$Var(\eta) = E(\eta^2) - E^2(\eta) = Var(S(\tau)) - E(\rho(\tau)). \tag{13}$$

Thus, we obtain the compatibility conditions

$$\begin{cases} E(\eta) = E(S(\tau)) \\ Var(S(\tau)) \geq E(\rho(\tau)). \end{cases} \tag{14}$$

If $Var(S(\tau)) < E(\rho(\tau))$, a solution to the IFPT problem does not exist. In the special case when $S(t) = a + bt$ $(a,\ b \geq 0)$ and $\rho(t) = t$, Equation (11) becomes $E(\eta) = a + bE(\tau)$ and Equation (13) becomes $Var(\eta) = b^2 Var(\tau) - E(\tau)$, while Equation (14) coincides with Equation (2.3) of [2]. By writing the Taylor's expansions at $\theta = 0$ of both members of Equation (4), and equaling the terms with the same order in θ, one gets the successive derivatives of $\widehat{g}(\theta)$ at $\theta = 0$; thus, one can write any moment of η in terms of the expectation of a function of τ; for instance, it is easy to see that

$$E(\eta^3) = E[(S(\tau))^3] - 3E[S(\tau)\rho(\tau)], \tag{15}$$

$$E(\eta^4) = E[(S(\tau)^4] - 6E[(S(\tau)^2\rho(\tau)] + 3E[(\rho(\tau)^2], \tag{16}$$

$$E(\eta^5) = E[15S(\tau)\rho^2(\tau) - 240S^3(\tau)\rho(\tau) + S^5(\tau)]. \tag{17}$$

2.1. *The Special Case* $S(t) = \alpha + \beta\rho(t)$

If $S(t) = \alpha + \beta\rho(t)$, with $\alpha,\ \beta \geq 0$, from Equation (4) we get

$$\widehat{g}(\theta) = E[e^{-\theta(\alpha+\beta\rho(\tau)) - \frac{\theta^2}{2}\rho(\tau)}] = e^{-\theta\alpha}E[e^{-\theta\rho(\tau)(\beta+\theta/2)}]. \tag{18}$$

Thus, setting $\widetilde{\tau} = \rho(\tau)$, we obtain (see Equation (5)):

$$\widehat{g}(\theta) = e^{-\theta\alpha}E[e^{-\theta(\beta+\theta/2)\widetilde{\tau}}] = e^{-\theta\alpha}\widehat{\widetilde{f}}(\theta(\beta+\theta/2)), \tag{19}$$

having denoted by $\widetilde{f}$ the density of $\widetilde{\tau}$. In this way, we reduce the IFPT problem of $X(t) = \eta + B(\rho(t))$ below the boundary $S(t) = \alpha + \beta\rho(t)$ to that of BM below the linear boundary $\alpha + \beta t$. For instance, taking $\rho(t) = t^3/3$, the solution to the IFPT problem of $X(t)$ through the cubic boundary $S(t) = \alpha + \frac{\beta}{3}t^3$, and the FPT density f, is nothing but the solution to the IFPT problem of BM through the linear boundary $\alpha + \beta t$, and the FPT density $\widetilde{f}$.

Under the assumption that $S(t) = \alpha + \beta\rho(t)$, with $\alpha,\ \beta \geq 0$, a number of explicit results can be obtained, by using the analogous ones which are valid for BM and a linear boundary (see [2]). As for the question of the existence of solutions to the IFPT problem, we have:

Proposition 1. *Let be* $S(t) = \alpha + \beta\rho(t)$, *with* $\alpha,\ \beta \geq 0$; *for* $\gamma,\ \lambda > 0$, *suppose that the FPT density* $f = F'$ *is given by*

$$f(t) = \begin{cases} \frac{\lambda^\gamma}{\Gamma(\gamma)}\rho(t)^{\gamma-1}e^{-\lambda\rho(t)}\rho'(t) & if\ t > 0 \\ 0 & otherwise \end{cases} \tag{20}$$

(namely the density $\widetilde{f}$ of $\widetilde{\tau}$ is the Gamma density with parameters (γ, λ)). Then, the IFPT problem has solution, provided that $\beta \geq \sqrt{2\lambda}$, and the Laplace transform of the density g of the initial position η is given by:

$$\widehat{g}(\theta) = \left[e^{-\alpha\theta/2} \frac{(\beta - \sqrt{\beta^2 - 2\lambda})^\gamma}{(\theta + \beta - \sqrt{\beta^2 - 2\lambda})^\gamma} \right] \cdot \left[e^{-\alpha\theta/2} \frac{(\beta + \sqrt{\beta^2 - 2\lambda})^\gamma}{(\theta + \beta + \sqrt{\beta^2 - 2\lambda})^\gamma} \right], \tag{21}$$

which is the Laplace transform of the sum of two independent random variables, Z_1 and Z_2, such that $Z_i - \alpha/2$ has distribution Gamma of parameters γ and λ_i $(i = 1, 2)$, where $\lambda_1 = \beta - \sqrt{\beta^2 - 2\lambda}$ and $\lambda_2 = \beta + \sqrt{\beta^2 - 2\lambda}$.

Remark 5. *If f is given by Equation (20), that is $\widetilde{f}$ is the Gamma density, the compatibility condition in Equation (14) becomes $\beta \geq \sqrt{\lambda}$, which is satisfied under the assumption $\beta \geq \sqrt{2\lambda}$ required by Proposition 1. In the special case when $\gamma = 1$, then η has the same distribution as $\alpha + Z_1 + Z_2$, where Z_i are independent and exponential with parameter λ_i, $i = 1, 2$.*

The following result also follows from Proposition 2.5 of [2].

Proposition 2. *Let be $S(t) = \alpha + \beta\rho(t)$, with α, $\beta \geq 0$; for $\beta > 0$, suppose that the Laplace transform of $\widetilde{f}$ has the form:*

$$\widehat{\widetilde{f}}(\theta) = \sum_{k=1}^{N} \frac{A_k}{(\theta + B_k)^{c_k}}, \tag{22}$$

for some $c_k > 0$, $A_k, B_k > 0$, $k = 1, \ldots, N$. Then, there exists a value $\beta^* > 0$ such that the solution to the IFPT problem exists, provided that $\beta \geq b^*$.

If $\beta = 0$ and the Laplace transform of $\widetilde{f}$ has the form:

$$\widehat{\widetilde{f}}(\theta) = \sum_{k=1}^{N} \frac{A_k}{(\sqrt{2\theta} + B_k)^{c_k}}, \tag{23}$$

then, the solution to the IFPT problem exists.

2.2. Approximate Solution to the IFPT Problem for Non Linear Boundaries

Now, we suppose that there exist $\alpha_1, \alpha_2, \beta_1, \beta_2$ with $0 \leq \alpha_1 \leq \alpha_2$ and $\beta_2 \geq \beta_1 \geq 0$, such that, for every $t \geq 0$:

$$\alpha_1 + \beta_1\rho(t) \leq S(t) \leq \alpha_2 + \beta_2\rho(t), \tag{24}$$

namely $S(t)$ is enveloped from above and below by the functions $S_{\alpha_2,\beta_2}(t) = \alpha_2 + \beta_2\rho(t)$ and $S_{\alpha_1,\beta_1}(t) = \alpha_1 + \beta_1\rho(t)$.

Then, by using Proposition (3.13) of [16] (see also [1]), we obtain the following:

Proposition 3. *Let $S(t)$ a continuous, increasing boundary satisfying Equation (24) and suppose that the FPT τ of $X(t) = \eta + B(\rho(t))$ $(\eta > S(0))$ below the boundary $S(t)$ has an assigned probability density f and that there exists a density g with support $(S(0), +\infty)$, which is solution to the IFPT problem for $X(t)$ and the boundary $S(t)$; as before, denote by $\widetilde{f}(t)$ the density of $\rho(\tau)$ and by $\widehat{\widetilde{f}}(\theta)$ its Laplace transform, for $\theta > 0$. Then:*

(i) *If $\alpha_2 > \alpha_1$ and the function $g \in L^p(S(0), \alpha_2)$ for some $p > 1$, its Laplace transform $\widehat{g}(\theta)$ must satisfy:*

$$e^{-\alpha_2(\theta + 2(\beta_2 - \beta_1))} \left[\widehat{\widetilde{f}}\left(\frac{\theta(\theta + 2\beta_2)}{2} \right) - (\alpha_2 - S(0))^{\frac{p-1}{p}} \left(\int_{S(0)}^{\alpha_2} g^p(x)dx \right)^{1/p} \right] \leq \widehat{g}(\theta)$$

$$\leq e^{-\alpha_1\theta} \widehat{\widetilde{f}}\left(\frac{\theta(\theta + 2\beta_1)}{2} \right); \tag{25}$$

(ii) *If* $\alpha_1 = \alpha_2 = S(0)$, *then Equation* (25) *holds without any further assumption on* g *(and the term* $(\alpha_2 - S(0))^{\frac{p-1}{p}} \left(\int_{S(0)}^{\alpha_2} g^p(x)dx\right)^{1/p}$ *vanishes).*

Remark 6. *The smaller* $\alpha_2 - \alpha_1$ *and* $\beta_2 - \beta_1$*, the better the approximation to the Laplace transform of* g. *Notice that, if* g *is bounded, then the term* $(\alpha_2 - S(0))^{\frac{p-1}{p}} \left(\int_{S(0)}^{\alpha_2} g^p(x)dx\right)^{1/p}$ *can be replaced with* $(\alpha_2 - S(0))||g||_\infty$.

2.3. The IFPT Problem for $\overline{X}(t) = \eta + B(\rho(t))+$ Large Jumps

As an application of the previous results, we consider now the piecewise-continuous process $\overline{X}(t)$, obtained by superimposing to $X(t)$ a jump process, namely we set $\overline{X}(t) = \eta + B(\rho(t))$ for $t < T$, where T is an exponential distributed time with parameter $\mu > 0$; we suppose that, for $t = T$ the process $\overline{X}(t)$ makes a downward jump and it crosses the continuous increasing boundary S, irrespective of its state before the occurrence of the jump. This kind of behavior is observed e.g. in the presence of a so called *catastrophes* (see e.g., [17]). For $\eta \geq S(0)$, we denote by $\overline{\tau}_S = \inf\{t > 0 : \overline{X}(t) \leq S(t)\}$ the FPT of $\overline{X}(t)$ below the boundary $S(t)$. The following holds:

Proposition 4. *If there exists a solution* $\overline{g}$ *to the IFPT problem of* $\overline{X}(t)$ *below* $S(t)$ *with* $\overline{X}(0) = \eta \geq S(0)$, *then its Laplace transform is given by*

$$\widehat{\overline{g}}(\theta) = E\left[e^{-\theta S(\tau) - \frac{\theta^2}{2}\rho(\tau) - \mu\tau}\right] + \mu \int_0^{+\infty} e^{-\theta S(t) - \frac{\theta^2}{2}\rho(t) - \mu t} \left(\int_t^{+\infty} f(s)ds\right) dt. \tag{26}$$

Proof. For $t > 0$, one has:

$$P(\overline{\tau}_S \leq t) = P(\overline{\tau}_S \leq t | t < T)P(t < T) + 1 \cdot P(t \geq T) = P(\tau_S \leq t)e^{-\mu t} + (1 - e^{-\mu t}). \tag{27}$$

Taking the derivative, one obtains the FPT density of $\overline{\tau}$:

$$\overline{f}(t) = e^{-\mu t} f(t) + \mu e^{-\mu t} \int_t^{+\infty} f(s)ds, \tag{28}$$

where f is the density of τ. Then, by the same arguments used in the proof of Theorem 1, we obtain

$$\begin{aligned}
\widehat{\overline{g}}(\theta) &= E\left[e^{-\theta S(\overline{\tau}) - \frac{\theta^2}{2}\rho(\overline{\tau})}\right] \\
&= \int_0^\infty e^{-\theta S(t) - \frac{\theta^2}{2}\rho(t)} \overline{f}(t)dt \\
&= \int_0^\infty e^{-\theta S(t) - \frac{\theta^2}{2}\rho(t)} \left[e^{-\mu t} f(t) + \mu e^{-\mu t} \int_t^\infty f(s)ds\right] dt \\
&= \int_0^\infty e^{-\theta S(t) - \frac{\theta^2}{2}\rho(t) - \mu t} f(t)dt + \mu \int_0^\infty e^{-\theta S(t) - \frac{\theta^2}{2}\rho(t) - \mu t} \left(\int_t^\infty f(s)ds\right) dt
\end{aligned}$$

that is Equation (26). □

Remark 7. (i) *For* $\mu = 0$, *namely when no jump occurs, Equation* (26) *becomes Equation* (4).
(ii) *If* τ *is exponentially distributed with parameter* λ, *then Equation* (26) *provides:*

$$\widehat{\overline{g}}(\theta) = \frac{\lambda + \mu}{\lambda} E\left[e^{-\theta S(\tau) - \frac{\theta^2}{2}\rho(\tau) - \mu\tau}\right]. \tag{29}$$

(iii) *In the special case when $S(t) = \alpha + \beta\rho(t)$ $(\alpha, \beta \geq 0)$, we can reduce to the FPT $\overline{\overline{\tau}}$ of BM + large jumps below the linear boundary $\alpha + \beta t$; then, it is possible to write $\widehat{\widehat{g}}$ in terms of the Laplace transform of $\overline{\overline{\tau}}$. Really, by using Proposition 3.10 of [16] one gets*

$$\widehat{\widehat{g}}(\theta) = e^{-\alpha\theta}\left[\left(1 - \frac{2\mu}{\theta(\theta + 2\beta)}\right)^{-1} \widehat{\widehat{f}}\left(\frac{\theta(\theta+2\beta)}{2} - \mu\right) - \frac{2\mu}{\theta(\theta+2\beta) - 2\mu}\right],$$

where, for simplicity of notation we have denoted again with $\widehat{\widehat{f}}$ the Laplace transform of $\overline{\overline{\tau}}$; of course, if $\rho(t) = t$, then $\widehat{\widehat{f}}$ is the Laplace transform of $\overline{\tau}$. Notice that, if $\mu = 0$ the last equation is nothing but Equation (5) with α, β in place of a, b.

3. Some Examples

Example 1. *If $S(t) = a + bt$, with a, $b \geq 0$, and $X(t) = B_t$ $(\rho(t) = 1)$, examples of solution to the IFPT problem, for $X(t)$ and various FPT densities f, can be found in [2].*

Example 2. *Let be $S(t) = \alpha + \beta\rho(t)$, with α, $\beta \geq 0$, and suppose that τ has density $f(t) = \lambda e^{-\rho(t)}\rho'(t)\mathbf{1}_{(0,+\infty)}(t)$ (that is, the density $\widetilde{f}$ of $\widetilde{\tau} = \rho(\tau)$ is exponential with parameter λ). By using Proposition 1 we get that $\eta = \alpha + Z_1 + Z_2$, where Z_i are independent random variable, such that $Z_i - \alpha/2$ has exponential distribution with parameter λ_i $(i = 1,2)$, where $\lambda_1 = \beta - \sqrt{\beta^2 - 2\lambda}$ and $\lambda_2 = \beta + \sqrt{\beta^2 - 2\lambda}$. Then, the solution g to the IFPT problem for $X(t) = \eta + B(\rho(t))$, the boundary S and the exponential FPT distribution, is:*

$$g(x) = \begin{cases} \frac{\lambda_1\lambda_2}{\lambda_2 - \lambda_1} e^{-\lambda_1(x-\alpha)} - e^{-\lambda_2(x-\alpha)}, \text{ if } b > \sqrt{2\lambda} \\ 2\lambda(x-\alpha)e^{-\sqrt{2\lambda}(x-a)}, \text{ if } b = \sqrt{2\lambda}. \end{cases} \quad (x \geq \alpha) \tag{30}$$

In general, for a given continuous increasing boundary $S(t)$ and an assigned distribution of τ, it is difficult to calculate explicitly the expectation on the right-hand member of Equation (4) to get the Laplace transform of η. Thus, a heuristic solution to the IFPT problem can be achieved by using Equation (4) to calculate the moments of η (those up to the fifth order are given by Equations (11), (12) and (15)–(17)). Of course, even if one was able to find the moments of η of any order, this would not determinate the distribution of η. However, this procedure is useful to study the properties of the distribution of η, provided that the solution to the IFPT problem exists.

Example 3. *Let be $S(t) = t^2$, $\rho(t) = t$ and suppose that τ is exponentially distributed with parameter λ; we search for a solution $\eta > 0$ to the IFPT problem by using the method of moments, described above. The compatibility condition in Equation (14) requires that $\lambda^3 < 20$ (for instance, one can take $\lambda = 1$). From Equations (11), (12) and (15)–(17), and calculating the moments of τ up to the eighth order, we obtain:*

$$E(\eta) = E(\tau^2) = \frac{2}{\lambda^2};\ E(\eta^2) = E(\tau^4) - E(\tau) = \frac{24 - \lambda^3}{\lambda^4};\ \sigma^2(\eta) = Var(\eta) = \frac{20 - \lambda^3}{\lambda^4};$$

$$E(\eta^3) = E(\tau^6) - 3E(\tau^3) = \frac{720 - 18\lambda^3}{\lambda^6};\ E(\eta^4) = E(\tau^8) - 6E(\tau^3) + 3E(\tau^2) = \frac{8! - 36\lambda^5 + 6\lambda^6}{\lambda^8}.$$

Notice that, under the condition $\lambda^3 < 20$ the first four moments of η are positive, as it must be. However, they do not match those of a Gamma distribution.

An information about the asymmetry is given by the skewness value

$$\frac{E(\eta - E(\eta))^3}{\sigma(\eta)^3} = -12\frac{24 - \lambda^3}{(20 - \lambda^3)^{3/2}} < 0,$$

meaning that the candidate η has an asymmetric distribution with a tail toward the left.

4. Conclusions and Final Remarks

We have dealt with the IFPT problem for a continuously time-changed Brownian motion $X(t)$ starting from a random position η. For a given continuous, increasing boundary $S(t)$ with $\eta \geq S(0) \geq 0$, and an assigned continuous distribution function F, the IFPT problem consists in finding the distribution, or the density g of η, such that the first-passage time τ of $X(t)$ below $S(t)$ has distribution F. In this note, we have provided some extensions of the results, already known in the case when $X(t)$ is BM and $S(t)$ is a straight line, and we have reported some explicit examples. Really, the process we considered has the form $X(t) = \eta + \int_0^t \sigma(s) dB_s$, where B_t is standard Brownian motion, and $\sigma(t)$ is a non random continuous function of time $t \geq 0$, such that the function $\rho(t) = \int_0^t \sigma^2(s) ds$ is increasing and it satisfies the condition $\rho(+\infty) = +\infty$. Thus, a standard BM $\widehat{B}$ exists such that $X(t) = \eta + \widehat{B}(\rho(t))$. Our main result states that

$$\widehat{g}(\theta) = E\left[e^{-\theta S(\tau) - \frac{\theta^2}{2}\rho(\tau)}\right], \tag{31}$$

where, for $\theta > 0$, $\widehat{g}(\theta)$ denotes the Laplace transform of the solution g to the IFPT problem.

Notice that the above result can be extended to diffusions which are more general than the process $X(t)$ considered, for instance to a process of the form

$$U(t) = w^{-1}(\widehat{B}(\rho(t)) + w(\eta)), \tag{32}$$

where w is a regular enough, increasing function; such a process U is obtained from BM by a space transformation and a continuous time-change (see e.g., the discussion in [2]). Since $w(U(t)) = w(\eta) + \widehat{B}(\rho(t))$, the IFPT problem for the process U, the boundary $S(t)$ and the FPT distribution F, is reduced to the analogous IFPT problem for $X(t) = \eta_1 + \widehat{B}(\rho(t))$, starting from $\eta_1 = w(\eta)$, instead of η, the boundary $S_1(t) = w(S(t))$ and the same FPT distribution F. When $\sigma(t) = 1$, i.e. $\rho(t) = t$, the process $U(t)$ is conjugated to BM, according to the definition given in [2]; two examples of diffusions conjugated to BM are the Feller process, and the Wright–Fisher like (or CIR) process, (see e.g., [2]). The process $U(t)$ given by Equation (32) is indeed a weak solution of the SDE:

$$dU(t) = -\frac{\rho'(t) w''(U(t))}{2(w'(U(t)))^3} dt + \frac{\sqrt{\rho'(t)}}{w'(U(t))} dB_t\,, \tag{33}$$

where $w'(x)$ and $w''(x)$ denote first and second derivative of $w(x)$.

Provided that the deterministic function $\rho(t)$ is replaced with a random function, the representation in Equation (32) is valid also for a time homogeneous one-dimensional diffusion driven by the SDE

$$dU(t) = \mu(U(t))dt + \sigma(U(t))dB_t,\; U(0) = \eta, \tag{34}$$

where the drift (μ) and diffusion coefficients (σ) satisfy the usual conditions (see e.g., [18]) for existence and uniqueness of the solution of Equation (34). In fact, let $w(x)$ be the *scale function* associated to the diffusion $U(t)$ driven by the SDE Equation (34), that is, the solution of $Lw(x) = 0$, $w(0) = 0$, $w'(0) = 1$, where L is the infinitesimal generator of U given by $Lh = \frac{1}{2}\sigma^2(x)\frac{d^2h}{dx^2} + \mu(x)\frac{dh}{dx}$. As easily seen, if the integral $\int_0^t \frac{2\mu(z)}{\sigma^2(z)}\, dz$ converges, the scale function is explicitly given by

$$w(x) = \int_0^x \exp\left(-\int_0^t \frac{2\mu(z)}{\sigma^2(z)}\, dz\right) dt. \tag{35}$$

If $\zeta(t) := w(U(t))$, by Itô's formula one obtains

$$\zeta(t) = w(\eta) + \int_0^t w'(w^{-1}(\zeta(s)))\sigma(w^{-1}(\zeta(s)))dB_s\,, \tag{36}$$

that is, the process $\zeta(t)$ is a local martingale, whose quadratic variation is

$$\rho(t) \doteq \langle \zeta \rangle_t = \int_0^t [w'(U(s))\sigma(U(s))]^2 ds, \ t \geq 0. \tag{37}$$

The (random) function $\rho(t)$ is differentiable and $\rho(0) = 0$; if it is increasing to $\rho(+\infty) = +\infty$, by the Dambis, Dubins–Schwarz theorem (see e.g., [3]) one gets that there exists a standard BM $\widehat{B}$ such that $\zeta(t) = \widehat{B}(\rho(t)) + w(\eta)$. Thus, since w is invertible, one obtains the representation in Equation (32).

Notice, however, that the IFPT problem for the process U given by Equation (32) cannot be addressed as in the case when ρ is a deterministic function. In fact, if $\rho(t)$ given by Equation (37) is random, it results that $\rho(t)$ and the FPT τ are dependent. Thus, in line of principle it would be possible to obtain information about the Laplace transform of g, only in the case when the joint distribution of $(\rho(t), \tau)$ was explicitly known.

Funding: This research was funded by the MIUR Excellence Department Project awarded tothe Department of Mathematics, University of Rome Tor Vergata, CUP E83C18000100006.

Acknowledgments: I would like to express particular thanks to the anonymous referees for their constructive comments and suggestions leading to improvements of the paper.

Conflicts of Interest: The author declares no conflict of interest.

References

1. Abundo, M. Some randomized first-passage problems for one-dimensional diffusion processes. *Sci. Math. Jpn.* **2013**, *76*, 33–46.
2. Abundo, M. An inverse first-passage problem for one-dimensional diffusion with random starting point. *Stat. Probab. Lett.* **2012**, *82*, 7–14. [CrossRef]
3. Revuz, D.; Yor, M. *Continous Martingales and Brownian Motion*; Springer: Berlin/Heidelberg, Germany, 1991.
4. Darling, D.A.; Siegert, A.J.F. The first passage problem for a continuous Markov process. *Ann. Math. Stat.* **1953**, *24*, 264–639. [CrossRef]
5. Hieber, P.; Scherer, M. A note on first-passage times of continuously time-changed Brownian motion. *Stat. Probab. Lett.* **2012**, *82*, 165–172. [CrossRef]
6. Hurd, T.R. Credit risk modeling using time-changed Brownian motion. *Int. J. Theor. Appl. Financ.* **2009**, *12*, 1213–1230. [CrossRef]
7. Jackson, K.; Kreinin, A.; Zhang, W. Randomization in the first hitting problem. *Stat. Probab. Lett.* **2009**, *79*, 2422–2428. [CrossRef]
8. Jaimungal, S.; Kreinin, A.; Valov, A. The generalized Shiryaev problem and Skorokhod embedding. *Theory Probab. Appl.* **2014**, *58*, 493–502. [CrossRef]
9. Lanska, V.; Smiths, C.E. The effect of a random initial value in neural first-passage-time models. *Math. Biosci.* **1989**, *93*, 191–215. [CrossRef]
10. Peskir, G. On integral equations arising in the first-passage problem for Brownian motion. *J. Integral Equat. Appl.* **2002**, *14*, 397–423. [CrossRef]
11. Sacerdote, L.; Zucca, C. Threshold shape corresponding to a Gamma firing distribution in an Ornstein-Uhlenbeck neuronal model. *Sci. Math. Jpn.* **2003**, *19*, 1319–1346.
12. Zucca, C.; Sacerdote, L. On the inverse first-passage-time problem for a Wiener process. *Ann. Appl. Probab.* **2009**, *8*, 375–385. [CrossRef]
13. Abundo, M. Limit at zero of the first-passage time density and the inverse problem for one-dimensional diffusions. *Stoch. Anal. Appl.* **2006**, *24*, 1119–1145. [CrossRef]
14. Klebaner, F.C. *Introduction to Stochastic Calculus With Applications*; Imperial College Press: London, UK, 2005.
15. Di Nardo, E.; Nobile, A.G.; Pirozzi, E.; Ricciardi, L.M. A computational approach to first-passage-time problems for Gauss-Markov processes. *Adv. Appl. Probab.* **2001**, *33*, 453–482. [CrossRef]
16. Abundo, M. An overview on inverse first-passage-time problems for one-dimensional diffusion processes. *Lect. Notes Semin. Interdiscip. Matematica* **2015**, *12*, 1–44. Available online: http://dimie.unibas.it/site/home/info/documento3012448.html (accessed on 28 May 2018)

17. Di Crescenzo, A.; Giorno, V.; Nobile, A.G.; Ricciardi, L.M. On the M/M/1 queue with catastrophes and its continuous approximation. *Queueing Syst.* **2003**, *43*, 329–347. [CrossRef]
18. Ikeda, N.; Watanabe, S. *Stochastic Differential Equations and Diffusion Processes*; North-Holland Publishing Company: Amsterdam, The Netherlands, 1981.

mathematics

Article

Stability Analysis of Cohen–Grossberg Neural Networks with Random Impulses

Ravi Agarwal [1,2], Snezhana Hristova [3,*], Donal O'Regan [4] and Peter Kopanov [3]

1 Department of Mathematics, Texas A&M University-Kingsville, Kingsville, TX 78363, USA; Ravi.Agarwal@tamuk.edu
2 Florida Institute of Technology, Melbourne, FL 32901, USA
3 Faculty of Mathematics, Plovdiv University, Tzar Asen 24, 4000 Plovdiv, Bulgaria; pkopanov@yahoo.com
4 School of Mathematics, Statistics and Applied Mathematics, National University of Ireland, H91 CF50 Galway, Ireland; donal.oregan@nuigalway.ie
* Correspondence: snehri@gmail.com

Received: 27 July 2018; Accepted: 17 August 2018; Published: 21 August 2018

Abstract: The Cohen and Grossberg neural networks model is studied in the case when the neurons are subject to a certain impulsive state displacement at random exponentially-distributed moments. These types of impulses significantly change the behavior of the solutions from a deterministic one to a stochastic process. We examine the stability of the equilibrium of the model. Some sufficient conditions for the mean-square exponential stability and mean exponential stability of the equilibrium of general neural networks are obtained in the case of the time-varying potential (or voltage) of the cells, with time-dependent amplification functions and behaved functions, as well as time-varying strengths of connectivity between cells and variable external bias or input from outside the network to the units. These sufficient conditions are explicitly expressed in terms of the parameters of the system, and hence, they are easily verifiable. The theory relies on a modification of the direct Lyapunov method. We illustrate our theory on a particular nonlinear neural network.

Keywords: Cohen and Grossberg neural networks; random impulses; mean square stability

1. Introduction

Artificial neural networks are important technical tools for solving a variety of problems in various scientific disciplines. Cohen and Grossberg [1] introduced and studied in 1983 a new model of neural networks. This model was extensively studied and applied in many different fields such as associative memory, signal processing and optimization problems. Several authors generalized this model [2] by including delays [3,4], impulses at fixed points [5,6] and discontinuous activation functions [7]. Furthermore, a stochastic generalization of this model was studied in [8]. The included impulses model the presence of the noise in artificial neural networks. Note that in some cases in the artificial neural network, the chaos improves the noise (see, for example, [9]).

To the best of our knowledge, there is only one published paper studying neural networks with impulses at random times [10]. However, in [10], random variables are incorrectly mixed with deterministic variables; for example $I_{[\xi_k,\xi_{k+1})}(t)$ for the random variables ξ_k, ξ_{k+1} is not a deterministic index function (it is a stochastic process), and it has an expected value labeled by E, which has to be taken into account on page 13 of [10]; in addition, in [10], one has to be careful since the expected value of a product of random variables is equal to the product of expected values only for independent random variables. We define the generalization of Cohen and Grossberg neural network with impulses at random times, briefly giving an explanation of the solutions being stochastic processes, and we study stability properties. Note that a brief overview of randomness in neural networks and some methods for their investigations are given in [11] where the models are stochastic ones. Impulsive perturbation is a common phenomenon in real-world systems, so it is also important to consider impulsive systems. Note that the stability of deterministic models with impulses for neural networks was studied in [12–18]. However, the occurrence of impulses at random times needs to be considered in real-world systems. The stability problem for the differential equation with impulses at random times was studied in [19–21]. In this paper, we study the general case of the time-varying potential (or voltage) of the cells, with the time-dependent amplification functions and behaved functions, as well as time-varying strengths of connectivity between cells and variable external bias or input from outside the network to the units. The study is based on an application of the Lyapunov method. Using Lyapunov functions, some stability sufficient criteria are provided and illustrated with examples.

2. System Description

We consider the model proposed by Cohen and Grossberg [1] in the case when the neurons are subject to a certain impulsive state displacement at random moments.

Let $T_0 \geq 0$ be a fixed point and the probability space $(\Omega, \mathcal{F}, P)$ be given. Let a sequence of independent exponentially-distributed random variables $\{\tau_k\}_{k=1}^{\infty}$ with the same parameter $\lambda > 0$ defined on the sample space Ω be given. Define the sequence of random variables $\{\xi_k\}_{k=0}^{\infty}$ by:

$$\xi_k = T_0 + \sum_{i=1}^{k} \tau_i, \quad k = 0, 1, 2, \ldots. \tag{1}$$

The random variable τ_k measures the waiting time of the k-th impulse after the $(k-1)$-th impulse occurs, and the random variable ξ_k denotes the length of time until k impulses occur for $t \geq T_0$.

Remark 1. *The random variable $\Xi = \sum_{i=1}^{k} \tau_i$ is Erlang distributed, and it has a pdf $f_{\Xi}(t) = \lambda e^{-\lambda t} \frac{(\lambda t)^{k-1}}{(k-1)!}$ and a cdf $F(t) = P(\Xi < t) = 1 - e^{-\lambda t} \sum_{j=0}^{k-1} \frac{(\lambda t)^j}{j!}$.*

Consider the general model of the Cohen–Grossberg neural networks with impulses occurring at random times (RINN):

$$\begin{aligned} x_i'(t) &= -a_i(x_i(t))\Big(b(x_i(t)) - \sum_{j=1}^{n} c_{ij}(t) f_j(x_j(t)) + I_i(t)\Big) \\ &\quad \text{for} \quad t \geq T_0, \ \xi_k < t < \xi_{k+1}, \ k = 0, 1, \ldots, \ \ i = 1, 2, \ldots n, \\ x_i(\xi_k + 0) &= \Phi_{k,i}(x_i(\xi_k - 0)) \qquad \text{for} \quad k = 1, 2, \ldots, \\ x_i(T_0) &= x_i^0, \end{aligned} \tag{2}$$

where n corresponds to the number of units in a neural network; $x_i(t)$ denotes the potential (or voltage) of cell i at time t, $x(t) = (x_1(t), x_2(t), \ldots, x_n(t)) \in \mathbb{R}^n$, $f_j(x_j(t))$ denotes the activation functions of the neurons at time t and represents the response of the j-th neuron to its membrane potential and $f(x) = (f_1(x_1), f_2(x_2), \ldots, f_n(x_n))$. Now, $a_i(.) > 0$ represents an amplification function; $b_i(.)$ represents an appropriately behaved function; the $n \times n$ connection matrix $C(t) = (c_{ij}(t))$ denotes the strengths of connectivity between cells at time t; and if the output from neuron j excites (resp., inhibits) neuron i, then $c_{ij}(t) \geq 0$ (resp., $c_{ij}(t) \leq 0$), and the functions $I_i(t)$, $I(t) = (I_1(t), I_2(t), \ldots, I_n(t)) \in \mathbb{R}^n$ correspond to the external bias or input from outside the network to the unit i at time t.

We list some assumptions, which will be used in the main results:

(H1) For all $i = 1, 2, \ldots, n$, the functions $a_i \in C(\mathbb{R}, (0, \infty))$, and there exist constants $A_i, B_i > 0$ such that $0 < A_i \leq a_i(u) \leq B_i$ for $u \in \mathbb{R}$.

(H2) There exist positive numbers $M_{i,j}$, $i, j = 1, 2, \ldots, n$ such that $|c_{i,j}(t)| \leq M_{i,j}$ for $t \geq 0$.

Remark 2. *In the case when the strengths of connectivity between cells are constants, then Assumption (H2) is satisfied.*

For the activation functions, we assume:

(H3) The neuron activation functions are Lipschitz, i.e., there exist positive numbers L_i, $i = 1,2,\ldots,n$, such that $|f_i(u) - f_i(v)| \le L_i|u - v|$ for $u, v \in \mathbb{R}$.

Remark 3. *Note that the activation functions satisfying Condition (H3) are more general than the usual sigmoid activation functions.*

2.1. Description of the Solutions of Model (2)

Consider the sequence of points $\{t_k\}_{k=1}^{\infty}$ where the point t_k is an arbitrary value of the corresponding random variable τ_k, $k = 1,2,\ldots$. Define the increasing sequence of points $\{T_k\}_{k=1}^{\infty}$ by:

$$T_k = T_0 + \sum_{i=1}^{k} t_k. \tag{3}$$

Note that T_k are values of the random variables ξ_k, $k = 1,2,\ldots$.

Consider the corresponding RINN (2) initial value problem for the system of differential equations with fixed points of impulses $\{T_k\}_{k=1}^{\infty}$ (INN):

$$\begin{aligned} x_i'(t) &= -a_i(x_i(t))\Big(b(x_i(t)) - \sum_{j=1}^{n} c_{ij}(t) f_j(x_j(t)) + I_i(t)\Big) \\ &\qquad \text{for } \ t \ge T_0,\ \ t \ne T_k,\ k = 0,1,\ldots,\ \ i = 1,2,\ldots n, \\ x_i(T_k + 0) &= \Phi_{k,i}(x_i(T_k - 0)) \qquad \text{for } \ k = 1,2,\ldots, \\ x_i(T_0) &= x_i^0. \end{aligned} \tag{4}$$

The solution of the differential equation with fixed moments of impulses (4) depends not only on the initial point (T_0, x^0), but on the moments of impulses T_k, $k = 1,2,\ldots$, i.e., the solution depends on the chosen arbitrary values t_k of the random variables τ_k, $k = 1,2,\ldots$. We denote the solution of the initial value problem (4) by $x(t; T_0, x^0, \{T_k\})$. We will assume that:

$$x(T_k; T_0, x^0, \{T_k\}) = \lim_{t \to T_k - 0} x(t; T_0, x^0, \{T_k\}) \quad \text{for any } k = 1,2,\ldots. \tag{5}$$

Remark 4. *Note that the limit (5) is well defined since $T_k, k = 1,2\ldots,$ are points from $\mathbb{R}$. This is different than* $\lim_{t \to \xi_k - 0} x(t)$ *because ξ_k is a random variable (see its incorrect use by the authors in [10]).*

The set of all solutions $x(t; T_0, x^0, \{T_k\})$ of the initial value problem for the impulsive fractional differential Equation (4) for any values t_k of the random variables τ_k, $k = 1,2,\ldots$ generates a stochastic process with state space $\mathbb{R}^n$. We denote it by $x(t; T_0, x^0, \{\tau_k\})$, and we will say that it is a solution of RINN (2).

Remark 5. *Note that $x(t; T_0, x^0, \{T_k\})$ is a deterministic function, but $x(t; T_0, x^0, \{\tau_k\})$ is a stochastic process.*

Definition 1. *For any given values t_k of the random variables τ_k, $k = 1,2,3,\ldots$, respectively, the solution $x(t; T_0, x^0, \{T_k\})$ of the corresponding initial value problem (IVP) for the INN (4) is called a sample path solution of the IVP for RINN (2).*

Definition 2. *A stochastic process $x(t; T_0, x^0, \{\tau_k\})$ with an uncountable state space $\mathbb{R}^n$ is said to be a solution of the IVP for the system of RINN (2) if for any values t_k of the random variables τ_k, $k = 1,2,\ldots$, the corresponding function $x(t; T_0, x^0, \{T_k\})$ is a sample path solution of the IVP for RINN (2).*

2.2. *Equilibrium of Model (2)*

We define an equilibrium of the model (2) assuming Condition (H1) is satisfied:

Definition 3. *A vector $x^* \in \mathbb{R}^n$, $x^* = (x_1^*, x_2^*, \dots, x_n^*)$ is an equilibrium point of RINN (2), if the equalities:*

$$0 = b(x_i^*) - \sum_{j=1}^{n} c_{ij}(t) f_j(x_i^*) + I_i(t) \text{ for } i = 1, 2, \dots, n \tag{6}$$

and

$$x_i^* = \Phi_{k,i}(x_i^*) \text{ for } t \geq 0, k = 1, 2, \dots, \; i = 1, 2, \dots, n \tag{7}$$

hold.

We assume the following:
(H4) Let RINN (2) have an equilibrium vector $x^* \in \mathbb{R}^n$.

If Assumption (H4) is satisfied, then we can shift the equilibrium point x^* of System (2) to the origin. The transformation $y(t) = x(t) - x^*$ is used to put System (2) in the following form:

$$\begin{aligned} y_i'(t) &= -p_i(y_i(t))\Big(q(y_i(t)) - \sum_{j=1}^{n} c_{ij}(t) F_j(y_j(t))\Big) \\ &\quad \text{for } \; t \geq T_0, \; \xi_k < t < \xi_{k+1}, \; k = 0, 1, \dots, \; i = 1, 2, \dots n, \\ y_i(\xi_k + 0) &= \phi_{k,i}(y(\xi_k - 0)) \qquad \text{for } \; k = 1, 2, \dots, \\ y_i(T_0) &= y_i^0, \end{aligned} \tag{8}$$

where $p_i(u) = a_i(u + x_i^*)$,$q_i(u) = b_i(u + x_i^*) - b_i(x_i^*)$, $F_j(u) = f_j(u + x_j^*) - f_j(x_j^*)$, $j = 1, 2, \dots, n$ and $\phi_{k,i}(u) = \Phi_{k,i}(u + x_i^*) - \Phi_{k,i}(x_i^*)$, $i = 1, 2, \dots, n$, $k = 1, 2, \dots, y_i^0 = x_i^0 - x_i^*$.

Remark 6. *If Assumption (H3) is fulfilled, then the function F in RINN (8) satisfies $|F_j(u)| \leq L_j|u|$, $j = 1, 2, \dots, n$, for $u \in \mathbb{R}$.*

Note that if the point $x^* \in \mathbb{R}^n$ is an equilibrium of RINN (2), then the point $y^* = 0$ is an equilibrium of RINN (8). This allows us to study the stability properties of the zero equilibrium of RINN (8).

3. Some Stability Results for Differential Equations with Impulses at Random Times

Consider the general type of initial value problem (IVP) for a system of nonlinear random impulsive differential equations (RIDE):

$$\begin{aligned} x'(t) &= g(t, x(t)) \text{ for } t \geq T_0, \; \xi_k < t < \xi_{k+1}, \\ x(\xi_k + 0) &= \Psi_k(x(\xi_k - 0)) \qquad \text{for } \; k = 1, 2, \dots, \\ x(T_0) &= x^0; \end{aligned} \tag{9}$$

with $x^0 \in \mathbb{R}^n$, random variables ξ_k, $k = 1, 2, \dots$ are defined by (1), $g \in C([T_0, \infty) \times \mathbb{R}^n, \mathbb{R}^n)$ and $\Psi_k : \mathbb{R}^n \to \mathbb{R}^n$.

Definition 4. *Let $p > 0$. Then, the trivial solution ($x^0 = 0$) of RIDE (9) is said to be* p-moment *exponentially stable if for any initial point $(T_0, y^0) \in \mathbb{R}_+ \times \mathbb{R}^n$, there exist constants $\alpha, \mu > 0$ such that $E[||y(t; T_0, y^0, \{\tau_k\})||^p] < \alpha ||y^0||^p e^{-\mu(t - T_0)}$ for all $t > T_0$, where $y(t; T_0, x^0, \{\tau_k\})$ is the solution of the IVP for RIDE (9).*

Definition 5. *Let $p > 0$. Then, the equilibrium x^* of RINN (2) is said to be* p-moment exponentially stable *if for any initial point $(T_0, x^0) \in \mathbb{R}_+ \times \mathbb{R}^n$, there exist constants $\alpha, \mu > 0$ such that $E[||x(t; T_0, x^0, \{\tau_k)\}) - x^*||^p] < \alpha||x^0 - x^*||^p e^{-\mu(t-T_0)}$ for all $t > T_0$, where $x(t; T_0, x^0, \{\tau_k)\}$ is the solution of the IVP for RINN (2).*

Remark 7. *We note that the two-moment exponential stability for stochastic equations is known as the mean square exponential stability, and in the case of $p = 1$, it is called mean exponential stability.*

Note that the p-moment exponential stability of RIDE (9) was studied in [20] by an application of Lyapunov functions from the class $\Lambda(J, \Delta)$, $J \subset \mathbb{R}_+$, $\Delta \subset \mathbb{R}^n$, $0 \in \Delta$ with:

$$\Lambda(J,\Delta) = \{V(t,x) \in C(J \times \Delta, \mathbb{R}_+) : \ V(t,0) \equiv 0, \\ V(t,x) \text{ is locally Lipschitzian with respect to x}\}.$$

We will use the Dini derivative of the Lyapunov function $V(t,x) \in \Lambda(J,\Delta)$ given by:

$$_{(9)}D_+V(t,x) = \limsup_{h \to 0^+} \frac{1}{h}\Big\{ V(t,x) - V(t-h, x - hg(t,x)) \Big\} \\ \text{for } \ t \in \ J, \ x \in \Delta. \tag{10}$$

Now, we will give a sufficient condition result:

Theorem 1 ([20]). *Let the following conditions be satisfied:*

1. *For $t \geq 0 : g(t,0) \equiv 0$ and $\Psi_k(0) = 0$, $k = 1, 2, \ldots$ and for any initial values (T_0, x^0), the corresponding IVP for the ordinary differential equation $x'(t) = g(t, x(t))$ has a unique solution.*
2. *The function $V \in \Lambda([T_0, \infty), \mathbb{R}^n)$, and there exist positive constants a, b such that:*
   (i) $a||x||^p \leq V(t,x) \leq b||x||^p$ for $t \geq T_0$, $x \in \mathbb{R}^n$;
   (ii) there exists a function $m \in C(\mathbb{R}_+, \mathbb{R}_+)$: $\inf_{t\geq 0} m(t) = L \geq 0$, and the inequality:
   $$_{(9)}D_+V(t,x) \leq -m(t)V(t,x), \quad \text{for } t \geq 0, \ \ x \in \mathbb{R}^n$$
   holds;
   (iii) for any $k = 1, 2, \ldots$, there exist constants w_k : $0 \leq w_k < 1 + \frac{L}{\lambda}$ for $t \geq 0$ such that:
   $$V(t, I_k(t,x)) \leq w_k V(t,x) \quad \text{ for } \ t \geq 0, \ x \in \mathbb{R}^n. \tag{11}$$

Then, the trivial solution of RIDE (9) is p-moment exponentially stable.

4. Stability Analysis of Neural Networks with Random Impulses

We will introduce the following assumptions:

(H5) For $i = 1, 2, \ldots, n$, the functions $b_i \in C(\mathbb{R}, \mathbb{R})$, and there exist constants $\beta_i > 0$ such that $u\Big(b_i(u + x_i^*) - b_i(x_i^*)\Big) \geq \beta_i u^2$ for any $u \in \mathbb{R}$ where $x^* \in \mathbb{R}^n$, $x^* = (x_1^*, x_2^*, \ldots, x_n^*)$, is the equilibrium from Condition (H4).

Remark 8. *If Condition (H5) is satisfied, then the inequality $uq(u) \geq \beta_i u^2, u \in \mathbb{R}$ holds for RINN (8).*

(H6) The inequality:

$$\nu = 2 \min_{i=\overline{1,n}} A_i\beta_i \ - \max_{i=\overline{1,n}} B_i\Big(\max_{i=\overline{1,n}} \sum_{j=1}^n M_{ij}L_j + (\sum_{i=1}^n \max_{j=\overline{1,n}} M_{ij}L_j)\Big) > 0 \tag{12}$$

holds.

(H7) For any $k = 1, 2, \dots$, there exists positive number $K_k < 1 + \frac{\nu}{\lambda}$ such that the inequalities:

$$\sum_{i=1}^{n} \Big(\Phi_{k,i}(x_i) - \Phi_{k,i}(x_i^*)\Big)^2 \leq K_k \sum_{i=1}^{n}(x_i - x_i^*)^2, \quad x_i \in \mathbb{R}, i = 1, 2, \dots, n,$$

hold where $x^* \in \mathbb{R}^n$, $x^* = (x_1^*, x_2^*, \dots, x_n^*)$, is the equilibrium from Condition (H4).

Remark 9. *If Assumption (H7) is fulfilled, then the impulsive functions ϕ_k, $k = 1, 2, \dots$ in RINN (8) satisfy the inequalities $\sum_{i=1}^{n} \phi_{k,i}^2(u_i) \leq K_k \sum_{i=1}^{n} u_i^2$.*

Theorem 2. *Let Assumptions (H1)–(H7) be satisfied. Then, the equilibrium point x^* of RINN (2) is mean square exponentially stable.*

Proof. Consider the quadratic Lyapunov function $V(t, x) = x^T x$, $x \in \mathbb{R}^n$. From Remarks 6, 8 and inequality $2|uv| \leq u^2 + v^2$, we get:

$$\begin{aligned}
{}_{(8)}D_+V(t,y) &\leq 2\sum_{i=1}^{n} y_i\Big(- p_i(y_i)\big(q(y_i) - \sum_{j=1}^{n} c_{ij}(t)F_j(y_j)\big)\Big) \\
&= -2\sum_{i=1}^{n} y_i p_i(y_i) q(y_i) + 2\sum_{i=1}^{n} y_i p_i(y_i) \sum_{j=1}^{n} c_{ij}(t)F_j(y_j) \\
&\leq -2\sum_{i=1}^{n} A_i\beta_i y_i^2 + 2\sum_{i=1}^{n} |y_i| B_i \sum_{j=1}^{n} M_{ij}L_j|y_j| \\
&\leq -2\sum_{i=1}^{n} A_i\beta_i y_i^2 + \sum_{i=1}^{n} B_i \sum_{j=1}^{n} M_{ij}L_j(y_i^2 + y_j^2) \\
&\leq -2\sum_{i=1}^{n} A_i\beta_i y_i^2 + \sum_{i=1}^{n} B_i y_i^2 \sum_{j=1}^{n} M_{ij}L_j + \sum_{i=1}^{n} B_i \sum_{j=1}^{n} M_{ij}L_j y_j^2 \\
&\leq -2\min_{i=\overline{1,n}} A_i\beta_i \sum_{i=1}^{n} y_i^2 \\
&\quad + \max_{i=\overline{1,n}} B_i\Big(\max_{i=\overline{1,n}} \sum_{j=1}^{n} M_{ij}L_j + \big(\sum_{i=1}^{n} \max_{j=\overline{1,n}} M_{ij}L_j\big)\Big) \sum_{i=1}^{n} y_i^2 \\
&= -\nu \sum_{i=1}^{n} y_i^2.
\end{aligned} \tag{13}$$

where the positive constant ν is defined by (12). Therefore, Condition 2(ii) of Theorem 1 is satisfied. Furthermore, from (H7), it follows that Condition 2(iii) of Theorem 1 is satisfied.

From Theorem 1, the zero solution of the system (9) is mean square exponentially stable, and therefore, the equilibrium point x^* of RINN (2) is mean square exponentially stable. □

Example 1. *Let $n = 3$, $t_0 = 0.1$, and the random variables $\tau_k, k = 1, 2, \dots$ are exponentially distributed with $\lambda = 1$. Consider the following special case of RINN (2):*

$$\begin{aligned}
x_i'(t) &= -a_i(x_i(t))\Big(2x_i(t) + \sum_{j=1}^{3} c_{ij}(t)f_j(x_j(t)) - \pi\Big) \\
&\qquad \textit{for } t \geq 0 \;\; \xi_k < t < \xi_{k+1}, \;\; i = 1, 2, 3, \\
x_i(\xi_k + 0) &= \Phi_{k,i}(x_i(\xi_k - 0)) \qquad \textit{for } k = 1, 2, \dots, \\
x_i(0.1) &= x_i^0,
\end{aligned} \tag{14}$$

with $a_i(u) = 2 + \frac{|u|}{1+|u|} \in [2,3), i = 1,2,3,\ f_i(u) = \alpha_i \cos(u),\ \alpha_1 = 0.1, \alpha_2 = 0.01, \alpha_2 = 2,\ \Phi_{k,i}(u) = u \sin k + (1 - \sin k)0.5\pi$, *and* $C = c_{ij}(t)$ *is given by:*

$$C(t) = \begin{pmatrix} -0.1\sin t & 0.4 & 0.3 \\ -\frac{t^2}{5t^2+1} & 0.3 & \frac{t}{5t+1} \\ \frac{t}{10t+1} & -0.2\cos t & -0.1\sin t \end{pmatrix}. \tag{15}$$

The point $x^* = (0.5\pi, 0.5\pi, 0.5\pi)$ is the equilibrium point of RINN (14), i.e., Condition (H4) is satisfied. Now, Assumption (H1) is satisfied with $A_i = 2, B_i = 3,\ i = 1,2,3$. In addition, Assumption (H5) is satisfied with $\beta_i = 2,\ i = 1,2,3$.

Furthermore, $|c_{ij}| \le M_{ij},\ i,j = 1,2,3,\ t \ge 0$ where $M = \{M_{ij}\}$, is given by:

$$M = \begin{pmatrix} 0.1 & 0.4 & 0.3 \\ 0.2 & 0.3 & 0.2 \\ 0.1 & 0.2 & 0.1 \end{pmatrix}. \tag{16}$$

Therefore, Assumption (H2) is satisfied. Note that Assumption (H3) is satisfied with Lipschitz constants $L_1 = 0.1, L_2 = 0.01, L_3 = 2$.

Then, the constant ν defined by (12) is $\nu = 8 - 3(1.814) = 2.558 > 0$. Next, Assumption (H7) is fulfilled with $K_k = 1$ because:

$$\begin{aligned} &\sum_{i=1}^{3} \Big(\Phi_{k,i}(x_i) - \Phi_{k,i}(x_i^*)\Big)^2 = \sum_{i=1}^{3} \Big(x_i \sin k + (1 - \sin k)0.5\pi - 0.5\pi\Big)^2 \\ &= \sum_{i=1}^{3} \Big((x_i - 0.5\pi)\sin k\Big)^2 \le \sum_{i=1}^{3} \Big(x_i - 0.5\pi\Big)^2,\ k = 1,2,\ldots. \end{aligned} \tag{17}$$

Therefore, according to Theorem 1, the equilibrium of RINN (14) is mean square exponentially stable.

Consider the system (14) without any kind of impulses. The equilibrium $x^* = (0.5\pi, 0.5\pi, 0.5\pi)$ is asymptotically stable (see Figures 1 and 2). Therefore, an appropriate perturbation of the neural networks by impulses at random times can keep the stability properties of the equilibrium.

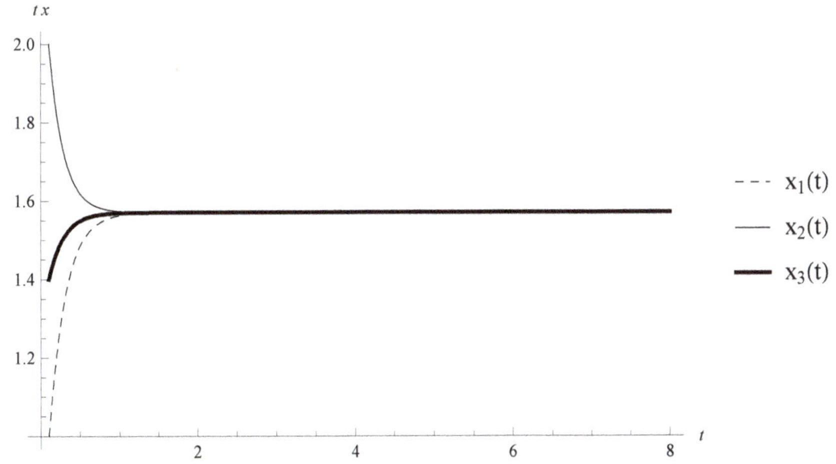

Figure 1. Example 1. Graph of the solution of the system ODE corresponding to (14) with $x_1^0 = 1, x_2^0 = 2, x_3^0 = 1.4$.

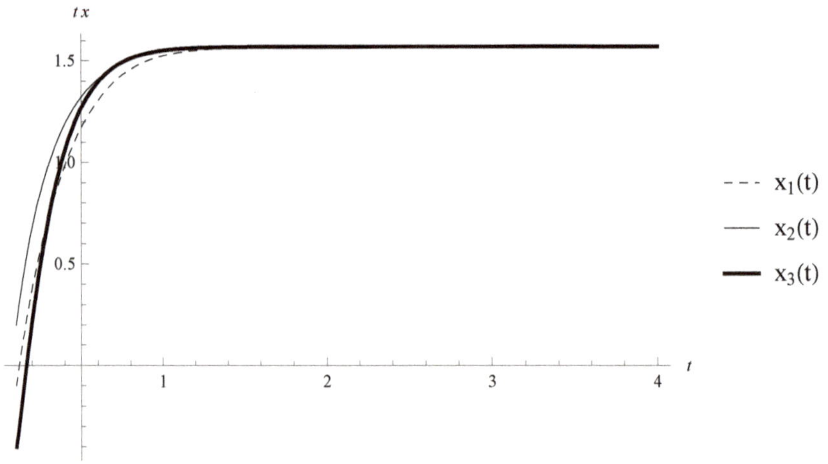

Figure 2. Example 1. Graph of the solution of the system ODE corresponding to (14) with $x_1^0 = -0.1, x_2^0 = 0.2, x_3^0 = -0.4$.

Remark 10. *Note that Condition (H7) is weaker than Condition (3.6) in Theorem 3.2 [16], and as a special case of Theorem 2, we obtain weaker conditions for exponential stability of the Cohen and Grossberg model without any type of impulses. For example, if we consider (14) according to Condition (3.6) [16], the inequality* $\delta = 2||M||2\frac{3}{4} = 1.0374 < 1$ *is not satisfied, and Theorem 3.2 [16] does not give us any result about stability (compare with Example 1).*

Now, consider the following assumption:
(H8) The inequality:

$$\nu = \min_{i=\overline{1,n}} \gamma_i - \sum_{i=1}^{n} \max_{j=\overline{1,n}} M_{ij} > 0 \tag{18}$$

holds.

Theorem 3. *Let Assumptions (H1)–(H5), (H7) and (H8) be satisfied. Then, the equilibrium point x^* of RINN (2) is mean exponentially stable.*

Proof. For any $u \in \mathbb{R}^n$, we define $V(u) = \sum_{i=1}^{n} \int_0^{u_i} \frac{sign(s)}{a_i(s)} ds$. Then:

$$V(u) \le \sum_{i=1}^{n} \int_0^{u_i} \frac{sign(s)}{A_i} ds = \sum_{i=1}^{n} \frac{1}{A_i} |u_i| \le A||u||$$

and:

$$V(u) \ge \sum_{i=1}^{n} \int_0^{u_i} \frac{sign(s)}{B_i} ds = \sum_{i=1}^{n} \frac{1}{B_i} |u_i| \ge B||u||$$

where $A = \max_{i=\overline{1,n}} \frac{1}{A_i}$, $B = \min_{i=\overline{1,n}} \frac{1}{B_i}$.

Then, for $t \geq 0$ and $y \in \mathbb{R}^n$ according to Remarks 6 and 8, we obtain:

$$\begin{aligned}
{}_{(8)}D_+V(y) &\leq \sum_{i=1}^{n} -sgn(y_i)\Big(q(y_i) - \sum_{j=1}^{n} c_{ij}(t)F_j(y_j)\Big) \\
&\leq \sum_{i=1}^{n}\Big(-\beta_i\,|y_i| + \sum_{j=1}^{n} M_{ij}|F_j(y_j)|\Big) \leq \sum_{i=1}^{n}\Big(-\beta_i\,|y_i| + \sum_{j=1}^{n} M_{ij}\,|y_j|\Big) \\
&= -\sum_{i=1}^{n}\beta_i\,|y_i| + \sum_{i=1}^{n}\sum_{j=1}^{n} M_{ij}|y_j| \leq -\min_{i=\overline{1,n}}\beta_i \sum_{i=1}^{n}|y_i| + \Big(\sum_{i=1}^{n}\max_{j=\overline{1,n}} M_{ij}\Big)\sum_{j=1}^{n}|y_j| \\
&\leq -\nu\sum_{i=1}^{n}|y_i| \leq -\frac{\nu}{B}V(u).
\end{aligned} \tag{19}$$

Furthermore, from (H7) and Remark 9, it follows that Condition 2(iii) of Theorem 1 is satisfied. From Theorem 1, we have that Theorem 3 is true. □

Example 2. *Let $n = 3$, $t_0 = 0.1$, and the random variables $\tau_k, k = 1, 2, \ldots$ are exponentially distributed with $\lambda = 1$. Consider the following special case of RINN (2):*

$$\begin{aligned}
x_i'(t) &= -a_i(x_i(t))\Big(2x_i(t) + \sum_{j=1}^{3} c_{ij}(t)f_j(x_j(t)) - 1\Big) \\
&\quad \text{for } t \geq 0 \;\; \xi_k < t < \xi_{k+1}, \;\; i = 1, 2, 3 \\
x_i(\xi_k + 0) &= \Phi_{k,i}(x_i(\xi_k - 0)) \qquad \text{for } k = 1, 2, \ldots \\
x_i(0.1) &= x_i^0,
\end{aligned} \tag{20}$$

with $a_i(u) = 2 + \frac{|u|}{1+|u|} \in [2, 3), i = 1, 2, 3$, $f_i(u) = log(\frac{u}{1-u})$, $\Phi_{k,i}(u) = u\sin k + (1 - \sin k)0.5$, and $C = c_{ij}(t)$ is given by (15).

The point $x^* = (0.5, 0.5, 0.5)$ is the equilibrium point of RINN (20), i.e., Condition (H4) is satisfied. Now, Assumption (H5) is satisfied with $\beta_i = 2$, $i = 1, 2, 3$.

Furthermore, $|c_{ij}| \leq M_{ij}$, $i, j = 1, 2, 3$, $t \geq 0$ where $M = \{M_{ij}\}$, is given by (16). Therefore, Assumption (H2) is satisfied. Then, the inequality $\min_{i=\overline{1,n}} \beta_i = 2 > \sum_{i=1}^{3} \max_{j=\overline{1,3}} M_{ij} = 0.4 + 0.3 + 0.2 = 0.9$ holds.

According to Theorem 3, the equilibrium of (20) is mean exponentially stable.

Consider the system (20) without any kind of impulses. The equilibrium $x^* = (0.5,\ 0.5,\ 0.5)$ is asymptotically stable (see Figures 3 and 4). Therefore, an appropriate perturbation of the neural networks by impulses at random times can keep the stability properties of the equilibrium.

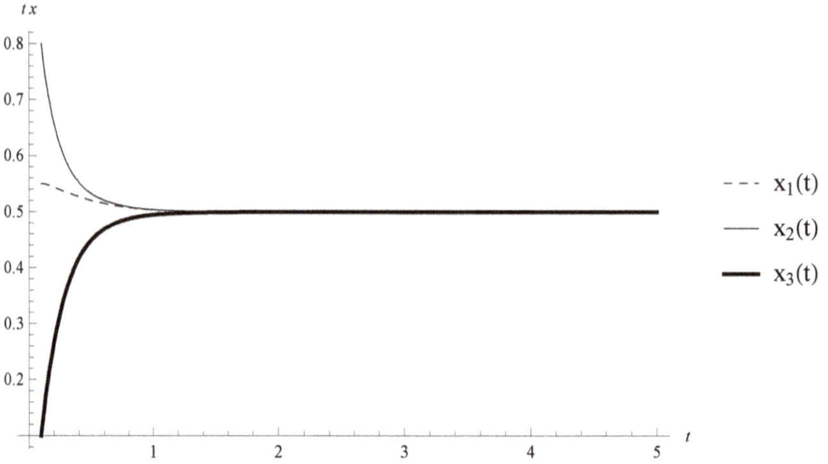

Figure 3. Example 2. Graph of the solution of the system ODE corresponding to (20) with $x_1^0 = 0.55, x_2^0 = 0.8, x_3^0 = 0.1$.

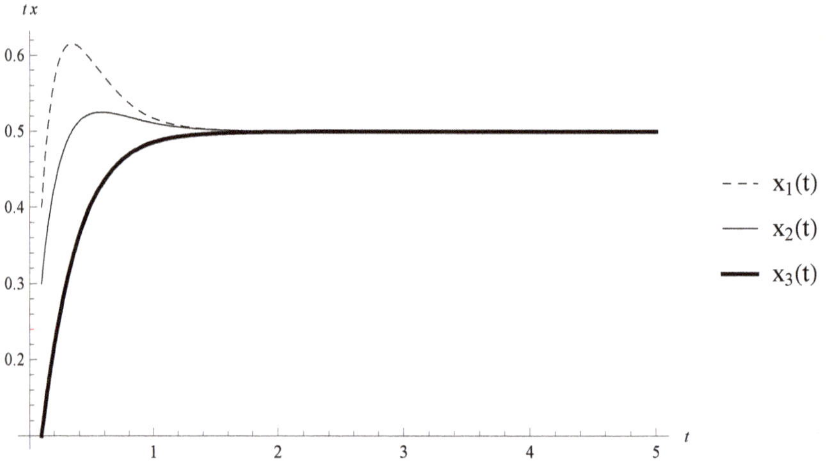

Figure 4. Example 2. Graph of the solution of the system ODE corresponding to (20) with $x_1^0 = 0.4, x_2^0 = 0.3, x_3^0 = 0.1$.

5. Conclusions

In this paper, we study stability properties of the equilibrium point of a generalization of the Cohen–Grossberg model of neural networks in the case when:

- the potential (or voltage) of any cell is perturbed instantaneously at random moments, i.e., the neural network is modeled by a deterministic differential equation with impulses at random times. This presence of randomness in the differential equation totally changes the behavior of the solutions (they are not deterministic functions, but stochastic processes).
- the random moments of the impulsive state displacements of neurons are exponentially distributed.

- the connection matrix $C = (c_{ij})$ is not a constant matrix which is usually the case in the literature (it is a matrix depending on time since the strengths of connectivity between cells could be changed in time).
- the external bias or input from outside the network to any unit is not a constant (it is variable in time).
- sufficient conditions for mean-square exponential stability and for mean exponential stability of the equilibrium are obtained.

Author Contributions: All authors contributed equally to the writing of this paper. All four authors read and approved the final manuscript.

Funding: The research was partially supported by Fund MU17-FMI-007, University of Plovdiv "Paisii Hilendarski".

Conflicts of Interest: The authors declare that they have no competing interests.

References

1. Cohen, M.; Grossberg, S. Stability and global pattern formation and memory storage by competitive neural networks . *IEEE Trans. Syst. Man Cyber* **1983**, *13*, 815–826. [CrossRef]
2. Guo, S.; Huang, L. Stability analysis of Cohen-Grossberg neural networks. *IEEE Trans. Neural Netw.* **2006**, *17*, 106–117. [CrossRef] [PubMed]
3. Bai, C. Stability analysis of Cohen–Grossberg BAM neural networks with delays and impulses. *Chaos Solitons Fractals* **2008**, *35*, 263–267. [CrossRef]
4. Cao, J.; Liang, J. Boundedness and stability for Cohen–Grossberg neural network with time-varying delays. *J. Math. Anal. Appl.* **2004**, *296*, 665–685. [CrossRef]
5. Aouiti, Ch.; Dridi, F. New results on impulsive Cohen–Grossberg neural networks. *Neural Process. Lett.* **2018**, *48*, 1–25. [CrossRef]
6. Liu, M.; Jiang, H.; Hu, C. Exponential Stability of Cohen-Grossberg Neural Networks with Impulse Time Window. *Discret. Dyn. Nat. Soc.* **2016**, *2016*, 2762960. [CrossRef]
7. Meng, Y.; Huang, L.; Guo, Z.; Hu, Q. Stability analysis of Cohen–Grossberg neural networks with discontinuous neuron activations. *Appl. Math. Model.* **2010**, *34*, 358–365. [CrossRef]
8. Huang, C.; Huang, L.; He, Y. Mean Square Exponential Stability of Stochastic Cohen-Grossberg Neural Networks with Unbounded Distributed Delays. *Discret. Dyn. Nat. Soc.* **2010**, *2010*, 513218. [CrossRef]
9. Bucolo, M.; Caponetto, R.; Fortuna, L.; Frasca, M.; Rizzo, A. Does chaos work better than noise? *IEEE Circuits Syst. Mag.* **2002**, *2*, 4–19 . [CrossRef]
10. Vinodkumar, A.; Rakkiyappan, R. Exponential stability results for fixed and random type impulsive Hopfield neural networks. *Int. J. Comput. Sci. Math.* **2016**, *7*, 1–19. [CrossRef]
11. Scardapane, S.; Wang, D. Randomness in neural networks: an overview. *WIREs Data Min. Knowl. Discov.* **2017**, *7*, 1–18. [CrossRef]
12. Gopalsamy, K. Stability of artificial neural networks with impulses. *Appl. Math. Comput.* **2004**, *154*, 783–813. [CrossRef]
13. Rakkiyappan, R.; Balasubramaiam, P.; Cao, J. Global exponential stability of neutral-type impulsive neural networks. *Nonlinear Anal. Real World Appl.* **2010**, *11*, 122–130. [CrossRef]
14. Song, X.; Zhao, P.; Xing, Z.; Peng, J. Global asymptotic stability of CNNs with impulses and multi-proportional delays. *Math. Methods Appl. Sci.* **2016**, *39*, 722–733.
15. Wu, Z.; Li, C. Exponential stability analysis of delayed neural networks with impulsive time window. In Proceedings of the 2017 Ninth International Conference on Advanced Computational Intelligence (ICACI), Doha, Qatar, 4–6 February 2017; pp. 37–42.
16. Wang, L.; Zou, X. Exponential stability of Cohen-Grossberg neural networks. *Neural Netw.* **2002**, *15*, 415–422. [CrossRef]
17. Yang, Z.; Xu, D. Stability analysis of delay neural networks with impulsive effects. *IEEE Trans. Circuits Syst. Express Briefs* **2015**, *52*, 517–521. [CrossRef]
18. Zhou, Q. Global exponential stability of BAM neural networks with distributed delays and impulses. *Nonlinear Anal. Real World Appl.* **2009**, *10*, 144–153. [CrossRef]

19. Agarwal, R.; Hristova, S.; O'Regan, D.; Kopanov, P. P-moment exponential stability of differential equations with random impulses and the Erlang distribution. *Mem. Differ. Equ. Math. Phys.* **2017**, *70*, 99–106. [CrossRef]
20. Agarwal, R.; Hristova, S.; O'Regan, D. Exponential stability for differential equations with random impulses at random times. *Adv. Differ. Equ.* **2013**, *372*, 12. [CrossRef]
21. Agarwal, R.; Hristova, S.; O'Regan, D.; Kopanov, P. Impulsive differential equations with Gamma distributed moments of impulses and p-moment exponential stability. *Acta Math. Sci.* **2017**, *37*, 985–997. [CrossRef]

mathematics

MDPI

Article

On Short-Term Loan Interest Rate Models: A First Passage Time Approach

Giuseppina Albano [1,*] and Virginia Giorno [2]

[1] Dipartimento di Scienze Economiche e Statistiche, University of Salerno, 84084 Fisciano, SA, Italy
[2] Dipartimento di Informatica, University of Salerno, 84084 Fisciano, SA, Italy; giorno@unisa.it
* Correspondence: pialbano@unisa.it; Tel.: +39-089-962-645

Received: 2 February 2018; Accepted: 25 April 2018; Published: 3 May 2018

Abstract: In this paper, we consider a stochastic diffusion process able to model the interest rate evolving with respect to time and propose a first passage time (FPT) approach through a boundary, defined as the "alert threshold", in order to evaluate the risk of a proposed loan. Above this alert threshold, the rate is considered at the risk of usury, so new monetary policies have been adopted. Moreover, the mean FPT can be used as an indicator of the "goodness" of a loan; i.e., when an applicant is to choose between two loan offers, s/he will choose the one with a higher mean exit time from the alert boundary. An application to real data is considered by analyzing the Italian average effect global rate by means of two widely used models in finance, the Ornstein-Uhlenbeck (Vasicek) and Feller (Cox-Ingersoll-Ross) models.

Keywords: loan interest rate regulation; diffusion model; first passage time (FPT)

1. Introduction

In recent decades, increasing attention has been paid to the study of the dynamics underlying the interest rates. The intrinsically stochastic nature of the interest rates has suggested the formulation of various models often based on stochastic differential equations (SDEs) (see, for example, [1,2] and references therein). More recently, further stochastic representations of non-usurious interest rates have been provided in order to obtain information concerning costs of loans. Most of them are simple and convenient time-homogeneous parametric models, attempting to capture certain features of observed dynamic movements, such as heteroschedasticity, long-run equilibrium, and other peculiarities (see, for example, [3–5]).

An interest rate is "usurious" if it is markedly above current market rates. France was the first European country to introduce an anti-usury law in 1966. In Italy, the first law of this nature (Law No. 108) was introduced in 1996. An inventory of interest rate restrictions against usury in the EU Member States was achieved at the end of 2010. In particular, the EU authorities' attention focused on the interest rate restrictions established on precise legal rules restricting credit price, both directly by fixed thresholds as well as indirectly by intervening on the calculation of compound interest (Directorate-General of the European Commission, 2011).

Since May 2011, the Italian law has governed interest rates in loans with new regulations, fixing a threshold above which interest rates applied in loans are considered usurious. The threshold rate is based on the actual global average rate of interest (TEGM) that is quarterly determined by the Italian Ministry of Economy and Finance (Ministero dell'Economia e delle Finanze), and it is a function of various types of homogeneous transactions. Specifically, the threshold rate is calculated as 125% of the reference TEGM plus 4%. Therefore,

$$\textit{Threshold rate} = 1.25\, TEGM + 0.04.$$

Moreover, the difference between the TEGM and the usury threshold cannot exceed 8%, so the maximum value admissible for TEGM cannot exceed 16%.

Note that the penal code (art. 644, comma 4, c.p.) establishes that the scheduling of the usury interest rate takes into account errands, wages, and costs, but not taxes related to the loan supply, but, to compute the TEGM, the Bank of Italy does not consider these items. Therefore, this difference between the principle stated by the legislature and the instructions of the Bank of Italy decreases both the average rates and the threshold rates. Therefore, another boundary that is lower than that established by the Bank of Italy should be introduced. This case has also been extended to other European countries.

The basic idea of the present work is to investigate the (random) time in which an interest rate reaches an "alert boundary", that is near the admitted limit of 0.16. To do this, we start with two classical models in the literature: Vasicek and Cox-Ingersoll-Ross (CIR) ([6,7]) since they provide good characterization of the short-term real rate process. In particular, the CIR model is able to capture the dependence of volatility on the level of interest rates ([8]).

We then investigated the first passage time (FPT) through a boundary generally depending on time. This approach is useful in economy since it suggests the time in which the trend of a loan interest rate can be considered at risk of usury, so it has to be modified from the owner of the loan service. Moreover, the mean first exit time through the alert boundary could be adopted as an indicator of the "goodness" of the loan, in the sense that an applicant choosing between two loan offers will choose the one with a higher mean exit time from the alert boundary. For the FPT analysis, we consider a constant boundary; clearly this kind of approach is applicable to other underlying models that are different from the Vasicek and CIR models and to boundaries generally depending on time, which is the case of time-dependent loan interest rate.

The layout of the paper is as follows. In Section 2, a brief review of diffusion models describing the dynamics of the interest rate is discussed. The FPT problem through a time-dependent threshold $S(t)$ is analyzed. In Section 3, we consider data of the TEGM published by Bank of Italy. In particular, we compare the Vasicek and CIR models in order to establish which model better fits our data. Moreover, a Chow test shows the presence of structural breaks. In Section 4, the FPT problem through a constant "alert boundary" is analyzed. Concluding remarks follow.

2. Mathematical Background

We denote by $\{X(t),\ t \geq t_0\}$ the stochastic process describing the dynamics of a loan interest rate. We assume that $X(t)$ is a time-homogeneous diffusion process defined in $I = (r_1, r_2)$ by the following SDE:

$$dX(t) = A_1[X(t)]dt + \sqrt{A_2[X(t)]}\, dW(t), \qquad X(t_0) = x_0 \ a.s., \tag{1}$$

where $A_1(x)$ and $A_2(x) > 0$ denote the drift and the infinitesimal variance of $X(t)$ and $W(t)$ is a standard Wiener process. The instantaneous drift $A_1(x)$ represents a force that keeps pulling the process towards its long-term mean, whereas $A_2(x)$ represents the amplitude of the random fluctuations. Let

$$h(x) = \exp\left\{-2\int^{x} \frac{A_1(z)}{A_2(z)}\, dz\right\}, \qquad s(x) = \frac{2}{A_2(x)\, h(x)} \tag{2}$$

be the scale function and speed density of $X(t)$, respectively. The transition probability density function (pdf) of $X(t)$, denoted by $f(x,t|y,\tau)$, is a solution of the Kolmogorov equation,

$$\frac{\partial f(x,t|y,\tau)}{\partial \tau} + A_1(y)\frac{\partial f(x,t|y,\tau)}{\partial y} + \frac{A_2(y)}{2}\frac{\partial^2 f(x,t|y,\tau)}{\partial y^2} = 0$$

and of the Fokker–Planck equation,

$$\frac{\partial f(x,t|y,\tau)}{\partial t} = -\frac{\partial}{\partial x}\Big[A_1(x)\, f(x,t|y,\tau)\Big] + \frac{\partial^2}{\partial x^2}\Big[\frac{A_2(x)}{2}\, f(x,t|y,\tau)\Big],$$

with the delta initial conditions:

$$\lim_{t\downarrow\tau} f(x,t|y,\tau) = \lim_{\tau\uparrow t} f(x,t|y,\tau) = \delta(x-y).$$

The above conditions assure the uniqueness of the transition pdf only when the endpoints of the diffusion interval are natural; otherwise, suitable boundary conditions may have to be imposed (cf., for istance, [9]).

Further, if $X(t)$ admits a steady-state behavior, then the steady-state pdf is

$$W(x) \equiv \lim_{t\to\infty} f(x,t|x_0,t_0) = \frac{s(x)}{\int_{-\infty}^{\infty} s(z)\,dz}.$$

Let

$$T_{X_0} = \inf_{t>t_0}\{t \,:\, X(t) > S(t) \mid X(t_0) = x_0\}$$

be the FPT variable of $X(t)$ through a time-dependent boundary $S(t)$ starting from x_0, and let $g[S(t),t|x_0,t_0] = dP(T_{x_0} < t)/dt$ be its pdf. In the following, we assume that $x_0 < S(t_0)$ since in our context x_0 represents the initial observed value of the interest rate. The FPT problem has far-reaching implications (see, for instance, [10,11]).

As shown in [12,13], if $S(t)$ is in $C^2[t_0,\infty)$, g can be obtained as a solution of the following second-kind Volterra integral equation:

$$g[S(t),t|y,\tau] = -2\Psi[S(t),t|y,\tau] + 2\int_{t_0}^{t} g[S(\vartheta),\vartheta|y,\tau]\,\Psi[S(t),t|S(\vartheta),\vartheta]\,d\vartheta \tag{3}$$

where

$$\Psi[S(t),t|y,\tau] = \frac{1}{2} f(x,t|y,\tau)\Big\{S'(t) - A_1[S(t)] + \frac{3}{4}A_2'[S(t)]\Big\} + \frac{A_2[S(t)]}{2}\,\frac{\partial f(x,t|y,\tau)}{\partial x}\Big|_{x=S(t)}.$$

If $A_1(x)$ and $A_2(x)$ are known, i.e., if the process is fixed, some closed form solution of (3) can be obtained for particular choices of the boundary $S(t)$. Further results have been obtained in [14–16]. Alternatively, a numerical algorithm can be successfully used; for example, the R package fptdApprox is also a useful instrument for the numerical evaluation of the FPT pdf (see [17,18]).

Further, if the FPT is a sure event and if $S(t) = S$ is time-independent, the moments of the FPT can be evaluated via a recursive Siegert-type formula (see, for instance, [9]):

$$t_n(S|x_0) = \int_0^{\infty} t^n\, g(S,t|x_0)\,dt = n\int_{x_0}^{S} dz\, h(z)\int_{-\infty}^{z} s(u)\, t_{n-1}(S|u)\,du \qquad n = 1,2,\ldots \tag{4}$$

where $t_0(S|x_0) = P(T_{x_0} < \infty) = 1$ and $h(x)$ and $s(x)$ given in Equation (2).

3. Modeling the Italian Loans

In this section, we consider two stochastic processes widely used in the financial literature, the Vasicek and CIR models (see [1,19]), for describing a historical series of Italian average rates on loans. We use the Akaike information criterion (AIC) as an indicator of the goodness of fit of the two models. Moreover, the presence of structural breaks is verified by means of a Chow test applied to the Euler discretization of the corresponding SDE. More precisely, the Chow test is sequentially applied for each instant in order to evaluate whether the coefficients of the Euler discretization made on each subinterval are equal to those including all observed time intervals.

TEGM values are quarterly settled and published by the Bank of Italy (see https://www.bancaditalia.it) for different types of credit transactions. We refer to the TEGM values for a particular credit transaction, "one-fifth of salary transfer", in the period from 1 July 1997 to 31 March 2015 (data are quarterly observed, so the number of observations is 72). Moreover, two amount classes are analyzed:

- *Dataset A:* up to 10 million lira (until 31 December 2001) and up to € 5000 (after 2002);
- *Dataset B:* above 10 million lire (until 31 December 2001) and above € 5000 (after 2002).

In Figure 1, Dataset A is shown on the left and Dataset B is on the right.

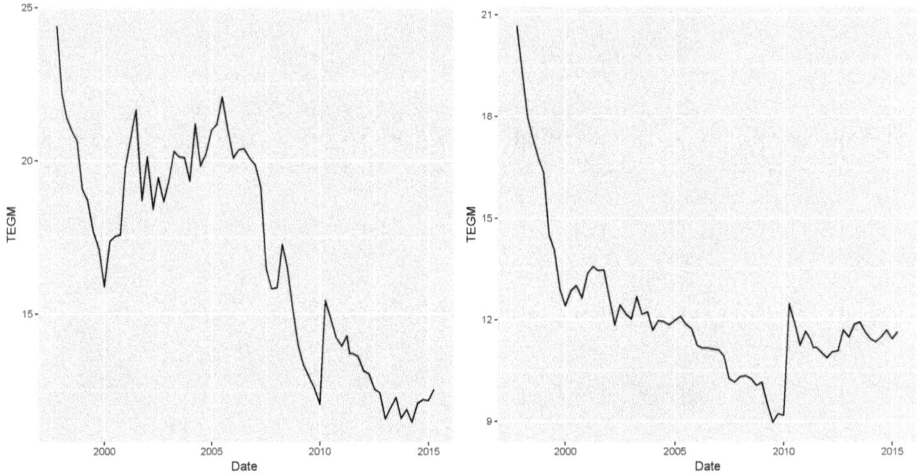

Figure 1. TEGM for one-fifth of salary transfer up to € 5000 (on the **left**) and above € 5000 (on the **right**).

We estimate the parameters for the Vasiceck and CIR models, maximizing the conditional likelihood function. Specifically, we assume that the process $X(t)$ is observed at n discrete time instants $t_1, \ldots, t_n$ with $t_i \geq t_0$ and denote by $x_1, \ldots, x_n$ the corresponding observations.

Let $\boldsymbol{\theta}$ be the vector of the unknown parameters and let us assume $P[X(t_1) = x_1] = 1$. The likelihood function is

$$L(x_1, \ldots, x_n; \boldsymbol{\theta}) = \prod_{i=2}^{n} f(x_i, t_i | x_{i-1}, t_{i-1}).$$

3.1. The Vasiceck Model

The Vasiceck model describes the short rate's dynamics. It can be used in the evaluation of interest rate derivatives and is more suitable for credit markets. It is specified by the following SDE:

$$dX(t) = [\theta_1 - \theta_2 X(t)]dt + \theta_3\, dW(t), \tag{5}$$

where $\theta_1, \theta_2, \theta_3$ are positive constants. The model (5) with $\theta_1 = 0$ was originally proposed by Ornstein and Uhlenbeck in 1930 in the physical context to describe the velocity of a particle moving in a fluid under the influence of friction and it was then generalized by Vasicek in 1977 to model loan interest rates. It is also used as a model and in physical and biological contexts (see, for instance, [20–23]).

We note that, for $\theta_2 > 0$, the process $X(t)$ is *mean reverting* oscillating around the equilibrium point θ_1/θ_2. The process is defined in $\mathbb{R}$ and the boundaries $\pm\infty$ are natural. The transition pdf of $X(t)$ is given by

$$f(x,t|x_0,t_0) = \frac{1}{\sqrt{2\pi V(t|t_0)}} \exp\left\{-\frac{[x - M(t|x_0,t_0)]^2}{2V(t|t_0)}\right\}, \tag{6}$$

where

$$M(t|x_0,t_0) = \frac{\theta_1}{\theta_2}\left[1 - e^{-\theta_2(t-t_0)}\right] + x_0 e^{-\theta_2(t-t_0)}, \qquad V(t|t_0) = \frac{\theta_3^2}{2\theta_2}\left[1 - e^{-2\theta_2(t-t_0)}\right]$$

represent the mean and the variance of $X(t)$ with the condition that $X(t_0) = x_0$, respectively.

Further, $X(t)$ has the following steady-state density:

$$W(x) = \frac{s(x)}{\int_{-\infty}^{\infty} s(z)\,dz} = \sqrt{\frac{\theta_2}{\pi\theta_3^2}} \exp\left\{-\frac{\theta_2}{\theta_3^2}\left(x - \frac{\theta_1}{\theta_2}\right)^2\right\},$$

which describes a Gaussian distribution with mean θ_1/θ_2 and variance $\theta_3^2/2\,\theta_2$.

Let $\boldsymbol{\theta} = (\theta_1, \theta_2, \theta_3)$ be the vector of the unknown parameters. The maximum likelihood estimate is obtained as $\widehat{\boldsymbol{\theta}} = \arg\max_{\boldsymbol{\theta}} \log L(x_1, \ldots, x_n; \boldsymbol{\theta})$. Implementing this method, making use of the R package sde (see [24,25]), the procedure produces the results shown in Table 1. In the last row of this table, the AIC, i.e.,

$$AIC = 6 - 2\log L(x_1, \ldots, x_n; \widehat{\boldsymbol{\theta}}),$$

is shown for the two datasets.

Table 1. ML estimates of Model (5) for Dataset A (on the left) and for Dataset B (on the right). The last row shows the AIC.

	Vasicek Model			
	Dataset A		**Dataset B**	
	estimate	standard error	estimate	standard error
$\widehat{\theta}_1$	0.9473455	0.62502358	1.9016919	0.41732799
$\widehat{\theta}_2$	0.0658379	0.03621181	0.1675858	0.03403613
$\widehat{\theta}_3$	1.0355084	0.08881145	0.5610075	0.04793382
AIC	207.8207		113.8432	

For Datasets A and B, the Chow test applied to the Euler discretization of Model (5) shows a structural break at time $t = 42$, corresponding to 1 January 2008 (p-value = 0.002726) for Dataset A and at time $t = 47$ corresponding to 1 January 2009 (p-value = 0.006231) for Dataset B. In Table 2, the ML estimates for Datasets A and B are shown considering separately the series before and after these dates. Precisely, we consider for Dataset A the following sub-periods:

- first period: 1 July 1997–1 October 2007;
- second period: 1 January 2008–31 March 2015;

for Dataset B, the sub-intervals are as follows:

- first period: 1 July 1997–1 October 2008;
- second period: 1 January 2009–31 March 2015.

The existence of a structural break is quite clear just looking at the data in Figure 1, but the Chow test permits us to establish the time at which the break verifies, and the AIC values confirm that the estimations evaluated in the two periods work better then the estimates on the whole dataset.

Table 2. ML estimates of Model (5) and the corresponding AIC for the periods indicated by Chow test for Dataset A (on the top) and for Dataset B (on the bottom).

The Vasicek Model				
Dataset A				
	First Period 1 July 1997–1 October 2007		*Second Period* 1 January 2008–31 March 2015	
	estimate	standard error	estimate	standard error
$\widehat{\theta}_1$	5.3865862	2.4412428	4.3464435	1.6780736
$\widehat{\theta}_2$	0.2862057	0.1245413	0.3411949	0.1264711
$\widehat{\theta}_3$	1.2012565	0.1489351	0.8262836	0.1179829
AIC	126.2129		67.89682	
Dataset B				
	First Period 1 July 1997–1 October 2008		*Second Period* 1 January 2009–31 March 2015	
	estimate	standard error	estimate	standard error
$\widehat{\theta}_1$	1.5003276	0.37401583	7.9272605	3.1400451
$\widehat{\theta}_2$	0.1380841	0.02919803	0.6949408	0.2786503
$\widehat{\theta}_3$	0.4386253	0.04613525	0.7898723	0.1424101
AIC	54.51816		48.00966	

In Figure 2, the steady state pdf are plotted for the two datasets, making use of the estimates of the parameter $\boldsymbol{\theta}$ given in Table 1 for the whole period and in Table 2 for the sub-intervals.

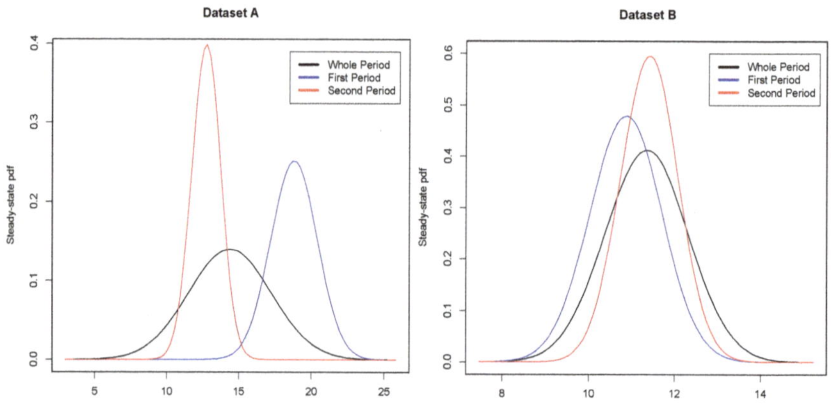

Figure 2. The Vasicek steady-state densities for Datasets A (on the **left**) and B (on the **right**) evaluated by using $\boldsymbol{\theta}$ given in Table 1 for the whole period: 1 July 1997–31 March 2015 and by using the parameters given in Table 2 for the sub-intervals identified by the Chow test.

3.2. The CIR Model

The CIR model, originally introduced by Feller as a model for population growth in 1951, was proposed by John C. Cox, Jonathan E. Ingersoll, and Stephen A. Ross as an extension of the valuation of interest rate derivatives. It describes the evolution of interest rates, and it is characterized by the following SDE:

$$dX(t) = [\theta_1 - \theta_2 X(t)]dt + \theta_3 \sqrt{X(t)}\, dW(t). \tag{7}$$

We point out that Model (7) has widely been used in the literature in the context of neuronal modeling (see, for example, [26–28]).

The process $X(t)$ in (7) is defined in $I = (0, +\infty)$. The nature of the boundaries 0 and $+\infty$ depends on the parameters of the process and establishes the conditions associated with the Kolmogorov and Fokker–Planck equations to determine the transition pdf. In particular, the lower boundary 0 is exit if $\theta_1 \leq 0$, regular if $0 < \theta_1 < \theta_3^2/2$, and entrance if $\theta_1 \geq \theta_3^2/2$, whereas the endpoint $+\infty$ is natural (see [29]). In the following, we assume that $\theta_1, \theta_2, \theta_3$ are positive constants and that $\theta_1 \geq \theta_3^2/2$. This last condition assures that $X(t)$ is strictly positive so that the zero state is unattainable. In this case, the 0 state is an entrance boundary, so that the transition pdf can be obtained solving the Kolmogorov and Fokker–Planck equations with the initial delta condition and a reflecting condition on the zero state. Specifically, denoting

$$h^{-1}(y) = y^{2\theta_1/\theta_3^2} e^{-2\theta_2 y/\theta_3^2}$$

the inverse of the scale function defined in (2), the reflecting condition for the Kolmogorov equation is

$$\lim_{y\to 0} h^{-1}(y) \frac{\partial}{\partial y} f(x, t|y, \tau) = 0,$$

whereas, for the Fokker–Planck equation, it is

$$\lim_{x\to 0} \left\{ \frac{\partial}{\partial x} \left[\frac{\theta_3^2 x}{2} f(x, t|y, \tau) \right] - (\theta_1 - \theta_2 x) f(x, t|y, \tau) \right\} = 0.$$

Therefore, for $\theta_1 \geq \theta_3^2/2$, one obtains

$$f(x, t|x_0, t_0) = \frac{2\theta_2}{\theta_3^2[1 - e^{-\theta_2 t}]} \exp\left\{ -\frac{2\theta_2(x + x_0 e^{-\theta_2 t})}{\theta_3^2[1 - e^{-\theta_2 t}]} \right\} \left(\frac{x}{x_0} e^{-\theta_2 t} \right)^{\theta_1/\theta_3^2 - 1/2}$$
$$\times I_{2\theta_1/\theta_3^2 - 1} \left[\frac{4\theta_2 (e^{\theta_2 t} x\, x_0)^{1/2}}{\theta_3^2 (e^{\theta_2 t} - 1)} \right] \tag{8}$$

where $I_\nu(z)$ denotes the modified Bessel function of the first kind:

$$I_\nu(z) = \sum_{k=0}^{\infty} \frac{(z/2)^{2k+\nu}}{k!\Gamma(\nu + k + 1)}$$

and Γ is the Euler Gamma function:

$$\Gamma(z) = \int_0^{+\infty} t^{z-1} e^{-t}\, dt.$$

The steady-state pdf for $X(t)$ is a Gamma distribution with shape parameter $2\theta_1/\theta_3^2$ and scale parameter $\theta_3^2/2\theta_2$, i.e.,

$$W(x) = \frac{1}{x\Gamma(2\theta_1/\theta_3^2)} \left(\frac{2\theta_2}{\theta_3^2} x \right)^{2\theta_1/\theta_3^2} \exp\left\{ -\frac{2\theta_2}{\theta_3^2} x \right\}.$$

For Model (7), in Table 3, the maximum likelihood estimates of the parameters and the standard errors and the AIC values are shown for Datasets A and B. Moreover, in the last row of this table, the AIC is shown for the two datasets.

Note that the Chow test applied to the Euler discretization of Model (7) produces the same results that the Vasiceck model does. Indeed, Models (5) and (7) show the same trend, but in the CIR model one assumes residuals heteroschedasticity that does not bias the parameter estimates; it only makes the standard errors incorrect. Moreover, in Table 4, the estimates of the parameters, the standard errors, and the AIC values are shown before and after the structural breaks indicated by the Chow test. In addition, in this case, the estimates for the two separated periods work better than the estimates using only one model for the whole period shown from the AIC values.

Table 3. ML estimates of Model (7) for Dataset A (on the left) and for Dataset B (on the right). Last row shows the AIC.

	CIR Model			
	Dataset A		Dataset B	
	estimate	standard error	estimate	standard error
$\widehat{\theta}_1$	0.87234371	0.57486263	0.71000000	0.69208186
$\widehat{\theta}_2$	0.06140606	0.03471271	0.06913925	0.05744829
$\widehat{\theta}_3$	0.24781675	0.02122343	0.16565230	0.01524389
AIC	204.3006		123.7508	

In Figure 3, the steady state pdf are plotted for the two datasets making use of the estimates of the parameter $\boldsymbol{\theta}$ given in Table 3 for the whole period and in Table 4 for the sub-intervals.

Table 4. ML estimates of Model (7) and the corresponding AIC for the periods indicated by the Chow test for Dataset A (on the top) and for Dataset B (on the bottom).

	CIR Model			
	Dataset A			
	before 1 January 2008		after 1 January 2008	
	estimate	standard error	estimate	standard error
$\widehat{\theta}_1$	0.70000000	1.96946686	0.50000000	1.5237235
$\widehat{\theta}_2$	0.04645403	0.10114005	0.05015628	0.1157965
$\widehat{\theta}_3$	0.25705245	0.02838514	0.21809718	0.0293771
AIC	130.8434		73.32189	
	Dataset B			
	before 1 January 2009		after 1 January 2009	
	estimate	standard error	estimate	standard error
$\widehat{\theta}_1$	0.55000000	0.7378984	0.53234000	2.02364571
$\widehat{\theta}_2$	0.06276713	0.0587019	0.03746054	0.18050398
$\widehat{\theta}_3$	0.12443707	0.0159996	0.21651927	0.03194828
AIC	58.07491		75.83153	

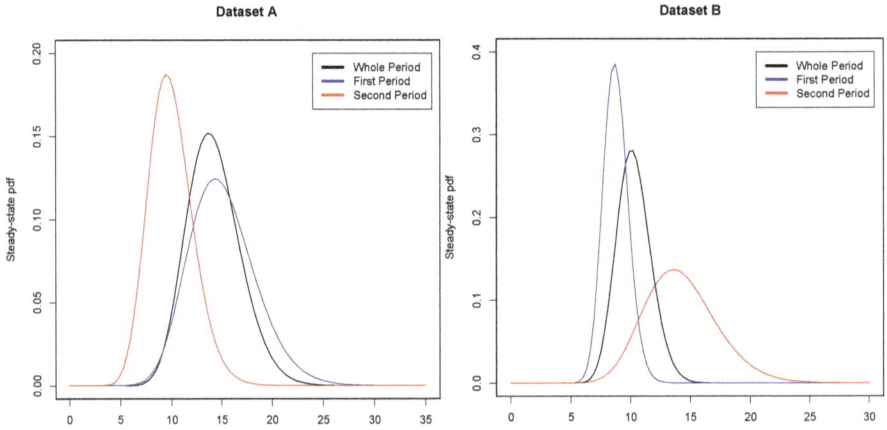

Figure 3. CIR steady-state densities for Datasets A (on the **left**) and B (on the **right**) evaluated by using θ given in Table 3 for the whole period: 1 July 1997–31 March 2015 and by using the parameters given in Table 4 for the sub-intervals identified by the Chow test.

4. FPT Analysis for TEGM

In this section, we consider the FPT analysis for the Vasicek model. This choice is motivated by the results of Section 3. Indeed, comparing the two models by looking at the AIC values, we can see that the Vasicek model better fits our datasets in all cases (only for Dataset A does the CIR model work better than the Vasicek model).

For Datasets A and B, due to the Markovianity of the process, we consider the estimates relative to the second periods (1 January 2008–31 March 2015 for Dataset A and 1 January 2009–31 March 2015 for Dataset B) shown in Table 2.

By using the recursive Equation (4), we obtain the estimates of FPT moments for Dataset A on the period 1 January 2008–31 March 2015. In Table 5, these estimates are shown for various values of S (on the top), with $x_0 = 13.28$ corresponding to the mean of the data in the considered period, and various values of the initial point x_0 (on the bottom) fixing the alert boundary $S = 15$.

Table 6 shows the analogous analysis of Table 5 for Dataset B, with $x_0 = 12.1636$ and $S = 14$.

We note that, by increasing the distance between S and x_0, the mean FPT increases. From an economic point of view, the choice of such a distance can be interpreted as a choice of "propensity of risk" of an available loan. Figure 4 shows the mean FPT (quarters starting from the loan deposit) as a function of the alert boundary (up) and as a function of the initial point x_0. Clearly, each applicant knows the initial point x_0 and can choose the "alert boundary" S.

Table 5. For Dataset A, second period, mean, second order moment, and variance of the random variable T_{x_0} through various values of the threshold S (on the top) and for various values of x_0 (on the bottom).

$x_0 = 13.28$	S	$t_1(S\|x_0)$	$t_2(S\|x_0)$	$Var(S\|x_0)$
	14.0	6.780026	1.312067×10^2	8.5238×10^1
	14.2	1.018325×10	2.612374×10^2	1.575387×10^2
	14.4	1.488234×10	5.135466×10^2	2.920625×10^2
	14.6	2.157931×10	1.021095×10^3	5.554285×10^2
	14.8	3.144937×10	2.089585×10^3	1.100522×10^3
	15	4.651822×10	4.462957×10^3	2.299013×10^3
	15.2	7.038275×10	1.00658×10^4	5.112068×10^3
	15.4	1.096303×10^2	2.421257×10^4	1.219378×10^4
	15.6	1.76713×10^2	6.262414×10^4	3.139667×10^4
	15.8	2.959467×10^2	1.752716×10^5	8.768709×10^4
	16	5.16416×10^2	5.332541×10^5	2.665686×10^5
$S = 15$	x_0	$t_1(S\|x_0)$	$t_2(S\|x_0)$	$Var(S\|x_0)$
	12.0	5.116115×10	4.937052×10^3	2.319589×10^3
	12.2	1.808844×10^2	6.413534×10^4	6.413534×10^4
	12.4	1.803493×10^2	6.394063×10^4	3.141475×10^4
	12.6	1.797354×10^2	6.371754×10^4	3.141272×10^4
	12.8	1.79022×10^2	6.345886×10^4	3.140997×10^4
	13	1.781812×10^2	6.315425×10^4	3.140572×10^4
	13.2	1.77174×10^2	6.279051×10^4	3.139988×10^4
	13.4	1.759456×10^2	6.234752×10^4	3.139065×10^4
	13.6	1.74417×10^2	6.179734×10^4	3.137605×10^4
	13.8	1.724714×10^2	6.109852×10^4	3.135214×10^4
	14	1.699329×10^2	6.018864×10^4	3.131143×10^4

Table 6. For Dataset B, second period, mean, second order moment, and variance of the random variable T_{x_0} through the threshold S for various values of S and of x_0.

$x_0 = 12.1636$	S	$t_1(S\|x_0)$	$t_2(S\|x_0)$	$Var(S\|x_0)$
	13	0.2653719×10^2	1.568046×10^3	8.638236×10^2
	13.2	0.5332441×10^2	5.985332×10^3	3.141839×10^3
	13.4	1.134217×10^2	2.631623×10^4	1.345175×10^4
	13.6	2.606306×10^2	1.371027×10^5	6.917443×10^4
	13.8	6.547357×10^2	8.602519×10^5	4.315731×10^5
	14	1.808475×10^3	6.548544×10^6	3.277963×10^6
	14.2	5.502628×10^3	6.05788×10^7	3.029989×10^7
	14.4	1.844072×10^4	6.801859×10^8	3.401257×10^8
	14.6	6.800683×10^4	9.250082×10^9	4.625154×10^9
	14.8	2.75718×10^5	1.520417×10^{11}	7.602126×10^{10}
	15	1.227843×10^6	3.015198×10^{12}	1.507601×10^{12}
$S = 14$	x_0	$t_1(S\|x_0)$	$t_2(S\|x_0)$	$Var(S\|x_0)$
	11	1.813076×10^3	6.565252×10^3	3.278009×10^6
	11.2	1.812688×10^3	6.563835×10^6	3.277998×10^6
	11.4	1.81221×10^3	6.562115×10^6	3.27801×10^6
	11.6	1.811604×10^3	6.559909×10^6	3.277999×10^6
	11.8	1.810807×10^3	6.557028×10^6	3.278004×10^6
	12.	1.809714×10^3	6.553067×10^6	3.278×10^6
	12.2	1.80814×10^3	6.547343×10^6	3.277975×10^6
	12.4	1.805734×10^3	6.538643×10^6	3.277966×10^6
	12.6	1.801816×10^3	6.52444×10^6	3.2779×10^6
	12.8	1.794949×10^3	6.499556×10^6	3.277714×10^6
	13	1.781937×10^3	6.452422×10^6	3.277121×10^6

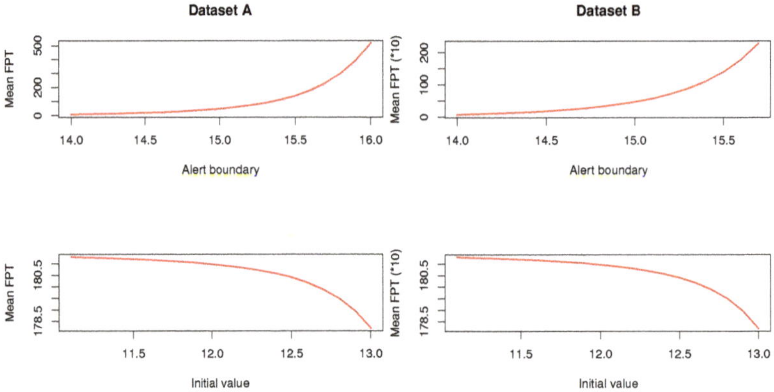

Figure 4. Mean FPT versus the alert boundary ($x_0 = 13.28$) and the initial value x_0 ($S = 15$) for Dataset A, second period (**left**), and for Dataset B, second period (**right**).

5. Conclusions

This paper addresses stochastic modeling of loan interest rate dynamics according to the current laws against usury. Such modeling states an upper bound, above which an interest rate is considered a usury rate and illegal. Here we focus on the Italian case and consider two models commonly used in short-term loan rates, i.e., the Vasicek and CIR models. We propose a strategy based on FPT through an alert boundary, above which the rate is considered at the risk of usury and hence has to be kept under control. Moreover, the mean first exit time through the alert boundary can be an indicator of the "goodness" of the loan, in the sense that an applicant, when he/she is choosing between two loan offers, should choose the one with a higher mean exit time from the alert boundary.

The procedure was applied to a historical series of Italian average rates on loans in the period from 1 July 1997 to 31 March 2015. We considered "one-fifth of salary transfer" and two amount classes were analyzed: (a) up to 10 million lira (until 31 December 2001) and up to € 5000 (after 2002); and (b) above 10 million lire (until 31 December 2001) and above € 5000 (after 2002). The model parameters were estimated by MLE, and a Chow test was applied to detect the presence of structural breaks in our datasets.

The model and proposed strategy are apt for further development. Indeed, we can extend the analysis to more general processes in which some parameters are time-dependent, or we can consider time-dependent thresholds to model varying loan interest rates. Further generalization can include analysis of FPT through two boundaries: the upper one describing an alert threshold and the lower one representing a favorable interest rate.

Author Contributions: Giuseppina Albano and Virginia Giorno contributed equally in the writing of this work.

Acknowledgments: This paper was supported partially by INDAM-GNCS.

Conflicts of Interest: The authors declare no conflict of interest.

References

1. Brigo, B.; Mercurio, F. *Interest Rate Models-Theory and Practice with Smile, Inflation and Credit*; Springer: Berlin, Germany, 2007.
2. Buetow, G.W., Jr.; Hanke, B.; Fabozzi, F.J. Impact of different interest rate models on bound value measures. *J. Fixed Income* **2001**, *3*, 41–53. [CrossRef]
3. Fan, J.; Jiang, J.; Zhang, C.; Zhou, Z. Time-dependent diffusion models for term structure dynamics. *Stat. Sin.* **2003**, *13*, 965–992.
4. Chen, W.; Xu, L.; Zhu, S.P. Stock loan valuation under a stochastic interest rate model. *Comput. Math. Appl.* **2015**, *70*, 1757–1771. [CrossRef]
5. Di Lorenzo, E.; Orlando, A.; Sibillo, M. A Stochastic model fro loan interest rates. *Banks Bank Syst.* **2013**, *8*, 94–99.
6. Cox, J.C.; Ingersoll, J.E.; Ross, S.A. A Theory of the Term Structure of Interest Rates. *Econometrica* **1985**, *53*, 385–407. [CrossRef]
7. Vasicek, O. An equilibrium characterization of the term structure. *J. Financ. Econ.* **1977**, *5*, 177–188. [CrossRef]
8. Khramov, V. *Estimating Parameters of Short-Term Real Interest Rate Models*; IMF Working Papers; International Monetary Fund: Washington, DC, USA, 2012.
9. Ricciardi, L.M.; Di Crescenzo, A.; Giorno, V.; Nobile, A.G. An outline of theoretical and algorithmic approaches to first passage time problems with applications to biological modeling. *Math. Jpn.* **1999**, *50*, 247–322.
10. Albano, G.; Giorno, V.; Román-Román, P.; Torres-Ruiz, F. On the effect of a therapy able to modify both the growth rates in a Gompertz stochastic model. *Math. Biosci.* **2013**, *245*, 12–21. [CrossRef] [PubMed]
11. Albano, G.; Giorno, V.; Román-Román, P.; Torres-Ruiz, F. On a Non-homogeneous Gompertz-Type Diffusion Process: Inference and First Passage Time. In *Computer Aided Systems Theory-EUROCAST*; Moreno Diaz, R., Pichler, F., Quesada Arencibia, A., Eds.; Lecture Notes in Computer Science; Springer: Cham, Switzerland, 2017; Volume 10672, pp. 47–54.
12. Buonocore, A.; Nobile, A.G.; Ricciardi, L.M. A new integral equation for the evaluation of first-passage-time probability densitie. *Adv. Appl. Prob.* **1987**, *19*, 784–800. [CrossRef]
13. Giorno, V.; Nobile, A.G.; Ricciardi, L.M., Sato, S. On the evaluation of firtst-passage-time probability densities via non-singular integral equations. *Adv. Appl. Prob.* **1989**, *21*, 20–36. [CrossRef]
14. Buonocore, A.; Caputo, L.; D'Onofrio, G.; Pirozzi, E. Closed-form solutions for the first-passage-time problem and neuronal modeling. *Ricerche di Matematica* **2015**, *64*, 421–439. [CrossRef]
15. Di Crescenzo, A.; Giorno, V.; Nobile, A.G. Analysis of reflected diffusions via an exponential time-based transformation. *J. Stat. Phys.* **2016**, *163*, 1425–1453. [CrossRef]

16. Abundo, M. The mean of the running maximum of an integrated Gauss–Markov process and the connection with its first-passage time. *Stoch. Anal. Appl.* **2017**, *35*, 499–510. [CrossRef]
17. Román-Román, P.; Serrano-Pérez, J.; Torres-Ruiz, F. More general problems on first-passage times for diffusion processes: A new version of the fptdapprox r package. *Appl. Math. Comput.* **2014**, *244*, 432–446. [CrossRef]
18. Román-Román, P.; Serrano-Pérez, J.; Torres-Ruiz, F. fptdapprox: Approximation of First-Passage-Time Densities for Diffusion Processes, R Package Version 2.1, 2015. Available online: http://cran.r-project.org/package=fptdApprox (accessed on 23 November 2017).
19. Linetsky, V. Computing hitting time densities for CIR and OU diffusions: Applications to mean reverting models. *J. Comput. Financ.* **2004**, *4*, 1–22. [CrossRef]
20. Albano, G.; Giorno, V. A stochastic model in tumor growth. *J. Theor. Biol.* **2006**, *242*, 229–236. [CrossRef] [PubMed]
21. Buonocore, A.; Caputo, L.; Nobile, A.G.; Pirozzi, E. Restricted Ornstein–Uhlenbeck process and applications in neuronal models with periodic input signals. *J. Comp. Appl. Math.* **2015**, *285*, 59–71. [CrossRef]
22. Dharmaraja, S.; Di Crescenzo, A.; Giorno, V.; Nobile, A.G. A continuous-time Ehrenfest model with catastrophes and its jump-diffusion approximation. *J. Stat. Phys.* **2015**, *161*, 326–345. [CrossRef]
23. Giorno, V.; Spina, S. On the return process with refractoriness for non-homogeneous Ornstein–Uhlenbeck neuronal model. *Math. Biosci. Eng.* **2014**, *11*, 285–302. [PubMed]
24. Iacus, S.M. *Simulation and Inference for Stochastic Differential Equations with R Examples*; Springer Series in Statistics: Berlin, Germany, 2008.
25. Iacus, S.M. sde-Manual and Help on Using the sde R Package. 2009. Available online: http://www.rdocumentation.org/packages/sde/versions/2.0.9 (accessed on 24 October 2017).
26. Giorno, V.; Lanský, P.; Nobile, A.G.; Ricciardi, L.M. Diffusion approximation and first-passage-time problem for a model neuron. III. A birth-and-death process approach. *Biol. Cybern.* **1988**, *58*, 387–404. [CrossRef] [PubMed]
27. Giorno, V.; Nobile, A.G.; Ricciardi, L.M. Single neuron's activity: On certain problems of modeling and interpretation. *BioSystems* **1997**, *40*, 65–74. [CrossRef]
28. Nobile, A.G.; Pirozzi, E. On time non-homogeneous Feller-type diffusion process in neuronal modeling. In *Computer Aided Systems Theory-EUROCAST 2017*; Moreno Diaz, R., Pichler, F., Quesada Arencibia, A., Eds.; Lecture Notes in Computer Science, Springer: Cham, Switzerland, 2015; Volume 9520, pp. 183–191.
29. Karlin, S.; Taylor, H.W. *A Second Course in Stochastic Processes*; Academic Press: New York, NY, USA, 1981.

mathematics

MDPI

Article

Fractional Queues with Catastrophes and Their Transient Behaviour

Giacomo Ascione [1], Nikolai Leonenko [2] and Enrica Pirozzi [1,*]

[1] Dipartimento di Matematica e Applicazioni "Renato Caccioppoli", Università degli Studi di Napoli Federico II, 80126 Napoli, Italy; giacomo.ascione@unina.it

[2] School of Mathematics, Cardiff University, Cardiff CF24 4AG, UK; LeonenkoN@cardiff.ac.uk

* Correspondence: enrica.pirozzi@unina.it

Received: 27 July 2018; Accepted: 31 August 2018; Published: 6 September 2018

Abstract: Starting from the definition of fractional M/M/1 queue given in the reference by Cahoy et al. in 2015 and M/M/1 queue with catastrophes given in the reference by Di Crescenzo et al. in 2003, we define and study a fractional M/M/1 queue with catastrophes. In particular, we focus our attention on the transient behaviour, in which the time-change plays a key role. We first specify the conditions for the global uniqueness of solutions of the corresponding linear fractional differential problem. Then, we provide an alternative expression for the transient distribution of the fractional M/M/1 model, the state probabilities for the fractional queue with catastrophes, the distributions of the busy period for fractional queues without and with catastrophes and, finally, the distribution of the time of the first occurrence of a catastrophe.

Keywords: fractional differential-difference equations; fractional queues; fractional birth-death processes; busy period

MSC: 60K25; 60J80

1. Introduction

Stochastic models for queueing systems have a wide range of applications in computer systems, sales points, telephone or telematic systems and also in several areas of science including biology, medicine and many others. The well known M/M/1 queueing model [1–5] constitutes the theoretical basis for building many other refined models for service systems.

Due to the Markov nature of its inter-arrival times of the customers and of its service times, the model can be mathematically treated in a simple manner, and, for this reason, it is widely used in many modeling contexts. Nevertheless, in the past few decades, the advent of fractional operators, such as fractional derivatives and integrals (see, for instance, [6] and [7] and references therein), has made it clear that different time scales, themselves random, that preserve memory (therefore not Markovian), allow the construction of more realistic stochastic models.

The introduction of the fractional Caputo derivative into the system of differential-difference equations for an M/M/1-type queue was done in [8], where, for a fractional M/M/1 queue, the state probabilities were determined. In this kind of queue model, the inter-arrival times and service times are characterized by Mittag–Leffler distributions [9]; in this case, the model does not have the property of memory loss that is typical of the exponential distributed times of the classical M/M/1 model. Indeed, a time-changed birth-death process [10,11], by means of an inverse stable subordinator [12], solves the corresponding fractional system of differential-difference equations and fractional Poisson processes [13] characterize the inter-arrival and service times.

The fractional M/M/1 model in [8] is an interesting and powerful model, not only because it is a generalization of the classical one, where the fractional order is set to 1, but also because its range of applications is extremely wide. Its importance can be further augmented by including in the model the occurrence of catastrophes, as it was considered in [14] for the classical M/M/1.

The catastrophe is a particular event that occurs in a random time leading to the instantaneous emptying of the system, or to a momentary inactivation of the system, as, for example, the action of a virus program that can make a computer system inactive [15]; other applications of models with catastrophes can be found in population dynamics and reliability contexts (see [16] and references therein).

Motivated by the mathematical need to enrich the fractional M/M/1 model of [8] with the inclusion of catastrophes, we study in this paper the above model; specifically, we determine the transient distribution, the distribution of the busy period (including that of the fractional M/M/1 queue of [8]) and the probability distribution of the time of the first occurrence of the catastrophe.

For these purposes, we need to guarantee the global uniqueness of the solution of the considered linear fractional Cauchy problem on Banach spaces. After recalling the definitions and known results in Section 2, we address the problem of uniqueness in Section 3. In Section 2, we also provide the transient distribution of the fractional M/M/1 model in an alternative form to that given in [8]. In Section 4, the distribution of the busy period for the fractional M/M/1 queue (without catastrophes) is obtained. Here, the time-changed birth-death process plays a key role to derive the results. In Section 5, we define the fractional queue with catastrophes; we are able to obtain the distribution of the transient state probabilities by following a strategy similar to that in [14]. We also found the distribution of the busy period and of the time of the first occurrence of the catastrophe starting from the empty system. Some special operators and functions used in this paper are specified in the Appendices A and B.

2. Definition of a Fractional Process Related to M/M/1 Queues

The classical M/M/1 queue process $N(t), t \geq 0$ can be described as continuous time Markov chain whose state space is the set $\{0, 1, 2, \dots\}$ and the state probabilities

$$p_n(t) = \mathbb{P}(N(t) = n | N(0) = 0),\ n = 0, 1, 2 \dots \tag{1}$$

satisfy the following differential-difference equations:

$$\begin{cases} D_t p_n(t) = -(\alpha + \beta) p_n(t) + \alpha p_{n-1}(t) + \beta p_{n+1}(t), & n \geq 1, \\ D_t p_0(t) = -\alpha p_0(t) + \beta p_1(t), & \\ p_n(0) = \delta_{n,0}, & n \geq 0, \end{cases} \tag{2}$$

where $\delta_{n,0}$ is the Kroeneker delta symbol, $D_t = \frac{d}{dt}$ and $\alpha, \beta > 0$ are the entrance and service rates, respectively.

Let $S_\nu(t), t \geq 0, \nu \in (0,1)$ be the Lévy ν-stable subordinator with Laplace exponent given by:

$$\log \mathbb{E} e^{-z S_\nu(t)} = -t z^\nu,\ z > 0.$$

Consider the inverse ν-stable subordinator

$$L_\nu(t) = \inf\{u \geq 0 :\ S_\nu(u) > t\},\ t \geq 0.$$

For $0 < \nu < 1$, the fractional M/M/1 queue process $N^\nu(t),\ t \geq 0$ is defined by a non-Markovian time change $L_\nu(t)$ independent of $N(t), t \geq 0$, i.e.,

$$N^\nu(t) = N(L_\nu(t)),\ t \geq 0. \tag{3}$$

This process was defined in [8] and it is non-Markovian with non-stationary and non-independent increments. For $\nu = 1$, by definition, $N^1(t) = N(t),\ t \geq 0$. Then, for a fixed $\nu \in (0,1]$, the state probabilities

$$p_n^\nu(t) = \mathbb{P}\{N^\nu(t) = n | N^\nu(0) = 0\},\ n = 0, 1, \dots \tag{4}$$

of the number of customers in the system at time t in the fractional M/M/1 queue are characterized by arrivals and services determined by fractional Poisson processes of order $\nu \in (0,1]$ [13] with parameters α and β. They are solutions of the following system of differential-difference equations

$$\begin{cases} {}_0^C D_t^\nu p_n^\nu(t) = -(\alpha+\beta)p_n^\nu(t) + \alpha p_{n-1}^\nu(t) + \beta p_{n+1}^\nu(t), & n \geq 1, \\ {}_0^C D_t^\nu p_0^\nu(t) = -\alpha p_0^\nu(t) + \beta p_1^\nu(t), & \\ p_n^\nu(0) = \delta_{n,0}, & n \geq 0, \end{cases} \tag{5}$$

where ${}_0^C D_t^\nu$ is the Caputo fractional derivative (see Appendix A).

Using Equation (5) and representation (3), the state probabilities are obtained in [8]:

$$\begin{aligned} p_n^\nu(t) &= \left(1 - \frac{\alpha}{\beta}\right)\left(\frac{\alpha}{\beta}\right)^n \\ &+ \left(\frac{\alpha}{\beta}\right)^n \sum_{k=0}^{+\infty} \sum_{m=0}^{n+k} \frac{k-m}{k+m} \binom{k+m}{k} \alpha^k \\ &\times \beta^{m-1} t^{\nu(k+m)-\nu} E_{\nu,\nu(k+m)-\nu+1}^{k+m}(-(\alpha+\beta)t^\nu), \end{aligned} \tag{6}$$

as well as its Laplace transform

$$\begin{aligned} \pi_n^\nu(z) = \int_0^{+\infty} e^{-zt} p_n^\nu(t) dt &= \left(1 - \frac{\alpha}{\beta}\right)\left(\frac{\alpha}{\beta}\right)^n \frac{1}{z} + \\ &+ \left(\frac{\alpha}{\beta}\right)^n \sum_{k=0}^{+\infty} \sum_{m=0}^{n+k} \frac{k-m}{k+m} \binom{k+m}{k} \alpha^k \\ &\times \beta^{m-1} \frac{z^{\nu-1}}{(z^\nu + \alpha + \beta)^{k+m}},\ z > 0. \end{aligned}$$

In Equation (6), the functions $E_{\nu,\mu}^\rho$ are generalized Mittag–Leffler functions (see Appendix B). Note that $p_n^\nu(t) \geq 0\ \forall n \geq 0$ and $\sum_{n=0}^{+\infty} p_n^\nu(t) = 1$.

Alternatively, let $h_\nu(t,x) = \dfrac{d}{dx}\mathbb{P}\{L_\nu(t) \leq x\},\ x \geq 0$, be the density of $L_\nu(t)$; then it is known (see, i.e., [17]) that

$$\int_0^{+\infty} e^{-sx} h_\nu(t,x) dx = E_\nu(-st^\nu),\ s \geq 0, \tag{7}$$

and (see, i.e., [18], Proposition 4.1)

$$h_\nu(t,x) = \frac{1}{\pi}\int_0^{+\infty} u^{\nu-1} e^{-tu - xu^\nu \cos(\nu\pi)} \sin\left(\pi\nu - xu^\nu \sin(\pi\nu)\right) du,\ x \geq 0. \tag{8}$$

Using (7) and an analytical expression for $p_n^\nu(t)$ given in [19], we can write down an alternative expression for (6) as

$$
\begin{aligned}
p_n^\nu(t) = & \left(\frac{p}{q}\right)^n \sum_{r=n}^{\infty} \frac{(\alpha+\beta)^r}{r!} t^{r\nu} E_\nu^{(r)}(-(\alpha+\beta)t^\nu) \\
& \times \sum_{r=0}^{\left[\frac{r-n}{2}\right]} \frac{r+1-2k}{r+1} \binom{r+1}{k} p^k q^{r-k},
\end{aligned} \tag{9}
$$

where $p = \dfrac{\alpha}{\alpha+\beta}$, $q = \dfrac{\beta}{\alpha+\beta}$, and $E_\nu^{(r)}(-(\alpha+\beta)t^\nu)$ is the $r-$th derivative of the function $E_\nu(z)$ evaluated at $z = -(\alpha+\beta)t^\nu$.

Actually, it is easy to see from (7) that

$$
\int_0^{+\infty} e^{-sx} x^r h_\nu(t,x)dx = E_\nu^{(r)}(-st^\nu)t^{\nu r};
$$

thus, using [19] and (3), we have

$$
\begin{aligned}
p_n^\nu(t) &= \int_0^{+\infty} p_n(s)h_\nu(t,s)ds \\
&= \left(\frac{p}{q}\right)^n \sum_{r=n}^{\infty} \frac{(\alpha+\beta)^r}{r!} \int_0^{+\infty} e^{-(\alpha+\beta)s} s^\nu h_\nu(t,s)ds \\
&\times \sum_{r=0}^{\left[\frac{r-n}{2}\right]} \frac{r+1-2k}{r+1} \binom{r+1}{k} p^k q^{r-k},
\end{aligned}
$$

and formula (9) follows. On the other hand, using (8), we have

$$
\begin{aligned}
p_n^\nu(t) &= \int_0^{+\infty} p_n(s)h_\nu(t,s)ds \\
&= \left(\frac{p}{q}\right)^n \sum_{r=n}^{\infty} \frac{(\alpha+\beta)^r}{r!} \int_0^{+\infty} u^{\nu-1} e^{-tu} F_{\nu,r}(u)du \\
&\times \sum_{r=0}^{\left[\frac{r-n}{2}\right]} \frac{r+1-2k}{r+1} \binom{r+1}{k} p^k q^{r-k},
\end{aligned}
$$

where

$$
F_{\nu,r}(u) = \frac{1}{\pi} \int_0^{+\infty} \exp\left\{-(\alpha+\beta)x - xu^\nu \cos(\nu\pi)\right\} x^r \sin\left(\pi\nu - xu^\nu \sin(\pi\nu)\right) dx.
$$

3. Linear Fractional Cauchy Problems on Banach Spaces

In order to describe the transient probabilities for our queues, we will need some uniqueness results for solutions of linear fractional Cauchy problems defined on Banach spaces. To do that, let us recall the following Theorem (Theorem 3.19 from [20]):

Theorem 1. *Let $(X, |\cdot|)$ be a Banach space and $J = [0,T]$ for some $T > 0$. Consider the ball $B_R = \{x \in X : |x| \leq R\}$. Let $\nu \in (0,1)$ and $f : J \times B_R \to X$ and consider the following Cauchy problem:*

$$
\begin{cases}
{}_0^C D_t^\nu x(t) = f(t,x(t)), \\
x(0) = x_0,
\end{cases} \tag{10}
$$

where ${}_0^C D_t^\nu$ is the Caputo derivative operator (see Appendix A).
Suppose that:

- *$f \in C(J \times B_R, X)$;*
- *There exists a constant $\overline{M}(R) > 0$ such that*

$$|f(t,x(t))| \leq \overline{M}(R)$$

for all $x \in B_R$ and $t \in J$ and such that

$$R \geq |x_0| + \frac{\overline{M}(R)T^{\nu}}{\Gamma(\nu+1)};$$

- *There exists a constant $\overline{L} > 0$ such that $\overline{L} \geq \frac{2\overline{M}(R)}{\Gamma(\nu+1)}$;*
- *There exists a constant $\overline{L}_0 > 0$ such that*

$$|f(t,x_1) - f(t,x_2)| \leq \overline{L}_0 |x_1 - x_2|$$

for all $x_1, x_2 \in B_R$ and $t \in J$;
- *There exist constants $\nu_1 \in (0,\nu)$ and $\tau > 0$ such that*

$$\overline{L}_A = \frac{\overline{L}_0}{\Gamma(\nu)} \frac{T^{(1+\beta)(1-\nu_1)}}{(1+\beta)^{1-\nu_1}} \left(\frac{\nu_1}{\tau}\right)^{\nu_1} < 1,$$

where $\beta = \frac{\nu-1}{1-\nu_1}$.

Then, if $x_0 \in B_R$, the problem (10) admits a unique solution $x^* \in C_\nu(J, B_R)$.

The previous theorem can be easily adapted to the case in which $J = [t_0, T + t_0]$ and the starting point of the derivative is t_0. Since we are interested in linear (eventually non-homogeneous) equations, let us show how the previous theorem can be adapted in such a case.

Corollary 1. *Consider the system (10) and suppose $f(t,x) = Ax + \xi$ where $A : X \to X$ is a linear and continuous operator and $\xi \in X$. Then, there exists a $R > |x_0|$ and $T > 0$ such that the system admits a unique solution $x^* \in C_\nu(J, B_R)$.*

Proof. Observe that, if $|x| \leq R$, then

$$|f(x)| \leq \|A\||x| + |\xi| \leq \|A\|R + |\xi|.$$

Let us choose T such that the conditions of Theorem 1 are verified. To do that, consider $\overline{M}(R) = \|A\|R + |\xi|$. Fix $R \geq |x_0|$ and define $\tilde{R} = R + \varepsilon$ for some $\varepsilon > 0$. Define then

$$T = \left[\frac{\varepsilon\Gamma(\nu+1)}{\overline{M}(\tilde{R})}\right]^{\frac{1}{\nu}}$$

and observe that

$$|x_0| + \frac{\overline{M}(\tilde{R})T^{\nu}}{\Gamma(\nu+1)} = |x_0| + \varepsilon \leq R + \varepsilon = \tilde{R}.$$

Thus, one can fix $\overline{L} = \frac{2\overline{M}(\tilde{R})}{\Gamma(\nu+1)}$ and $\overline{L}_0 = \|A\|$. Moreover, since for fixed $\nu_1 \in (0,\nu)$ the function $\tau \mapsto \overline{L}_A(\tau)$ is decreasing and $\lim_{\tau \to 0} \overline{L}_A(\tau) = 0$, then one can easily find a $\tau > 0$ such that $\overline{L}_A(\tau) < 1$. Since we are under the hypotheses of Theorem 1, then we have shown the local existence and uniqueness of a solution $x^* \in C_\nu(J, B_{\tilde{R}})$. □

However, using such corollary, we can only afford local uniqueness. Global uniqueness of the solution of the Cauchy problem (10) can be obtained with the additional hypothesis that such solution is uniformly bounded:

Corollary 2. *Suppose we are under the hypotheses of Corollary 1. If there exists a solution $x^* \in C([0,+\infty[,X)$ and a constant $k > 0$ such that for any $t \geq 0$ we have $|x^*(t)| \leq k$, then such solution is unique.*

Proof. Observe that $|x_0| \leq k$ and then fix $\tilde{R} = k + \varepsilon$. Define

$$\Delta T = \left[\frac{\varepsilon\Gamma(\nu+1)}{\overline{M}(\tilde{R})}\right]^{\frac{1}{\nu}}.$$

Fix $T_1 = \Delta T$ and observe that, by using Corollary 1, there exists a unique solution in $[0, T_1]$. Since x^* is a solution of such problem, we have that x^* is unique. Suppose we have defined T_{n-1} such that x^* is the unique solution of system (10) in $[0, T_{n-1}]$. Consider the problem

$$\begin{cases} {}^{C}_{T_{n-1}}D^{\nu}_{t}x(t) = f(x(t)), \\ x(T_{n-1}) = x^*(T_{n-1}). \end{cases} \tag{11}$$

Define then $T_n = T_{n-1} + \Delta T$ and observe that, since $|x^*(T_{n-1})| \leq k$, by using Corollary 1, there exists a unique solution in $[T_{n-1}, T_n]$.

By using a change of variables, it is easy to show that

$${}^{C}_{T_{n-1}}D^{\nu}_{t}x = {}^{C}_{0}D^{\nu}_{t-T_{n-1}}\tilde{x},$$

where $\tilde{x} : t \mapsto x(t + T_{n-1})$. By using such relation, we have that system (11) is equivalent to

$$\begin{cases} {}^{C}_{0}D^{\nu}_{t}\tilde{x}(t) = f(\tilde{x}(t)), \\ \tilde{x}(0) = x^*(T_{n-1}), \end{cases}$$

whose unique solution is $\tilde{x}(t) = x^*(t + T_{n-1})$ so that $x(t) = x^*(t)$ and $x^*(t)$ is the unique solution of system (10) in $[0, T_n]$. Since $T_n \to +\infty$ as $n \to +\infty$, we have global uniqueness of limited solutions. □

4. The Fractional M/M/1 Queue

Let us consider again the fractional M/M/1 process $N^{\nu}(t)$, $t \geq 0$ defined by (3) with state probabilites in (6).

Consider the Hilbert space $(l^2(\mathbb{R}), |\cdot|_2)$ with the norm $|x|_2^2 = \sum_{k=0}^{+\infty} x_k^2$ and let $C_{\nu}([0,T], l^2(\mathbb{R}))$ be the space of the ν-Hölder continuous functions from $[0,T]$ to $l^2(\mathbb{R})$. One can rewrite the system (5) in $l^2(\mathbb{R})$ as follows:

$$\begin{cases} {}^{C}_{0}D^{\nu}_{t}\boldsymbol{p}^{\nu}(t) = A_0\boldsymbol{p}^{\nu}(t), \\ \boldsymbol{p}^{\nu}(0) = (\delta_{n,0})_{n\geq 0}, \end{cases} \tag{12}$$

where $\boldsymbol{p}^{\nu}(t) = (p^{\nu}_n(t))_{n\geq 0} \in C([0,T], l^2(\mathbb{R}))$ and

$$A_0 = \begin{pmatrix} -\alpha & \beta & 0 & 0 & 0 & \cdots \\ \alpha & -(\alpha+\beta) & \beta & 0 & 0 & \cdots \\ 0 & \alpha & -(\alpha+\beta) & \beta & 0 & \cdots \\ 0 & 0 & \alpha & -(\alpha+\beta) & \beta & \cdots \\ \vdots & \vdots & \vdots & \ddots & \ddots & \ddots \end{pmatrix}$$

is an infinite tridiagonal matrix with $A_0 = (a_{i,j})_{i,j\geq 0}$. Let us show the following:

Lemma 1. *The linear operator A_0 is continuous and $\|A_0\| \leq 2(\alpha+\beta)$.*

Proof. To show that A_0 is continuous, let us use Schur's test (Theorem 5.2 in [21]). Observe that

$$\sum_{k=0}^{+\infty} |a_{k,0}| = 2\alpha, \qquad \sum_{k=0}^{+\infty} |a_{k,j}| = 2(\alpha+\beta) \text{ for } j \neq 0$$

so that, in general,

$$\sum_{k=0}^{+\infty} |a_{k,j}| \le 2(\alpha+\beta).$$

Moreover,

$$\sum_{k=0}^{+\infty} |a_{0,k}| = \alpha+\beta, \qquad \sum_{k=0}^{+\infty} |a_{j,k}| = 2(\alpha+\beta) \text{ for } j \neq 0,$$

so that, in general,

$$\sum_{k=0}^{+\infty} |a_{j,k}| \le 2(\alpha+\beta).$$

By Schur's test, we have that A_0 is a bounded operator on l^2 and

$$\|A_0\| \le 2(\alpha+\beta).$$

□

Thus, by Corollary 1, we obtain local existence and uniqueness of the solution of system (5). Global uniqueness can be obtained a posteriori, since the solutions of such system are known.

Let us also observe that the distributions of the inter-arrival times are Mittag–Leffler distributions. To do that, consider the system, for fixed $n \ge 0$

$$\begin{cases} {}_0^C D_t^\nu b_n^\nu(t) = -\alpha b_n^\nu(t), \\ {}_0^C D_t^\nu b_{n+1}^\nu(t) = \alpha b_n^\nu(t), \\ b_n^\nu(0) = 1, \\ b_{n+1}^\nu(0) = 0, \end{cases}$$

which are the state probabilities of a queue with null death rate, fixed birth rate, starting with n customers and with an absorbent state $n+1$. Under such assumptions, $b_{n+1}^\nu(t)$ is the probability that a customer arrives before t. Moreover, the normalizing condition becomes

$$b_n^\nu(t) + b_{n+1}^\nu(t) = 1.$$

One can solve the first equation (see Appendix A) to obtain

$$b_n^\nu(t) = E_\nu(-\alpha t^\nu),$$

where E_ν is the one-parameter Mittag–Leffler function (see Appendix B), and then, by using the normalizing condition, we have

$$b_{n+1}^\nu(t) = 1 - E_\nu(-\alpha t^\nu).$$

In a similar way, let us show that the distributions of the service times are Mittag–Leffler distributions. To show that, consider the system, for fixed $n \geq 0$,

$$\begin{cases} {}^C_0 D^\nu_t d^\nu_n(t) = \beta d^\nu_{n+1}(t), \\ {}^C_0 D^\nu_t d^\nu_{n+1}(t) = -\beta d^\nu_{n+1}(t), \\ d^\nu_n(t) = 0, \\ d^\nu_{n+1}(t) = 1, \end{cases}$$

which are the state probabilities of a queue with null birth rate, fixed death rate, starting with $n+1$ customers with an absorbent state n. Under such assumption, $d^\nu_n(t)$ is the probability that a customer is served before t. Moreover, the normalizing condition becomes

$$d^\nu_n(t) + d^\nu_{n+1}(t) = 1.$$

One can solve the second equation to obtain

$$d^\nu_{n+1}(t) = E_\nu(-\beta t^\nu), \; t \geq 0$$

and then, by using the normalizing condition, we have

$$d^\nu_n(t) = 1 - E_\nu(-\beta t^\nu), \; t \geq 0.$$

Moreover, since we know that $\forall t \geq 0$ $p^\nu_n(t) \geq 0$ and $\sum_{n=0}^{\infty} p^\nu_n(t) = 1$, by the continuous inclusion $l^1(\mathbb{R}) \subseteq l^2(\mathbb{R})$ (see [22]), $(p_n(t))_{n \geq 0}$ is uniformly bounded in $l^2(\mathbb{R})$ and then, by Corollary 2, it is the (global) unique solution of system (5).

Distribution of the Busy Period

We want to determine the probability distribution $K^\nu(t)$ of the busy period K^ν of a fractional M/M/1 queue. To do this, we will follow the lines of the proof given in [1] and [4].

Theorem 2. *Let K^ν be the random variable describing the duration of the busy period of a fractional M/M/1 queue $N^\nu(t)$ and consider $K^\nu(t) = \mathbb{P}(K^\nu \leq t)$. Then,*

$$K^\nu(t) = 1 - \sum_{n=1}^{+\infty} \sum_{m=0}^{+\infty} C_{n,m} t^{\nu(n+2m-1)} E^{n+2m}_{\nu,\nu(n+2m-1)+1}(-(\alpha+\beta)t^\nu), \tag{13}$$

where

$$C_{n,m} = \binom{n+2m}{m} \frac{n}{n+2m} \alpha^{n+m-1} \beta^m. \tag{14}$$

Proof. Let us first define a queue $\overline{N}^\nu(t)$ such that $\mathbb{P}(\overline{N}^\nu(0) = 1) = 1$ and $\overline{N}^\nu(t)$ behaves like $N^\nu(t)$ except for the state 0 being an absorbent state. Thus, state probability functions are solution of the following system

$$\begin{cases} {}^C_0 D^\nu_t \overline{p}^\nu_0 = \beta \overline{p}^\nu_1(t), \\ {}^C_0 D^\nu_t \overline{p}^\nu_1 = -(\alpha+\beta)\overline{p}^\nu_1(t) + \beta \overline{p}^\nu_2(t), \\ {}^C_0 D^\nu_t \overline{p}^\nu_n = -(\alpha+\beta)\overline{p}^\nu_n(t) + \alpha \overline{p}^\nu_{n-1}(t) + \beta \overline{p}^\nu_{n+1}(t), & n \geq 2, \\ \overline{p}^\nu_n(0) = \delta_{n,1}, & n \geq 0. \end{cases} \tag{15}$$

First, we want to show that, if we consider $L_\nu(t)$, the inverse of a ν-stable subordinator that is independent from $\overline{N}^1(t)$, then $\overline{N}^\nu(t) \overset{d}{=} \overline{N}^1(L_\nu(t))$. To do that, consider the probability generating function $G^\nu(z,t)$ of $\overline{N}^\nu(t)$ defined as

$$G^\nu(z,t) = \sum_{k=0}^{+\infty} z^k \overline{p}_k^\nu(t). \tag{16}$$

From system (15), we know that $G^\nu(z,t)$ solves the following fractional Cauchy problem:

$$\begin{cases} z {}_0^C D_t^\nu G^\nu(z,t) = [\alpha z^2 - (\alpha+\beta)z + \beta][G^\nu(z,t) - \overline{p}_0^\nu(t)], \\ G^\nu(z,0) = z, \end{cases} \tag{17}$$

which, for $\nu = 1$, becomes

$$\begin{cases} z \frac{d}{dt} G^1(z,t) = [\alpha z^2 - (\alpha+\beta)z + \beta][G^1(z,t) - \overline{p}_0^1(t)], \\ G^1(z,0) = z. \end{cases} \tag{18}$$

Taking the Laplace transform in Equation (17) and using Equation (A1), we have

$$z[s^\nu \widetilde{G}^\nu(z,s) - z s^{\nu-1}] = [\alpha z^2 - (\alpha+\beta)z + \beta][\widetilde{G}^\nu(z,s) - \overline{\pi}_0^\nu(s)], \tag{19}$$

where $\widetilde{G}^\nu(z,s)$ and $\overline{\pi}_0^\nu(s)$ are Laplace transforms of $G^\nu(z,t)$ and $\overline{p}_0^\nu(t)$.

We know that $\overline{N}^\nu(t) \overset{d}{=} \overline{N}^1(L_\nu(t))$ if and only if

$$\overline{p}_n^\nu(t) = \mathbb{P}(\overline{N}^\nu(t) = n) = \mathbb{P}(\overline{N}^1(L_\nu(t)) = n) = \int_0^{+\infty} \overline{p}_n^1(y)\, \mathbb{P}(L_\nu(t) \in dy) \tag{20}$$

and then if and only if, by Equation (16),

$$G^\nu(z,t) = \int_0^{+\infty} G^1(z,y)\, \mathbb{P}(L_\nu(t) \in dy). \tag{21}$$

Taking the Laplace transform in Equations (20) and (21) for $n = 0$ and by using (see, i.e., Equation (10) in [12])

$$\mathcal{L}[\mathbb{P}(L_\nu(t) \in dy)](s) = s^{\nu-1} e^{-ys} dy, \tag{22}$$

we know we have to show that

$$\overline{\pi}_0^\nu(s) = \int_0^{+\infty} \overline{p}_n^1(y) s^{\nu-1} e^{-ys^\nu} dy \tag{23}$$

and

$$\widetilde{G}^\nu(z,s) = \int_0^{+\infty} G^1(z,y) s^{\nu-1} e^{-ys^\nu} dy. \tag{24}$$

Since Equation (17) admits a unique solution, then we only need to show that the right-hand sides of Equations (23) and (24) solve Equation (19), that is to say that we have to verify

$$z\left[s^\nu \int_0^{+\infty} G^1(z,y) e^{-ys^\nu} dy - z\right] = [\alpha z^2 - (\alpha+\beta)z + \beta]\left[\int_0^{+\infty} G^1(z,y) e^{-ys^\nu} dy - \int_0^{+\infty} \overline{p}_0^1(y) e^{-ys^\nu} dy\right]. \tag{25}$$

To do that, consider the right-hand side of the previous equation and, recalling that $G^1(z,t)$ is solution of Equation (18):

$$\int_0^{+\infty} [\alpha z^2 - (\alpha+\beta)z + \beta][G^1(z,y) - \overline{p}_0^1(y)]e^{-ys^\nu} dy = \int_0^{+\infty} \left(\frac{d}{dy} G^1(z,y)\right) e^{-ys^\nu} dy$$

and then, by integrating by parts, we have Equation (25).

Now remark that $\overline{p}_0^\nu(t) = B^\nu(t)$. Thus, we want to determine $\overline{p}_0^\nu(t)$. To do that, let us recall, from [1,4] that

$$\overline{p}_n^1(t) = nt^{-1}\alpha^{\frac{n}{2}-1}\beta^{-\frac{n}{2}}e^{-(\alpha+\beta)t}I_n(2\sqrt{\alpha\beta}t) \text{ for } n \geq 1$$

from which, explicitly writing $I_n(2\sqrt{\alpha\beta}t)$, we have

$$\overline{p}_n^1(t) = \sum_{m=0}^{+\infty} \binom{n+2m}{m} \frac{n}{n+2m} \frac{1}{(n+2m-1)!} \alpha^{n+m-1}\beta^m t^{n+2m-1} e^{-(\alpha+\beta)t} \text{ for } n \geq 1.$$

Posing $C_{n,m} = \binom{n+2m}{m}\frac{n}{n+2m}\alpha^{n+m-1}\beta^m$, we have

$$\overline{p}_n^1(t) = \sum_{m=0}^{+\infty} \frac{C_{n,m}}{(n+2m-1)!} t^{n+2m-1} e^{-(\alpha+\beta)t} \text{ for } n \geq 1$$

and then

$$\overline{p}_0^1(t) = 1 - \sum_{n=1}^{+\infty}\sum_{m=0}^{+\infty} \frac{C_{n,m}}{(n+2m-1)!} t^{n+2m-1} e^{-(\alpha+\beta)t}. \tag{26}$$

Since $\overline{N}^\nu(t) = \overline{N}^1(L_\nu(t))$, we have

$$\overline{p}_0^\nu(t) = \int_0^{+\infty} \overline{p}_0^1(y)\,\mathbb{P}(L_\nu(t) \in dy)$$

and then, using Equation (26), we have

$$\overline{p}_0^\nu(t) = 1 - \sum_{n=1}^{+\infty}\sum_{m=0}^{+\infty} \frac{C_{n,m}}{(n+2m-1)!} \int_0^{+\infty} y^{n+2m-1} e^{-(\alpha+\beta)y}\,\mathbb{P}(L_\nu(t) \in dy). \tag{27}$$

Taking the Laplace transform in Equation (27), using Equation (22), we have

$$\overline{\pi}_0^\nu(s) = \frac{1}{s} - \sum_{n=1}^{+\infty}\sum_{m=0}^{+\infty} \frac{C_{n,m}}{(n+2m-1)!} s^{\nu-1} \int_0^{+\infty} y^{n+2m-1} e^{-(\alpha+\beta+s^\nu)y} dy$$

and then integrating

$$\overline{\pi}_0^\nu(s) = \frac{1}{s} - \sum_{n=1}^{+\infty}\sum_{m=0}^{+\infty} C_{n,m} \frac{s^{\nu-1}}{(\alpha+\beta+s^\nu)^{n+2m}}.$$

Finally, using formula (A2), we have

$$\overline{p}_0^\nu(s) = 1 - \sum_{n=1}^{+\infty}\sum_{m=0}^{+\infty} C_{n,m} t^{\nu(n+2m-1)} E^{n+2m}_{\nu,\nu(n+2m-1)+1}(-(\alpha+\beta)t^\nu)$$

□

Remark 1. *As $\nu \to 1$ we obtain, by using*

$$E^{n+2m}_{1,n+2m}(-(\alpha+\beta)t) = \frac{e^{-(\alpha+\beta)t}}{(n+2m-1)!},$$

that $\overline{p}_0^\nu(t) \to \overline{p}_0^1(t)$ *and then* $K^\nu(t) \to K^1(t)$.

5. The Fractional M/M/1 Queue with Catastrophes

Let us consider a classical M/M/1 queue with FIFO discipline and subject to catastrophes whose effect is to instantaneously empty the queue [14] and let $N_\xi^1(t)$ be the number of customers in the system at time t with state probabilities

$$p_n^{1,\xi}(t) = \mathbb{P}(N_\xi^1(t) = n | N_\xi^1(0) = 0),\ n = 0, 1, \ldots$$

Then, the function $p_n^{1,\xi}$ satisfy the following differential-difference equations:

$$\begin{cases} D_t p_0^{1,\xi}(t) = -(\alpha+\xi)p_0^{1,\xi}(t) + \beta p_1^{1,\xi}(t) + \xi, \\ D_t p_n^{1,\xi}(t) = -(\alpha+\beta+\xi)p_n^{1,\xi}(t) + \alpha p_{n-1}^{1,\xi}(t) + \beta p_{n+1}^{1,\xi}(t), & n \geq 1, \\ p_n^{1,\xi}(0) = \delta_{n,0}, & n \geq 0, \end{cases} \tag{28}$$

where $\delta_{n,0}$ is the Kroeneker delta symbol, $D_t = \frac{d}{dt}$, $\alpha, \beta > 0$ are the entrance and service rates, respectively, and $\xi > 0$ is the rate of the catastrophes when the system is not empty.

For $\nu \in (0,1)$, we define the fractional M/M/1 queue process with catastrophes as

$$N_\xi^\nu(t) = N_\xi^1(L_\nu(t)),\ t \geq 0,$$

where L_ν is an inverse ν-stable subordinator that is independent of $N_\xi^1(t)$, $t \geq 0$ (see Section 2).

We will show that the state probabilities

$$p_n^{\nu,\xi} := \mathbb{P}(N_\xi^\nu(t) = n | N_\xi^\nu(0) = 0)$$

satisfy the following differential-difference fractional equations:

$$\begin{cases} {}_0^C D_t^\nu p_0^{\nu,\xi}(t) = -(\alpha+\xi)p_0^{\nu,\xi}(t) + \beta p_1^{\nu,\xi}(t) + \xi, \\ {}_0^C D_t^\nu p_n^{\nu,\xi}(t) = -(\alpha+\beta+\xi)p_n^{\nu,\xi}(t) + \alpha p_{n-1}^{\nu,\xi}(t) + \beta p_{n+1}^{\nu,\xi}(t), & n \geq 1, \\ p_n^{\nu,\xi}(0) = \delta_{n,0}, & n \geq 0, \end{cases} \tag{29}$$

where ${}_0^C D_t^\nu$ is the Caputo fractional derivative (see Appendix A).

In the classical case, catastrophes occur according to a Poisson process with rate ξ if the system is not empty. In our case, consider for a fixed $n > 0$,

$$\begin{cases} {}_0^C D_t^\nu c_0^\nu(t) = \xi(1 - c_0^\nu(t)), \\ {}_0^C D_t^\nu c_n^\nu(t) = -\xi c_n^\nu(t), \\ c_0^\nu(0) = 0, \\ c_n^\nu(0) = 1, \end{cases}$$

which describes the state probabilities of an initially non empty system with null birth and death rate but positive catastrophe rate. In such case, $c_0^\nu(t)$ is the probability a catastrophe occurs before time t. Moreover, the normalization property becomes

$$c_0^\nu(t) + c_n^\nu(t) = 1.$$

In such case, we can solve the second equation to obtain

$$c_n^{\nu}(t) = E_{\nu}(-\xi t^{\nu}),\ t \geq 0.$$

Using the normalization property, we finally obtain

$$c_0^{\nu}(t) = 1 - E_{\nu}(-\xi t^{\nu}),\ t \geq 0 \tag{30}$$

and then the distributions of the inter-occurrence of the catastrophes are Mittag–Leffler distributions. We can conclude that, in the fractional case, catastrophes occur according to a fractional Poisson process ([10,11,13]) with rate ξ if the system is not empty. Since the operators ${}_0^C D_t^{\nu}$ are Caputo fractional derivatives, we expect the stationary behaviour of the queue to be the same as the classic one. Denoting with N_{ξ}^1 the number of customers in the system at the steady state of a classical M/M/1 with catastrophes and defining the state probabilities

$$q_n = \mathbb{P}(N_{\xi}^1 = n),\ n \geq 0,$$

we can use the results obtained in [15] to observe that

$$q_n = \left(1 - \frac{1}{z_1}\right)\left(\frac{1}{z_1}\right)^n,\ n \geq 0, \tag{31}$$

where z_1 is the solution of

$$\alpha z^2 - (\alpha + \beta + \xi)z + \beta = 0 \tag{32}$$

such that $z_1 > 1$. Let us call z_2 the other solution of Equation (32) and observe that $0 < z_2 < 1 < z_1$. Some properties coming from such equations that will be useful hereafter are

$$\alpha + \beta + \xi = \alpha z_i + \frac{\beta}{z_i} \tag{33}$$

and

$$\alpha z_i^2 = (\alpha + \beta + \xi)z_i - \beta \tag{34}$$

with $i = 1, 2$.

5.1. Alternative Representation of the Fractional M/M/1 Queue with Catastrophes

We want to obtain an alternative representation of the fractional M/M/1 queue with catastrophes in a way which is similar to Lemma 2.1 in [14]. To do that, we firstly need to assure that system (29) admits a unique uniformly bounded solution. To do that, let us write system (29) in the form

$$\begin{cases} {}_0^C D_t^{\nu} \boldsymbol{p}^{\nu,\xi}(t) = f(\boldsymbol{p}^{\nu,\xi}(t)), \\ \boldsymbol{p}^{\nu,\xi}(t) = (\delta_{n,0})_{n \geq 0}, \end{cases} \tag{35}$$

where $\boldsymbol{p}^{\nu,\xi}(t) = (p_n^{\nu,\xi}(t))_{n \geq 0} \in C([0,T], l^2(\mathbb{R}))$, $f(\boldsymbol{x}) = A_{\xi}\boldsymbol{x} + \boldsymbol{\xi}$, $\boldsymbol{\xi} = (\xi, 0, \dots, 0, \dots)$ and

$$A_{\xi} = \begin{pmatrix} -(\alpha+\xi) & \beta & 0 & 0 & 0 & \cdots \\ \alpha & -(\alpha+\beta+\xi) & \beta & 0 & 0 & \cdots \\ 0 & \alpha & -(\alpha+\beta+\xi) & \beta & 0 & \cdots \\ 0 & 0 & \alpha & -(\alpha+\beta+\xi) & \beta & \cdots \\ \vdots & \vdots & \vdots & \ddots & \ddots & \ddots \end{pmatrix}$$

is an infinite tridiagonal matrix with $A_{\xi} = (a_{i,j})_{i,j \geq 0}$. We need to show the following:

Lemma 2. *The linear operator A_ξ is continuous and $\|A_\xi\| \le 2(\alpha+\beta)+\xi$.*

Proof. To obtain an estimate of the norm of A_ξ, let us use Schur's test. Observe that

$$\sum_{k=0}^{+\infty} |a_{k,0}| = 2\alpha+\xi, \qquad \sum_{k=0}^{+\infty} |a_{k,j}| = 2\alpha+2\beta+\xi \text{ with } j \neq 0,$$

so that, in general,

$$\sum_{k=0}^{+\infty} |a_{k,j}| \le 2\alpha+2\beta+\xi.$$

Moreover,

$$\sum_{k=0}^{+\infty} |a_{0,k}| = \alpha+\beta+\xi, \qquad \sum_{k=0}^{+\infty} |a_{j,k}| = 2\alpha+2\beta+\xi \text{ for } j \neq 0$$

so that, in general,

$$\sum_{k=0}^{+\infty} |a_{j,k}| \le 2\alpha+2\beta+\xi.$$

By Schur's test, we have that A_ξ is a bounded operator on l^2 and

$$\|A_\xi\| \le 2(\alpha+\beta)+\xi.$$

□

Observe that, if $\xi = 0$, the operator A_0 is the same of system (12). Let us also observe that by Corollary 1 there locally exists a unique solution. Moreover, if we show that a solution is uniformly bounded, such solution is unique. Now, we are ready to adapt Lemma 2.1 of [14] to the fractional case.

Theorem 3. *Let $\widetilde{N}^\nu(t)$ be the number of customers in a fractional M/M/1 queue with arrival rate αz_1 and service rate $\frac{\beta}{z_1}$ such that $\mathbb{P}(\widetilde{N}^\nu(0)=0)=1$ and consider N a random variable independent from $\widetilde{N}^\nu(t)$ whose state probabilities q_n are defined in Equation (31). Define*

$$M^\nu(t) := \min\{\widetilde{N}^\nu(t), N\},\ t \ge 0.$$

Then, the state probabilities of $M^\nu(t)$ are the unique solutions of (29).

Moreover, $M^\nu(t) \overset{d}{=} N_\xi^\nu(t)$, where $\overset{d}{=}$ is the equality in distribution, and then $p_n^{\nu,\xi}(t)$, $n = 0,1,\dots$ are the unique solutions of (29).

Proof. Define $p_n^{*,\nu}(t) = \mathbb{P}(M^\nu(t)=n)$ and $\widetilde{p}_n^\nu(t) = \mathbb{P}(\widetilde{N}^\nu(t)=n)$. Since $\widetilde{N}^\nu(t)$ and N are independent, then

$$p_n^{*,\nu}(t) = \mathbb{P}(N=n)\,\mathbb{P}(\widetilde{N}^\nu(t) \ge n) + \mathbb{P}(\widetilde{N}^\nu(t)=n)\,\mathbb{P}(N>n),$$

which, by using the definitions of $\widetilde{p}_n^\nu(t)$ and q_n, becomes

$$p_n^{*,\nu}(t) = q_n \sum_{k=n}^{+\infty} \widetilde{p}_k^\nu(t) + \left(\sum_{k=n+1}^{+\infty} q_n\right) \widetilde{p}_n^\nu(t). \tag{36}$$

Moreover, by using Equation (31), we have

$$\sum_{k=n+1}^{+\infty} q_n = \left(1-\frac{1}{z_1}\right) \sum_{k=n+1}^{+\infty} \left(\frac{1}{z_1}\right)^k = \left(1-\frac{1}{z_1}\right)\left(\frac{1}{z_1}\right)^{n+1} \sum_{k=0}^{+\infty} \left(\frac{1}{z_1}\right)^k = \left(\frac{1}{z_1}\right)^{n+1} \tag{37}$$

and then, substituting Equation (37) in (36), we obtain

$$p_n^{*,\nu}(t) = q_n \sum_{k=n}^{+\infty} \widetilde{p}_k^{\nu}(t) + \left(\frac{1}{z_1}\right)^{n+1} \widetilde{p}_n^{\nu}(t). \tag{38}$$

We want to show that $M^{\nu}(t) = N^{\nu}(t)$. Since, by definition, $p_n^{*,\nu}(t)$ are non-negative and $\sum_{n=0}^{+\infty} p_n^{*,\nu}(t) = 1$, they are uniformly bounded in $l^2(\mathbb{R})$. Thus, we only need to show that $\boldsymbol{p}^{*,\nu}(t) = (p_n^{*,\nu}(t))_{n\geq 0}$ solves system (35).
The initial conditions are easily verified, so we only need to verify the differential relations. Observe that

$$p_0^{*,\nu}(t) = q_0 + \frac{1}{z_1}\widetilde{p}_0^{\nu}(t)$$

and then, applying the Caputo derivative operator, we obtain

$${}_0^C D_t^{\nu} p_0^{*,\nu}(t) = \frac{1}{z_1} {}_0^C D_t^{\nu} \widetilde{p}_0^{\nu}(t).$$

Since $\widetilde{p}_0^{\nu}(t)$ is a solution of system (5) with rates αz_1 and $\frac{\beta}{z_1}$, we have

$${}_0^C D_t^{\nu} p_0^{*,\nu}(t) = -\alpha \widetilde{p}_0^{\nu}(t) + \frac{\beta}{z_1^2}\widetilde{p}_1^{\nu}(t).$$

Observe also that

$$p_1^{*,\nu}(t) = q_1(1-\widetilde{p}_0^{\nu}(t)) + \left(\frac{1}{z_1}\right)^2 \widetilde{p}_1^{\nu}(t)$$

so we have

$$\begin{aligned} &-(\alpha+\xi)p_0^{*,\nu}(t) + \beta p_1^{*,\nu}(t) + \xi \\ &\qquad = -(\alpha+\xi)\left(q_0 + \left(\frac{1}{z_1}\right)\widetilde{p}_0^{\nu}(t)\right) + \beta\left[q_1(1-\widetilde{p}_0^{\nu}(t)) + \left(\frac{1}{z_1}\right)^2 \widetilde{p}_1^{\nu}(t)\right] + \xi. \end{aligned}$$

After some calculations, we obtain

$$-(\alpha+\xi)p_0^{*,\nu}(t) + \beta p_1^{*,\nu}(t) + \xi = -(\alpha+\xi)q_0 - \frac{\alpha+\xi}{z_1}\widetilde{p}_0^{\nu}(t) + \beta q_1 - \beta q_1 \widetilde{p}_0^{\nu}(t) + \frac{\beta}{z_1^2}\widetilde{p}_1^{\nu}(t) + \xi.$$

Let us remark that

$$q_0 = 1 - \frac{1}{z_1}, \qquad q_1 = \left(1-\frac{1}{z_1}\right)\left(\frac{1}{z_1}\right),$$

so we have

$$\begin{aligned} &-(\alpha+\xi)p_0^{*,\nu}(t) + \beta p_1^{*,\nu}(t) + \xi \\ &\qquad = \frac{-\alpha z_1^2 + (\alpha+\beta+\xi)z_1 - \beta}{z_1^2} + \frac{(-(\alpha+\beta+\xi)z_1 + \beta)\widetilde{p}_0^{\nu}(t)}{z_1^2} + \frac{\beta}{z_1^2}\widetilde{p}_1^{\nu}(t). \end{aligned}$$

By using Equations (32) and (34), we obtain

$$-(\alpha+\xi)p_0^{*,\nu}(t) + \beta p_1^{*,\nu}(t) + \xi = -\alpha\widetilde{p}_0^{\nu}(t) + \frac{\beta}{z_1^2}\widetilde{p}_1^{\nu}(t) = {}_0^C D_t^{\nu} p_0^{*,\nu}(t).$$

Rewrite now Equation (38) in the form

$$p_n^{*,\nu}(t) = q_n\left(1-\sum_{k=0}^{+\infty}\widetilde{p}_k^{\nu}(t)\right)+\left(\frac{1}{z_1}\right)^{n+1}\widetilde{p}_n^{\nu}(t) \tag{39}$$

and then apply a Caputo derivative operator to obtain

$${}_0^C D_t^{\nu} p_n^{*,\nu}(t) = -q_n\sum_{k=1}^{+\infty}{}_0^C D_t^{\nu}\widetilde{p}_k^{\nu}(t) - q_n {}_0^C D_t^{\nu}\widetilde{p}_0^{\nu}(t)+\left(\frac{1}{z_1}\right)^{n+1}{}_0^C D_t^{\nu}\widetilde{p}_n^{\nu}(t).$$

Since $\widetilde{p}_n^{\nu}(t)$ is a solution of system (5) with birth rate αz_1 and death rate $\frac{\beta}{z_1}$, then we have

$$\begin{aligned}
{}_0^C D_t^{\nu} p_n^{*,\nu}(t) = {} & q_n\left(\alpha z_1+\frac{\beta}{z_1}\right)\sum_{k=1}^{n-1}\widetilde{p}_k^{\nu} - q_n\alpha z_1\sum_{k=0}^{n-2}\widetilde{p}_k^{\nu}(t)-\frac{\beta}{z_1}q_n\sum_{k=2}^{n}\widetilde{p}_k^{\nu}(t)\\
&+\alpha z_1 q_n\widetilde{p}_0(t)-\frac{\beta}{z_1}q_n\widetilde{p}_1^{\nu}-\left(\frac{1}{z_1}\right)^{n+1}\left(\alpha z_1+\frac{\beta}{z_1}\right)\widetilde{p}_n^{\nu}(t)\\
&+\alpha\left(\frac{1}{z_1}\right)^{n}\widetilde{p}_{n-1}^{\nu}(t)+\beta\left(\frac{1}{z_1}\right)^{n+2}\widetilde{p}_{n+1}^{\nu}(t).
\end{aligned}$$

Remark that, by using Equation (39),

$$\begin{aligned}
-(\alpha+\beta+\xi)&p_n^{*,\nu}(t)+\alpha p_{n-1}^{*,\nu}(t)+\beta p_{n+1}^{*,\nu}(t) =\\
&-(\alpha+\beta+\xi)\left(q_n\left(1-\sum_{k=0}^{n-1}\widetilde{p}_k^{\nu}(t)\right)+\left(\frac{1}{z_1}\right)^{n+1}\widetilde{p}_n^{\nu}(t)\right)\\
&+\alpha\left(q_{n-1}\left(1-\sum_{k=0}^{n-2}\widetilde{p}_k^{\nu}(t)\right)+\left(\frac{1}{z_1}\right)^{n}\widetilde{p}_{n-1}^{\nu}(t)\right)\\
&+\beta\left(q_{n+1}\left(1-\sum_{k=0}^{n}\widetilde{p}_k^{\nu}(t)\right)+\left(\frac{1}{z_1}\right)^{n+2}\widetilde{p}_{n+1}^{\nu}(t)\right).
\end{aligned}$$

Then, recalling that by definition $q_{n-1} = z_1 q_n$ and $q_{n+1} = \frac{q_n}{z_1}$ and doing some calculations, we have

$$\begin{aligned}
-(\alpha+\beta+\xi)&p_n^{*,\nu}(t)+\alpha p_{n-1}^{*,\nu}(t)+\beta p_{n+1}^{*,\nu}(t) =\\
&(\alpha+\beta+\xi)q_n\sum_{k=1}^{n-1}\widetilde{p}_k^{\nu}(t)-\alpha z_1 q_n\sum_{k=0}^{n-2}\widetilde{p}_k^{\nu}(t)-\frac{\beta}{z_1}q_n\sum_{k=2}^{n}\widetilde{p}_k^{\nu}(t)\\
&+\frac{(\alpha+\beta+\xi)z_1-\beta}{z_1}q_n\widetilde{p}_0^{\nu}(t)-\frac{\beta}{z_1}q_n\widetilde{p}_1^{\nu}(t)-(\alpha+\beta+\xi)\left(\frac{1}{z_1}\right)^{n+1}\widetilde{p}_n^{\nu}(t)\\
&+\alpha\left(\frac{1}{z_1}\right)^{n}\widetilde{p}_{n-1}^{\nu}(t)+\beta\left(\frac{1}{z_1}\right)^{n+2}\widetilde{p}_{n+1}^{\nu}(t)+\frac{\alpha z_1^2-(\alpha+\beta+\xi)z_1+\beta}{z_1}q_n.
\end{aligned}$$

Finally, by using Equations (32), (33) and (34), we have

$$\begin{aligned}
&-(\alpha+\beta+\xi)p_n^{*,\nu}(t)+\alpha p_{n-1}^{*,\nu}(t)+\beta p_{n+1}^{*,\nu}(t)= \\
&\quad\left(\alpha z_1+\frac{\beta}{z_1}\right)q_n\sum_{k=1}^{n-1}\widetilde{p}_k^{\nu}(t)-\alpha z_1 q_n\sum_{k=0}^{n-2}\widetilde{p}_k^{\nu}(t)-\frac{\beta}{z_1}q_n\sum_{k=2}^{n}\widetilde{p}_k^{\nu}(t) \\
&\quad+\alpha z_1 q_n\widetilde{p}_0^{\nu}(t)-\frac{\beta}{z_1}q_n\widetilde{p}_1^{\nu}(t)-\left(\alpha z_1+\frac{\beta}{z_1}\right)\left(\frac{1}{z_1}\right)^{n+1}\widetilde{p}_n^{\nu}(t) \\
&\quad+\alpha\left(\frac{1}{z_1}\right)^{n}\widetilde{p}_{n-1}^{\nu}(t)+\beta\left(\frac{1}{z_1}\right)^{n+2}\widetilde{p}_{n+1}^{\nu}(t)={}_0^C D_t^{\nu}p_n^{*,\nu}(t).
\end{aligned}$$

We have shown that the state probabilities $p_n^{*,\nu}(t)$ of $M^{\nu}(t)$ are the unique solutions of system (29). Now, we need to show that $M^{\nu}(t)\overset{d}{=}N_{\xi}^{\nu}(t)$. To do this, consider $\widetilde{N}^1(t)$ a classical M/M/1 queue with arrival rate αz_1 and service rate $\frac{\beta}{z_1}$, N a random variable independent from $\widetilde{N}^{\nu}(t)$ and $\widetilde{N}^1(t)$ with probability masses q_n and finally $L_{\nu}(t)$ the inverse of a ν-stable subordinator which is independent from N and $\widetilde{N}^1(t)$. Define also $M^1(t)=\min\{\widetilde{N}^1(t),N\}$. By Lemma 2.1 of [14], we know that $M^1(t)\overset{d}{=}N_{\xi}^1(t)$, so $M^1(L_{\nu}(t))\overset{d}{=}N_{\xi}^1(L_{\nu}(t))\overset{d}{=}N_{\xi}^{\nu}(t)$. However, by definition, we know that $\widetilde{N}^1(L_{\nu}(t))\overset{d}{=}\widetilde{N}^{\nu}(t)$, thus finally

$$M^{\nu}(t)\overset{d}{=}M^1(L_{\nu}(t))\overset{d}{=}N_{\xi}^1(L_{\nu}(t))\overset{d}{=}N_{\xi}^{\nu}(t).$$

□

5.2. State Probabilities for the Fractional M/M/1 with Catastrophes

Since we have defined $N_{\xi}^{\nu}(t)\overset{d}{=}N_{\xi}^1(L_{\nu}(t))$, where $L_{\nu}(t)$ is the inverse of a ν-stable subordinator, which is independent from $N_{\xi}^1(t)$, we can use such definition and Theorem 3 with the results obtained in [14] to study the state probabilities of $N_{\xi}^{\nu}(t)$. In particular, we refer to the formula

$$\begin{aligned}
p_n^{1,\xi}(t)=q_n&+\sum_{m=0}^{+\infty}\sum_{r=0}^{m+n}\frac{C_{n,m,r}^1}{(m+r-1)!}t^{m+r-1}e^{-(\alpha+\beta+\xi)t} \\
&+\sum_{m=0}^{+\infty}\sum_{r=m+n+1}^{+\infty}\frac{C_{n,m,r}^2}{(m+r-1)!}t^{m+r-1}e^{-(\alpha+\beta+\xi)t},
\end{aligned}\tag{40}$$

where

$$\begin{aligned}
C_{n,m,r}^1&=\frac{z_1-1}{(z_1-z_2)z_1^{n+m+1-r}}\binom{m+r}{r}\frac{m-r}{m+r}\beta^m\alpha^{r-1}, \\
C_{n,m,r}^2&=\frac{1-z_2}{(z_1-z_2)z_2^{n+m+1-r}}\binom{m+r}{r}\frac{r-m}{r+m}\beta^m\alpha^{r-1}.
\end{aligned}\tag{41}$$

By using such formula, we can show the following:

Theorem 4. *For any $t>0$ and $n=0,1,\dots$, we have*

$$\begin{aligned}
p_n^{\nu,\xi}(t)=q_n&+\sum_{m=0}^{+\infty}\sum_{r=0}^{m+n}C_{n,m,r}^1 t^{\nu(m+r-1)}E_{\nu,\nu(m+r-1)+1}^{m+r}(-(\alpha+\beta+\xi)t^{\nu}) \\
&+\sum_{m=0}^{+\infty}\sum_{r=m+n+1}^{+\infty}C_{n,m,r}^2 t^{\nu(m+r-1)}E_{\nu,\nu(m+r-1)+1}^{m+r}(-(\alpha+\beta+\xi)t^{\nu}),
\end{aligned}\tag{42}$$

where $C_{n,m,r}^i$ are defined in (41).

Proof. From $N_\xi^\nu(t) \stackrel{d}{=} N_\xi^1(L_\nu(t))$, we have

$$p_n^{\nu,\xi}(t) = \int_0^{+\infty} p_n^{1,\xi}(y)\,\mathbb{P}(L_\nu(t) \in dy)$$

and then, by using formula (40),

$$\begin{aligned} p_n^{\nu,\xi}(t) = q_n &+ \sum_{m=0}^{+\infty}\sum_{r=0}^{m+n} \frac{C^1_{n,m,r}}{(m+r-1)!}\int_0^{+\infty} y^{m+r-1}e^{-(\alpha+\beta+\xi)y}\,\mathbb{P}(L_\nu(t)\in dy) \\ &+ \sum_{m=0}^{+\infty}\sum_{r=m+n+1}^{+\infty} \frac{C^2_{n,m,r}}{(m+r-1)!}\int_0^{+\infty} y^{m+r-1}e^{-(\alpha+\beta+\xi)y}\,\mathbb{P}(L_\nu(t)\in dy). \end{aligned}$$

Taking the Laplace transform and using Equation (22), we obtain

$$\begin{aligned} \pi_n^{\nu,\xi}(s) = q_n &+ \sum_{m=0}^{+\infty}\sum_{r=0}^{m+n} \frac{C^1_{n,m,r}}{(m+r-1)!}s^{\nu-1}\int_0^{+\infty} y^{m+r-1}e^{-(\alpha+\beta+\xi+s^\nu)y}dy \\ &+ \sum_{m=0}^{+\infty}\sum_{r=m+n+1}^{+\infty} \frac{C^2_{n,m,r}}{(m+r-1)!}s^{\nu-1}\int_0^{+\infty} y^{m+r-1}e^{-(\alpha+\beta+\xi+s^\nu)y}dy \end{aligned}$$

and then, integrating

$$\begin{aligned} \pi_n^{\nu,\xi}(s) = q_n &+ \sum_{m=0}^{+\infty}\sum_{r=0}^{m+n} C^1_{n,m,r}\frac{s^{\nu-1}}{(\alpha+\beta+\xi+s^\nu)^{m+r}} \\ &+ \sum_{m=0}^{+\infty}\sum_{r=m+n+1}^{+\infty} C^2_{n,m,r}\frac{s^{\nu-1}}{(\alpha+\beta+\xi+s^\nu)^{m+r}}. \end{aligned}$$

Finally, by using Equation (A2), we obtain

$$\begin{aligned} p_n^{\nu,\xi}(t) = q_n &+ \sum_{m=0}^{+\infty}\sum_{r=0}^{m+n} C^1_{n,m,r}t^{\nu(m+r-1)}E^{m+r}_{\nu,\nu(m+r-1)+1}(-(\alpha+\beta+\xi)t^\nu) \\ &+ \sum_{m=0}^{+\infty}\sum_{r=m+n+1}^{+\infty} C^2_{n,m,r}t^{\nu(m+r-1)}E^{m+r}_{\nu,\nu(m+r-1)+1}(-(\alpha+\beta+\xi)t^\nu). \end{aligned}$$

□

Remark 2. *From formula (42), we can easily see that* $\lim_{t\to+\infty} p_n^{\nu,\xi}(t) = q_n$ *so, as we expected, the steady-state probabilities are the same as the classical ones. For such reason, we can say that the fractional behaviour is influential only in the transient state of the queue.*

Remark 3. *As* $\nu \to 1$*, by using*

$$E^{m+r}_{1,m+r}(-(\alpha+\beta+\xi)t) = \frac{e^{-(\alpha+\beta+\xi)t}}{(m+r-1)!},$$

we obtain that $\lim_{\nu\to1} p_n^{\nu,\xi}(t) = p_n^{1,\xi}(t)$.

Remark 4. *If* $\alpha < \beta$ *and* $\xi = 0$*, then* $z_1 = \frac{\beta}{\alpha}$ *and* $z_2 = 1$*. For such reason,* $q_n = \left(1-\frac{\alpha}{\beta}\right)\left(\frac{\alpha}{\beta}\right)^n$, $C^1_{n,m,r} = \left(\frac{\alpha}{\beta}\right)^n \frac{m-r}{m+r}\binom{m+r}{m}\alpha^m\beta^{r-1}$ *and* $C^2_{n,m,r} = 0$*. Then, we have that* $p_n^{\nu,\xi}(t)$ *of Equation (42) has the form of Equation (6).*

If $\alpha > \beta$ and $\xi \to 0$ then $z_1 = 1$ and $z_2 = \frac{\beta}{\alpha}$. In such case, $q_n = 0$, $C^1_{n,m,r} = 0$ and $C^2_{n,m,r} = \alpha^{n+m}\beta^{r-n-1}\binom{m+r}{m}\frac{r-m}{m+r}$. For such case, we have

$$\lim_{\xi \to 0} p_n^{\nu,\xi}(t) = \left(\frac{\alpha}{\beta}\right)^n \sum_{m=0}^{+\infty} \sum_{r=m+n+1}^{+\infty} \alpha^m \beta^{r-1} \binom{m+r}{m} \frac{r-m}{m+r} t^{\nu(m+r-1)} E^{m+r}_{\nu,\nu(m+r-1)+1},$$

which is not recognizable as a previously obtained formula. This is due to the fact that the formula

$$\lim_{\xi \to 0} p_n^{1,\xi}(t) = \frac{e^{-(\alpha+\beta)t}}{\beta t} \left(\frac{\alpha}{\beta}\right)^n \sum_{r=n+1}^{+\infty} r \left(\frac{\beta}{\alpha}\right)^{\frac{r}{2}} I_r(2\sqrt{\alpha\beta}t) \tag{43}$$

(which is the one that is obtained from (42) as $\nu = 1$ and $\alpha > \beta$, as done in [14]) has no known equivalent in the fractional case. It is also interesting to observe that in [8] another representation of the Laplace transform of $p^\nu_n(t)$ is given in formula 2.40, which is not easily invertible, but has been obtained by using (43) instead of Sharma's representation of $p^1_n(t)$ ([2])

$$p_n^1(t) = \left(1 - \frac{\alpha}{\beta}\right)\left(\frac{\alpha}{\beta}\right)^n + e^{-(\alpha+\beta)t}\left(\frac{\alpha}{\beta}\right)^n \sum_{r=0}^{+\infty} \frac{(\alpha t)^r}{r!} \sum_{m=0}^{k+r} (r-m) \frac{(\beta t)^{m-1}}{m!}.$$

5.3. Distribution of the Busy Period

Let B^ν denote the duration of the busy period and $B^\nu(t) = \mathbb{P}(B^\nu \le t)$ be its probability distribution function. Let us observe that, if we pose $N^\nu(0) = 1$, then the queue empties within t if and only if a catastrophe occurs within t or otherwise the queue empties without catastrophes within t. Let us remark that, if there is no occurrence of catastrophes, the queue behaves as a fractional M/M/1. Let us define K^ν as the duration of a busy period for a fractional M/M/1 queue without catastrophes, Ξ^ν the time of first occurrence of a catastrophe for a non empty queue and $K^\nu(t) = \mathbb{P}(K^\nu \le t)$ and $\Xi^\nu(t) = \mathbb{P}(\Xi^\nu \le t)$ their probability distribution functions. Thus, we have

$$B^\nu(t) = \Xi^\nu(t) + (1 - \Xi^\nu(t))K^\nu(t). \tag{44}$$

Remark 5. *If we denote with $b^\nu(t)$, $\xi^\nu(t)$ and $k^\nu(t)$ the probability density functions of B^ν, Ξ^ν and K^ν, we have, by deriving formula (44),*

$$b^\nu(t) = \xi^\nu(t)(1 - K^\nu(t)) + (1 - \Xi^\nu(t))k^\nu(t),$$

which, for $\nu = 1$, is formula (17) of [14].

By using formula (44), we can finally show:

Theorem 5. *Let B^ν be the duration of the busy period of a fractional M/M/1 queue with catastrophes and $B^\nu(t) = \mathbb{P}(B^\nu \le t)$. Then,*

$$B^\nu(t) = 1 - E_\nu(-\xi t^\nu) \sum_{n=1}^{+\infty} \sum_{m=0}^{+\infty} C_{n,m} t^{\nu(n+2m-1)} E^{n+2m}_{\nu,\nu(n+2m-1)+1}(-(\alpha+\beta)t^\nu), \tag{45}$$

where $C_{n,m}$ is given in (14).

Proof. Observe that, by formula (30), we have

$$\Xi^\nu(t) = c_0^\nu(t) = 1 - E_\nu(-\xi t^\nu)$$

and by formula (13) we also have a closed form of $K^\nu(t)$. Thus, by using formula (44), we obtain Equation (45). □

5.4. Distribution of the Time of the First Occurrence of a Catastrophe

We already know that if the queue starts from a non-empty state, then the occurrence of the catastrophes is a Mittag–Leffler distribution. However, we are interested in such distribution as the queue starts being empty. To do that, we will need some auxiliary discrete processes.

Theorem 6. *Let $\mathcal{D}^\nu$ be the time of first occurrence of a catastrophe as $\mathbb{P}(N^\nu(0)=0)=1$ and let $\mathcal{D}^\nu(t)=\mathbb{P}(\mathcal{D}^\nu \le t)$. Then,*

$$\mathcal{D}^\nu(t) = 1 - \sum_{j=1}^{+\infty}\sum_{m=0}^{+\infty} C_{m,j} t^{\nu(2m+j-1)} E^{2m+j}_{\nu,\nu(2m+j-1)+1}[-(\alpha+\beta+\xi)t^\nu], \tag{46}$$

where

$$C_{m,j} = \frac{j}{2m+j}\frac{(\beta+\xi)^j-\alpha^j}{\beta+\xi-\alpha}\binom{2m+j}{m}(\alpha\beta)^m.$$

Proof. Following the lines of [14], let us consider the process $\overline{N}^\nu(t)$ with state space $\{-1,0,1,2,\dots\}$ such that $\mathbb{P}(\overline{N}^\nu(t)=0)=1$ and posing $r_n(t)=\mathbb{P}(\overline{N}^\nu(t)=n)$, $n\ge -1$ as its state probability, we have

$$\begin{cases} {}^C_0D^\nu_t r^\nu_{-1}(t) = \xi[1-r^\nu_{-1}(t)-r^\nu_0(t)], \\ {}^C_0D^\nu_t r^\nu_0(t) = -\alpha r^\nu_0(t)+\beta r^\nu_1(t), \\ {}^C_0D^\nu_t r^\nu_n(t) = -(\alpha+\beta+\xi)r^\nu_n(t)+\alpha r^\nu_{n-1}(t)+\beta r^\nu_{n+1}(t), & n\ge 1, \\ r^\nu_n(0)=\delta_{n,0}, & n\ge -1. \end{cases} \tag{47}$$

Let us remark that such process represents our queue until a catastrophe occurs: in such case, instead of emptying the queue, the state of the process becomes -1, which is an absorbent state. With such interpretation, we can easily observe that $\mathcal{D}^\nu(t)=r^\nu_{-1}(t)$.

In order to determine $r^\nu_n(t)$, we will first show that $\overline{N}^\nu(t) \overset{d}{=} \overline{N}^1(L_\nu(t))$ where $L_\nu(t)$ is the inverse of a ν-stable subordinator which is independent from $\overline{N}^1$. To do that, let us consider $\overline{N}^\nu(t)+1$ instead of $\overline{N}^\nu(t)$. Let us remark that $\mathbb{P}(\overline{N}^\nu(t)+1=n)=r^\nu_{n-1}(t)$. Let $G^\nu(z,t)=\sum_{n=0}^{+\infty} z^n r^\nu_{n-1}(t)$ be the probability generating function of $\overline{N}^\nu(t)+1$. Multiplying the third sequence of equations in (47) with z^{n+1} and then, summing all these equations, we have

$${}^C_0D^\nu_t\left(\sum_{n=2}^{+\infty} z^n r^\nu_{n-1}(t)\right) = -(\alpha+\beta+\xi)\sum_{n=2}^{+\infty} z^n r^\nu_{n-1}(t) + \alpha\sum_{n=2}^{+\infty} z^n r^\nu_{n-2}(t) + \beta\sum_{n=2}^{+\infty} z^n r^\nu_n(t). \tag{48}$$

Now observe that

$$\sum_{n=2}^{+\infty} z^n r^\nu_{n-1}(t) = \sum_{n=0}^{+\infty} z^n r^\nu_{n-1}(t) - r^\nu_{-1}(t) - zr^\nu_0(t) = G^\nu(z,t) - r^\nu_{-1}(t) - zr^\nu_0(t); \tag{49}$$

moreover,

$$\begin{aligned}\sum_{n=2}^{+\infty} z^n r^\nu_{n-2}(t) &= \sum_{n=1}^{+\infty} z^{n+1} r^\nu_{n-1}(t) = z\sum_{n=1}^{+\infty} z^n r^\nu_{n-1}(t) \\ &= z[G^\nu(z,t)-r^\nu_{-1}(t)] = z[G^\nu(z,t)-r^\nu_{-1}(t)-zr^\nu_0(t)] + z^2 r^\nu_0(t); \end{aligned} \tag{50}$$

finally,

$$\sum_{n=2}^{+\infty} z^n r_n^\nu(t) = \sum_{n=3}^{+\infty} z^{n-1} r_{n-1}^\nu(t) = \frac{1}{z}\sum_{n=3}^{+\infty} z^n r_{n-1}^\nu(t)$$
$$= \frac{1}{z}[G^\nu(z,t) - r_{-1}^\nu(t) - z r_0^\nu(t) - z^2 r_1^\nu(t)]$$
$$= \frac{1}{z}[G^\nu(z,t) - r_{-1}^\nu(t) - z r_0^\nu(t)] - z r_1^\nu(t). \quad (51)$$

Using Equations (49),(50) and (51) in Equation (48), we obtain

$${}_0^C D_t^\nu [G^\nu(z,t) - r_{-1}^\nu(t) - z r_0^\nu(t)]$$
$$= \left[\alpha z - (\alpha+\beta+\xi) + \frac{\beta}{z}\right][G^\nu(z,t) - r_{-1}^\nu(t) - z r_0^\nu(t)]$$
$$+ \alpha z^2 r_0^\nu(t) - \beta z r_1^\nu(t). \quad (52)$$

Finally, by using the first and the second equation of Equation (47) in Equation (52), we obtain

$${}_0^C D_t^\nu G^\nu(z,t) = \left[\alpha z - (\alpha+\beta+\xi) + \frac{\beta}{z}\right][G^\nu(z,t) - r_{-1}^\nu(t) - z r_0^\nu(t)]$$
$$+ \alpha z(z-1) r_0^\nu(t) + \xi[1 - r_{-1}^\nu(t) - r_0^\nu(t)].$$

We have obtained that the probability generating function $G^\nu(z,t)$ of $\overline{N}^\nu(t)+1$ solves the Cauchy problem

$$\begin{cases} z\,{}_0^C D_t^\nu G^\nu(z,t) = \left[\alpha z^2 - (\alpha+\beta+\xi)z + \beta\right][G^\nu(z,t) - r_{-1}^\nu(t) - z r_0^\nu(t)] \\ \qquad + \alpha z^2(z-1) r_0^\nu(t) + \xi z[1 - r_{-1}^\nu(t) - r_0^\nu(t)], \\ G^\nu(z,0) = z, \end{cases} \quad (53)$$

that, for $\nu = 1$, becomes

$$\begin{cases} z\frac{d}{dt} G^1(z,t) = \left[\alpha z^2 - (\alpha+\beta+\xi)z + \beta\right][G^1(z,t) - r_{-1}^1(t) - z r_0^1(t)] \\ \qquad + \alpha z^2(z-1) r_0^1(t) + \xi z[1 - r_{-1}^1(t) - r_0^1(t)], \\ G^1(z,0) = z. \end{cases} \quad (54)$$

Let $\widetilde{G}^\nu(z,s)$, $\widetilde{r}_0^\nu(s)$ and $\widetilde{r}_{-1}^\nu(s)$ be the Laplace transforms of $G^\nu(z,t)$, $r_0^\nu(t)$ and $r_{-1}^\nu(t)$ and let us take the Laplace transform in Equation (53) to obtain

$$z[s^\nu \widetilde{G}^\nu(z,s) - s^{\nu-1} z] = \left[\alpha z^2 - (\alpha+\beta+\xi)z + \beta\right][\widetilde{G}^\nu(z,s) - \widetilde{r}_{-1}^\nu(s) - z\widetilde{r}_0^\nu(s)]$$
$$+ \alpha z^2(z-1)\widetilde{r}_0^\nu(s) + \xi z\left[\frac{1}{s} - \widetilde{r}_{-1}^\nu(s) - \widetilde{r}_0^\nu(s)\right]. \quad (55)$$

Now, let us remark that $\overline{N}^\nu(t)+1 \overset{d}{=} \overline{N}^1(L_\nu(t))+1$ if and only if for all $n \ge 0$:

$$r_{n-1}^\nu(t) = \int_0^{+\infty} r_{n-1}^1(y)\,\mathbb{P}(L_\nu(t) \in dy) \quad (56)$$

that is to say if and only if

$$G^\nu(z,t) = \int_0^{+\infty} G^1(z,y)\,\mathbb{P}(L_\nu(t) \in dy).$$

Taking Laplace transform and using Equation (22), we obtain

$$\begin{aligned}\widetilde{G}^{\nu}(z,s) &= s^{\nu-1}\int_0^{+\infty} G^1(z,y)e^{-ys^{\nu}}dy,\\ \widetilde{r}^{\nu}_{-1}(s) &= s^{\nu-1}\int_0^{+\infty} r^1_{-1}(y)e^{-ys^{\nu}}dy,\\ \widetilde{r}^{\nu}_{0}(s) &= s^{\nu-1}\int_0^{+\infty} r^1_{0}(y)e^{-ys^{\nu}}dy.\end{aligned} \tag{57}$$

Thus, by substituting the formulas (57) in (55), we obtain

$$\begin{aligned} z\left[s^{\nu}s^{\nu-1}\int_0^{+\infty} G^1(z,y)e^{-ys^{\nu}}dy - s^{\nu-1}z\right] &= [\alpha z^2-(\alpha+\beta+\xi)z+\beta]\\ &\times\left[s^{\nu-1}\int_0^{+\infty} G^1(z,y)e^{-ys^{\nu}}dy - s^{\nu-1}\int_0^{+\infty} r^1_{-1}(y)e^{-ys^{\nu}}dy - zs^{\nu-1}\int_0^{+\infty} r^1_0(y)e^{-ys^{\nu}}dy\right]\\ &+\alpha z^2(z-1)s^{\nu-1}\int_0^{+\infty} r^1_0(y)e^{-ys^{\nu}}dy\\ &+\xi z\left[\frac{1}{s} - s^{\nu-1}\int_0^{+\infty} r^1_{-1}(y)e^{-ys^{\nu}}dy - s^{\nu-1}\int_0^{+\infty} r^1_0(y)e^{-ys^{\nu}}dy\right].\end{aligned}$$

Finally, multiplying with $\frac{1}{s^{\nu-1}}$, we have

$$\begin{aligned} z\left[s^{\nu}\int_0^{+\infty} G^1(z,y)e^{-ys^{\nu}}dy - z\right] &= [\alpha z^2-(\alpha+\beta+\xi)z+\beta]\\ &\times\left[\int_0^{+\infty} G^1(z,y)e^{-ys^{\nu}}dy - \int_0^{+\infty} r^1_{-1}(y)e^{-ys^{\nu}}dy - z\int_0^{+\infty} r^1_0(y)e^{-ys^{\nu}}dy\right]\\ &+\alpha z^2(z-1)\int_0^{+\infty} r^1_0(y)e^{-ys^{\nu}}dy\\ &+\xi z\left[\frac{1}{s^{\nu}} - \int_0^{+\infty} r^1_{-1}(y)e^{-ys^{\nu}}dy - \int_0^{+\infty} r^1_0(y)e^{-ys^{\nu}}dy\right].\end{aligned} \tag{58}$$

Now we know that $\overline{N}^{\nu}(t) \overset{d}{=} \overline{N}^1(L_{\nu}(t))$ if and only if Equation (58) is verified. For this reason, we only need to show such equation. To do that, remarking that $\int_0^{+\infty} e^{-ys^{\nu}}dy = \frac{1}{s^{\nu}}$, consider the right-hand side of Equation (58) and observe that

$$\begin{aligned} &[\alpha z^2-(\alpha+\beta+\xi)z+\beta]\left[\int_0^{+\infty} G^1(z,y)e^{-ys^{\nu}}dy - \int_0^{+\infty} r^1_{-1}(y)e^{-ys^{\nu}}dy - z\int_0^{+\infty} r^1_0(y)e^{-ys^{\nu}}dy\right]\\ &\quad+\alpha z^2(z-1)\int_0^{+\infty} r^1_0(y)e^{-ys^{\nu}}dy + \xi z\left[\frac{1}{s^{\nu}} - \int_0^{+\infty} r^1_{-1}(y)e^{-ys^{\nu}}dy - \int_0^{+\infty} r^1_0(y)e^{-ys^{\nu}}dy\right]\\ &\quad= \int_0^{+\infty}([\alpha z^2-(\alpha+\beta+\xi)z+\beta][G^1(z,y) - r^1_{-1}(y) - zr^1_0(y)]\\ &\quad+\alpha z^2(z-1)r^1_0(y) + \xi z[1 - r^1_{-1}(y) - r^1_0(y)])e^{-ys^{\nu}}dy.\end{aligned}$$

Thus, by using Equation (54), we have

$$\begin{aligned}
&[\alpha z^2-(\alpha+\beta+\xi)z+\beta]\left[\int_0^{+\infty}G^1(z,y)e^{-ys^\nu}dy-\int_0^{+\infty}r^1_{-1}(y)e^{-ys^\nu}dy-z\int_0^{+\infty}r^1_0(y)e^{-ys^\nu}dy\right]\\
&\quad+\alpha z^2(z-1)\int_0^{+\infty}r^1_0(y)e^{-ys^\nu}dy+\xi z\left[\frac{1}{s^\nu}-\int_0^{+\infty}r^1_{-1}(y)e^{-ys^\nu}dy-\int_0^{+\infty}r^1_0(y)e^{-ys^\nu}dy\right]\\
&\quad=z\int_0^{+\infty}\left(\frac{d}{dt}G^1(z,y)\right)e^{-ys^\nu}dy\\
&\quad=z\left[\int_0^{+\infty}G^1(z,y)e^{-ys^\nu}dy-z\right],
\end{aligned}$$

concluding the proof of our first claim.

From Theorem 3.1 of [14], we know that

$$r^1_{-1}(t)=1-\sum_{j=1}^{+\infty}\sum_{m=0}^{+\infty}\frac{C_{m,j}}{(2m+j-1)!}t^{2m+j-1}e^{-(\alpha+\beta+\xi)t} \tag{59}$$

and, since we know that $\overline{N}^\nu(t)\overset{d}{=}\overline{N}^1(L_\nu(t))$, we can use (59) in (56) with $n=0$ to obtain:

$$r^\nu_{-1}(t)=1-\sum_{j=1}^{+\infty}\sum_{m=0}^{+\infty}\frac{C_{m,j}}{(2m+j-1)!}\int_0^{+\infty}y^{2m+j-1}e^{-(\alpha+\beta+\xi)y}\,\mathbb{P}(L_\nu(t)\in dy). \tag{60}$$

Taking the Laplace transform in (60) and using formula (22), we obtain

$$\widetilde{r}^\nu_{-1}(s)=\frac{1}{s}-\sum_{j=1}^{+\infty}\sum_{m=0}^{+\infty}\frac{C_{m,j}}{(2m+j-1)!}s^{\nu-1}\int_0^{+\infty}y^{2m+j-1}e^{-(\alpha+\beta+\xi+s^\nu)y}dy$$

and then integrate

$$\widetilde{r}^\nu_{-1}(s)=\frac{1}{s}-\sum_{j=1}^{+\infty}\sum_{m=0}^{+\infty}C_{m,j}\frac{s^{\nu-1}}{(\alpha+\beta+\xi+s^\nu)^{2m+j}}. \tag{61}$$

Finally, applying the inverse Laplace transform on Equation (61) and using formula (A2), we complete the proof. □

6. Conclusions

Our work focused on the transient behaviour of a fractional M/M/1 queue with catastrophes, deriving formulas for the state probabilities, the distribution of the busy period and the distribution of the time of the first occurrence of a catastrophe. This is a non-Markov generalization of the classical M/M/1 queue with catastrophes, obtained through a time-change. The introduction of fractional dynamics in the equations that master the behaviour of the queue led to a sort of transformation of the time scale. Fractional derivatives are global operators, so the state probabilities preserve memory of their past, eventually slowing down the entire dynamics. Indeed, we can see how Mittag–Leffler functions take place where in the classical case we expected to see exponentials. However, such fractional dynamic seems to affect only the transient behaviour, since we have shown in Remark 2 that the limit behaviour is the same.

The main difficulty that is linked with fractional queues (or in general time-changed queues) is the fact that one has to deal with non-local derivative operators, such as the Caputo derivative, losing Markov property. However, fractional dynamics and fractional processes are gaining attention, due to their wide range of applicability, from physics to finance, from computer science to biology. Moreover, time-changed processes have formed a thriving field of application in mathematical finance.

Future works will focus on an extension of such results to $E_k/M/1$ and $M/E_k/1$ queues, or even to a generalization of fractional M/M/1 queue to a time-changed M/M/1 queue by using the inverse of any subordinator.

Author Contributions: The three authors have participated equally in the development of this work. The paper was also written and reviewed cooperatively.

Funding: This research was partially funded by GNCS and MANM. Nikolai Leonenko was supported in parts by the project MTM2015–71839–P(cofounded with Feder funds), of the DGI, MINECO, and the Australian Research Council's Discovery Projects funding scheme (project number DP160101366).

Acknowledgments: We want to thank V. V. Ahn of QUT, Brisbane, Australia for his support.

Conflicts of Interest: The authors declare no conflict of interest.

Appendix A. Fractional Integrals and Derivatives

Let us recall the definition of fractional integral [7]. Given a function $x : [t_0, t_1] \subseteq \mathbb{R} \to \mathbb{R}$, its fractional integral of order $\nu > 0$ in $[t_0, t]$ for $t_0 \le t \le t_1$ is given by:

$$ {}_{t_0}\mathcal{I}_t^{\nu} x = \frac{1}{\Gamma(\nu)} \int_{t_0}^{t} (t-\tau)^{\nu-1} x(\tau) d\tau. $$

The Riemann–Liouville fractional derivative operator is defined as:

$$ {}_{t_0}^{RL}D_t^{\nu} = \frac{d^m}{dt^m} {}_{t_0}\mathcal{I}_t^{m-\nu} $$

while the Caputo fractional derivative operator is defined as:

$$ {}_{t_0}^{C}D_t^{\nu} = {}_{t_0}\mathcal{I}_t^{m-\nu} \frac{d^m}{dt^m} $$

whenever $m-1 < \nu < m$. Obviously, such operators are linear. It is interesting to remark that

$$ {}_0^{RL}D_t^{\nu} 1 = \frac{t^{-\nu}}{\Gamma(1-\nu)}, \qquad {}_0^{C}D_t^{\nu} 1 = 0. $$

Note that, for a function $x(t), t \ge 0$ and $\nu \in (0,1)$, the Caputo fractional derivative is defined as:

$$ \begin{aligned} {}_0^{C}D_t^{\nu} x &= \frac{1}{\Gamma(1-\nu)} \int_0^t \frac{d}{dt} x(t-s) \frac{ds}{s^{\nu}} \\ &= {}_0^{RL}D_t^{\nu} x - \frac{x(0)}{\Gamma(1-\nu) t^{\nu}}, \end{aligned} $$

where

$$ {}_0^{RL}D_t^{\nu} x = \frac{1}{\Gamma(1-\nu)} \frac{d}{dt} \int_0^t x(t-s) \frac{ds}{s^{\nu}}, $$

and for its Laplace transform, denoting by $\widetilde{x}(z)$ the Laplace transform of x,

$$ \mathcal{L}[{}_0^{C}D_t^{\nu} x](z) = z^{\nu} \widetilde{x}(z) - z^{\nu-1} x(0). \tag{A1} $$

Moreover, for $\nu \in (0,1)$, a well-posed fractional Cauchy problem with Riemann–Liouville derivatives is given in the form

$$ \begin{cases} {}_{t_0}^{RL}D_t^{\nu} x = f(t, x(t)), \\ \left[{}_{t_0}I_t^{1-\nu} x \right]_{|t=t_0} = x_0, \end{cases} $$

in which the initial condition is given in terms of fractional integrals, while if we use Caputo derivatives we have:

$$\begin{cases} {}^{C}_{t_0}D^{\nu}_{t}x = f(t, x(t)), \\ x(t_0) = x_0, \end{cases}$$

in which the initial condition is related only with the initial value of the function. For such reason, we will prefer adopting Caputo derivatives as fractional derivatives in this paper.

Finally, let us remark that the definition of fractional integral and derivative can be also considered for functions $x : [t_0, t_1] \subseteq \mathbb{R} \to B$ where B is a Banach space and all the involved integrals are Bochner integrals ([23]).

Appendix B. Some Special Functions

We recall the definitions of some special functions we use in such text.

Gamma funcion

The Gamma function is defined as:

$$\Gamma(z) := \int_0^{\infty} t^{z-1}e^{-t}dt.$$

In particular, we have $\Gamma(z+1) = z\Gamma(z)$ and, for $z = n \in \mathbb{N}$, $\Gamma(n+1) = n!$.

The modified Bessel function ([24]) of the first kind can be defined by its power series expansion as:

$$I_r(x) = \sum_{m=0}^{+\infty} \frac{\left(\frac{x}{2}\right)^{2m+r}}{m!\Gamma(m+r+1)}.$$

One-parameter Mittag–Leffler functions ([6]) are defined by their power series expansion as:

$$E_{\nu}(z) = \sum_{k=0}^{\infty} \frac{z^k}{\Gamma(\nu k+1)}, \ \nu > 0, \ z \in \mathbb{C}.$$

Two-parameters Mittag–Leffler functions are also defined by their power series expansion as:

$$E_{\nu,\mu}(z) = \sum_{k=0}^{\infty} \frac{z^k}{\Gamma(\nu k+\mu)}, \ \nu > 0, \ \mu > 0, \ z \in \mathbb{C}.$$

Remark that $E_{\nu,1}(t) = E_{\nu}(t)$.

Generalized Mittag–Leffler functions are defined by their power series expansion as:

$$E^{\rho}_{\nu,\mu}(z) = \sum_{k=0}^{+\infty} \frac{(\rho)_k}{\Gamma(\nu k+\mu)} \frac{z^k}{k!}, \ \nu > 0, \ \mu > 0, \ \rho > 0, \ z \in \mathbb{C},$$

where $(\rho)_k$ is the Pochhammer symbol

$$(\rho)_k = \rho(\rho+1)(\rho+2)\cdots(\rho+k-1).$$

An alternative way to define the Generalized Mittag–Leffler function is:

$$E^{\rho}_{\alpha,\beta}(z) = \sum_{k=0}^{+\infty} \frac{z^k\Gamma(\rho+k)}{k!\Gamma(\alpha k+\beta)\Gamma(\rho)}, \ z \in \mathbb{C}.$$

Remark also that $E^1_{\alpha,\beta} = E_{\alpha,\beta}$. Functions with similar series expansions are also involved in the study of the asymptotic behaviour of some integrals, which arise from a Feynman path integral approach to some financial problems (see, i.e., [25] Section 4).
Recall also the following Laplace transform formula [9]

$$\mathcal{L}[z^{\gamma-1}E^{\delta}_{\nu,\gamma}(wz^{\nu})](s) = \frac{s^{\nu\delta-\gamma}}{(s^{\nu}-w)^{\delta}}, \ \gamma,\nu,\delta,w \in \mathbb{C}, \ \Re(\gamma),\Re(\nu),\Re(\delta) > 0, \ s \in \mathbb{C}, \ |ws^{\nu}| < 1. \tag{A2}$$

Finally let us remark, for $\nu \in (0,1)$ that the Cauchy problem

$$\begin{cases} {}^{C}_{0}D^{\nu}_{t}x = \lambda x, \\ x(0) = x_0, \end{cases}$$

admits as unique solution $x(t) = x_0 E_{\nu}(\lambda t^{\nu})$ ([6], p. 295).

References

1. Conolly, B.W. *Lecture Notes on Queueing Systems;* E. Horwood Limited: Ann Arbor, MI, USA, 1975; ISBN 0470168579, 9780470168578.
2. Conolly, B.W.; Langaris, C. On a new formula for the transient state probabilities for M/M/1 queues and computational implications. *J. Appl. Probab.* **1993**, *30*, 237–246. [CrossRef]
3. Kleinrock, L. *Queueing Systems: Theory;* Wiley: Hoboken, NJ, USA, 1975; ISBN 0471491101.
4. Lakatos, L.; Szeidl, L.; Telek, M. *Introduction to Queueing Systems with Telecommunication Applications;* Springer Science & Business Media: Berlin, Germany, 2012; ISBN 978-1-4614-5316-1.
5. Parthasarathy, P.R. A transient solution to an M/M/1 queue: A simple approach. *Adv. Appl. Probab.* **1987**, *19*, 997–998, doi:10.2307/1427113. [CrossRef]
6. Kilbas, A.A.; Srivastava, H.M.; Trujillo, J.J. Theory and Applications of Fractional Differential Equations. *North-Holland Math. Stud.* **2006**, *204*, 7–10.
7. Li, C.; Qian, D.; Chen, Y.Q. On Riemann-Liouville and Caputo derivatives. *Discrete Dyn. Nat. Soc.* **2011**, *2011*. [CrossRef]
8. Cahoy, D.O.; Polito, F.; Phoha,V. Transient behavior of fractional queues and related processes. *Methodol. Comput. Appl.* **2015**, *17*, 739–759. [CrossRef]
9. Haubold, H.J.; Mathai, A.M.; Saxena, R.K. Mittag–Leffler functions and their applications. *J. Appl. Math.* **2011**, *2011*. [CrossRef]
10. Meerschaert, M.M.; Nane, E.; Vellaisamy, P. The fractional Poisson process and the inverse stable subordinator. *Electron. J. Probab.* **2011**, *16*, 1600–1620. [CrossRef]
11. Laskin, N. Fractional Poisson process. *Commun. Nonlinear Sci.* **2003**, *8*, 201–213. [CrossRef]
12. Meerschaert, M.M.; Straka, P. Inverse stable subordinators. *Math. Model. Nat. Phenom.* **2013**, *8*, 1–16. [CrossRef] [PubMed]
13. Aletti, G.; Leonenko, N.; Merzbach, E. Fractional Poisson fields and martingales. *J. Stat. Phys.* **2018**, *170*, 700–730. [CrossRef]
14. Di Crescenzo, A.; Giorno, V.; NobileL, A.; Ricciardi, G.M. On the M/M/1 queue with catastrophes and its continuous approximation. *Queueing Syst.* **2003**, *43*, 329–347.:1023261830362. [CrossRef]
15. Krishna Kumar, B.; Arivudainambi, D. Transient solution of an M/M/1 queue with catastrophes. *Comput. Math. Appl.* **2000**, *40*, 1233–1240. [CrossRef]
16. Giorno, V.; Nobile, A.; Pirozzi, E. A state-dependent queueing system with asymptotic logarithmic distribution. *J. Math. Anal. Appl.* **2018**, *458*, 949–966. [CrossRef]
17. Bingham, N.H. Limit theorems for occupation times of Markov processes. *Zeitschrift für Wahrscheinlichkeitstheorie und Verwandte Gebiete* **1971**, *17*, 1–22. [CrossRef]
18. Kataria, K.K.; Vellaisamy, P. On densities of the product, quotient and power of independent subordinators. *J. Math. Anal. Appl.* **2018**, *462*, 1627–1643. [CrossRef]
19. Leguesdron, P.; Pellaumail, J.; Sericola, B. Transient analysis of the M/M/1 queue. *Adv. Appl. Probab.* **1993**, *25*, 702–713. [CrossRef]

20. Zhou, Y. *Basic Theory of Fractional Differential Equations*; World Scientific: Singapore, 2014; ISBN 978-981-3148-16-1.
21. Halmos, P.R.; Sunder, V.S. *Bounded Integral Operators on L^2 Spaces*; Springer Science & Business Media: Berlin, Germany, 2012; ISBN:978-3-642-67018-3.
22. Villani, A. Another note on the inclusion $L^p(\mu) \subset L^q(\mu)$. *Am. Math. Mon.* **1985**, *92*, 485–C76. [CrossRef]
23. Yosida, K. *Functional Analysis*; Springer: Berlin, Germany, 1978; ISBN 978-3-642-61859-8.
24. Korenev, B.G. *Bessel Functions and Their Applications*; CRC Press: Boca raton, FL, USA, 2003; ISBN 9780415281300.
25. Issaka, A.; SenGupta, I. Feynman path integrals and asymptotic expansions for transition probability densities of some Lévy driven financial markets. *J. Appl. Math. Comput.* **2017**, *54*, 159–182. [CrossRef]

mathematics

MDPI

Article

An $M^{[X]}/G(a,b)/1$ Queueing System with Breakdown and Repair, Stand-By Server, Multiple Vacation and Control Policy on Request for Re-Service

G. Ayyappan and S. Karpagam *

Department of Mathematics, Pondicherry Engineering College, Puducherry 605014, India; ayyappan@pec.edu
* Correspondence: karpagammaths19@gmail.com

Received: 29 March 2018; Accepted: 29 May 2018; Published: 14 June 2018

Abstract: In this paper, we discuss a non-Markovian batch arrival general bulk service single-server queueing system with server breakdown and repair, a stand-by server, multiple vacation and re-service. The main server's regular service time, re-service time, vacation time and stand-by server's service time are followed by general distributions and breakdown and repair times of the main server with exponential distributions. There is a stand-by server which is employed during the period in which the regular server remains under repair. The probability generating function of the queue size at an arbitrary time and some performance measures of the system are derived. Extensive numerical results are also illustrated.

Keywords: non-Markovian queue; general bulk service; multiple vacation; breakdown and repair; stand-by server; re-service

Mathematics Subject Classification: 60K25; 90B22; 68M20

1. Introduction

Queueing systems with general bulk service and vacations have been studied by many researchers because they deal with effective utilization of the server's idle time for secondary jobs. Such queueing systems have a wide range of application in many real-life situations such as production line systems, inventory systems, digital communications and computer networks. Doshi [1] and Takagi [2] have made a comprehensive survey of queueing systems with vacations. A batch arrival $M^{[X]}/G/1$ queueing system with multiple vacations was first studied by Baba [3]. Krishna Reddy et al. [4] have discussed an $M^{[X]}/G(a,b)/1$ model with an N-policy, multiple vacations and setup times. Jeyakumar and Senthilnathan [5] analyzed the bulk service queueing system with multiple working vacations and server breakdown.

The first work on re-service was done by Madan [6]. He consider an $M/G/1$ queueing model, in which the server performs the first essential service for all arriving customers. As soon as the first service is executed, they may leave the system with probability $(1-\theta)$, and the second optional service is provided with θ. Madan et al. [7] considered a bulk arrival queue with optional re-service. Jeyakumar and Arumuganathan [8] discussed a bulk queue with multiple vacation and a control policy on request for re-service. Recently, Haridass and Arumuganathan [9] analyzed a batch service queueing system with multiple vacations, setup times and server choice of admitting re-service.

No system is found to be perfect in the real world, since all the devices fail more or less frequently. Thus, the random failures and systematic repair of components of a machining system have a significant impact on the output and the productivity of the machining system. A detailed survey on queues with interruptions was undertaken by Krishnamoorthy et al. [10]. Ayyappan and Shyamala [11] derived the transient solution to an $M^{[X]}/G/1$ queueing system with feedback, random breakdowns, Bernoulli schedule server vacation and random setup time. An $M/G/1$ queue with two phases of service subject to random breakdown and delayed repair was examined by Choudhury and Tadj [12]. Senthilnathan and Jeyakumar [13] studied the behavior of the server breakdown without interruption in an $M^{[X]}/G(a,b)/1$ queueing system with multiple vacations and closedown time. An $M/G/1$ two-phase multi-optional retrial queue with Bernoulli feedback, non-persistent customers and breakdown and repair was analyzed by Lakshmi and Ramanath [14]. Recently, a discrete time queueing system with server breakdowns and changes in the repair times was investigated by Atencia [15].

The operating machine may fail in some cases, but due to the standby machines of the queueing machining system, it remains operative and continues to perform the assigned job. The provision of stand-by and repairmen support to the queueing system maintains the smooth functioning of the system. In the field of computer and communications systems, distribution and service systems, production/manufacturing systems, etc., the applications of queueing models with standby support may be noted.

This paper is organized as follows. A literature survey is given in Section 2. In Section 3, the queuing problem is defined. The system equations are developed in Sections 4. The Probability Generating Function (PGF) of the queue length distribution in the steady state is obtained in Section 5. Various performance measures of the queuing system are derived in Section 6. A computational study is illustrated in Section 7. Conclusions are given in Section 8.

2. Literature Survey

Various authors have analyzed queueing problems of server vacation with several combinations. A batch arrival queue with a vacation time under a single vacation policy was analyzed by Choudhury [16]. Jeyakumar and Arumuganathan [17] have discussed steady state analysis of an $M^{[X]}]/G/1$ queue with two service modes and multiple vacation, in which they obtained PGF of the queue size and some performance measures. Balasubramanian et al. [18] discussed steady state analysis of a non-Markovian bulk queueing system with overloading and multiple vacations. Haridass and Arumuganathan [19] discussed a batch arrival general bulk service queueing system with a variant threshold policy for secondary jobs. Recently, Choudhury and Deka [20] discussed a batch arrival queue with an unreliable server and delayed repair, with two phases of service and Bernoulli vacation under multiple vacation policy.

Queueing systems, where the service discipline involves more than one service, have been receiving much attention recently. They are said to have an additional service channel, or to have feedback, or to have optional re-service, or to have two phases of heterogeneous service. Madan [21] analyzed a queueing system with feedback. Madan [22], generalized his previous model by incorporating server vacation. Medhi [23] discussed a single server Poisson input queue with a second optional channel. Arumugananathan and Maliga [24] also examined a bulk queue with re-service of the service station and setup time. Baruah et al. [25] studied a batch arrival queue with two types of service, balking, re-service and vacation. Ayyappan and Sathiya [11] derived the PGF of the non-Markovian queue with two types of service and optional re-service with a general vacation distribution.

One can find an enormous amount of work done on queueing systems with breakdowns. For some papers on random breakdowns in queueing systems, the reader may see Aissani et al. [26], Maraghi et al. [27] and Fadhil et al. [28]. Rajadurai et al. [29] analyzed an $M^{[X]}/G/1$ retrial queue with two phases of service under Bernoulli vacation and random breakdown. Jiang et al. [30] have made a computational analysis of a queue with working breakdown and delayed repair.

The operating system may fail in some cases, but due to stand-by machines, it remains operative and continuous to perform the assigned job. Madan [31] studied the steady state behavior of a queuing system with a stand-by server to serve customers only during the repair period. In that work, repair times were assumed to follow an exponential distribution. Khalaf [32] examined the queueing system with four different main servers' interruption and a stand-by server. Jain et al. [33] have made a cost analysis of the machine repair problem with standby, working vacation and server breakdown. Kumar et al. [34] discussed a bi-level control of a degraded machining system with two unreliable servers, multiple standbys, startup and vacation. Murugeswari et al. [35] analyzed the bulk arrival queueing model with a stand-by server and compulsory server vacation. Recently, we provided an excellent survey on standby by Kolledath et al. [36].

3. Model Description

This paper deals with a queueing model whose arrival follows a compound Poisson process with intensity rate λ. The main server and stand-by servers serve the customers under the general bulk service rule. The general bulk service rule was first introduced by Neuts [37]. The general bulk service rule states that the server will start to provide service only when at least 'a' units are present in the queue, and the maximum service capacity is 'b' (b > a). On completion of a batch service, if less than 'a' customers are present in the queue, then the server has to wait until the queue length reaches the value 'a'. If less than or equal to 'b' and greater than or equal to 'a' customers are in the queue, then all the existing customers are taken into service. If greater than or equal to 'b' customers are in the queue, then 'b' customers are taken into service. The main server may breakdown at any time during regular service with exponential rate α, and in such cases, the main server immediately goes for a repair, which follows an exponential distribution with rate η, while the service to the current batch is interrupted. Such a batch of customers is changed to the stand-by server, which starts service to that batch afresh. The stand-by server remains in the system until the main server's repair is completed. At the instant of repair completion, if the stand-by server is busy, then the current batch of customers is exchanged to the main server, which starts that batch service afresh. At the completion of a regular service (by the main server), the leaving batch may request for a re-service with probability π. However, the re-service is rendered only when the number of customers waiting in the queue is less than a. If no request for re-service is made after the completion of a regular service and the number of customers in the queue is less than a, then the server will avail itself of a vacation of a random length. The server takes a sequence of vacations until the queue size reaches at least a. In addition, we assume that the service time of the main server and stand-by server, re-service and vacation time of the main server are independent of each other and follow a general (arbitrary) distribution.

Notations

Let X be the group size random variable of arrival, g_k be the probability of 'k' customers arriving in a batch and $X(z)$ be its PGF. $S_b(.)$, $R(.)$, $S_s(.)$ and $V(.)$ represent the Cumulative Distribution Functions (CDF) of the regular service and re-service time of the main server, the service time of the stand-by server and the vacation time of the main server with corresponding probability density functions of $s_b(x)$, $r(x)$, $s_s(x)$ and $v(x)$, respectively. $S_b^0(t)$, $R^0(t)$, $S_s^0(t)$ and $V^{(0)}(t)$ represent the remaining regular service and re-service time of service given by the main server, the remaining service time of service given by the stand-by server and the remaining vacation time of the main server at time 't', respectively. $\tilde{S}_b(\theta)$, $\tilde{R}(\theta)$, $\tilde{S}_s(\theta)$ and $\tilde{V}(\theta)$ represent the Laplace–Stieltjes Transform (LST) of S_b, R, S_s and V, respectively.

For further development of the queueing system, let us define the following:

$\varepsilon(t)$ = 1, 2, 3, 4, 5 and 6 at time t; the main server is in regular service, re-service and vacation, and at time t, the stand-by server is in service and idle, respectively.

$Z(t) = j$, if the server is on the j-th vacation.

$N_s(t)$ = number of customers in service station at time t.

$N_q(t)$ = number of customers in the queue at time t.

Define the probabilities:

$$T_n(t)\Delta t = Pr\{N_q(t) = n,\ \varepsilon(t) = 5\},\ 0 \le n \le a-1,$$
$$P_{m,n}(x,t)\Delta t = Pr\{N_s(t) = m,\ N_q(t) = n,\ x \le S_b^0(t) \le x+\Delta t,\ \varepsilon(t) = 1\},\quad a \le m \le b,\ n \ge 0,$$
$$R_n(x,t)\Delta t = Pr\{N_s(t) = m,\ N_q(t) = n,\ x \le R^0(t) \le x+\Delta t,\ \varepsilon(t) = 2\},\quad a \le m \le b,\ n \ge 0,$$
$$B_{m,n}(x,t)\Delta t = Pr\{N_s(t) = m,\ N_q(t) = n,\ x \le S_s^0(t) \le x+\Delta t,\ \varepsilon(t) = 4\},\quad a \le m \le b,\ n \ge 0,$$
$$Q_{l,j}(x,t)\Delta t = Pr\{Z(t) = l,\ N_q(t) = j,\ x \le V^0(t) \le x+\Delta t,\ \varepsilon(t) = 3\},\quad l \ge 1, j \ge 0.$$

4. Queue Size Distribution

From the above-defined probabilities, we can easily construct the following steady state equations:

$$(\lambda+\eta)T_0 = \sum_{m=a}^{b} B_{m,0}(0), \tag{1}$$

$$(\lambda+\eta)T_n = \sum_{m=a}^{b} B_{m,n}(0) + \sum_{k=1}^{n} T_{n-k}\lambda g_k,\ 1 \le n \le a-1, \tag{2}$$

$$-P'_{i,0}(x) = -(\lambda+\alpha)P_{i,0}(x) + \sum_{m=a}^{b} P_{m,i}(0)s_b(x) + \eta\int_0^\infty B_{i,0}(y)\,dy s_b(x) + R_i(0)s_b(x) + \sum_{l=1}^{\infty} Q_{l,i}(0)s_b(x),\ a \le i \le b, \tag{3}$$

$$-P'_{i,j}(x) = -(\lambda+\alpha)P_{i,j}(x) + \eta\int_0^\infty B_{i,j}(y)\,dy s_b(x) + \sum_{k=1}^{j} P_{i,j-k}(x)\lambda g_k,\quad j \ge 1,\ a \le i \le b-1, \tag{4}$$

$$-P'_{b,j}(x) = -(\lambda+\alpha)P_{b,j}(x) + \sum_{m=a}^{b} P_{m,b+j}(0)s_b(x) + \eta\int_0^\infty B_{b,j}(y)\,dy s_b(x) + R_{b+j}(0)s_b(x) + \sum_{l=1}^{\infty} Q_{l,b+j}(0)s_b(x) + \sum_{k=1}^{j} P_{b,j-k}(x)\lambda g_k,\ j \ge 1, \tag{5}$$

$$-B'_{i,0}(x) = -(\lambda+\eta)B_{i,0}(x) + \sum_{m=a}^{b} B_{m,i}(0)s_s(x) + \alpha\int_0^\infty P_{i,0}(y)\,dy s_s(x) + \sum_{k=0}^{a-1} T_k\lambda g_{i-k}s_s(x),\ a \le i \le b, \tag{6}$$

$$-B'_{i,j}(x) = -(\lambda+\eta)B_{i,j}(x) + \alpha\int_0^\infty P_{i,j}(y)\,dy s_s(x) + \sum_{k=1}^{j} B_{i,j-k}(x)\lambda g_k,\quad j \ge 1,\ a \le i \le b-1, \tag{7}$$

$$-B'_{b,j}(x) = -(\lambda+\eta)B_{b,j}(x) + \sum_{m=a}^{b} B_{m,b+j}(0)s_s(x) + \sum_{k=1}^{j} B_{b,j-k}(x)\lambda g_k + \alpha\int_0^\infty P_{b,j}(y)\,dy\, s_s(x) + \sum_{k=0}^{a-1} T_k\lambda g_{b+j-k}s_s(x),\ j \ge 1, \tag{8}$$

$$-R_0'(x) = -\lambda R_0(x) + \pi \sum_{m=a}^{b} P_{m,0}(0) r(x), \tag{9}$$

$$-R_n'(x) = -\lambda R_n(x) + \pi \sum_{m=a}^{b} P_{m,n}(0) r(x) + \sum_{k=1}^{n} R_{n-k}(x)\lambda g_k,\ 1 \le n \le a-1, \tag{10}$$

$$-R_n'(x) = -\lambda R_n(x) + \sum_{k=1}^{n} R_{n-k}(x)\lambda g_k,\ n \ge a, \tag{11}$$

$$-Q_{1,0}'(x) = -\lambda Q_{1,0}(x) + (1-\pi) \sum_{m=a}^{b} P_{m,0}(0) v(x) + R_0(0) v(x) + \eta T_0 v(x), \tag{12}$$

$$\begin{aligned} -Q_{1,n}'(x) &= -\lambda Q_{1,n}(x) + (1-\pi) \sum_{m=a}^{b} P_{m,n}(0) v(x) + R_n(0) v(x) + \eta T_n v(x) \\ &\quad + \sum_{k=1}^{n} Q_{1,n-k}(x)\lambda g_k,\ 1 \le n \le a-1, \end{aligned} \tag{13}$$

$$-Q_{1,n}'(x) = -\lambda Q_{1,n}(x) + \sum_{k=1}^{n} Q_{1,n-k}(x)\lambda g_k,\ n \ge a, \tag{14}$$

$$-Q_{j,0}'(x) = -\lambda Q_{j,0}(x) + Q_{j-1,0}(0) v(x),\ j \ge 2, \tag{15}$$

$$\begin{aligned} -Q_{j,n}'(x) &= -\lambda Q_{j,n}(x) + Q_{j-1,n}(0) v(x) + \sum_{k=1}^{n} Q_{j,n-k}(x)\lambda g_k,\ j \ge 2, \\ &\qquad 1 \le n \le a-1, \end{aligned} \tag{16}$$

$$-Q_{j,n}'(x) = -\lambda Q_{j,n}(x) + \sum_{k=1}^{n} Q_{j,n-k}(x)\lambda g_k,\ j \ge 2,\ n \ge a. \tag{17}$$

Taking the LST on both sides of Equations (3)–(17), we get,

$$\begin{aligned} \theta \tilde{P}_{i,0}(\theta) - P_{i,0}(0) &= (\lambda+\alpha)\tilde{P}_{i,0}(\theta) - \sum_{m=a}^{b} P_{m,i}(0)\tilde{S}_b(\theta) - \eta \int_0^\infty B_{i,0}(y)\,dy \tilde{S}_b(\theta) \\ &\quad - R_i(0)\tilde{S}_b(\theta) - \sum_{l=1}^{\infty} Q_{l,i}(0)\tilde{S}_b(\theta),\ a \le i \le b, \end{aligned} \tag{18}$$

$$\begin{aligned} \theta \tilde{P}_{i,j}(\theta) - P_{i,j}(0) &= (\lambda+\alpha)\tilde{P}_{i,j}(\theta) - \eta \int_0^\infty B_{i,j}(y)\,dy\ \tilde{S}_b(\theta) - \sum_{k=1}^{j} \tilde{P}_{i,j-k}(\theta)\lambda g_k, \\ &\qquad a \le i \le b-1,\ j \ge 1, \end{aligned} \tag{19}$$

$$\begin{aligned} \theta \tilde{P}_{b,j}(\theta) - P_{b,j}(0) &= (\lambda+\alpha)\tilde{P}_{b,j}(\theta) - \sum_{m=a}^{b} P_{m,b+j}(0)\tilde{S}_b(\theta) - \eta \int_0^\infty B_{b,j}(y)\,dy \tilde{S}_b(\theta) \\ &\quad - R_{b+j}(0)\tilde{S}_b(\theta) - \sum_{l=1}^{\infty} Q_{l,b+j}(0)\tilde{S}_b(\theta) - \sum_{k=1}^{j} \tilde{P}_{b,j-k}(\theta)\lambda g_k,\ j \ge 1, \end{aligned} \tag{20}$$

$$\begin{aligned} \theta \tilde{B}_{i,0}(\theta) - B_{i,0}(0) &= (\lambda+\eta)\tilde{B}_{i,0}(\theta) - \sum_{m=a}^{b} B_{m,i}(0)\tilde{S}_s(\theta) - \alpha \int_0^\infty P_{i,0}(y)\,dy \tilde{S}_s(\theta) \\ &\quad - \sum_{k=0}^{a-1} T_k \lambda g_{i-k}\tilde{S}_s(\theta),\ a \le i \le b, \end{aligned} \tag{21}$$

$$\begin{aligned} \theta \tilde{B}_{i,j}(\theta) - B_{i,j}(0) &= (\lambda+\eta)\tilde{B}_{i,j}(\theta) - \alpha \int_0^\infty P_{i,j}(y)\,dy\ \tilde{S}_s(\theta) - \sum_{k=1}^{j} \tilde{B}_{i,j-k}(\theta)\lambda g_k, \\ &\qquad j \ge 1,\ a \le i \le b-1, \end{aligned} \tag{22}$$

$$\theta\tilde{B}_{b,j}(\theta) - B_{b,j}(0) = (\lambda+\eta)\tilde{B}_{b,j}(\theta) - \sum_{m=a}^{b} B_{m,b+j}(0)\tilde{S}_s(\theta) - \sum_{k=1}^{j}\tilde{B}_{b,j-k}(\theta)\lambda g_k$$
$$- \alpha\int_0^\infty P_{b,j}(y)\,dy\,\tilde{S}_s(\theta) - \sum_{k=0}^{a-1} T_k\lambda g_{b+j-k}\tilde{S}_s(\theta),\ j\geq 1, \tag{23}$$

$$\theta\tilde{R}_0(\theta) - R_0(0) = \lambda\tilde{R}_0(\theta) - \pi\sum_{m=a}^{b} P_{m,0}(0)\tilde{R}(\theta), \tag{24}$$

$$\theta\tilde{R}_n(\theta) - R_n(0) = \lambda\tilde{R}_n(\theta) - \pi\sum_{m=a}^{b} P_{m,n}(0)\tilde{R}(\theta) - \sum_{k=1}^{n}\tilde{R}_{n-k}(\theta)\lambda g_k,$$
$$1\leq n\leq a-1, \tag{25}$$

$$\theta\tilde{R}_n(\theta) - R_n(0) = \lambda\tilde{R}_n(\theta) - \sum_{k=1}^{n}\tilde{R}_{n-k}(\theta)\lambda g_k,\ n\geq a, \tag{26}$$

$$\theta\tilde{Q}_{1,0}(\theta) - Q_{1,0}(0) = \lambda\tilde{Q}_{1,0}(\theta) - (1-\pi)\sum_{m=a}^{b} P_{m,0}(0)\tilde{V}(\theta) - R_0(0)\tilde{V}(\theta)$$
$$-\eta T_0\tilde{V}(\theta), \tag{27}$$

$$\theta\tilde{Q}_{1,n}(\theta) - Q_{1,n}(0) = \lambda\tilde{Q}_{1,n}(\theta) - (1-\pi)\sum_{m=a}^{b} P_{m,n}(0)\tilde{V}(\theta) - R_n(0)\tilde{V}(\theta)$$
$$-\eta T_n\tilde{V}(\theta) - \sum_{k=1}^{n}\tilde{Q}_{1,n-k}(\theta)\lambda g_k,\ 1\leq n\leq a-1, \tag{28}$$

$$\theta\tilde{Q}_{1,n}(\theta) - Q_{1,n}(0) = \lambda\tilde{Q}_{1,n}(\theta) - \sum_{k=1}^{n}\tilde{Q}_{1,n-k}(\theta)\lambda g_k,\ n\geq a, \tag{29}$$

$$\theta\tilde{Q}_{j,0}(\theta) - Q_{j,0}(0) = \lambda\tilde{Q}_{j,0}(\theta) - Q_{j-1,0}(0)\tilde{V}(\theta),\ j\geq 2, \tag{30}$$

$$\theta\tilde{Q}_{j,n}(\theta) - Q_{j,n}(0) = \lambda\tilde{Q}_{j,n}(\theta) - Q_{j-1,n}(0)\tilde{V}(\theta) - \sum_{k=1}^{n}\tilde{Q}_{j,n-k}(\theta)\lambda g_k,\ j\geq 2,$$
$$1\leq n\leq a-1, \tag{31}$$

$$\theta\tilde{Q}_{j,n}(\theta) - Q_{j,n}(0) = \lambda\tilde{Q}_{j,n}(\theta) - \sum_{k=1}^{n}\tilde{Q}_{j,n-k}(\theta)\lambda g_k,\ j\geq 2,\ n\geq a. \tag{32}$$

To find the Probability Generating Function (PGF) for the queue size, we define the following PGFs:

$$\tilde{P}_i(z,\theta) = \sum_{j=0}^{\infty}\tilde{P}_{i,j}(\theta)z^j,\ \ P_i(z,0) = \sum_{j=0}^{\infty} P_{i,j}(0)z^j,\ a\leq i\leq b,$$
$$\tilde{R}(z,\theta) = \sum_{j=0}^{\infty}\tilde{R}_j(\theta)z^j,\ \ R(z,0) = \sum_{j=0}^{\infty} R_j(0)z^j,$$
$$\tilde{B}_i(z,\theta) = \sum_{j=0}^{\infty}\tilde{B}_{i,j}(\theta)z^j,\ \ B_i(z,0) = \sum_{j=0}^{\infty} B_{i,j}(0)z^j,\ a\leq i\leq b, \tag{33}$$
$$\tilde{Q}_l(z,\theta) = \sum_{j=0}^{\infty}\tilde{Q}_{l,j}(\theta)z^j\ \ Q_l(z,0) = \sum_{j=0}^{\infty} Q_{l,j}(0)z^j,\ l\geq 1.$$

By multiplying Equations (18)–(32) with suitable power of z^n and summing over n ($n = 0$ to ∞) and using Equation (33), we get:

$$(\theta - u(z))\tilde{P}_i(z,\theta) = P_i(z,0) - \tilde{S}_b(\theta)\Big[\sum_{m=a}^{b} P_{m,i}(0) + R_i(0) + \sum_{j=0}^{\infty} Q_{l,i}(0) + \eta\tilde{B}_i(z,0)\Big],$$
$$a \le i \le b-1, \quad (34)$$

$$z^b(\theta - u(z))\tilde{P}_b(z,\theta) = (z^b - \tilde{S}_b(\theta))P_b(z,0) - \tilde{S}_b(\theta)\Big[\sum_{m=a}^{b-1} P_m(z,0) + R(z,0) + \sum_{l=1}^{\infty} Q_l(z,0) + z^b\eta\tilde{B}_b(z,0) - \sum_{j=0}^{b-1}\Big(\sum_{m=a}^{b} P_{m,j}(0)z^j + R_j(0)z^j + \sum_{j=0}^{\infty} Q_{l,j}(0)z^j\Big)\Big], \quad (35)$$

$$(\theta - v(z))\tilde{B}_i(z,\theta) = B_i(z,0) - \tilde{S}_s(\theta)\Big[\alpha\tilde{P}_i(z,0) + \sum_{m=a}^{b} B_{m,i}(0) + \sum_{k=0}^{a-1} T_k\lambda g_{i-k}\Big],$$
$$a \le i \le b-1, \quad (36)$$

$$z^b(\theta - v(z))\tilde{B}_b(z,\theta) = (z^b - \tilde{S}_s(\theta))B_b(z,0) - \tilde{S}_s(\theta)\Big[\sum_{m=a}^{b-1} B_m(z,0) + z^b\alpha\tilde{P}_b(z,0) + \lambda\sum_{k=0}^{a-1}\sum_{j=b}^{\infty} T_k z^k g_{j-k} z^{j-k} - \sum_{j=0}^{b-1}\sum_{m=a}^{b} B_{m,j}(0)z^j\Big], \quad (37)$$

$$(\theta - w(z))\tilde{R}(z,\theta) = R(z,0) - \pi\tilde{R}(\theta)\sum_{n=0}^{a-1}\sum_{m=a}^{b} P_{m,n}(0)z^n, \quad (38)$$

$$(\theta - w(z))\tilde{Q}_1(z,\theta) = Q_1(z,0) - \tilde{V}(\theta)\sum_{n=0}^{a-1}\Big[(1-\pi)\sum_{m=a}^{b} P_{m,n}(0)z^n + R_n(0)z^n + \eta T_n z^n\Big], \quad (39)$$

$$(\theta - w(z))\tilde{Q}_j(z,\theta) = Q_j(z,0) - \tilde{V}(\theta)\sum_{n=0}^{a-1} Q_{j-1,n}(0)z^n,\ j \ge 2, \quad (40)$$

where
$u(z) = \lambda + \alpha - \lambda X(z)$, $v(z) = \lambda + \eta - \lambda X(z)$, $w(z) = \lambda - \lambda X(z)$.
Substitute $\theta = u(z)$ in (34) and (35), we get,

$$P_i(z,0) = \tilde{S}_b(u(z))\Big[\sum_{m=a}^{b} P_{m,i}(0) + R_i(0) + \sum_{j=0}^{\infty} Q_{l,i}(0) + \eta\tilde{B}_i(z,0)\Big],$$
$$a \le i \le b-1, \quad (41)$$

$$P_b(z,0) = \frac{\tilde{S}_b(u(z))}{(z^b - \tilde{S}_b(u(z)))}\Big[\sum_{m=a}^{b-1} P_m(z,0) + R(z,0) + \sum_{l=1}^{\infty} Q_l(z,0) + z^b\eta\tilde{B}_b(z,0) - \sum_{j=0}^{b-1}\Big(\sum_{m=a}^{b} P_{m,j}(0)z^j + R_j(0)z^j + \sum_{j=0}^{\infty} Q_{l,j}(0)z^j\Big)\Big], \quad (42)$$

substitute $\theta = v(z)$ in (36) and (37), we get,

$$B_i(z,0) = \tilde{S}_s(v(z))\left[\alpha\tilde{P}_i(z,0) + \sum_{m=a}^{b} B_{m,i}(0) + \sum_{k=0}^{a-1} T_k\lambda g_{i-k}\right],\ a \le i \le b-1, \tag{43}$$

$$B_b(z,0) = \frac{\tilde{S}_s(v(z))}{(z^b - \tilde{S}_s(v(z)))}\Bigg[\sum_{m=a}^{b-1} B_m(z,0) + z^b\alpha\tilde{P}_b(z,0) + \lambda\sum_{k=0}^{a-1}\sum_{j=b}^{\infty} T_k z^k g_{j-k} z^{j-k} - \sum_{j=0}^{b-1}\sum_{m=a}^{b} B_{m,j}(0)z^j\Bigg], \tag{44}$$

substitute $\theta = w(z)$ in (38) to (40), we get

$$R(z,0) = \pi\tilde{R}(w(z))\sum_{n=0}^{a-1}\sum_{m=a}^{b} P_{m,n}(0)z^n, \tag{45}$$

$$Q_1(z,0) = \tilde{V}(w(z))\sum_{n=0}^{a-1}\left[(1-\pi)\sum_{m=a}^{b} P_{m,n}(0)z^n + R_n(0)z^n + \eta T_n z^n\right], \tag{46}$$

$$Q_j(z,0) = \tilde{V}(w(z))\sum_{n=0}^{a-1} Q_{j-1,n}(0)z^n,\ j \ge 2. \tag{47}$$

Substitute Equations (41)–(47) in Equations (34)–(40) after simplification, and we get,

$$(\theta - u(z))\tilde{P}_i(z,\theta) = (\tilde{S}_b(u(z)) - \tilde{S}_b(\theta))\Bigg[\sum_{m=a}^{b} P_{m,i}(0) + R_i(0) + \sum_{j=0}^{\infty} Q_{l,i}(0) + \eta\tilde{B}_i(z,0)\Bigg],\ a \le i \le b-1, \tag{48}$$

$$(\theta - u(z))\tilde{P}_b(z,\theta) = \frac{(\tilde{S}_b(u(z)) - \tilde{S}_b(\theta))}{(z^b - \tilde{S}_b(u(z)))}\Bigg[\sum_{m=a}^{b-1} P_m(z,0) + R(z,0) + \sum_{l=1}^{\infty} Q_l(z,0) + z^b\eta\tilde{B}_b(z,0) - \sum_{j=0}^{b-1}\Big(\sum_{m=a}^{b} P_{m,j}(0)z^j + R_j(0)z^j + \sum_{j=0}^{\infty} Q_{l,j}(0)z^j\Big)\Bigg], \tag{49}$$

$$(\theta - v(z))\tilde{B}_i(z,\theta) = (\tilde{S}_s(v(z)) - \tilde{S}_s(\theta))\left[\alpha\tilde{P}_i(z,0) + \sum_{m=a}^{b} B_{m,i}(0) + \sum_{k=0}^{a-1} T_k\lambda g_{i-k}\right],\ a \le i \le b-1, \tag{50}$$

$$(\theta - v(z))\tilde{B}_b(z,\theta) = \frac{(\tilde{S}_s(v(z)) - \tilde{S}_s(\theta))}{(z^b - \tilde{S}_s(v(z)))}\Bigg[\sum_{m=a}^{b-1} B_m(z,0) + z^b\alpha\tilde{P}_b(z,0) + \lambda\sum_{k=0}^{a-1}\sum_{j=b}^{\infty} T_k z^k g_{j-k} z^{j-k} - \sum_{j=0}^{b-1}\sum_{m=a}^{b} B_{m,j}(0)z^j\Bigg], \tag{51}$$

$$(\theta - w(z))\tilde{R}(z,\theta) = (\tilde{R}(w(z)) - \tilde{R}(\theta))\pi\sum_{n=0}^{a-1}\sum_{m=a}^{b} P_{m,n}(0)z^n, \tag{52}$$

$$(\theta - w(z))\tilde{Q}_1(z,\theta) = (\tilde{V}(w(z)) - \tilde{V}(\theta))\sum_{n=0}^{a-1}\Bigg[(1-\pi)\sum_{m=a}^{b} P_{m,n}(0)z^n + R_n(0)z^n + \eta T_n z^n\Bigg], \tag{53}$$

$$(\theta - w(z))\tilde{Q}_j(z,\theta) = (\tilde{V}(w(z)) - \tilde{V}(\theta))\sum_{n=0}^{a-1} Q_{j-1,n}(0)z^n,\ j \ge 2. \tag{54}$$

5. Probability Generating Function of the Queue Size

5.1. The PGF of the Queue Size at an Arbitrary Time Epoch

Let $P(z)$ be the PGF of the queue size at an arbitrary time epoch. Then,

$$P(z) = \sum_{i=a}^{b} \tilde{P}_i(z,0) + \sum_{i=a}^{b} \tilde{B}_i(z,0) + \tilde{R}(z,0) + \sum_{l=1}^{\infty} \tilde{Q}_l(z,0) + T(z). \tag{55}$$

By substituting $\theta = 0$ in Equations (48)–(54), then Equation (55) becomes:

$$P(z) = \frac{\begin{array}{l} K_1(z) \sum_{i=a}^{b-1} (z^b - z^i) c_i + (1 - \tilde{V}(w(z))) K_3(z) \sum_{n=0}^{a-1} c_n z^n \\ + K_2(z) \sum_{i=a}^{b-1} (z^b - z^i) d_i + (\tilde{V}(w(z)) - \tilde{R}(w(z))) K_3(z) \sum_{n=0}^{a-1} \pi p_n z^n \\ + \left[\eta [Y_1(z) - \tilde{V}(w(z)) K_3(z)] - v(z) K_2(z) + w(z) Y_1(z) \right] \sum_{k=0}^{a-1} T_k z^k \end{array}}{w(z) Y_1(z)} \tag{56}$$

where $p_i = \sum_{m=a}^{b} P_{m,i}(0)$, $v_i = \sum_{l=1}^{\infty} Q_{l,i}(0)$, $q_i = \sum_{m=a}^{b} B_{m,i}(0), R_i(0) = r_i$, $c_i = p_i + v_i + r_i$ and $d_i = q_i + \sum_{k=0}^{a-1} T_k \lambda g_{i-k}$ and the expressions for $K_1(z)$, $K_2(z)$, $K_3(z)$ and $Y_1(z)$ are defined in Appendix A.

5.2. Steady State Condition

The probability generating function has to satisfy $P(1) = 1$. In order to satisfy this condition, applying L'Hopital's rule and evaluating $\lim_{z \to 1} P(z)$, then equating the expression to one, we have, $H = (-\lambda X_1) F_1$, where the expressions H and F_1 are defined in Appendix B.

Since p_i, c_i, d_i and T_i are probabilities of $'i'$ customers being in the queue, it follows that H must be positive. Thus, $P(1) = 1$ is satisfied iff $(-\lambda X_1) F_1 > 0$. If:

$$\rho = \frac{\lambda X_1 (\alpha + \eta)(1 - \tilde{S}_b(\alpha))(1 - \tilde{S}_s(\eta))}{b \alpha \eta [\tilde{S}_b(\alpha)(1 - \tilde{S}_s(\eta)) + \tilde{S}_s(\eta)(1 - \tilde{S}_b(\alpha))]}$$

then $\rho < 1$ is the condition for the existence of the steady state for the model under consideration.

5.3. Computational Aspects

Equation (56) has $2b + a$ unknowns $c_0, c_1, ..., c_{b-1}, d_a, ..., d_{b-1}, p_0, p_1, ..., p_{a-1}$ and $T_0, T_1, ..., T_{a-1}$. Now, Equation (56) gives the PGF of the number of customers involving only $2b + a$ unknowns. We can express $c_i (0 \leq i \leq a-1)$ in terms of p_i and T_i in such a way that the numerator has only $2b$ constants. Now, Equation (56) gives the PGF of the number of customers involving only $2b$ unknowns. By Rouche's theorem, it can be proven that $Y_1(z)$ has $2b - 1$ zeros inside and one on the unit circle $|z| = 1$. Since $P(z)$ is analytic within and on the unit circle, the numerator must vanish at these points, which gives $2b$ equations in $2b$ unknowns. We can solve these equations by any suitable numerical technique.

5.4. Result 1

The probability that $n (0 \leq n \leq a-1)$ customers are in queue during the main server's re-service completion r_n can be expressed as the probability of n customers in the queue during the main server's regular busy period p_n as,

$$r_n = \pi \sum_{k=0}^{n} \gamma_{n-k} p_k, \; n = 0, 1, 2..., a-1 \tag{57}$$

where γ_n are the probabilities of n customers arriving during the main server's re-service time.

5.5. Result 2

The probability that $n(0 \leq n \leq a-1)$ customers are in queue c_n can be expressed as the sum of the probability of n customers in the queue during the main server's busy period and the stand-by server's idle time p_n and T_n as,

$$c_n = \sum_{k=0}^{n} \tau_{n-k}^{(1)} p_k + \tau_{n-k}^{(2)} T_k, \; 0 \leq n \leq a-1. \tag{58}$$

where:

$$\begin{aligned} \tau_n^{(1)} &= \frac{\pi(\gamma_n - \beta_n) + \sum_{k=1}^{n} \beta_k \tau_{n-k}^{(1)}}{1 - \beta_0} \\ \tau_n^{(2)} &= \frac{\eta \beta_n + \sum_{k=1}^{n} \beta_k \tau_{n-k}^{(2)}}{1 - \beta_0} \end{aligned} \tag{59}$$

γ_n, β_n are the probabilities of n customers arriving during the main server's re-service and vacation time, respectively.

5.6. Particular Case

Case 1:

When there is no breakdown and re-service, then Equation (65) reduces to:

$$P(z) = \frac{(\tilde{S}_b(w(z)) - 1) \sum_{i=a}^{b-1} (z^b - z^i) c_i + (z^b - 1)(\tilde{V}(w(z)) - 1) \sum_{n=0}^{a-1} c_n z^n}{(-w(z))(z^b - \tilde{S}_b(w(z)))} \tag{60}$$

which coincides with the PGF of Senthilnathan et al. [19] without closedown.

Case 2:

When there is no breakdown, then Equation (65) reduces to:

$$P(z) = \frac{\begin{aligned} &(1 - \tilde{S}_b(w(z))) \sum_{i=a}^{b-1} (z^b - z^i) c_i + (z^b - 1)(1 - \tilde{V}(w(z))) \sum_{n=0}^{a-1} c_n z^n \\ &+ (z^b - 1)(\tilde{V}(w(z)) - \tilde{R}(w(z))) \sum_{n=0}^{a-1} \pi p_n z^n \end{aligned}}{(w(z))(z^b - \tilde{S}_b(w(z)))} \tag{61}$$

which is the PGF of Jeyakumar et al. [38].

5.7. PGF of the Queue Size in Various Epochs

5.7.1. PGF of the Queue Size in the Main Server's Service Completion Epoch

The probability generating function of the main server's service completion epoch $M(z)$ is obtained from Equations (48) and (49):

$$M(z) = \frac{\begin{array}{l}(1-\tilde{S}_b(u(z)))\Big[v(z)(z^b-\tilde{S}_s(v(z)))\Big(\sum_{i=a}^{b-1}(z^b-z^i)c_i-\sum_{k=0}^{a-1}c_kz^k \\ +\pi\tilde{R}(w(z))\sum_{n=0}^{a-1}p_nz^n+\tilde{V}(w(z))\sum_{n=0}^{a-1}((1-\pi)p_n+r_n+\eta T_n+v_n)z^n\Big) \\ +z^b\eta(1-\tilde{S}_s(v(z)))\Big(\sum_{i=a}^{b-1}(z^b-z^i)(q_i+r_i^{(2)}+\sum_{k=0}^{a-1}T_k\lambda g_{i-k})\Big)-v(z)T(z)\Big]\end{array}}{Y_1(z)} \tag{62}$$

5.7.2. PGF of the Queue Size in the Vacation Completion Epoch

The PGF of the main server's vacation completion epoch $V(z)$ is obtained from Equations (53) and (54); we get,

$$V(z) = \frac{(1-\tilde{V}(w(z)))\sum_{n=0}^{a-1}((1-\pi)p_n+r_n+\eta T_n+v_n)z^n}{w(z)} \tag{63}$$

5.7.3. PGF of the Queue Size in the Main Server's Re-Service Completion Epoch

The PGF of the main server's re-service completion epoch $R(z)$ is obtained from Equation (52); we get,

$$R(z) = \frac{(1-\tilde{R}(w(z)))\sum_{n=0}^{a-1}\pi p_nz^n}{w(z)} \tag{64}$$

5.7.4. PGF of the Queue Size in the Stand-by Server's Service Completion Epoch

The probability generating function of the stand-by server's service completion epoch $N(z)$ is derived from Equations (50) and (51); we get,

$$N(z) = \frac{\begin{array}{l}(1-\tilde{S}_s(v(z)))\Big[z^b\alpha(1-\tilde{S}_b(u(z)))\Big(\sum_{i=a}^{b-1}(z^b-z^i)c_i-\sum_{k=0}^{a-1}c_kz^k \\ +\pi\tilde{R}(w(z))\sum_{n=0}^{a-1}p_nz^n+\tilde{V}(w(z))\sum_{n=0}^{a-1}((1-\pi)p_n+r_n+\eta T_n+v_n)z^n\Big) \\ +u(z)(z^b-\tilde{S}_b(u(z)))\Big(\sum_{i=a}^{b-1}(z^b-z^i)(q_i+\sum_{k=0}^{a-1}T_k\lambda g_{i-k})-v(z)T(z)\Big)\Big]\end{array}}{Y_1(z)} \tag{65}$$

6. Some Performance Measures

6.1. The Main Server's Expected Length of Idle Period

Let K be the random variable denoting the 'idle period due to multiple vacation processes'. Let Y be the random variable defined by:

$$Y = \begin{cases} 0 & \text{if the server finds at least '}a\text{' customers after the first vacation} \\ 1 & \text{if the server finds less than '}a\text{' customers after the first vacation} \end{cases}$$

Now,

$$\begin{aligned} E(K) &= E(K/Y=0)P(Y=0)+E(K/Y=1)P(Y=1) \\ &= E(V)P(Y=0)+(E(V)+E(K))P(Y=1), \end{aligned}$$

Solving for $E(K)$, we get:

$$E(K) = \frac{E(V)}{(1 - P(Y = 1))} = \frac{E(V)}{\left(1 - \sum_{n=0}^{a-1} \sum_{i=0}^{n} \left[\beta_i [p_n + r_n + \eta T_n]\right]\right)}$$

6.2. Expected Queue Length

The mean number of customers waiting in the queue $E(Q)$ in an arbitrary time epoch is obtained by differentiating $P(z)$ at $z = 1$ and is given by:

$$E(Q) = \frac{\begin{array}{l} f_1(X, S_b, S_s)\left[\sum_{i=a}^{b-1}[b(b-1) - i(i-1)]c_i\right] \\ + f_1(X, S_b, S_s)\left[\sum_{i=a}^{b-1}(b(b-1) - i(i-1))d_i\right] \\ + f_2(X, S_b, S_s)\left[\sum_{i=a}^{b-1}(b-i)c_i\right] + f_3(X, S_b, S_s)\sum_{i=a}^{b-1}(b-i)d_i \\ + f_4(X, S_b, S_s, V)\sum_{n=0}^{a-1} nc_n + f_5(X, S_b, S_s, V)\sum_{n=0}^{a-1} c_n \\ + f_6(X, S_b, S_s, R, V)\sum_{n=0}^{a-1} \pi n p_n + f_7(X, S_b, S_s, R, V)\sum_{n=0}^{a-1} \pi p_n \\ + f_8(X, S_b, S_s, V)\sum_{n=0}^{a-1} nT_n + f_9(X, S_b, S_s, V)\sum_{n=0}^{a-1} T_n \end{array}}{3(F_{12})^2}, \tag{66}$$

the expressions for $f_i (i = 1, 2, ..., 9)$ are defined in Appendix B.

6.3. Expected Waiting Time

The expected waiting time is obtained using Little's formula as:

$$E(W) = \frac{E(Q)}{\lambda E(X)} \tag{67}$$

where $E(Q)$ is given in Equation (66).

7. Numerical Example

A numerical example of our model is analyzed for a particular case with the following assumptions:

1. The batch size distribution of the arrival is geometric with mean 2.
2. Take $a = 5$ and $b = 8$, and the service time distribution is Erlang-2 (both servers).
3. The vacation and re-service time of the main server follow an exponential distribution with parameter $\omega = 5, \epsilon = 3$, respectively.
4. Let m_1 be the service rate for the main server.
5. Let m_2 be the service rate for the stand-by server.

The unknown probabilities of the queue size distribution are computed using numerical techniques. The zeros of the function $Y_1(z)$ are obtained (see Figure 1), and simultaneous equations are solved by using MATLAB. The values which are satisfies the stability condition (see Figure 2) are used for calculating the table values.

The expected queue length $E(Q)$ and the expected waiting time $E(W)$ are calculated for various arrival rate sand service rates, and the results are tabulated.

From Tables 1–4, the following observations can be made.

1. As arrival rate λ increases, the expected queue size and expected waiting time are also increase.
2. When the main server's and stand-by server's service rate increases, the expected queue size and expected waiting time decrease.
3. When the main server's vacation rate increases, the expected queue size increases.

Table 1. Arrival rate vs. expected queue length and expected waiting time for the values $m_1 = 10, m_2 = 9.5$, $\alpha = 1$, $\eta = 2$, $\pi = 0.3$, $\epsilon = 3$, and $\omega = 5$.

λ	ρ	$E(Q)$	$E(W)$
5.00	0.131407	8.657374	0.865737
5.25	0.137978	9.539724	0.908545
5.50	0.144548	10.375816	0.943256
5.75	0.151119	11.149076	0.969485
6.00	0.157689	11.843026	0.986919
6.25	0.164259	12.441064	0.995285
6.50	0.170830	12.927026	0.994387
6.75	0.177400	13.285663	0.984123
7.00	0.183970	13.502456	0.964461
7.25	0.190541	13.563622	0.935422
7.50	0.197111	13.456834	0.897122

Table 2. Main server's service rate vs. expected queue length and expected waiting time for the values $\lambda = 5$, $m_2 = 5$, $\alpha = 1$, $\eta = 2$, $\pi = 0.3$, $\epsilon = 3$, and $\omega = 5$.

m_1	ρ	$E(Q)$	$E(W)$
5.25	0.257410	26.824821	2.682482
5.50	0.248885	26.082729	2.608273
5.75	0.240905	25.388555	2.538855
6.00	0.233420	24.735574	2.473557
6.25	0.226385	24.117976	2.411798
6.50	0.219761	23.531054	2.353105
6.75	0.213513	22.971202	2.297120
7.00	0.207610	22.435611	2.243561
7.25	0.202024	21.921511	2.192151
7.50	0.196731	21.427009	2.142701
7.75	0.191707	20.950354	2.095035
8.00	0.186934	20.490051	2.049005

Table 3. Stand-by server's service rate vs. expected queue length and expected waiting time for the values $\lambda = 5$, $m_1 = 10$, $\alpha = 1$, $\eta = 2$, $\pi = 0.3$, $\epsilon = 3$, and $\omega = 5$.

m_2	ρ	$E(Q)$	$E(W)$
4.0	0.260103	65.007246	4.062953
4.5	0.254645	59.415276	3.713455
5.0	0.249401	53.808260	3.363016
5.5	0.244361	48.320593	3.020037
6.0	0.239515	43.028238	2.689265
6.5	0.234853	37.970425	2.373152
7.0	0.230366	33.162455	2.072653
7.5	0.226045	28.605406	1.787838
8.0	0.22188	24.292435	1.518277
8.5	0.217865	20.212075	1.263255
9.0	0.213991	16.351005	1.021938

Table 4. The effect of the main server's vacation rate on expected queue length for the values $\lambda = 5$, $m_1 = 10$, $m_2 = 9.5$, $\alpha = 1$, $\eta = 2$, $\pi = 0.3$, and $\epsilon = 3$.

ω	Erlang	Exponential
5.00	8.657374	8.279153
5.25	8.808004	8.448114
5.50	8.950939	8.607757
5.75	9.086640	8.758748
6.00	9.215535	8.901685
6.25	9.338068	9.037158
6.50	9.454625	9.165674
6.75	9.565595	9.287730
7.00	9.671330	9.403767
7.25	9.772165	9.514200
7.50	9.868410	9.619407
7.75	9.960351	9.719734
8.00	10.048247	9.815494

```
clc
clear all
syms z X Y1 Y2;
f=0.25;
for Lam = 5:f:7.5
    Lam
m1 = 10;
m2 = 9.5;
a=5;
b=8;
Eta = 2;
Alp = 1;
X = -3/(10*((7*z)/10 - 1));
u = Lam - Lam*X+Alp;
v = Lam - Lam*X+Eta;
S1 = ((2*m1)/(2*m1+u))^2;
B1 = ((2*m2)/(2*m2+v))^2;
Y1=u*v*(z^b-S1)*(z^b-B1)-z^b*z^b*Alp*Eta*(1-S1)*(1-B1);
Y2 = solve(Y1)
end
```

Figure 1. MATLAB code for finding the roots.

```
clc
clear all
f=0.25;
for Lam =5.00:f:7.5
   Lam;
    fprintf('Lam=%g \n', Lam);
m1 = 10;
m2 = 9.5;
a=5;
b=8;
pi= 0.3;
Eta = 2;
Alp = 1;
X1=2;
S1 = ((2*m1)/(2*m1+Alp))^2;
B1 = ((2*m2)/(2*m2+Eta))^2;
N=(Lam*X1*(Alp+Eta)*(1-S1)*(1-B1));
D=b*Alp*Eta*(B1*(1-S1)+S1*(1-B1));
fprintf('rho=%g \n', (N/D))
end
```

Figure 2. MATLAB code for finding the rho value.

8. Conclusions

In this paper, a batch arrival general bulk service queueing system with breakdown and repair, stand-by server, multiple vacation and control policy on request for re-service is analyzed. The probability generating function of the queue size distribution at an arbitrary time is obtained. Some performance measures are calculated. The particular cases of the model are also deduced. From the numerical results, it is observed that when the arrival rate increases, the expected queue length and waiting time of the customers are also increase; if the service rate increases (for both server's), then the expected queue length and expected waiting time decrease. It is also observed that, if the main server's vacation rate increases, then the expected queue length increases.

Author Contributions: G.A.: To describe the model. S.K.: Convert the theoretical model into mathematical model and solving.

Conflicts of Interest: There is no conflict of interest by the author to publish this paper.

Appendix A

The expressions used in Equation (56) are defined as follows:

$$K_1(z) = w(z)(1-\tilde{S}_b(u(z)))A_1(z),$$
$$K_2(z) = w(z)(1-\tilde{S}_s(v(z)))A_2(z),$$
$$K_3(z) = Y_1(z) - w(z)A_1(z)(1-\tilde{S}_b(u(z))),$$

where

$$A_1(z) = v(z)(z^b - \tilde{S}_s(v(z))) + z^b\alpha(1-\tilde{S}_s(v(z)))$$
$$A_2(z) = u(z)(z^b - \tilde{S}_b(u(z))) + z^b\eta(1-\tilde{S}_b(u(z)))$$
$$Y_1(z) = u(z)v(z)(z^b - \tilde{S}_b(u(z)))(z^b - \tilde{S}_s(v(z))) - z^{2b}\alpha\eta(1-\tilde{S}_b(u(z)))(1-\tilde{S}_s(v(z))).$$

Appendix B

The expressions for f_i's in (66) are defined as follows:

$$
\begin{aligned}
&f_1(X, S_b, S_s) = 3E_1F_{12},\\
&f_2(X, S_b, S_s) = 3F_7F_{12} - 2E_1F_{13},\\
&f_3(X, S_b, S_s) = 3F_9F_{12} - 2E_1F_{13},\\
&f_4(X, S_b, S_s, V) = 6b\alpha\eta V_1E_2F_{12},\\
&f_5(X, S_b, S_s, V) = 3F_{12}[b\alpha\eta E_2V_2 + V_1(F_7 - F_2)] - 2b\alpha\eta E_2V_1F_{13},\\
&f_6(X, S_b, S_s, R_1, V) = 6b\alpha\eta(R_1 - V_1)E_2F_{12},\\
&f_7(X, S_b, S_s, R_1, V) = 3F_{12}[b\alpha\eta E_2(R_2 - V_2) + (R_1 - V_1)(F_7 - F_2)] - 2b\alpha\eta E_2(R_1 + V_1)F_{13},\\
&f_8(X, S_b, S_s, V) = 6F_{11}F_{12},\\
&f_9(X, S_b, S_s, V) = 3E_{10}F_{12} - 2F_{11}F_{13},
\end{aligned}
$$

where:

$$
\begin{aligned}
&E_1 = -\lambda X_1(\alpha+\eta)(1 - \tilde{S}_b(\alpha))(1 - \tilde{S}_s(\eta)),\\
&E_2 = \tilde{S}_b(\alpha)(\tilde{S}_s(\eta) - 1) + \tilde{S}_s(\eta)(\tilde{S}_b(\alpha) - 1),\\
&E_3 = S_{b1}(1 - \tilde{S}_s(\eta)) + S_{s1}(1 - \tilde{S}_b(\alpha)),\\
&E_4 = -\lambda X_1(1 - \tilde{S}_s(\eta)),\\
&E_5 = -\lambda X_2(1 - \tilde{S}_s(\eta)) + 2\lambda X_1S_{s1},\\
&E_6 = -\lambda X_3(1 - \tilde{S}_s(\eta)) + 3\lambda X_2S_{s1} + 3\lambda X_1S_{s2},\\
&E_7 = -\lambda X_1(1 - \tilde{S}_b(\alpha)),\\
&E_8 = -\lambda X_2(1 - \tilde{S}_b(\alpha)) + 2\lambda X_1S_{b1},\\
&E_9 = -\lambda X_3(1 - \tilde{S}_b(\alpha)) + 3\lambda X_2S_{b1} + 3\lambda X_1S_{b2},\\
&E_{10} = \eta[b\alpha\eta E_2V_2 - V_1(F_2 - F_7)] + \lambda X_2(E_1 - F_1) + \lambda X_1(F_9 - F_2) - (\eta/3)(F_8 + F_{10}),\\
&F_1 = E_1 - b\alpha\eta E_2,\\
&F_2 = (1 - \tilde{S}_b(\alpha))(1 - \tilde{S}_s(\eta))[2(\lambda X_1)^2 - \lambda X_2(\alpha+\eta) - 2b(2b-1)\alpha\eta]\\
&\quad + [b(b-1)\alpha\eta - 2b\lambda X_1(\alpha+\eta)][(1 - \tilde{S}_b(\alpha)) + (1 - \tilde{S}_s(\eta))]\\
&\quad + 2[2b\alpha\eta + \lambda X_1(\alpha+\eta)]E_3 + 2b\alpha\eta(b - S_{b1} - S_{s1})\\
&F_3 = (\alpha+\eta)(b - S_{s1}) + E_4 - b\alpha\tilde{S}_s(\eta),\\
&F_4 = (\alpha+\eta)(b(b-1) - S_{s2}) - b\alpha[(b-1)\tilde{S}_s(\eta) + 2S_{s1}] - 2b\lambda X_1 + E_5,\\
&F_5 = (\alpha+\eta)(b - S_{b1}) + E_7 - b\eta\tilde{S}_b(\alpha),\\
&F_6 = (\alpha+\eta)(b(b-1) - S_{b2}) - b\eta[(b-1)\tilde{S}_b(\alpha) + 2S_{b1}] - 2b\lambda X_1 + E_8,\\
&F_7 = E_8(\alpha+\eta)(1 - \tilde{S}_s(\eta)) + 2E_7F_3,\\
&F_8 = E_9(\alpha+\eta)(1 - \tilde{S}_s(\eta)) + 3E_8F_3 + 3E_7F_4,\\
&F_9 = E_5(\alpha+\eta)(1 - \tilde{S}_b(\alpha)) + 2E_4F_5,\\
&F_{10} = E_6(\alpha+\eta)(1 - \tilde{S}_b(\alpha)) + 3E_5F_5 + 3E_4F_6,\\
&F_{11} = b\alpha\eta E_2(\eta V_1 + \lambda X_1) + \eta\lambda X_1[(\alpha+\eta)E_3 - (1 - \tilde{S}_b(\alpha))F_3 + (1 - \tilde{S}_s(\eta))F_5],\\
&F_{12} = -2\lambda X_1F_1,\\
&F_{13} = -3[\lambda X_1F_2 + \lambda X_2F_1].
\end{aligned}
$$

$$
H = \left\{ \begin{array}{l} 2E_1\sum_{i=a}^{b-1}(b-i)c_i + 2E_1\sum_{i=a}^{b-1}(b-i)d_i + 2b\alpha\eta E_2V_1\sum_{n=0}^{a-1}c_n \\ +2b\alpha\eta E_2(R_1+V_1)\sum_{n=0}^{a-1}\pi p_n + 2F_{11}\sum_{n=0}^{a-1}T_n, \end{array} \right.
$$

where:

$$S_{b1} = -\lambda X_1 \tilde{S}_b'(\alpha),\ S_{b2} = \tilde{S}_b''(\alpha)(-\lambda X_1)^2 - \lambda X_2 \tilde{S}_b'(\alpha),$$
$$S_{s1} = -\lambda X_1 \tilde{S}_s'(\eta),\ S_{s2} = \tilde{S}_s''(\eta)(-\lambda X_1)^2 - \lambda X_2 \tilde{S}_s'(\eta),$$
$$R_1 = \lambda X_1 E(R),\ R_2 = \lambda X_2 E(R) + \lambda^2 X_1^2 E(R^2)$$
$$V_1 = \lambda X_1 E(V),\ V_2 = \lambda X_2 E(V) + \lambda^2 X_1^2 E(V^2),\ X_1 = E(X),\ X_2 = E(X^2).$$

References

1. Doshi, B.T. Queueing systems with vacations—A survey. *Queueing Syst.* **1986**, *1*, 29–66. [CrossRef]
2. Takagi, H. Vacation and Priority Systems. *Queuing Analysis: A Foundation of Performance Evaluation;* North-Holland: Amsterdam, The Netherlands, 1991; Volume I.
3. Baba, Y. On the $M^{[X]}/G/1$ queue with vacation time. *Oper. Res. Lett.* **1986**, *5*, 93–98. [CrossRef]
4. Krishna Reddy, G.V.; Nadarajan, R.; Arumuganathan, R. Analysis of a bulk queue with N policy multiple vacations and setup times. *Comput. Oper. Res.* **1998**, *25*, 957–967. [CrossRef]
5. Jeyakumar, S.; Senthilnathan, B. Modelling and analysis of a bulk service queueing model with multiple working vacations and server breakdown. *RAIRO-Oper. Res.* **2017**, *51*, 485–508. [CrossRef]
6. Madan, K.C. An M/G/1 queue with second optional service. *Queueing Syst.* **2000**, *34*, 37–46. [CrossRef]
7. Madan, K.C. On $M^{[X]}/(G1, G2)/1$ queue with optional re-service. *Appl. Math. Comput.* **2004**, *152*, 71–88.
8. Jeyakumar, S.; Arumuganathan, R. A Non-Markovian Bulk Queue with Multiple Vacations and Control Policy on Request for Re-Service. *Qual. Technol. Quant. Manag.* **2011**, *8*, 253–269. [CrossRef]
9. Haridass, M.; Arumuganathan, R. A batch service queueing system with multiple vacations, setup time and server's choice of admitting re-service. *Int. J. Oper. Res.* **2012**, *14*, 156–186. [CrossRef]
10. Krishnamoorthy, A.; Pramod, P.K.; Chakravarthy, S.R. Queues with interruptions: A survey. *Oper. Res. Decis. Theory* **2014**, *22*, 290–320. [CrossRef]
11. Ayyappan, G.; Shyamala, S. Transient solution of an $M^{[X]}/G/1$ queueing model with feedback, random breakdowns, Bernoulli schedule server vacation and random setup time. *Int. J. Oper. Res.* **2016**, *25*, 196–211. [CrossRef]
12. Choudhury, G.; Tadj, L. An M/G/1 queue with two phases of service subject to the server breakdown and delayed repair. *Appl. Math. Model.* **2009**, *33*, 2699–2709. [CrossRef]
13. Senthilnathan, B.; Jeyakumar, S. A study on the behaviour of the server breakdown without interruption in a $M^{[X]}/G(a,b)/1$ queueing system with multiple vacations and closedown time. *Appl. Math. Comput.* **2012**, *219*, 2618–2633.
14. Lakshmi, K.; Kasturi Ramanath, K. An M/G/1 two phase multi-optional retrial queue with Bernoulli feedback, non-persistent customers and breakdown and repair. *Int. J. Oper. Res.* **2014**, *19*, 78–95. [CrossRef]
15. Atencia, I. A discrete-time queueing system with server breakdowns and changes in the repair times. *Ann. Oper. Res.* **2015**, *235*, 37–49. [CrossRef]
16. Choudhury, G. A batch arrival queue with a vacation time under single vacation policy. *Comput. Oper. Res.* **2002**, *29*, 1941–1955. [CrossRef]
17. Jeyakumar, S.; Arumuganathan, R. Analysis of Single Server Retrial Queue with Batch Arrivals, Two Phases of Heterogeneous Service and Multiple Vacations with N-Policy. *Int. J. Oper. Res.* **2008**, *5*, 213–224.
18. Balasubramanian, M.; Arumuganathan. R.; Senthil Vadivu. A. Steady state analysis of a non-Markovian bulk queueing system with overloading and multiple vacations. *Int. J. Oper. Res.* **2010**, *9*, 82–103. [CrossRef]
19. Haridass, M.; Arumuganathan, R. Analysis of a batch arrival general bulk service queueing system with variant threshold policy for secondary jobs. *Int. J. Math. Oper. Res.* **2011**, *3*, 56–77. [CrossRef]
20. Choudhury, G.; Deka, M. A batch arrival unreliable server delaying repair queue with two phases of service and Bernoulli vacation under multiple vacation policy. *Qual. Technol. Quant. Manag.* **2018**, *15*, 157–186. [CrossRef]
21. Madan, K.C. A cyclic queueing system with three servers and optional two-way feedback. *Microelectron. Reliabil.* **1988**, *28*, 873–875. [CrossRef]
22. Madan, K.C. On a single server queue with two-stage heterogeneous service and deterministic server vacations. *Int. J. Syst. Sci.* **2001**, *32*, 837–844. [CrossRef]

23. Medhi, J. A Single Server Poisson Input Queue with a Second Optional Channel. *Queueing Syst.* **2002**, *42*, 239–242. [CrossRef]
24. Arumuganathan, R.; Malliga, T.J. Analysis of a bulk queue with repair of service station and setup time. *Int. J. Can. Appl. Math. Quart.* **2006**, *13*, 19–42.
25. Baruah, M.; Madan, K.C.; Eldabi, T. A Two Stage Batch Arrival Queue with Reneging during Vacation and Breakdown Periods. *Am. J. Oper. Res.* **2013**, *3*, 570–580. [CrossRef]
26. Aissani, A.; Artalejo, J. On the single server retrial queue subject to breakdowns. *Queueing Syst.* **1988**, *30*, 309–321. [CrossRef]
27. Maraghi, F.A.; Madan, K.C.; Darby-Dowman, K. Bernoulli schedule vacation queue with batch arrivals and random system breakdowns having general repair time distribution. *Int. J. Oper. Res.* **2010**, *7*, 240–256. [CrossRef]
28. Fadhil, R.; Madan, K.C.; Lukas, A.C. An M(X)/G/1 Queue with Bernoulli Schedule General Vacation Times, Random Breakdowns, General Delay Times and General Repair Times. *Appl. Math. Sci.* **2011**, *5*, 35–51.
29. Rajadurai, P.; Varalakshmi, M.; Saravanarajan, M.C.; Chandrasekaran, V.M. Analysis of $M^{[X]}/G/1$ retrial queue with two phase service under Bernoulli vacation schedule and random breakdown. *Int. J. Math. Oper. Res.* **2015**, *7*, 19–41. [CrossRef]
30. Jiang, T.; Xin, B. Computational analysis of the queue with working breakdowns and delaying repair under a Bernoulli-schedule-controlled policy. *J. Commun. Stat. Theory Methods* **2018**, 1–16. [CrossRef]
31. Madan, K.C. A bulk input queue with a stand-by. *S. Afr. Stat. J.* **1995**, *29*, 1–7.
32. Khalaf, R.F. Queueing Systems With Four Different Main Server's Interruptions and a Stand-By Server. *Int. J. Oper. Res.* **2014**, *3*, 49–54. [CrossRef]
33. Jain, M.; Preeti. Cost analysis of a machine repair problem with standby, working vacation and server breakdown. *Int. J. Math. Oper. Res.* **2014**, *6*, 437–451. [CrossRef]
34. Kumar, K.; Jain, M. Bi-level control of degraded machining system with two unreliable servers, multiple standbys, startup and vacation. *Int. J. Oper. Res.* **2014**, *21*, 123–142. [CrossRef]
35. Murugeswari, N.; Maragatha Sundari, S. A Standby server bulk arrival Queuing model of Compulsory server Vacation. *Int. J. Eng. Dev. Res.* **2017**, *5*, 337–341.
36. Kolledath, S.; Kumar, K.; Pippal, S. Survey on queueing models with standbys support. *Yugoslav J. Oper. Res.* **2018**, *28*, 3–20. [CrossRef]
37. Neuts, M.F. A general class of bulk queues with poisson input. *The Annals of Mathematical Statistics*, **1967**, *38*, 759–770. [CrossRef]
38. Ayyappan, G.; Sathiya, S. Non Markovian Queue with Two Types service Optional Re-service and General Vacation Distribution. *Appl. Appl. Math. Int. J.* **2016**, *11*, 504–526.

mathematics

Article

Fusion Estimation from Multisensor Observations with Multiplicative Noises and Correlated Random Delays in Transmission

Raquel Caballero-Águila [1,*], Aurora Hermoso-Carazo [2] and Josefa Linares-Pérez [2]

1 Department of Statistics, University of Jaén, Paraje Las Lagunillas, 23071 Jaén, Spain
2 Department of Statistics, University of Granada, Avda. Fuentenueva, 18071 Granada, Spain; ahermoso@ugr.es (A.H.-C.); jlinares@ugr.es (J.L.-P.)
* Correspondence: raguila@ujaen.es; Tel.: +34-953-212-926

Received: 20 July 2017; Accepted: 29 August 2017; Published: 4 September2017

Abstract: In this paper, the information fusion estimation problem is investigated for a class of multisensor linear systems affected by different kinds of stochastic uncertainties, using both the distributed and the centralized fusion methodologies. It is assumed that the measured outputs are perturbed by one-step autocorrelated and cross-correlated additive noises, and also stochastic uncertainties caused by multiplicative noises and randomly missing measurements in the sensor outputs are considered. At each sampling time, every sensor output is sent to a local processor and, due to some kind of transmission failures, one-step correlated random delays may occur. Using only covariance information, without requiring the evolution model of the signal process, a local least-squares (LS) filter based on the measurements received from each sensor is designed by an innovation approach. All these local filters are then fused to generate an optimal distributed fusion filter by a matrix-weighted linear combination, using the LS optimality criterion. Moreover, a recursive algorithm for the centralized fusion filter is also proposed and the accuracy of the proposed estimators, which is measured by the estimation error covariances, is analyzed by a simulation example.

Keywords: fusion estimation; sensor networks; random parameter matrices; multiplicative noises; random delays

1. Introduction

Over the past decades, the use of sensor networks has experienced a fast development encouraged by the wide range of potential applications in many areas, since they usually provide more information than traditional single-sensor communication systems. So, important advances have been achieved concerning the estimation problem in networked stochastic systems and the design of multisensor fusion techniques [1]. Many of the existing fusion estimation algorithms are related to conventional systems (see e.g., [2–5], and the references therein), where the sensor measured outputs are affected only by additive noises and each sensor transmits its outputs to the fusion center over perfect connections.

However, in a network context, usually the restrictions of the physical equipment or the uncertainties in the external environment, inevitably cause problems in both the sensor outputs and the transmission of such outputs, that can worsen dramatically the quality of the fusion estimators designed without considering these drawbacks [6]. Multiplicative noise uncertainties and missing measurements are some of the random phenomena that usually arise in the sensor measured outputs and motivate the design of new estimation algorithms (see e.g., [7–11], and references therein).

Furthermore, when the sensors send their measurements to the processing center via a communication network some additional network-induced phenomena, such as random delays or measurement losses, inevitably arise during this transmission process, which can spoil the fusion estimators performance and motivate the design of fusion estimation algorithms for systems with one (or even several) of the aforementioned uncertainties (see e.g., [12–24], and references therein). All the above cited papers on signal estimation with random transmission delays assume independent random delays at each sensor and mutually independent delays between the different sensors; in [25] this restriction was weakened and random delays featuring correlation at consecutive sampling times were considered, thus allowing to deal with some common practical situations (e.g., those in which two consecutive observations cannot be delayed).

It should be also noted that, in many real-world problems, the measurement noises are usually correlated; this occurs, for example, when all the sensors operate in the same noisy environment or when the sensor noises are state-dependent. For this reason, the fairly conservative assumption that the measurement noises are uncorrelated is commonly weakened in many of the aforementioned research papers on signal estimation. Namely, the optimal Kalman filtering fusion problem in systems with noise cross-correlation at consecutive sampling times is addressed, for example, in [19]; also, under different types of noise correlation, centralized and distributed fusion algorithms for systems with multiplicative noise are obtained in [11,20], and for systems where the measurements might have partial information about the signal in [7].

In this paper, covariance information is used to address the distributed and centralized fusion estimation problems for a class of linear networked stochastic systems with multiplicative noises and missing measurements in the sensor measured outputs, subject to transmission random one-step delays. It is assumed that the sensor measurement additive noises are one-step autocorrelated and cross-correlated, and the Bernoulli variables describing the measurement delays at the different sensors are correlated at the same and consecutive sampling times. As in [25], correlated random delays in the transmission are assumed to exist, with different delay rates at each sensor; however, the proposed observation model is more general than that considered in [25] since, besides the random delays in the transmission, multiplicative noises and missing phenomena in the measured outputs are considered; also cross-correlation between the different sensor additive noises is taken into account. Unlike [7–11] where multiplicative noise uncertainties and/or missing measurements are considered in the sensor measured outputs, in this paper random delays in the transmission are also assumed to exist. Hence, a unified framework is provided for dealing simultaneously with missing measurements and uncertainties caused by multiplicative noises, along with random delays in the transmission and, hence, the proposed fusion estimators have wide applicability. Recursive algorithms for the optimal linear distributed and centralized filters under the least-squares (LS) criterion are derived by an innovation approach. Firstly, local estimators based on the measurements received from each sensor are obtained and then the distributed fusion filter is generated as the LS matrix-weighted linear combination of the local estimators. Also, a recursive algorithm for the optimal linear centralized filter is proposed. Finally, it is important to note that, even though the state augmentation method has been largely used in the literature to deal with the measurement delays, such method leads to a significant rise of the computational burden, due to the increase of the state dimension. In contrast to such approach, the fusion estimators proposed in the current paper are obtained without needing the state augmentation; so, the dimension of the designed estimators is the same as that of the original state, thus reducing the computational cost compared with the existing algorithms based on the augmentation method.

The rest of the paper is organized as follows. The multisensor measured output model with multiplicative noises and missing measurements, along with the transmission random one-step delay model, are presented in Section 2. The distributed fusion estimation algorithm is derived in Section 3, and a recursive algorithm for the centralized LS linear filtering estimator is proposed in Section 4. The effectiveness of the proposed estimation algorithms is analyzed in Section 5 by a simulation example and some conclusions are drawn in Section 6.

Notation: The notation throughout the paper is standard. $\mathbb{R}^n$ and $\mathbb{R}^{m\times n}$ denote the n-dimensional Euclidean space and the set of all $m \times n$ real matrices, respectively. For a matrix A, the symbols A^T and A^{-1} denote its transpose and inverse, respectively; the notation $A \otimes B$ represents the Kronecker product of the matrices A, B. If the dimensions of vectors or matrices are not explicitly stated, they are assumed to be compatible with algebraic operations. In particular, I denotes the identity matrix of appropriate dimensions. The notation $a \wedge b$ indicates the minimum value of two real numbers a, b. For any function $G_{k,s}$, depending on the time instants k and s, we will write $G_k = G_{k,k}$ for simplicity; analogously, $F^{(i)} = F^{(ii)}$ will be written for any function $F^{(ij)}$, depending on the sensors i and j. Moreover, for an arbitrary random vector $\alpha_k^{(i)}$, we will use the notation $\overline{\alpha}_k^{(i)} \equiv E\left[\alpha_k^{(i)}\right]$, where $E[.]$ is the mathematical expectation operator. Finally, $\delta_{k,s}$ denotes the Kronecker delta function.

2. Problem Formulation and Model Description

This paper is concerned with the LS linear filtering estimation problem of discrete-time stochastic signals from randomly delayed observations coming from networked sensors using the distributed and centralized fusion methods. The signal measurements at the different sensors are affected by multiplicative and additive noises, and the additive sensor noises are assumed to be correlated and cross-correlated at the same and consecutive sampling times. Each sensor output is transmitted to a local processor over imperfect network connections and, due to network congestion or some other causes, random one-step delays may occur during this transmission process; in order to model different delay rates in the transmission from each sensor to the local processor, different sequences of correlated Bernoulli random variables with known probability distributions are used.

In the distributed fusion method, each local processor produces the LS linear filter based on the measurements received from the sensor itself; afterwards, these local estimators are transmitted to the fusion center over perfect connections, and the distributed fusion filter is generated by a matrix-weighted linear combination of the local LS linear filtering estimators using the mean squared error as optimality criterion. In the centralized fusion method, all measurement data of the local processors are transmitted to the fusion center, also over perfect connections, and the LS linear filter based on all the measurements received is obtained by a recursive algorithm.

Next, we present the observation model and the hypotheses on the signal and noise processes necessary to address the estimation problem.

2.1. Signal Process

The distributed and centralized fusion filtering estimators will be obtained under the assumption that the evolution model of the signal to be estimated is unknown and only information about its mean and covariance functions is available; specifically, the following hypothesis is required:

Hypothesis 1. *The n_x-dimensional signal process $\{x_k\}_{k\geq 1}$ has zero mean and its autocovariance function is expressed in a separable form, $E[x_k x_s^T] = A_k B_s^T$, $s \leq k$, where $A_k, B_s \in \mathbb{R}^{n_x \times n}$ are known matrices.*

Note that, when the system matrix Φ in the state-space model of a stationary signal is available, the signal autocovariance function is $E[x_k x_s^T] = \Phi^{k-s} E[x_s x_s^T]$, $s \leq k$, and Hypothesis 1 is clearly satisfied taking, for example, $A_k = \Phi^k$ and $B_s = E[x_s x_s^T](\Phi^{-s})^T$. Similarly, if $x_k = \Phi_{k-1} x_{k-1} + w_{k-1}$, the covariance function can be expressed as $E[x_k x_s^T] = \Phi_{k,s} E[x_s x_s^T]$, $s \leq k$, where $\Phi_{k,s} = \Phi_{k-1} \cdots \Phi_s$, and Hypothesis 1 is also satisfied taking $A_k = \Phi_{k,0}$ and $B_s = E[x_s x_s^T](\Phi_{s,0}^{-1})^T$. Furthermore, Hypothesis 1 covers even situations where the system matrix in the state-space model is singular, although a different factorization must be used in those cases (see e.g., [21]). Hence, Hypothesis 1 on the signal autocovariance function covers both stationary and non-stationary signals, providing a unified context to deal with a large number of different situations and avoiding the derivation of specific algorithms for each of them.

2.2. Multisensor Measured Outputs

Consider m sensors, whose measurements obey the following equations:

$$z_k^{(i)} = \theta_k^{(i)} \left(H_k^{(i)} + \varepsilon_k^{(i)} C_k^{(i)} \right) x_k + v_k^{(i)}, \quad k \geq 1, \quad i = 1, \dots, m \tag{1}$$

where $z_k^{(i)} \in \mathbb{R}^{n_z}$ is the measured output of the i-th sensor at time k, which is transmitted to a local processor by unreliable network connections, and $H_k^{(i)}, C_k^{(i)}$ are known time-varying matrices of suitable dimensions. For each sensor $i = 1, \dots, m$, $\{\theta_k^{(i)}\}_{k \geq 1}$ is a Bernoulli process describing the missing phenomenon, $\{\varepsilon_k^{(i)}\}_{k \geq 1}$ is a scalar multiplicative noise, and $\{v_k^{(i)}\}_{k \geq 1}$ is the measurement noise.

The following hypotheses on the observation model given by Equation (1) are required:

Hypothesis 2. *The processes $\{\theta_k^{(i)}\}_{k \geq 1}$, $i = 1, \dots, m$, are independent sequences of independent Bernoulli random variables with know probabilities $P(\theta_k^{(i)} = 1) = \overline{\theta}_k^{(i)}$, $k \geq 1$.*

Hypothesis 3. *The multiplicative noises $\{\varepsilon_k^{(i)}\}_{k \geq 1}$, $i = 1, \dots, m$, are independent sequences of independent scalar random variables with zero means and known second-order moments; we will denote $\sigma_k^{(i)} \equiv E\left[(\varepsilon_k^{(i)})^2 \right]$, $k \geq 1$.*

Hypothesis 4. *The sensor measurement noises $\{v_k^{(i)}\}_{k \geq 1}$, $i = 1, \dots, m$, are zero-mean sequences with known second-order moments defined by:*

$$E\left[v_k^{(i)} v_s^{(j)T} \right] = R_k^{(ij)} \delta_{k,s} + R_{k,k-1}^{(ij)} \delta_{k-1,s}, \quad s \leq k; \quad i, j = 1, \dots, m$$

From Hypothesis 2, different sequences of independent Bernoulli random variables with known probabilities are used to model the phenomenon of missing measurements at each sensor; so, when $\theta_k^{(i)} = 1$, which occurs with known probability $\overline{\theta}_k^{(i)}$, the state x_k is present in the measurement $z_k^{(i)}$ coming from the i-th sensor at time k; otherwise, $\theta_k^{(i)} = 0$ and the state is missing in the measured output from the i-th sensor at time k, which means that such observation only contains additive noise $v_k^{(i)}$ with probability $1 - \overline{\theta}_k^{(i)}$. Although these variables are assumed to be independent from sensor to sensor, such condition is not necessary to deduce either the centralized estimators or the local estimators, but only to obtain the cross-covariance matrices of the local estimation errors, which are necessary to determine the matrix weights of the distributed fusion estimators. Concerning Hypothesis 3, it should be noted that the multiplicative noises involved in uncertain systems are usually gaussian noises. Finally, note that the conservative hypothesis of independence between different sensor measurement noises has been weakened in Hypothesis 4, since such independence assumption may be a limitation in many real-world problems; for example, when all the sensors operate in the same noisy environment, the noises are usually correlated, or even some sensors may have the same measurement noises.

2.3. Observation Model with Random One-Step Delays

For each $k \geq 1$, assume that the measured outputs of the different sensors, $z_k^{(i)}$, $i = 1, \dots, m$, are transmitted to the local processors through unreliable communication channels and, due to network congestion or some other causes, random one-step delays with different rates are supposed to exist in these transmissions. Assuming that the first measurement is always available and considering different sequences of Bernoulli random variables, $\{\gamma_k^{(i)}\}_{k\geq 2}$, $i = 1, \ldots, m$, to model the random delays, the observations used in the estimation are described by:

$$y_k^{(i)} = (1-\gamma_k^{(i)})z_k^{(i)} + \gamma_k^{(i)} z_{k-1}^{(i)}, \ k \geq 2; \ \ y_1^{(i)} = z_1^{(i)}; \ \ i = 1, \ldots, m \tag{2}$$

From Equation (2) it is clear that $\gamma_k^{(i)} = 0$ means that $y_k^{(i)} = z_k^{(i)}$; that is, the local processor receives the data from the i-th sensor at the sampling time k. When $\gamma_k^{(i)} = 1$, then $y_k^{(i)} = z_{k-1}^{(i)}$, meaning that the measured output at time k is delayed and the previous one $z_{k-1}^{(i)}$ is used for the estimation. These Bernoulli random variables modelling the delays are assumed to be one-step correlated, thus covering many practical situations; for example, those in which consecutive observations transmitted through the same channel cannot be delayed, or situations where there are some sort of links between the different communications channels. Specifically, the following hypothesis is assumed:

Hypothesis 5. *$\{\gamma_k^{(i)}\}_{k\geq 2}$, $i = 1, \ldots, m$, are sequences of Bernoulli random variables with known means, $\overline{\gamma}_k^{(i)} \equiv E[\gamma_k^{(i)}]$, $k \geq 2$. It is assumed that $\gamma_k^{(i)}$ and $\gamma_s^{(j)}$ are independent for $|k-s| \geq 2$, and the second-order moments, $\overline{\gamma}_{k,s}^{(i,j)} \equiv E[\gamma_k^{(i)}\gamma_s^{(j)}]$, $s = k-1, k$, and $i, j = 1, \ldots, m$, are also known.*

Finally, the following independence hypothesis is also required:

Hypothesis 6. *For $i = 1, \ldots, m$, the processes $\{x_k\}_{k\geq 1}$, $\{\theta_k^{(i)}\}_{k\geq 1}$, $\{\varepsilon_k^{(i)}\}_{k\geq 1}$, $\{v_k^{(i)}\}_{k\geq 1}$ and $\{\gamma_k^{(i)}\}_{k\geq 2}$ are mutually independent.*

In the following proposition, explicit expressions for the autocovariance functions of the transmitted and received measurements, that will be necessary for the distributed fusion estimation algorithm, are derived.

Proposition 1. *For $i, j = 1, \ldots, m$, the autocovariance functions $\Sigma_{k,s}^{z^{(ij)}} \equiv E[z_k^{(i)} z_s^{(j)T}]$ and $\Sigma_{k,s}^{y^{(i)}} \equiv E[y_k^{(i)} y_s^{(i)T}]$ are given by:*

$$\begin{aligned}
\Sigma_k^{z^{(i)}} &= \overline{\theta}_k^{(i)} H_k^{(i)} A_k B_k^T H_k^{(i)T} + \sigma_k^{(i)} C_k^{(i)} A_k B_k^T C_k^{(i)T} + R_k^{(i)}, \ k \geq 1 \\
\Sigma_{k,s}^{z^{(ij)}} &= \overline{\theta}_k^{(i)} \overline{\theta}_s^{(j)} H_k^{(i)} A_k B_s^T H_s^{(j)T} + R_{k,k-1}^{(ij)} \delta_{k-1,s} + R_k^{(ij)} \delta_{k,s}, \ \ i \neq j, \ or \ s < k \\
\Sigma_{k,s}^{y^{(ij)}} &= (1 - \overline{\gamma}_k^{(i)} - \overline{\gamma}_s^{(j)} + \overline{\gamma}_{k,s}^{(i,j)}) \Sigma_{k,s}^{z^{(ij)}} + (\overline{\gamma}_s^{(j)} - \overline{\gamma}_{k,s}^{(i,j)}) \Sigma_{k,s-1}^{z^{(ij)}} \\
&\quad + (\overline{\gamma}_k^{(i)} - \overline{\gamma}_{k,s}^{(i,j)}) \Sigma_{k-1,s}^{z^{(ij)}} + \overline{\gamma}_{k,s}^{(i,j)} \Sigma_{k-1,s-1}^{z^{(ij)}}, \ \ k, s \geq 2 \\
\Sigma_{2,1}^{y^{(ij)}} &= (1 - \overline{\gamma}_2^{(i)}) \Sigma_{2,1}^{z^{(ij)}} + \overline{\gamma}_2^{(i)} \Sigma_1^{z^{(ij)}}; \quad \Sigma_1^{y^{(ij)}} = \Sigma_1^{z^{(ij)}}
\end{aligned} \tag{3}$$

Proof. From Equations (1) and (2), taking into account Hypotheses 1–6, the expressions given in Equation (3) are easily obtained. □

3. Distributed Fusion Linear Filter

In this section, we address the distributed fusion linear filtering problem of the signal from the randomly delayed observations defined by Equations (1) and (2), using the LS optimality criterion. In the distributed fusion method, each local processor provides the LS linear filter of the signal x_k based on the measurements from the corresponding sensor, which will be denoted by $\widehat{x}_{k/k}^{(i)}$; afterwards, these local filters are transmitted to the fusion center where the distributed filter, $\widehat{x}_{k/k}^{(D)}$, is designed as a matrix-weighted linear combination of such local filters. First, in Section 3.1, for each $i = 1, \ldots, m$, a recursive algorithm for the local LS linear filter, $\widehat{x}_{k/k}^{(i)}$, will be deduced. Then, in Section 3.2, the derivation of the cross-correlation matrices between any two local filters, $\widehat{\Sigma}^{(ij)}_{k/k} = E\big[\widehat{x}^{(i)}_{k/k}\widehat{x}^{(j)T}_{k/k}\big]$, $i,j = 1,\ldots,m$, will be detailed. Finally, in Section 3.3, the distributed fusion filter weighted by matrices, $\widehat{x}^{(D)}_{k/k}$, will be generated from the local filters by applying the LS optimality criterion.

3.1. Local LS Linear Filtering Recursive Algorithm

To obtain the signal LS linear filters based on the available observations from each sensor, we will use an innovation approach. For each sensor $i = 1,\ldots,m$, the innovation at time k, which represents the new information provided by the k-th observation, is defined by $\mu^{(i)}_k = y^{(i)}_k - \widehat{y}^{(i)}_{k/k-1}$, $k \geq 1$, where $\widehat{y}^{(i)}_{k/k-1}$ is the LS linear estimator of y_k based on the previous observations, $y^{(i)}_s$, $s \leq k-1$, with $\widehat{y}^{(i)}_{1/0} = E[y^{(i)}_1] = 0$.

As it is known (see e.g., [26]), the innovations, $\{\mu^{(i)}_k\}_{k\geq 1}$, constitute a zero-mean white process, and the LS linear estimator of any random vector α_k based on the observations $y^{(i)}_1,\ldots,y^{(i)}_L$, denoted by $\widehat{\alpha}^{(i)}_{k/L}$, can be calculated as a linear combination of the corresponding innovations, $\mu^{(i)}_1,\ldots,\mu^{(i)}_L$; namely,

$$\widehat{\alpha}^{(i)}_{k/L} = \sum_{h=1}^{L} E\big[\alpha_k \mu^{(i)T}_h\big]\Pi^{(i)-1}_h \mu^{(i)}_h \tag{4}$$

where $\Pi^{(i)}_h \equiv E\big[\mu^{(i)}_h \mu^{(i)T}_h\big]$ denotes the covariance matrix of $\mu^{(i)}_h$.

This general expression for the LS linear estimators along with the Orthogonal Projection Lemma (OPL), which guarantees that the estimation error is uncorrelated with all the observations or, equivalently, that it is uncorrelated with all the innovations, are the essential keys to derive the proposed recursive local filtering algorithm.

Taking into account Equation (4), the first step to obtain the signal estimators is to find an explicit formula for the innovation $\mu^{(i)}_h$ or, equivalently, for the observation predictor $\widehat{y}^{(i)}_{h/h-1}$.

Using the following alternative expression for the observations $y^{(i)}_k$ given by Equation (2),

$$\begin{aligned} y^{(i)}_k &= (1-\overline{\gamma}^{(i)}_k)\theta^{(i)}_k(H^{(i)}_k + \varepsilon^{(i)}_k C^{(i)}_k)x_k + \overline{\gamma}^{(i)}_k\overline{\theta}^{(i)}_{k-1}H^{(i)}_{k-1}x_{k-1} + w^{(i)}_k, \quad k \geq 2 \\ w^{(i)}_k &= \overline{\gamma}^{(i)}_k(\theta^{(i)}_{k-1} - \overline{\theta}^{(i)}_{k-1})H^{(i)}_{k-1}x_{k-1} + \overline{\gamma}^{(i)}_k\theta^{(i)}_{k-1}\varepsilon^{(i)}_{k-1}C^{(i)}_{k-1}x_{k-1} \\ &\quad + (1-\overline{\gamma}^{(i)}_k)v^{(i)}_k + \overline{\gamma}^{(i)}_k v^{(i)}_{k-1} - (\gamma^{(i)}_k - \overline{\gamma}^{(i)}_k)(z^{(i)}_k - z^{(i)}_{k-1}), \quad k \geq 2 \end{aligned} \tag{5}$$

and taking into account the independence hypotheses on the model, it is easy to see that:

$$\widehat{y}^{(i)}_{k/k-1} = (1-\overline{\gamma}^{(i)}_k)\overline{\theta}^{(i)}_k H^{(i)}_k \widehat{x}^{(i)}_{k/k-1} + \overline{\gamma}^{(i)}_k\overline{\theta}^{(i)}_{k-1}H^{(i)}_{k-1}\widehat{x}^{(i)}_{k-1/k-1} + \widehat{w}^{(i)}_{k/k-1}, \quad k \geq 2$$

Now, taking into account that $w^{(i)}_h$ is uncorrelated with $y^{(i)}_s$ for $s \leq h-2$, and using Equation (4) for $\widehat{w}^{(i)}_{k/k-1}$, we obtain that:

$$\begin{aligned} \widehat{y}^{(i)}_{k/k-1} &= (1-\overline{\gamma}^{(i)}_k)\overline{\theta}^{(i)}_k H^{(i)}_k \widehat{x}^{(i)}_{k/k-1} + \overline{\gamma}^{(i)}_k\overline{\theta}^{(i)}_{k-1}H^{(i)}_{k-1}\widehat{x}^{(i)}_{k-1/k-1} \\ &\quad + \sum_{h=1}^{(k-1)\wedge 2} \mathcal{W}^{(i)}_{k,k-h}\Pi^{(i)-1}_{k-h}\mu^{(i)}_{k-h}, \quad k \geq 2 \end{aligned} \tag{6}$$

where $\mathcal{W}^{(i)}_{k,k-h} \equiv E\big[w^{(i)}_k \mu^{(i)T}_{k-h}\big]$, $h = 1,2$.

Equation (6) for the one-stage observation predictor is the starting point to derive the local recursive filtering algorithm presented in Theorem 1; this algorithm provide also the filtering error covariance matrices, $P^{(i)}_{k/k} \equiv E\big[(x_k - \widehat{x}^{(i)}_{k/k})(x_k - \widehat{x}^{(i)}_{k/k})^T\big]$, which measure the accuracy of the estimators $\widehat{x}^{(i)}_{k/k}$ when the LS optimality criterion is used.

Theorem 1. *Under Hypotheses 1–6, for each single sensor node $i = 1, \ldots, m$, the local LS linear filter, $\widehat{x}^{(i)}_{k/k}$, and the corresponding error covariance matrix, $P^{(i)}_{k/k}$, are given by:*

$$\widehat{x}^{(i)}_{k/k} = A_k O^{(i)}_k, \quad k \geq 1 \tag{7}$$

and:

$$P^{(i)}_{k/k} = A_k \left(B_k - A_k r^{(i)}_k\right)^T, \quad k \geq 1 \tag{8}$$

where the vectors $O^{(i)}_k$ and the matrices $r^{(i)}_k = E[O^{(i)}_k O^{(i)T}_k]$ are recursively obtained from:

$$O^{(i)}_k = O^{(i)}_{k-1} + J^{(i)}_k \Pi^{(i)-1}_k \mu^{(i)}_k, \quad k \geq 1; \quad O^{(i)}_0 = 0 \tag{9}$$

$$r^{(i)}_k = r^{(i)}_{k-1} + J^{(i)}_k \Pi^{(i)-1}_k J^{(i)T}_k, \quad k \geq 1; \quad r^{(i)}_0 = 0 \tag{10}$$

and the matrices $J^{(i)}_k = E[O^{(i)}_k \mu^{(i)T}_k]$ satisfy:

$$J^{(i)}_k = \mathcal{H}^{(i)T}_{B_k} - r^{(i)}_{k-1}\mathcal{H}^{(i)T}_{A_k} - \sum_{h=1}^{(k-1)\wedge 2} J^{(i)}_{k-h}\Pi^{(i)-1}_{k-h}\mathcal{W}^{(i)T}_{k,k-h}, \quad k \geq 2; \quad J^{(i)}_1 = \mathcal{H}^{(i)T}_{B_1} \tag{11}$$

The innovations $\mu^{(i)}_k$, and their covariance matrices, $\Pi^{(i)}_k$, are given by:

$$\mu^{(i)}_k = y^{(i)}_k - \mathcal{H}^{(i)}_{A_k} O^{(i)}_{k-1} - \sum_{h=1}^{(k-1)\wedge 2} \mathcal{W}^{(i)}_{k,k-h}\Pi^{(i)-1}_{k-h}\mu^{(i)}_{k-h}, \quad k \geq 2; \quad \mu^{(i)}_1 = y^{(i)}_1 \tag{12}$$

and:

$$\begin{aligned} \Pi^{(i)}_k &= \Sigma^{y^{(i)}}_k - \mathcal{H}^{(i)}_{A_k}(\mathcal{H}^{(i)T}_{B_k} - J^{(i)}_k) - \sum_{h=1}^{(k-1)\wedge 2} \mathcal{W}^{(i)}_{k,k-h}\Pi^{(i)-1}_{k-h}(\mathcal{H}^{(i)}_{A_k} J^{(i)}_{k-h} + \mathcal{W}^{(i)}_{k,k-h})^T, \quad k \geq 2 \\ \Pi^{(i)}_1 &= \Sigma^{y^{(i)}}_1 \end{aligned} \tag{13}$$

The coefficients $\mathcal{W}^{(i)}_{k,k-h} = E[w^{(i)}_k \mu^{(i)T}_{k-h}]$, $h = 1, 2$, are calculated as:

$$\begin{aligned} \mathcal{W}^{(i)}_{k,k-1} &= \Sigma^{y^{(i)}}_{k,k-1} - \mathcal{H}^{(i)}_{A_k}\mathcal{H}^{(i)T}_{B_{k-1}} - \mathcal{W}^{(i)}_{k,k-2}\Pi^{(i)-1}_{k-2}(\mathcal{H}^{(i)}_{A_{k-1}} J^{(i)}_{k-2} + \mathcal{W}^{(i)}_{k-1,k-2})^T, \quad k \geq 3 \\ \mathcal{W}^{(i)}_{2,1} &= \Sigma^{y^{(i)}}_{2,1} - \mathcal{H}^{(i)}_{A_2}\mathcal{H}^{(i)T}_{B_1} \\ \mathcal{W}^{(i)}_{k,k-2} &= \overline{\gamma}^{(i)}_k (1 - \overline{\gamma}^{(i)}_{k-2}) R^{(i)}_{k-1,k-2}, \quad k \geq 4; \quad \mathcal{W}^{(i)}_{3,1} = \overline{\gamma}^{(i)}_3 R^{(i)}_{2,1} \end{aligned} \tag{14}$$

Finally, the matrices $\Sigma^{y^{(i)}}_{k,s}$ are given in Equation (3) and $\mathcal{H}^{(i)}_{\Psi_s}$, $\Psi = A, B$, $s = k-1, k$, are obtained by:

$$\mathcal{H}^{(i)}_{\Psi_s} = (1 - \overline{\gamma}^{(i)}_s)\overline{\theta}^{(i)}_s H^{(i)}_s \Psi_s + \overline{\gamma}^{(i)}_s \overline{\theta}^{(i)}_{s-1} H^{(i)}_{s-1}\Psi_{s-1}, \quad s \geq 2; \quad \mathcal{H}^{(i)}_{\Psi_1} = \overline{\theta}^{(i)}_1 H^{(i)}_1 \Psi_1 \tag{15}$$

Proof. The local filter $\widehat{x}^{(i)}_{k/k}$ will be obtained from the general expression given in Equation (4), starting from the computation of the coefficients:

$$\mathcal{X}^{(i)}_{k,h} = E[x_k \mu^{(i)T}_h] = E[x_k y^{(i)T}_h] - E[x_k \widehat{y}^{(i)T}_{h/h-1}], \quad 1 \leq h \leq k$$

The independence hypotheses and the separable structure of the signal covariance assumed in Hypothesis 1 lead to $E[x_k y_h^{(i)T}] = A_k \mathcal{H}_{B_h}^{(i)T}$, with $\mathcal{H}_{B_h}^{(i)}$ given by Equation (15). From Equation (6) for $\widehat{y}_{h/h-1}^{(i)}$, $h \geq 2$, we have:

$$E\big[x_k \widehat{y}_{h/h-1}^{(i)T}\big] = (1-\overline{\gamma}_h^{(i)})\overline{\theta}_h^{(i)} E\big[x_k \widehat{x}_{h/h-1}^{(i)T}\big] H_h^{(i)T} + \overline{\gamma}_h^{(i)}\overline{\theta}_{h-1}^{(i)} E\big[x_k \widehat{x}_{h-1/h-1}^{(i)T}\big] H_{h-1}^{(i)T} + \sum_{j=1}^{(h-1)\wedge 2} \mathcal{X}_{k,h-j}^{(i)} \Pi_{h-j}^{(i)-1} \mathcal{W}_{h,h-j}^{(i)T}$$

Hence, using now Equation (4) for $\widehat{x}_{h/h-1}^{(i)}$ and $\widehat{x}_{h-1/h-1}^{(i)}$, the filter coefficients are expressed as:

$$\begin{aligned} \mathcal{X}_{k,h}^{(i)} &= A_k \mathcal{H}_{B_h}^{(i)T} - \sum_{j=1}^{h-1} \mathcal{X}_{k,j}^{(i)} \Pi_j^{(i)-1} \Big((1-\overline{\gamma}_h^{(i)})\overline{\theta}_h^{(i)} \mathcal{X}_{h,j}^{(i)T} H_h^{(i)T} + \overline{\gamma}_h^{(i)}\overline{\theta}_{h-1}^{(i)} \mathcal{X}_{h-1,j}^{(i)T} H_{h-1}^{(i)T} \Big) \\ &\quad - \sum_{j=1}^{(h-1)\wedge 2} \mathcal{X}_{k,h-j}^{(i)} \Pi_{h-j}^{(i)-1} \mathcal{W}_{h,h-j}^{(i)T}, \quad 2 \leq h \leq k \\ \mathcal{X}_{k,1}^{(i)} &= A_k \mathcal{H}_{B_1}^{(i)T} \end{aligned}$$

which guarantees that $\mathcal{X}_{k,h}^{(i)} = A_k J_h^{(i)}$, $1 \leq h \leq k$, with $J_h^{(i)}$ given by:

$$\begin{aligned} J_h^{(i)} &= \overline{\mathcal{H}}_{B_h}^{(i)T} - \sum_{j=1}^{h-1} J_j^{(i)} \Pi_j^{(i)-1} J_j^{(i)T} \overline{\mathcal{H}}_{A_h}^{(i)T} - \sum_{j=1}^{(h-1)\wedge 2} J_{h-j}^{(i)} \Pi_{h-j}^{(i)-1} \mathcal{W}_{h,h-j}^{(i)T}, \quad h \geq 2 \\ J_1^{(i)} &= \overline{\mathcal{H}}_{B_1}^{(i)T} \end{aligned} \tag{16}$$

Therefore, by defining $O_k^{(i)} = \sum_{h=1}^{k} J_h^{(i)} \Pi_h^{(i)-1} \mu_h^{(i)}$ and $r_k^{(i)} = E\big[O_k^{(i)} O_k^{(i)T}\big]$, Equation (7) for the filter follows immediately from Equation (4), and Equation (8) is obtained by using the OPL to express $P_{k/k}^{(i)} = E\big[x_k x_k^T\big] - E\big[\widehat{x}_{k/k}^{(i)} \widehat{x}_{k/k}^{(i)T}\big]$, and applying Hypothesis 1 and Equation (7).

The recursive Equations (9) and (10) are directly obtained from the corresponding definitions, taking into account that $r_k^{(i)} = \sum_{h=1}^{k} J_h^{(i)} \Pi_h^{(i)-1} J_h^{(i)T}$ which, in turn, from Equation (16), leads to Equation (11) for $J_k^{(i)}$.

From now on, using that $\widehat{x}_{k/k-1}^{(i)} = A_k O_{k-1}^{(i)}$, $\widehat{x}_{k-1/k-1}^{(i)} = A_{k-1} O_{k-1}^{(i)}$ and Equation (15), the expression for the observation predictor given by Equation (6) will be rewritten as follows:

$$\widehat{y}_{k/k-1}^{(i)} = \mathcal{H}_{A_k}^{(i)} O_{k-1}^{(i)} + \sum_{h=1}^{(k-1)\wedge 2} \mathcal{W}_{k,k-h}^{(i)} \Pi_{k-h}^{(i)-1} \mu_{k-h}^{(i)}, \quad k \geq 2 \tag{17}$$

From Equation (17), Equation (12) for the innovation is directly obtained and, applying the OPL to express its covariance matrix as $\Pi_k^{(i)} = E\big[y_k^{(i)} y_k^{(i)T}\big] - E\big[\widehat{y}_{k/k-1}^{(i)} \widehat{y}_{k/k-1}^{(i)T}\big]$, the following identity holds:

$$\begin{aligned} \Pi_k^{(i)} &= \Sigma_k^{y^{(i)}} - \mathcal{H}_{A_k}^{(i)} E\big[O_{k-1}^{(i)} \widehat{y}_{k/k-1}^{(i)T}\big] - \textstyle\sum_{h=1}^{(k-1)\wedge 2} \mathcal{W}_{k,k-h}^{(i)} \Pi_{k-h}^{(i)-1} E\big[\mu_{k-h}^{(i)} \widehat{y}_{k/k-1}^{(i)T}\big], \quad k \geq 2 \\ \Pi_1^{(i)} &= \Sigma_1^{y^{(i)}} \end{aligned}$$

Now, using again Equation (17), and taking Equation (11) into account, it is deduced that $E\big[O_{k-1}^{(i)} \widehat{y}_{k/k-1}^{(i)T}\big] = \mathcal{H}_{B_k}^{(i)T} - J_k^{(i)}$ and, since $E\big[\widehat{y}_{k/k-1}^{(i)} \mu_{k-h}^{(i)T}\big] = \mathcal{H}_{A_k}^{(i)} E\big[O_{k-1}^{(i)} \mu_{k-h}^{(i)T}\big] + \mathcal{W}_{k,k-h}^{(i)}$ and $E\big[O_{k-1}^{(i)} \mu_{k-h}^{(i)T}\big] = J_{k-h}^{(i)}$, $h = 1, 2$, Equation (13) for $\Pi_k^{(i)}$ is obtained.

To complete the proof, the expressions for $\mathcal{W}^{(i)}_{k,k-h} = E\big[w^{(i)}_k \mu^{(i)T}_{k-h}\big]$, $h = 1,2$, with $w^{(i)}_k$ given in Equation (5), are derived using that $w^{(i)}_k$ is uncorrelated with $y^{(i)}_h$, $h \leq k-3$. Consequently, $\mathcal{W}^{(i)}_{k,k-2} = E\big[w^{(i)}_k y^{(i)T}_{k-2}\big]$, and Equation (14) for $\mathcal{W}^{(i)}_{k,k-2}$ is directly obtained from Equations (1), (2) and (5), using the hypotheses stated on the model.

Next, using Equation (4) for $\widehat{y}^{(i)}_{k-1/k-2}$ in $\mathcal{W}^{(i)}_{k,k-1} = E\big[w^{(i)}_k y^{(i)T}_{k-1}\big] - E\big[w^{(i)}_k \widehat{y}^{(i)T}_{k-1/k-2}\big]$, we have:

$$\mathcal{W}^{(i)}_{k,k-1} = E\big[w^{(i)}_k y^{(i)T}_{k-1}\big] - \mathcal{W}^{(i)}_{k,k-2}\Pi^{(i)-1}_{k-2}\left(E\big[y^{(i)}_{k-1}\mu^{(i)T}_{k-2}\big]\right)^T \tag{18}$$

To compute the first expectation involved in this formula, we write:

$$w^{(i)}_k = y^{(i)}_k - (1-\overline{\gamma}^{(i)}_k)\theta^{(i)}_k(H^{(i)}_k + \varepsilon^{(i)}_k C^{(i)}_k)x_k - \overline{\gamma}^{(i)}_k\overline{\theta}^{(i)}_{k-1}H^{(i)}_{k-1}x_{k-1}$$

and we apply the OPL to rewrite $E\big[x_s y^{(i)T}_{k-1}\big] = E\big[\widehat{x}^{(i)}_{s/k-1}y^{(i)T}_{k-1}\big]$, $s = k, k-1$, thus obtaining that $E\big[w^{(i)}_k y^{(i)T}_{k-1}\big] = \Sigma^{y^{(i)}}_{k,k-1} - \mathcal{H}^{(i)}_{A_k}E\big[O^{(i)}_{k-1}y^{(i)T}_{k-1}\big]$; then, by expressing $E\big[O^{(i)}_{k-1}y^{(i)T}_{k-1}\big] = E\big[O^{(i)}_{k-1}\mu^{(i)T}_{k-1}\big] + E\big[O^{(i)}_{k-1}\widehat{y}^{(i)T}_{k/k-1}\big]$ and using Equations (11) and (17), it follows that $E\big[w^{(i)}_k y^{(i)T}_{k-1}\big] = \Sigma^{y^{(i)}}_{k,k-1} - \mathcal{H}^{(i)}_{A_k}\mathcal{H}^{(i)T}_{B_{k-1}}$.

The second expectation in Equation (18) is easily computed taking into account that, from the OPL, it is equal to $E\big[\widehat{y}^{(i)}_{k-1/k-2}\mu^{(i)T}_{k-2}\big]$ and using Equation (17).

So the proof of Theorem 1 is completed. □

3.2. Cross-Correlation Matrices between Any Two Local Filters

To obtain the distributed filtering estimator, the cross-correlation matrices between any pair of local filters must be calculated; a recursive formula for such matrices is derived in the following theorem (the notation in this theorem is the same as that used in Theorem 1).

Theorem 2. *Under Hypotheses 1–6, the cross-correlation matrices between two local filters,* $\Sigma^{(ij)}_{k/k} = E\big[\widehat{x}^{(i)}_{k/k}\widehat{x}^{(j)T}_{k/k}\big]$, $i,j = 1,\dots,m$, *are calculated by:*

$$\widehat{\Sigma}^{(ij)}_{k/k} = A_k r^{(ij)}_k A^T_k, \ k \geq 1 \tag{19}$$

with $r^{(ij)}_k = E\big[O^{(i)}_k O^{(j)T}_k\big]$ *satisfying:*

$$r^{(ij)}_k = r^{(ij)}_{k-1} + J^{(ij)}_{k-1,k}\Pi^{(j)-1}_k J^{(j)T}_k + J^{(i)}_k\Pi^{(i)-1}_k J^{(ji)T}_k, \ k \geq 1; \ \ r^{(ij)}_0 = 0 \tag{20}$$

where $J^{(ij)}_{k-1,k} = E\big[O^{(i)}_{k-1}\mu^{(j)T}_k\big]$ *are given by:*

$$\begin{aligned} J^{(ij)}_{k-1,k} &= (r^{(i)}_{k-1} - r^{(ij)}_{k-1})\mathcal{H}^{(j)T}_{A_k} + \sum_{h=1}^{(k-1)\wedge 2} J^{(i)}_{k-h}\Pi^{(i)-1}_{k-h}\mathcal{W}^{(ji)T}_{k,k-h} \\ &\quad - \sum_{h=1}^{(k-1)\wedge 2} J^{(ij)}_{k-1,k-h}\Pi^{(j)-1}_{k-h}\mathcal{W}^{(j)T}_{k,k-h}, \ k \geq 2 \\ J^{(ij)}_{0,1} &= 0 \end{aligned} \tag{21}$$

and $J^{(ij)}_{k,s} = E\big[O^{(i)}_k \mu^{(j)T}_s\big]$, *for* $s = k-1, k$, *satisfy:*

$$J^{(ij)}_{k,s} = J^{(ij)}_{k-1,s} + J^{(i)}_k\Pi^{(i)-1}_k\Pi^{(ij)}_{k,s}, \ k \geq 2; \ \ J^{(ij)}_1 = J^{(i)}_1\Pi^{(i)-1}_1\Pi^{(ij)}_1 \tag{22}$$

The innovation cross-covariance matrices $\Pi_k^{(ij)} = E\big[\mu_k^{(i)}\mu_k^{(j)T}\big]$ *are obtained as:*

$$\begin{aligned}\Pi_k^{(ij)} &= \Sigma_k^{y^{(ij)}} - \mathcal{H}_{A_k}^{(i)}\big(\mathcal{H}_{B_k}^{(j)T} - J_k^{(j)} - J_{k-1,k}^{(ij)}\big) - \sum_{h=1}^{(k-1)\wedge 2} \mathcal{W}_{k,k-h}^{(i)}\Pi_{k-h}^{(i)-1}\Pi_{k-h,k}^{(ij)} \\ &\quad - \sum_{h=1}^{(k-1)\wedge 2} \mathcal{W}_{k,k-h}^{(ij)}\Pi_{k-h}^{(j)-1}\big(\mathcal{H}_{A_k}^{(j)}J_{k-h}^{(j)} + \mathcal{W}_{k,k-h}^{(j)}\big)^T, \ k \geq 2 \\ \Pi_1^{(ij)} &= \Sigma_1^{y^{(ij)}}\end{aligned} \tag{23}$$

where $\Pi_{k,s}^{(ij)} = E\big[\mu_k^{(i)}\mu_s^{(j)T}\big]$, $s = k-2, k-1$, *are given by:*

$$\Pi_{k,s}^{(ij)} = \mathcal{H}_{A_k}^{(i)}\big(J_s^{(j)} - J_{k-1,s}^{(ij)}\big) + \mathcal{W}_{k,s}^{(ij)} - \sum_{h=1}^{(k-1)\wedge 2} \mathcal{W}_{k,k-h}^{(i)}\Pi_{k-h}^{(i)-1}\Pi_{k-h,s}^{(ij)}, \ k \geq 2 \tag{24}$$

The coefficients $\mathcal{W}_{k,k-h}^{(ij)} = E\big[w_k^{(i)}\mu_{k-h}^{(j)T}\big]$, $h = 1, 2$, *are computed by:*

$$\begin{aligned}\mathcal{W}_{k,k-1}^{(ij)} &= \Sigma_{k,k-1}^{y^{(ij)}} - \mathcal{H}_{A_k}^{(i)}\mathcal{H}_{B_{k-1}}^{(j)T} - \mathcal{W}_{k,k-2}^{(ij)}\Pi_{k-2}^{(j)-1}\big(\mathcal{H}_{A_{k-1}}^{(j)}J_{k-2}^{(j)} + \mathcal{W}_{k-1,k-2}^{(j)}\big)^T, \ k \geq 3 \\ \mathcal{W}_{2,1}^{(ij)} &= \Sigma_{2,1}^{y^{(ij)}} - \mathcal{H}_{A_2}^{(i)}\mathcal{H}_{B_1}^{(j)T} \\ \mathcal{W}_{k,k-2}^{(ij)} &= \overline{\gamma}_k^{(i)}\big(1-\overline{\gamma}_{k-2}^{(j)}\big)R_{k-1,k-2}^{(ij)}, \ k \geq 4; \quad \mathcal{W}_{3,1}^{(ij)} = \overline{\gamma}_3^{(i)}R_{2,1}^{(ij)}\end{aligned} \tag{25}$$

Finally, the matrices $\Sigma_{k,s}^{y^{(ij)}}$, *and* $\mathcal{H}_{A_s}^{(l)}$, $\mathcal{H}_{B_s}^{(l)}$, $s = k-1, k$, $l = i, j$, *are given in Equations (3) and (15), respectively.*

Proof. Equation (19) for $\widehat{\Sigma}_{k/k}^{(ij)}$ is directly obtained using Equation (7) for the local filters and defining $r_k^{(ij)} = E\big[O_k^{(i)}O_k^{(j)T}\big]$.

Next, we derive the recursive formulas to obtain the matrices $r_k^{(ij)}$, which clearly satisfy Equation (20) just by using Equation (9) and defining $J_{s,k}^{(ij)} = E\big[O_s^{(i)}\mu_k^{(j)T}\big]$, $s = k-1, k$.

For later derivations, the following expression of the one-stage predictor of $y_k^{(j)}$ based on the observations of sensor i will be used; this expression is obtained from Equation (5), taking into account that $\widehat{x}_{k/s}^{(i)} = A_k O_s^{(i)}$, $s = k-1, k$, and defining $\mathcal{W}_{k,k-h}^{(ji)} = E\big[w_k^{(j)}\mu_{k-h}^{(i)T}\big]$, $h = 1, 2$:

$$\widehat{y}_{k/k-1}^{(j/i)} = \mathcal{H}_{A_k}^{(j)}O_{k-1}^{(i)} + \sum_{h=1}^{(k-1)\wedge 2} \mathcal{W}_{k,k-h}^{(ji)}\Pi_{k-h}^{(i)-1}\mu_{k-h}^{(i)}, \ k \geq 2 \tag{26}$$

As Equation (17) is a particular case of Equation (26), for $i = j$, hereafter we will also refer to it for the local predictors $\widehat{y}_{k/k-1}^{(i)}$, $k \geq 2$.

By applying the OPL, it is clear that $E\big[O_{k-1}^{(i)}y_k^{(j)T}\big] = E\big[O_{k-1}^{(i)}\widehat{y}_{k/k-1}^{(j/i)T}\big]$ and, consequently, we can rewrite $J_{k-1,k}^{(ij)} = E\big[O_{k-1}^{(i)}\big(\widehat{y}_{k/k-1}^{(j/i)} - \widehat{y}_{k/k-1}^{(j)}\big)^T\big]$; then, using Equation (26) for both predictors, Equation (21) is easily obtained. Also, Equation (22) for $J_{k,s}^{(ij)}$, $s = k-1, k$, is immediately deduced from Equation (9), just defining $\Pi_{k,s}^{(ij)} = E\big[\mu_k^{(i)}\mu_s^{(j)T}\big]$.

To obtain Equation (23), first we apply the OPL to express $\Pi_k^{(ij)} = \Sigma_k^{y^{(ij)}} - E\big[\widehat{y}_{k/k-1}^{(i/j)}\widehat{y}_{k/k-1}^{(j)T}\big] - E\big[\widehat{y}_{k/k-1}^{(i)}\mu_k^{(j)T}\big]$. Then, using Equation (26) for $\widehat{y}_{k/k-1}^{(i/j)}$ and $\widehat{y}_{k/k-1}^{(i)}$, and the definitions of $J_{k-1,k}^{(ij)}$ and $\Pi_{k-h,k}^{(ij)}$, we have:

$$E[\widehat{y}^{(i/j)}_{k/k-1}\widehat{y}^{(j)T}_{k/k-1}] = \mathcal{H}^{(i)}_{A_k}E[O^{(j)}_{k-1}\widehat{y}^{(j)T}_{k/k-1}] + \sum_{h=1}^{(k-1)\wedge 2}\mathcal{W}^{(ij)}_{k,k-h}\Pi^{(j)-1}_{k-h}E[\mu^{(j)}_{k-h}\widehat{y}^{(j)T}_{k/k-1}]$$

$$E[\widehat{y}^{(i)}_{k/k-1}\mu^{(j)T}_{k}] = \overline{\mathcal{H}}^{(i)}_{A_k}J^{(ij)}_{k-1,k} + \sum_{h=1}^{(k-1)\wedge 2}\mathcal{W}^{(i)}_{k,k-h}\Pi^{(i)-1}_{k-h}\Pi^{(ij)}_{k-h,k}$$

so Equation (23) is obtained taking into account that $E[O^{(j)}_{k-1}\widehat{y}^{(j)T}_{k/k-1}] = \mathcal{H}^{(j)T}_{B_k} - J^{(j)}_k$ and $E[\widehat{y}^{(j)}_{k/k-1}\mu^{(j)T}_{k-h}] = \mathcal{H}^{(j)}_{A_k}J^{(j)}_{k-h} + \mathcal{W}^{(j)}_{k,k-h}$, as it has been shown in the proof of Theorem 1.

Equation (24) for $\Pi^{(ij)}_{k,s} = E[y^{(i)}_k\mu^{(j)T}_s] - E[\widehat{y}^{(i)}_{k/k-1}\mu^{(j)T}_s]$, with $s = k-2, k-1$, is obtained from $E[y^{(i)}_k\mu^{(j)T}_s] = \mathcal{H}^{(i)}_{A_k}J^{(j)}_s + \mathcal{W}^{(ij)}_{k,s}$, and using Equation (26) in $E[\widehat{y}^{(i)}_{k/k-1}\mu^{(j)T}_s]$.

Finally, the reasoning to obtain Equation (25) for the coefficients $\mathcal{W}^{(ij)}_{k,k-h} = E[w^{(i)}_k\mu^{(j)T}_{k-h}]$, $h = 1,2$, is also similar to that used to derive $\mathcal{W}^{(i)}_{k,k-h}$ in Theorem 1, so it is omitted and the proof of Theorem 2 is then completed. □

3.3. *Derivation of the Distributed LS Fusion Linear Filter*

As it has been mentioned previously, a matrix-weighted fusion linear filter is now generated from the local filters by applying the LS optimality criterion. The distributed fusion filter at any time k is hence designed as a product, $\mathcal{F}_k\widehat{X}_{k/k}$, where $\widehat{X}_{k/k} = (\widehat{x}^{(1)T}_{k/k}, \dots, \widehat{x}^{(m)T}_{k/k})^T$ is the vector constituted by the local filters, and $\mathcal{F}_k \in \mathbb{R}^{n_x \times mn_x}$ is the matrix obtained by minimizing the mean squared error, $E[(x_k - \mathcal{F}_k\widehat{X}_{k/k})^T(x_k - \mathcal{F}_k\widehat{X}_{k/k})]$.

As it is known, the solution of this problem is given by $\mathcal{F}^{opt}_k = E[x_k\widehat{X}^T_{k/k}]\left(E[\widehat{X}_{k/k}\widehat{X}^T_{k/k}]\right)^{-1}$ and, consequently, the proposed distributed filter is expressed as:

$$\widehat{x}^{(D)}_{k/k} = E[x_k\widehat{X}^T_{k/k}]\widehat{\Sigma}^{-1}_{k/k}\widehat{X}_{k/k} \tag{27}$$

with $\widehat{\Sigma}_{k/k} \equiv E[\widehat{X}_{k/k}\widehat{X}^T_{k/k}] = \left(\widehat{\Sigma}^{(ij)}_{k/k}\right)_{i,j=1,\dots,m}$, where $\widehat{\Sigma}^{(ij)}_{k/k}$ are the cross-correlation matrices between any two local filters given in Theorem 2.

The distributed fusion linear filter weighted by matrices is presented in the following theorem.

Theorem 3. *Let $\widehat{X}_{k/k} = (\widehat{x}^{(1)T}_{k/k}, \dots, \widehat{x}^{(m)T}_{k/k})^T$ denote the vector constituted by the local LS filters given in Theorem 1, and $\widehat{\Sigma}_{k/k} = \left(\widehat{\Sigma}^{(ij)}_{k/k}\right)_{i,j=1,\dots,m}$, with $\widehat{\Sigma}^{(ij)}_{k/k} = E[\widehat{x}^{(i)}_{k/k}\widehat{x}^{(j)T}_{k/k}]$ given in Theorem 2. Then, the distributed filtering estimator, $\widehat{x}^{(D)}_{k/k}$, and the error covariance matrix, $P^{(D)}_{k/k}$, are given by:*

$$\widehat{x}^{(D)}_{k/k} = \left(\widehat{\Sigma}^{(1)}_{k/k}, \dots, \widehat{\Sigma}^{(m)}_{k/k}\right)\widehat{\Sigma}^{-1}_{k/k}\widehat{X}_{k/k}, \quad k \geq 1 \tag{28}$$

and:

$$P^{(D)}_{k/k} = A_kB^T_k - \left(\widehat{\Sigma}^{(1)}_{k/k}, \dots, \widehat{\Sigma}^{(m)}_{k/k}\right)\widehat{\Sigma}^{-1}_{k/k}\left(\widehat{\Sigma}^{(1)}_{k/k}, \dots, \widehat{\Sigma}^{(m)}_{k/k}\right)^T, \quad k \geq 1 \tag{29}$$

Proof. As it has been discussed previously, Equation (28) is immediately derived from Equation (27), since the OPL guarantees that $E[x_k\widehat{X}^T_{k/k}] = \left(E[\widehat{x}^{(1)}_{k/k}\widehat{x}^{(1)T}_{k/k}], \dots, E[\widehat{x}^{(m)}_{k/k}\widehat{x}^{(m)T}_{k/k}]\right) = \left(\widehat{\Sigma}^{(1)}_{k/k}, \dots, \widehat{\Sigma}^{(m)}_{k/k}\right)$. Equation (29) is obtained from $P^{(D)}_{k/k} = E[x_kx^T_k] - E[\widehat{x}^{(D)}_{k/k}\widehat{x}^{(D)T}_{k/k}]$, using Hypothesis 1 and Equation (28). Then, Theorem 3 is proved. □

4. Centralized LS Fusion Linear Filter

In this section, using an innovation approach, a recursive algorithm is designed for the LS linear centralized fusion filter of the signal, x_k, which will be denoted by $\widehat{x}^{(C)}_{k/k}$.

4.1. Stacked Observation Model

In the centralized fusion filtering, the observations of the different sensors are jointly processed at each sampling time to yield the filter $\widehat{x}_{k/k}^{(C)}$. To carry out this process, at each sampling time $k \geq 1$ we will deal with the vector constituted by the observations from all sensors, $y_k = \left(y_k^{(1)T}, \ldots y_k^{(m)T}\right)^T$, which, from Equation (2), can be expressed as:

$$y_k = (I - \Gamma_k) z_k + \Gamma_k z_{k-1}, \; k \geq 2; \; y_1 = z_1 \tag{30}$$

where $z_k = \left(z_k^{(1)T}, \ldots z_k^{(m)T}\right)^T$ is the vector constituted by the sensor measured outputs given in Equation (1), and $\Gamma_k = Diag\left(\gamma_k^{(1)}, \ldots, \gamma_k^{(m)}\right) \otimes I$.

Let us note that the stacked vector z_k is affected by random matrices $\Theta_k = Diag\left(\theta_k^{(1)}, \ldots, \theta_k^{(m)}\right) \otimes I$ and $\mathcal{E}_k = Diag\left(\varepsilon_k^{(1)}, \ldots, \varepsilon_k^{(m)}\right) \otimes I$, and by a measurement additive noise $v_k = \left(v_k^{(1)T}, \ldots v_k^{(m)T}\right)^T$; so, denoting $H_k = \left(H_k^{(1)T}, \ldots H_k^{(m)T}\right)^T$ and $C_k = \left(C_k^{(1)T}, \ldots C_k^{(m)T}\right)^T$, we have:

$$z_k = \Theta_k \left(H_k + \mathcal{E}_k C_k\right) x_k + v_k, \; k \geq 1 \tag{31}$$

Hence, the problem is to obtain the LS linear estimator of the signal, x_k, based on the observations $y_1, \ldots, y_k$, and this problem requires the statistical properties of the processes involved in Equations (30) and (31), which are easily inferred from the model Hypotheses 1–6:

Property 1. *$\{\Theta_k\}_{k\geq 1}$ is a sequence of independent random parameter matrices whit known means $\overline{\Theta}_k \equiv E[\Theta_k] = Diag\left(\overline{\theta}_k^{(1)}, \ldots, \overline{\theta}_k^{(m)}\right) \otimes I$.*

Property 2. *$\{\mathcal{E}_k\}_{k\geq 1}$ is a sequence of independent random parameter matrices whose entries have zero means and known second-order moments.*

Property 3. *The noise $\{v_k\}_{k\geq 1}$ is a zero-mean sequence with known second-order moments defined by the matrices $R_{k,s} \equiv \left(R_{k,s}^{(ij)}\right)_{i,j=1,\ldots,m}$.*

Property 4. *The matrices $\{\Gamma_k\}_{k\geq 2}$ have known means, $\overline{\Gamma}_k \equiv E[\Gamma_k] = Diag\left(\overline{\gamma}_k^{(1)}, \ldots, \overline{\gamma}_k^{(m)}\right) \otimes I$, $k \geq 2$, and Γ_k and Γ_s are independent for $|k - s| \geq 2$.*

Property 5. *The processes $\{x_k\}_{k\geq 1}$, $\{\Theta_k\}_{k\geq 1}$, $\{\mathcal{E}_k\}_{k\geq 1}$, $\{v_k\}_{k\geq 1}$ and $\{\Gamma_k\}_{k\geq 2}$ are mutually independent.*

4.2. Recursive Filtering Algorithm

In view of Equations (30) and (31) and the above properties, the study of the LS linear filtering problem based on the stacked observations is completely similar to that of the local filtering problem carried out in Section 3. Therefore, the centralized filtering algorithm described in the following theorem is derived by an analogous reasoning to that used in Theorem 1 and its proof is omitted.

Theorem 4. *The centralized LS linear filter, $\widehat{x}_{k/k}^{(C)}$, is given by:*

$$\widehat{x}_{k/k}^{(C)} = A_k O_k, \; k \geq 1$$

where the vectors O_k and the matrices $r_k = E\left[O_k O_k^T\right]$ are recursively obtained from:

$$O_k = O_{k-1} + J_k \Pi_k^{-1} \mu_k, \; k \geq 1; \; O_0 = 0$$

$$r_k = r_{k-1} + J_k \Pi_k^{-1} J_k^T, \; k \geq 1; \; r_0 = 0$$

The matrices $J_k = E[O_k\mu_k^T]$ *satisfy:*

$$J_k = \mathcal{H}_{B_k}^T - r_{k-1}\mathcal{H}_{A_k}^T - \sum_{h=1}^{(k-1)\wedge 2} J_{k-h}\Pi_{k-h}^{-1}\mathcal{W}_{k,k-h}^T, \quad k \geq 2; \quad J_1 = \mathcal{H}_{B_1}^T.$$

The innovations, μ_k*, and their covariance matrices,* Π_k*, are given by:*

$$\mu_k = y_k - \mathcal{H}_{A_k}O_{k-1} - \sum_{h=1}^{(k-1)\wedge 2} \mathcal{W}_{k,k-h}\Pi_{k-h}^{-1}\mu_{k-h}, \quad k \geq 2; \quad \mu_1 = y_1,$$

and:

$$\Pi_k = \Sigma_k^y - \mathcal{H}_{A_k}(\mathcal{H}_{B_k}^T - J_k) - \sum_{h=1}^{(k-1)\wedge 2} \mathcal{W}_{k,k-h}\Pi_{k-h}^{-1}(\mathcal{H}_{A_k}J_{k-h} + \mathcal{W}_{k,k-h})^T, \quad k \geq 2; \quad \Pi_1 = \Sigma_1^y,$$

respectively, and the coefficients $\mathcal{W}_{k,k-h} = E[w_k\mu_{k-h}^T]$, $h = 1, 2$, *satisfy:*

$$\begin{array}{l}
\mathcal{W}_{k,k-1} = \Sigma_{k,k-1}^y - \mathcal{H}_{A_k}\mathcal{H}_{B_{k-1}}^T - \mathcal{W}_{k,k-2}\Pi_{k-2}^{-1}(\mathcal{H}_{A_{k-1}}J_{k-2} + \mathcal{W}_{k-1,k-2})^T, \quad k \geq 3 \\
\mathcal{W}_{2,1} = \Sigma_{2,1}^y - \mathcal{H}_{A_2}\mathcal{H}_{B_1}^T \\
\mathcal{W}_{k,k-2} = \overline{\Gamma}_k R_{k-1,k-2}(I - \overline{\Gamma}_{k-2}), \quad k \geq 4; \quad \mathcal{W}_{3,1} = \overline{\Gamma}_3 R_{2,1}.
\end{array}$$

In the above formulas, the matrices Σ_k^y *and* $\Sigma_{k,k-1}^y$ *are computed by* $\Sigma_{k,s}^y = (\Sigma_{k,s}^{y^{(ij)}})_{i,j=1,\ldots,m}$, $s = k, k-1$, *with* $\Sigma_{k,s}^{y^{(ij)}}$ *given in Equation (3), and* $\mathcal{H}_{\Psi_k} = (\mathcal{H}_{\Psi_k}^{(1)T}, \ldots, \mathcal{H}_{\Psi_k}^{(m)T})^T$, *with* $\mathcal{H}_{\Psi_k}^{(i)}$ *defined in Equation (15).*

The performance of the LS linear filters $\widehat{x}_{k/k}^{(C)}$, $k \geq 1$, is measured by the error covariance matrices $P_{k/k}^{(C)} = E[x_k x_k^T] - E[\widehat{x}_{k/k}^{(C)}\widehat{x}_{k/k}^{(C)T}]$, whose computation, not included in Theorem 3, is immediate from Hypothesis 1 and expression $\widehat{x}_{k/k}^{(C)} = A_k O_k$ of the filter:

$$P_{k/k}^{(C)} = A_k(B_k - A_k r_k)^T, \quad k \geq 1$$

Note that these matrices only depend on the matrices A_k and B_k, which are known, and the matrices r_k, which are recursively calculated and do not depend on the current set of observations. Hence, the filtering error covariance matrices provide a measure of the estimators performance even before we get any observed data.

5. Numerical Simulation Example

In this section, a numerical example is shown to examine the performance of the proposed distributed and centralized filtering algorithms and how the estimation accuracy is influenced by the missing and delay probabilities. Let us consider that the system signal to be estimated is a zero-mean scalar process, $\{x_k\}_{k\geq 1}$, with autocovariance function $E[x_k x_j] = 1.025641 \times 0.95^{k-j}$, $j \leq k$, which is factorizable according to Hypothesis 1 just taking, for example, $A_k = 1.025641 \times 0.95^k$ and $B_k = 0.95^{-k}$.

Sensor measured outputs. The measured outputs of this signal are assumed to be provided by three different sensors and described by Equation (1):

$$z_k^{(i)} = \theta_k^{(i)}\left(H_k^{(i)} + \varepsilon_k^{(i)}C_k^{(i)}\right)x_k + v_k^{(i)}, \quad k \geq 1, \quad i = 1, 2, 3,$$

where

- $H_k^{(1)} = H_k^{(2)} = 1$, $H_k^{(3)} = 0.75$ and $C_k^{(1)} = C_k^{(2)} = 0$, $C_k^{(3)} = 0.95$.

- The processes $\{\theta_k^{(i)}\}_{k\geq 1}$, $i = 1,2,3$, are independent sequences of independent Bernoulli random variables with constant and identical probabilities for the three sensors $P(\theta_k^{(i)} = 1) = \overline{\theta}$.
- $\{\varepsilon_k^{(3)}\}_{k\geq 1}$ is a zero-mean Gaussian white process with unit variance.
- The additive noises $\{v_k^{(i)}\}_{k\geq 1}$, $i = 1,2,3$, are defined as $v_k^{(i)} = c_i(\eta_k + \eta_{k+1})$, $i = 1,2,3$, where $c_1 = 0.75$, $c_2 = 1$, $c_3 = 0.5$, and $\{\eta_k\}_{k\geq 1}$ is a zero-mean Gaussian white process with unit variance.

Note that there are only *missing measurements* in sensors 1 and 2, and both *missing measurements and multiplicative noise* in sensor 3. Also, it is clear that the additive noises $\{v_k^{(i)}\}_{k\geq 1}$, $i = 1,2,3$, are only correlated at the same and consecutive sampling times, with $R_k^{(ij)} = 2c_ic_j$, $R_{k,k-1}^{(ij)} = c_ic_j$, $i,j = 1,2,3$.

Observations with transmission random one-step delays. Next, according to our theoretical observation model, it is supposed that, at any sampling time $k \geq 2$, the data transmissions are subject to random one-step delays with different rates and such delays are correlated at consecutive sampling times. More precisely, let us assume that the available measurements $y_k^{(i)}$ are given by:

$$y_k^{(i)} = (1-\gamma_k^{(i)})z_k^{(i)} + \gamma_k^{(i)}z_{k-1}^{(i)}, \quad k \geq 2, \quad i = 1,2,3$$

where the variables $\{\gamma_k^{(i)}\}_{k\geq 2}$ modeling this type of correlated random delays are defined using two independent sequences of independent Bernoulli random variables, $\{\lambda_k^{(i)}\}_{k\geq 1}$, $i = 1,2$, with constant probabilities, $P[\lambda_k^{(i)} = 1] = \overline{\lambda}^{(i)}$, for all $k \geq 1$; specifically, we define $\gamma_k^{(i)} = \lambda_{k+1}^{(i)}(1-\lambda_k^{(i)})$, for $i = 1,2$, and $\gamma_k^{(3)} = \lambda_k^{(1)}(1-\lambda_{k+1}^{(1)})$.

It is clear that the sensor delay probabilities are time-invariant: $\overline{\gamma}^{(i)} = \overline{\lambda}^{(i)}(1-\overline{\lambda}^{(i)})$, for $i = 1,2$, and $\overline{\gamma}^{(3)} = \overline{\gamma}^{(1)}$. Moreover, the independence of the sequences $\{\lambda_k^{(i)}\}_{k\geq 1}$, $i = 1,2$, together with the independence of the variables within each sequence, guarantee that the random variables $\gamma_k^{(i)}$ and $\gamma_s^{(j)}$ are independent if $|k-s| \geq 2$, for any $i,j = 1,2,3$. Also, it is clear that, at each sensor, the variables $\{\gamma_k^{(i)}\}_{k\geq 2}$ are correlated at consecutive sampling times and $\overline{\gamma}_{k,s}^{(i,i)} = 0$, for $i = 1,2,3$ and $|k-s| = 1$. Finally, we have that $\{\gamma_k^{(3)}\}_{k\geq 2}$ is independent of $\{\gamma_k^{(2)}\}_{k\geq 2}$, but correlated with $\{\gamma_k^{(1)}\}_{k\geq 2}$ at consecutive sampling times, with $\overline{\gamma}_{k,k-1}^{(1,3)} = \overline{\gamma}^{(1)}\overline{\lambda}^{(1)}$ and $\overline{\gamma}_{k,k-1}^{(3,1)} = \overline{\gamma}^{(1)}(1-\overline{\lambda}^{(1)})$.

Let us observe that, for each sensor $i = 1,2,3$, if $\gamma_k^{(i)} = 1$, then $\gamma_{k+1}^{(i)} = 0$; this fact guarantees that, when the measurement at time k is delayed, the available measurement at time $k+1$ is well-timed. Therefore, this correlation model covers those situations where the possibility of consecutive delayed observations at the same sensor is avoided.

To illustrate the feasibility and analyze the effectiveness of the proposed filtering estimators, the algorithms were implemented in MATLAB, and a hundred iterations were run. In order to measure the estimation accuracy, the error variances of both distributed and centralized fusion estimators were calculated for different values of the probability $\overline{\theta}$ of the Bernoulli random variables which model the missing measurements phenomena, and for several values of the delay probabilities, $\overline{\gamma}^{(i)}$, $i = 1,2,3$, obtained from several values of $\overline{\lambda}^{(i)}$. Let us observe that the delay probabilities, $\overline{\gamma}^{(i)} = \overline{\lambda}^{(i)}(1-\overline{\lambda}^{(i)})$, for $i = 1,2$, are the same if $1-\overline{\lambda}^{(i)}$ is used instead of $\overline{\lambda}^{(i)}$; for this reason, only the case $\overline{\lambda}^{(i)} \leq 0.5$ was analyzed.

Performance of the local and fusion filtering algorithms. Let us assume that $\overline{\theta} = 0.5$, and consider the same delay probabilities, $\overline{\gamma}^{(i)} = 0.21$, for the three sensors obtained when $\overline{\lambda}^{(i)} = 0.3$, $i = 1,2$. In Figure 1, the error variances of the local, distributed and centralized filters are compared; this figure shows that the error variances of the distributed fusion filtering estimator are lower than those of every local estimator, but slightly greater than those of the centralized one. However, this slight difference is compensated by the fact that the distributed fusion structure reduces the computational cost and has better robustness and fault tolerance. Analogous results are obtained for other values of the probabilities $\overline{\theta}$ and $\overline{\gamma}^{(i)}$.

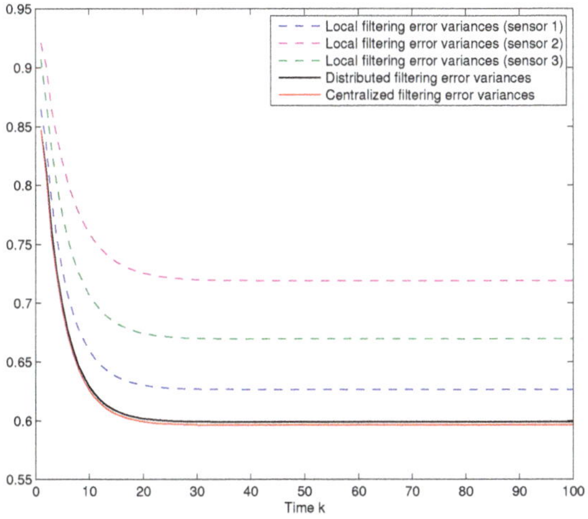

Figure 1. Filtering error variances for $\overline{\theta} = 0.5$, and $\overline{\gamma}^{(i)} = 0.21$, $i = 1, 2, 3$.

Influence of the missing measurements. Considering again $\overline{\gamma}^{(i)} = 0.21$, $i = 1, 2, 3$, in order to show the effect of the missing measurements phenomena, the distributed and centralized filtering error variances are displayed in Figure 2 for different values of the probability $\overline{\theta}$; specifically, when $\overline{\theta}$ is varied from 0.1 to 0.9. In this figure, both graphs (corresponding to the distributed and centralized fusion filters, respectively) show that the performance of the filters becomes poorer as $\overline{\theta}$ decrease, which means that, as expected, the performance of both filters improves as the probability of missing measurements, $1 - \overline{\theta}$, decreases. This figure also confirms that both methods, distributed and centralized, have approximately the same accuracy for the different values of the missing probabilities, thus corroborating the previous comments.

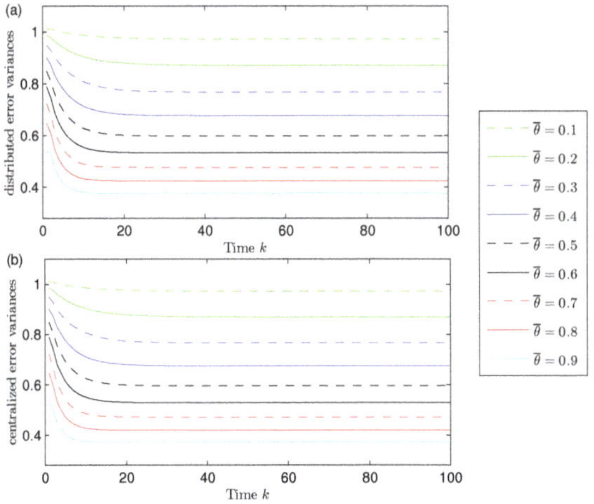

Figure 2. (**a**) Distributed and (**b**) centralized filtering error variances for different values of $\overline{\theta}$, when $\overline{\gamma}^{(i)} = 0.21$, $i = 1, 2, 3$.

Influence of the transmission delays. For $\overline{\theta} = 0.5$, different values for the probabilities $\overline{\gamma}^{(i)}$, $i = 1,2,3$, of the Bernoulli variables modelling the one-step delay phenomenon in the transmissions from the sensors to the local processors have been considered to analyze its influence on the performance of the distributed and centralized fusion filters. Since the behavior of the error variances is analogous for all the iterations, only the results at a specific iteration ($k = 100$) are displayed here. Specifically, Figure 3 shows a comparison of the filtering error variances at $k = 100$ in the following cases:

(I) Error variances versus $\overline{\lambda}^{(1)}$, when $\overline{\lambda}^{(2)} = 0.5$. In this case, the values $\overline{\lambda}^{(1)} = 0.1, 0.2, 0.3, 0.4$ and 0.5, lead to the values $\overline{\gamma}^{(1)} = \overline{\gamma}^{(3)} = 0.09, 0.16, 0.21, 0.24$ and 0.25, respectively, for the delay probabilities of sensors 1 and 3, whereas the delay probability of sensor 2 is constant and equal to 0.25.

(II) Error variances versus $\overline{\lambda}^{(1)}$, when $\overline{\lambda}^{(2)} = \overline{\lambda}^{(1)}$. Now, as in Figure 2, the delay probabilities of the three sensors are equal, and they all take the aforementioned values.

Figure 3 shows that the performance of the distributed and centralized estimators is indeed influenced by the probability $\overline{\lambda}^{(i)}$ and, as expected, better estimations are obtained as $\overline{\lambda}^{(i)}$ becomes smaller, due to the fact that the delay probabilities, $\overline{\gamma}^{(i)}$, decrease with $\overline{\lambda}^{(i)}$. Moreover, this figure shows that the error variances in case (II) are less than those of case (I). This is due to the fact that, while the delay probabilities of the three sensors are varied in case (II), only two sensors vary their delay probabilities in case (I); since the constant delay probability of the other sensor is assumed to take its greatest possible value, this figure confirms that the estimation accuracy improves as the delay probabilities decrease.

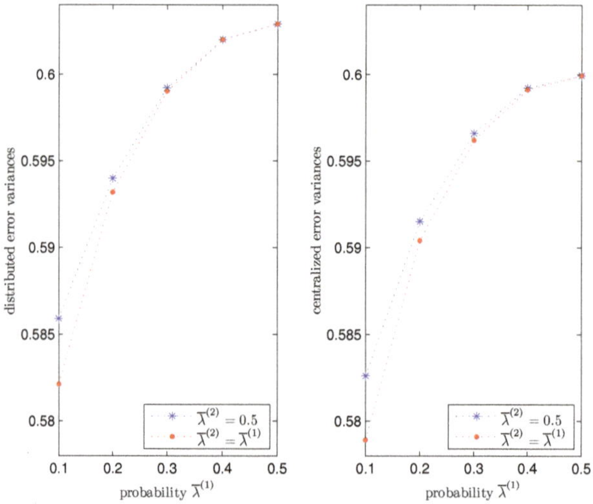

(a) (b)

Figure 3. (**a**) Distributed and (**b**) centralized filtering filtering error variances at $k = 100$, versus $\overline{\lambda}^{(1)}$.

Comparison. Next, we present a comparative analysis of the proposed centralized filter and the following ones:

- The centralized Kalman-type filter [4] for systems without uncertainties.
- The centralized filter [8] for systems with missing measurements.
- The centralized filter [25] for systems with correlated random delays.

Assuming the same probabilities $\overline{\theta} = 0.5$ and $\overline{\gamma}^{(i)} = 0.21$ as in Figure 1, and using one thousand independent simulations, the different centralized filtering estimates are compared using the mean square error (MSE) at each sampling time k, which is calculated as $\text{MSE}_k = \frac{1}{1000}\sum_{s=1}^{1000}\left(x_k^{(s)} - \widehat{x}_{k/k}^{(s)}\right)^2$, where $\left\{x_k^{(s)};\ 1 \leq k \leq 100\right\}$ denotes the s-th set of artificially simulated data and $\widehat{x}_{k/k}^{(s)}$ is the filter at the sampling time k in the s-th simulation run. The results are displayed in Figure 4, which shows that: (a) the proposed centralized filtering algorithm provides better estimations than the other filtering algorithms since the possibility of different simultaneous uncertainties in the different sensors is considered; (b) the centralized filter [8] outperforms the filter [25] since, even though the latter accommodates the effect of the delays during transmission, it does not take into account the missing measurement phenomenon in the sensors; (c) the filtering algorithm in [4] provides the worst estimations, a fact that was expected since neither the uncertainties in the measured outputs nor the delays during transmission are taken into account.

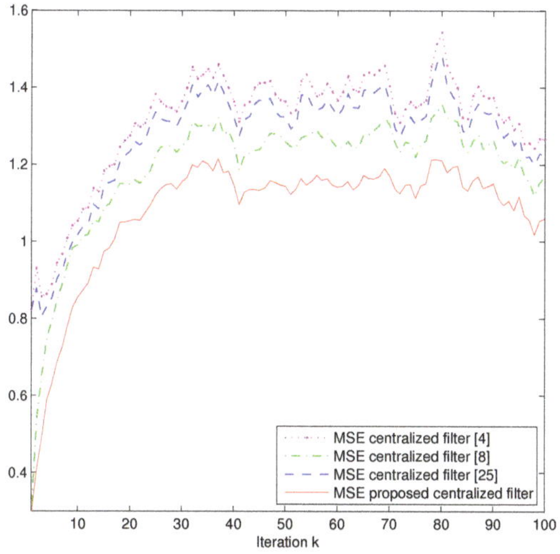

Figure 4. Filtering mean square errors when $\overline{\theta} = 0.5$ and $\overline{\gamma}^{(i)} = 0.21$.

Six-sensor network. Finally, according to the anonymous reviewers suggestion, the feasibility of the proposed estimation algorithms is tested for a larger number of sensors. More specifically, three additional sensors are considered with the same characteristics as the previous ones, but a probability $\overline{\theta^*} = P\big(\theta_k^{(i)} = 1\big) = 0.75$, $i = 4, 5, 6$, for the Bernoulli random variables modelling the missing measurements phenomena. The results are shown in Figure 5, from which similar conclusions to those from Figure 1 are deduced.

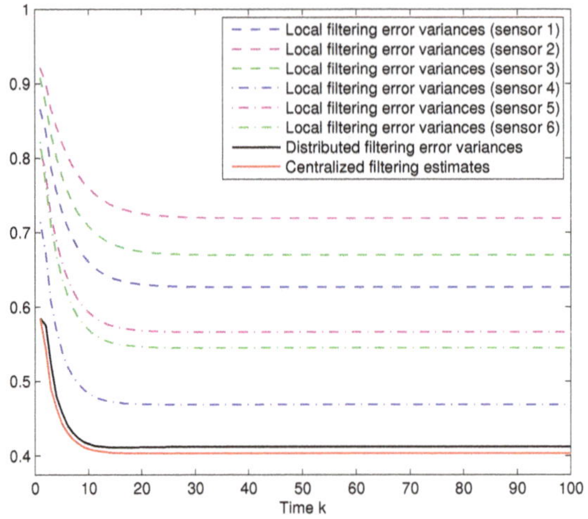

Figure 5. Filtering error variances when $\overline{\theta} = 0.5$, $\overline{\theta^{*}} = 0.75$ and $\overline{\gamma}^{(i)} = 0.21$.

6. Conclusions

In this paper, distributed and centralized fusion filtering algorithms have been designed in multi-sensor systems from measured outputs with both multiplicative and additive noises, assuming correlated random delays in transmissions. The main outcomes and results can be summarized as follows:

- *Covariance information approach.* The evolution model generating the signal process is not required to design the proposed distributed and centralized fusion filtering algorithms; nonetheless, they are also applicable to the conventional formulation using the state-space model.
- *Measured outputs with multiplicative and additive noises.* The sensor measured outputs are assumed to be affected by different stochastic uncertainties (namely, missing measurements and multiplicative noises), besides cross-correlation between the different sensor additive noises.
- *Random one-step transmission delays.* The fusion estimation problems are addressed assuming that random one-step delays may occur during the transmission of the sensor outputs through the network communication channels; the delays have different characteristics at the different sensors and they are assumed to be correlated and cross-correlated at consecutive sampling times. This correlation assumption covers many situations where the common assumption of independent delays is not realistic; for example, networked systems with stand-by sensors for the immediate replacement of a failed unit, thus avoiding the possibility of two successive delayed observations.
- *Distributed and centralized fusion filtering algorithms.* As a first step, a recursive algorithm for the local LS linear signal filter based on the measured output data coming from each sensor has been designed by an innovation approach; the computational procedure of the local algorithms is very simple and suitable for online applications. After that, the matrix-weighted sum that minimizes the mean-squared estimation error is proposed as distributed fusion estimator. Also, using covariance information, a recursive centralized LS linear filtering algorithm, with an analogous structure to that of the local algorithms, is proposed. The accuracy of the proposed fusion estimators, obtained under the LS optimality criterion, is measured by the error covariance matrices, which can be calculated offline as they do not depend on the current observed data set.

Acknowledgments: This research is supported by Ministerio de Economía y Competitividad and Fondo Europeo de Desarrollo Regional FEDER (grant No. MTM2014-52291-P).

Author Contributions: All the authors contributed equally to this work. Raquel Caballero-Águila, Aurora Hermoso-Carazo and Josefa Linares-Pérez provided original ideas for the proposed model and collaborated in the derivation of the estimation algorithms; they participated equally in the design and analysis of the simulation results; and the paper was also written and reviewed cooperatively.

Conflicts of Interest: The authors declare no conflict of interest.

References

1. Li, W.; Wang, Z.; Wei, G.; Ma, L.; Hu, J.; Ding, D. A Survey on multisensor fusion and consensus filtering for sensor networks. *Discret. Dyn. Nat. Soc.* **2015**, *2015*, 683701.
2. Ran, C.; Deng, Z. Self-tuning weighted measurement fusion Kalman filtering algorithm. *Comput. Stat. Data Anal.* **2012**, *56*, 2112–2128.
3. Feng, J.; Zeng, M. Optimal distributed Kalman filtering fusion for a linear dynamic system with cross-correlated noises. *Int. J. Syst. Sci.* **2012**, *43*, 385–398.
4. Yan, L.; Li, X.R.; Xia, Y.; Fu, M. Optimal sequential and distributed fusion for state estimation in cross-correlated noise. *Automatica* **2013**, *49*, 3607–3612.
5. Feng, J.; Wang, Z.; Zeng, M. Distributed weighted robust Kalman filter fusion for uncertain systems with autocorrelated and cross-correlated noises. *Inf. Fusion* **2013**, *14*, 78–86.
6. Hu, J.; Wang, Z.; Chen, D.; Alsaadi, F.E. Estimation, filtering and fusion for networked systems with network-induced phenomena: New progress and prospects. *Inf. Fusion* **2016**, *31*, 65–75.
7. Liu, Y.; He, X.; Wang, Z.; Zhou, D. Optimal filtering for networked systems with stochastic sensor gain degradation. *Automatica* **2014**, *50*, 1521–1525.
8. Caballero-Águila, R.; García-Garrido, I.; Linares-Pérez, J. Information fusion algorithms for state estimation in multi-sensor systems with correlated missing measurements. *Appl. Math. Comput.* **2014**, *226*, 548–563.
9. Peng, F.; Sun, S. Distributed fusion estimation for multisensor multirate systems with stochastic observation multiplicative noises. *Math. Probl. Eng.* **2014**, *2014*, 373270.
10. Pang, C.; Sun, S. Fusion predictors for multi-sensor stochastic uncertain systems with missing measurements and unknown measurement disturbances. *IEEE Sens. J.* **2015**, *15*, 4346–4354.
11. Tian, T.; Sun, S.; Li, N. Multi-sensor information fusion estimators for stochastic uncertain systems with correlated noises. *Inf. Fusion* **2016**, *27*, 126–137.
12. Shi, Y.; Fang, H. Kalman filter based identification for systems with randomly missing measurements in a network environment. *Int. J. Control* **2010**, *83*, 538–551.
13. Li, H.; Shi, Y. Robust H_∞ filtering for nonlinear stochastic systems with uncertainties and random delays modeled by Markov chains. *Automatica* **2012**, *48*, 159–166.
14. Li, N.; Sun, S.; Ma, J. Multi-sensor distributed fusion filtering for networked systems with different delay and loss rates. *Digit. Signal Process.* **2014**, *34*, 29–38.
15. Sun, S.; Ma, J. Linear estimation for networked control systems with random transmission delays and packet dropouts. *Inf. Sci.* **2014**, *269*, 349–365.
16. Caballero-Águila, R.; Hermoso-Carazo, A.; Linares-Pérez, J. Covariance-based estimation from multisensor delayed measurements with random parameter matrices and correlated noises. *Math. Probl. Eng.* **2014**, *2014*, 958474.
17. Chen, B.; Zhang, W.; Yu, L. Networked fusion Kalman filtering with multiple uncertainties. *IEEE Trans. Aerosp. Electron. Syst.* **2015**, *51*, 2332–2349.
18. Caballero-Águila, R.; Hermoso-Carazo, A.; Linares-Pérez, J. Optimal state estimation for networked systems with random parameter matrices, correlated noises and delayed measurements. *Int. J. Gen. Syst.* **2015**, *44*, 142–154.
19. Wang, S.; Fang, H.; Tian, X. Recursive estimation for nonlinear stochastic systems with multi-step transmission delays, multiple packet dropouts and correlated noises. *Signal Process.* **2015**, *115*, 164–175.
20. Chen, D.; Yu, Y.; Xu, L.; Liu, X. Kalman filtering for discrete stochastic systems with multiplicative noises and random two-step sensor delays. *Discret. Dyn. Nat. Soc.* **2015**, *2015*, 809734.
21. Caballero-Águila, R.; Hermoso-Carazo, A.; Linares-Pérez, J. Fusion estimation using measured outputs with random parameter matrices subject to random delays and packet dropouts. *Signal Process.* **2016**, *127*, 12–23.

22. Chen, D.; Xu, L.; Du, J. Optimal filtering for systems with finite-step autocorrelated process noises, random one-step sensor delay and missing measurements. *Commun. Nonlinear Sci. Numer. Simul.* **2016**, *32*, 211–224.
23. Wang, S.; Fang, H.; Tian, X. Minimum variance estimation for linear uncertain systems with one-step correlated noises and incomplete measurements. *Digit. Signal Process.* **2016**, *49*, 126–136.
24. Gao, S.; Chen, P.; Huang, D.; Niu, Q. Stability analysis of multi-sensor Kalman filtering over lossy networks. *Sensors* **2016**, *16*, 566.
25. Caballero-Águila, R.; Hermoso-Carazo, A.; Linares-Pérez, J. Linear estimation based on covariances for networked systems featuring sensor correlated random delays. *Int. J. Syst. Sci.* **2013**, *44*, 1233–1244.
26. Kailath, T.; Sayed, A.H.; Hassibi, B. *Linear Estimation*; Prentice Hall: Upper Saddle River, NJ, USA, 2000.

mathematics

MDPI

Article

Forecast Combinations in the Presence of Structural Breaks: Evidence from U.S. Equity Markets

Davide De Gaetano [1,2]

[1] University of Roma Tre, Via Silvio D'Amico, 77–00145 Rome, Italy

[2] SOSE—Soluzioni per il Sistema Economico S.p.A., Via Mentore Maggini, 48/C–00143 Rome, Italy; ddegaetano@sose.it

Received: 24 January 2018; Accepted: 20 February 2018; Published: 1 March 2018

Abstract: Realized volatility, building on the theory of a simple continuous time process, has recently received attention as a nonparametric ex-post estimate of the return variation. This paper addresses the problem of parameter instability due to the presence of structural breaks in realized volatility in the context of three HAR-type models. The analysis is conducted on four major U.S. equity indices. More specifically, a recursive testing methodology is performed to evaluate the null hypothesis of constant parameters, and then, the performance of several forecast combinations based on different weighting schemes is compared in an out-of-sample variance forecasting exercise. The main findings are the following: (i) the hypothesis of constant model parameters is rejected for all markets under consideration; (ii) in all cases, the recursive forecasting approach, which is appropriate in the absence of structural changes, is outperformed by forecast combination schemes; and (iii) weighting schemes that assign more weight in most recent observations are superior in the majority of cases.

Keywords: realized volatility; forecast combinations; structural breaks

1. Introduction

Modeling and forecasting volatility comprise an important issue in empirical finance. Traditional approaches are based on the univariate GARCH class of models or stochastic volatility models. Realized Volatility (RV) has lately become very popular; it uses improved measures of ex-post volatility constructed from high frequency data and provides an efficient estimate of the unobserved volatility of financial markets. In contrast with the GARCH approach, in which the volatility is treated as a latent variable, RV can be considered as an observable proxy, and as a consequence, it can be used in time series models to generate forecasts.

Many authors, staring from [1], have highlighted the importance of structural breaks in RV. Their presence in the data-generating process can induce instability in the model parameters. Ignoring structural breaks and wrongly assuming that the structure of a model remains fixed over time have clear adverse implications. The first finding is the inconsistency of the parameter estimates. Moreover, structural changes are likely to be responsible for most major forecast failures of time-invariant series models. Recently, Kumar [2] has found that volatility transmission from crude oil to equity sectors is structurally unstable and exhibits structural breaks; Gong and Lin [3] have examined whether structural breaks contain incremental information for forecasting the volatility of copper futures, and they have argued that considering structural breaks can improve the performance of most of the existing heterogeneous autoregressive-type models; Ma et al. [4] have introduced Markov regime switching to forecast the realized volatility of the crude oil futures market; in the same context, Wang et al. [5] have found that time-varying parameter models can significantly outperform their constant-coefficient counterparts for longer forecasting horizons.

In this paper, three different model specifications of the log-RV have been considered. The first is the Heterogeneous Autoregressive model (HAR-RV) proposed in [6], which is able to capture many of the features of volatility including long memory, fat tails and self-similarity. The second is the Leverage Heterogeneous Autoregressive model (LHAR-RV) proposed in [7], which is able to approximate both long-range dependence and the leverage effect. The last is the Asymmetric Heterogeneous Autoregressive model (AHAR-RV), which is a simplified version of the model proposed in [1]. In the spirit of the EGARCH model, the AHAR-RV allows for asymmetric effects from positive and negative returns. These models, which have become very popular in the econometric literature on RV, have very parsimonious linear structures, and as a consequence, they are extremely easy to implement and to estimate. Moreover, they have good performance in approximating many features that characterize the dynamics of RV [8], and in the forecasting context, they seem to provide results that are at least as good as more sophisticated models that consider additional components of the realized variance, such as semivariance and jumps [9,10].

The aim of this paper is to empirically investigate the relevance of structural breaks for forecasting RV of a financial time series. The presence of structural breaks in the considered RV-representations has been investigated and verified by resorting to a fluctuation test for parameter instability in a regression context. In particular, attention has been focused on the recursive estimates test [11]. This choice is particularly motivated in those cases where no particular pattern of the deviation from the null hypothesis of constant parameters is assumed. Furthermore, the proposal does not require the specification of the locations of the break points.

In order to handle parameter instability, some specific forecast combinations have been introduced and discussed. They are based on different estimation windows with alternative weighting schemes. These forecast combinations, proposed in a regression setting, are employed in financial time series, highlighting, also in this context, their usefulness in the presence of structural breaks. Moreover, all of them are feasible for a high sample size; they do not explicitly incorporate the estimation of the break dates; and as shown by [12], they do not suffer from this estimation uncertainty.

The forecasting performance of the proposed forecast combinations for the three different specifications of RV models has been compared in terms of two loss functions, the Mean Squared Error (MSE) and the Quasi-Likelihood (QLIKE) described below. These are the loss functions most widely used to compare volatility forecasting performance, and according to [13], they provide robust ranking of the models.

In order to statistically assess if the differences in the forecasting performance of the considered forecast combinations are relevant, the model confidence set, proposed in [14], has been used.

The empirical analysis has been conducted on four U.S. stock market indices: S&P 500, Dow Jones Industrial Average, Russell 2000 and Nasdaq 100. For all the series, the 5-min RV has been considered; it is one of the most used proxies of volatility, and as shown in [15], it favorably compares to more sophisticated alternatives in terms of estimation accuracy of asset price variation.

The structure of this paper is as follows. Section 2 introduces the empirical models for RV and briefly illustrates the problem of structural breaks. In Section 3, some of the most used procedures to test parameters' instability in the regression framework are reviewed. Attention has been focused on the class of fluctuation tests, and in particular, the recursive estimates test has been discussed. Section 4 introduces the problem of forecasting in the presence of structural breaks and discusses some forecast combinations able to take into account parameters' instability. In Section 5, the empirical results on the four U.S. stock market indices are reported and discussed. Some final remarks close the paper.

2. Realized Volatility Models

Let $p(s)$ be the log-price of a financial asset at time s, $\sigma^2(s)$ the instantaneous or spot volatility and $w(s)$ the standard Brownian motion. Define a simple continuous time process:

$$dp(s) = \sigma(s)dw(s) \tag{1}$$

and assume that $\sigma^2(s)$ has locally square integrable sample paths, stochastically independent of $w(s)$. The integrated volatility for day t is defined as the integral of $\sigma^2(s)$ over the interval $(t, t+1)$:

$$IV_t = \int_t^{t+1} \sigma^2(s)\, ds \tag{2}$$

where a full twenty four-hour day is represented by Time Interval 1. The integrated volatility is not observable, but it can be estimated using high frequency asset returns.

If m intraday returns are available for each day t, $\{r_{t,i}\}$ $i = 1, \ldots, m$, it is possible to define a precise volatility measure, called a realized volatility, as the squared sum of them over day t:

$$RV_t = \sum_{i=1}^{m} r_{t,i}^2 \tag{3}$$

If there were no market microstructure noise, the realized volatility would provide a consistent estimator of the integrated volatility, that is as the time interval approaches zero or equivalently m goes to infinity:

$$RV_t \to IV_t \tag{4}$$

In this paper, we focus on 5-min realized volatility; this choice is justified on the grounds of past empirical findings that show that at this frequency, there is no evidence of micro-structure noise [16]. Moreover, as shown in [15], 5-min RV favorably compares to more sophisticated alternatives in terms of estimation accuracy.

In the econometric literature, many approaches have been developed to model and forecast realized volatility with the aim of reproducing the main empirical features of financial time series such as long memory, fat tails and self-similarity. In this paper, attention has been focused on the classic Heterogeneous Autoregressive model of Realized Volatility (HAR-RV) and on some of its extensions.

The HAR-RV model, proposed in [6], has a very simple and parsimonious structure; moreover, empirical analysis [8] shows remarkably good forecasting performance. In this model, lags of RV are used at daily, weekly and monthly aggregated periods.

More precisely, let $v_t = log(RV_t)$ where RV_t is the realized volatility at time $t = 1, 2, \ldots, T$. The logarithmic version of the HAR-RV similar to that implemented by [17] is defined as:

$$v_t = \beta_0 + \beta_1 v_{t-1} + \beta_2 v_t^{(5)} + \beta_3 v_t^{(22)} + \epsilon_t \tag{5}$$

where $\epsilon_t \sim NID(0, \sigma^2)$ and $v_t^{(5)}$ and $v_t^{(22)}$ are defined, respectively, as:

$$v_t^{(5)} = \frac{v_{t-1} + v_{t-2} + \ldots + v_{t-5}}{5} \tag{6}$$

$$v_t^{(22)} = \frac{v_{t-1} + v_{t-2} + \ldots + v_{t-22}}{22} \tag{7}$$

The HAR-RV model is able to capture some well-known features of financial returns such as long memory and fat tails [8].

The first extension of this model is the Leverage Heterogeneous Autoregressive model of Realized Volatility (LHAR-RV) proposed in [7]. This model is defined as:

$$\begin{aligned} v_t = \beta_0 + \beta_1 v_{t-1} + \beta_2 v_t^{(5)} + \beta_3 v_t^{(22)} + \beta_4 r_{t-1}^{-} + \beta_5 r_t^{(5)-} + \\ + \beta_6 r_t^{(22)-} + \beta_7 r_{t-1}^{+} + \beta_8 r_t^{(5)+} + \beta_9 r_t^{(22)+} + \epsilon_t \end{aligned} \tag{8}$$

where $\epsilon_t \sim NID(0, \sigma^2)$, r_t are the daily returns and:

$$r_t^{(5)-} = \frac{r_{t-1} + r_{t-2} + \ldots + r_{t-5}}{5} I_{\{(r_{t-1}+r_{t-2}+\ldots+r_{t-5})<0\}} \tag{9}$$

$$r_t^{(5)+} = \frac{r_{t-1} + r_{t-2} + \ldots + r_{t-5}}{5} I_{\{(r_{t-1}+r_{t-2}+\ldots+r_{t-5})>0\}} \tag{10}$$

$$r_t^{(22)-} = \frac{r_{t-1} + r_{t-2} + \ldots + r_{t-22}}{22} I_{\{(r_{t-1}+r_{t-2}+\ldots+r_{t-22})<0\}} \tag{11}$$

$$r_t^{(22)+} = \frac{r_{t-1} + r_{t-2} + \ldots + r_{t-22}}{22} I_{\{(r_{t-1}+r_{t-2}+\ldots+r_{t-22})>0\}} \tag{12}$$

where I is the indicator function. The LHAR-RV model approximates both long-range dependence and the leverage effect. Some authors ([18]) suggest including only the negative part of heterogeneous returns since the estimates of the coefficients of the positive ones are usually not significant.

The second extension is the Asymmetric Heterogeneous Autoregressive model of Realized Volatility (AHAR-RV), which is a simplified version of the model proposed in [1]. It is defined as:

$$v_t = \beta_0 + \beta_1 v_{t-1} + \beta_2 v_t^{(5)} + \beta_3 v_t^{(22)} + \beta_4 \frac{|r_{t-1}|}{\sqrt{RV_{t-1}}} + \beta_5 \frac{|r_{t-1}|}{\sqrt{RV_{t-1}}} I_{\{r_{t-1}<0\}} + \epsilon_t \tag{13}$$

The last two terms allow for asymmetric effects from positive and negative returns in the spirit of the EGARCH model.

All the considered models can be rewritten in a standard regression framework:

$$y_t = \mathbf{x}_t' \boldsymbol{\beta} + \epsilon_t \tag{14}$$

where $y_t = v_t$, $\mathbf{x}_t$ is the $p \times 1$ vector of the regressors at time t and $\boldsymbol{\beta}$ is the $p \times 1$ vector of the corresponding coefficients. Of course, the number p and the specification of the vector $\mathbf{x}_t$ are different for each model.

Many studies (see, for example, [1]) agree on the existence of structural breaks in RV. If structural breaks are present in the data-generating process, they could induce instability in the model parameters. Ignoring them in the specification of the model could provide the wrong modeling and forecasting for the RV.

To deal with structural breaks, the linear regression model (14) is assumed to have time-varying coefficients, and so, it may be expressed as:

$$y_t = \mathbf{x}_t' \boldsymbol{\beta}_t + \epsilon_t \qquad t = 1, 2, \ldots T \tag{15}$$

In many applications, it is reasonable to assume that there are m breakpoints at the date $\tau_1, \tau_2, \cdots, \tau_m$ in which the coefficients shift from one stable regression relationship to a different one. Thus, there are $m+1$ segments in which the regression coefficients are constant. Model (15) can be rewritten as:

$$y_t = \mathbf{x}_t' \boldsymbol{\beta}_{\tau_{j-1}+1:\tau_j} + \epsilon_t \qquad t = 1, 2, \ldots T \qquad j = 1, 2, \ldots m+1 \tag{16}$$

and, by convention, $\tau_0 = 1$ and $\tau_{m+1} = T$.

3. Testing for Structural Changes

The presence of structural breaks can be tested through the null hypothesis that the regression coefficients remain constant over time, that is:

$$H_0 : \boldsymbol{\beta}_t = \boldsymbol{\beta} \qquad t = 1, 2, \ldots T \tag{17}$$

against the alternative that at least one coefficient varies over time.

In the statistical and econometric literature, testing for parameters' instability in a regression framework has been treated using different approaches. The classical test for structural change is the well-known Chow test [19]. This testing procedure splits the sample into two sub-samples, estimates the parameters for each sub-sample, and then, using a classic F statistic, a test on the equality of the two sets of parameters is performed. For a review, which includes also some extensions in different contexts, see [20]. The principal issue of the Chow test is the assumption that the break-date must be known a priori. Generally, this procedure is used by fixing an arbitrary candidate break-date or by selecting it on the basis of some known feature of the data. However, the results can be highly sensitive to these arbitrary choices; in the first case, the Chow test may be uninformative, and in the second case, it can be misleading [21].

More recently, the literature has focused on a more realistic problem in which the number of break points and their locations are supposed to be unknown (see [22], for a survey). In this context, one of the major contributions is the strategy proposed in Bai and Perron ([23–25]) who developed an iterative procedure that allows consistent estimation of the number and the location of the break points together with the unknown regression coefficients in each regime. In their procedure, the breaks are considered deterministic parameters, and so, the specification of their underlying generating process is not required. The number of breaks can be sequentially determined by testing for $q+1$ against q or using a global approach of testing for q against no breaks. However, the procedure needs the specification of some restrictions such as the minimum distance between breaks and their maximum number.

Another approach to change point testing is based on the generalized fluctuation tests (for a survey, see [26]). Such an approach has the advantage of not assuming a particular pattern of deviation from the null hypothesis. Moreover, although it is possible in principal to carry out the location of the break points, this method is commonly used only to verify their presence; with this aim, the fluctuation tests will be used in this paper. The general idea is to fit a regression model to the data and derive the empirical process that captures the fluctuation in the residuals or in the parameter estimates. Under the null hypothesis of constant regression coefficients, fluctuations are governed by functional central limit theorems ([27]), and therefore, boundaries can be found that are crossed by the corresponding limiting processes with fixed probability α. When the fluctuation of the empirical process increases, there is evidence of structural changes in the parameters. Moreover, its trajectory may also highlight the type of deviation from the null hypothesis, as well as the dating of the structural breaks. As previously pointed out, the generalized fluctuation tests can be based on the residuals or on the parameter estimates of the regression model. The first class includes the classical CUSUM based on Cumulative Sums of recursive residuals [28], the CUSUM test based on OLS residuals [29] and the Moving Sums (MOSUM) tests based on the recursive and OLS residuals [30]. The second class includes the Recursive Estimates (RE) test [11] and the Moving Estimates (ME) test [31]. In both, the vector of unknown parameters is estimated recursively with a growing number of observations, in the RE test, or with a moving data window, in the ME test, and then compared to the estimates obtained by using the whole sample.

Define:

$$\mathbf{y}'_{1:t} = (y_1, y_2, \ldots, y_t) \tag{18}$$

$$\mathbf{X}'_{1:t} = (\mathbf{x}_1, \mathbf{x}_2, \ldots, \mathbf{x}_t)' \tag{19}$$

and let:

$$\widehat{\beta}_{1:t} = (\mathbf{X}'_{1:t}\mathbf{X}_{1:t})^{-1}\mathbf{X}'_{1:t}\mathbf{y}_{1:t} \qquad t = p, p+1, \ldots, T \tag{20}$$

be the Ordinary Least Squares (OLS) estimate of the regression coefficients based on the observations up to t.

The basic idea is to reject the null hypothesis of parameter constancy if these estimates fluctuate too much. Formally, the test statistic is defined as:

$$S^{(T)} = \sup_{0 \leq z \leq 1} \left\| B^{(T)}(z) \right\|_{\infty} \quad (21)$$

with:

$$B^{(T)}(z) = \frac{\phi(z)}{\widehat{\sigma} T} (\mathbf{X}'_{1:T} \mathbf{X}_{1:T})^{1/2} (\widehat{\beta}_{1:\phi(z)} - \widehat{\beta}_{1:T}) \quad (22)$$

where:

$$\widehat{\sigma} = \left[\frac{1}{T-p} \sum_{t=1}^{T} (y_t - \mathbf{x}'_t \widehat{\beta}_{1:T}) \right]^{1/2} \quad (23)$$

and $\phi(z)$ is the largest integer less than or equal to $p + z(T - p)$ and $\|.\|_{\infty}$ the maximum norm. As proven in [11], $B^{(T)}(z)$ is a p-dimensional stochastic process such that

$$B^{(T)}(z) \xrightarrow{\mathcal{D}} B(z) \quad (24)$$

where $B(z)$ is a Brownian bridge. The distribution of $\sup_{0 \leq z \leq 1} \|B(z)\|_{\infty}$ is given in [32]; in particular, it is:

$$P\left(\sup_{0 \leq z \leq 1} \|B(z)\|_{\infty} \leq x \right) = \left[1 + 2 \sum_{i=1}^{\infty} (-1)^i e^{-2i^2 x^2} \right]^p \qquad x \geq 0 \quad (25)$$

4. Forecasting Methods in the Presence of Structural Breaks

Once the parameter instability due to the presence of structural breaks has been detected, the problem is how to account for it when generating forecasts. Indeed, parameter instability could cause forecast failures in macroeconomic and financial time series (for a survey, see [33]).

When it is possible to identify the exact date of the last break, the standard solution is to use only observations over the post-break period. In practice, the dates of the break points are not known a priori, and an estimation procedure has to be used. It could produce imprecise values, which negatively affect the specification of the forecasting model and, as a consequence, poor performance of the forecasts. Furthermore, even if the last break date is correctly estimated, the forecasts generated by this scheme are likely to be unbiased and may not minimize the mean square forecast error [12]. Moreover, if the last break is detected close to the boundaries of the data sample, the parameters of the forecasting model are estimated with a relatively short sample, and the estimation uncertainty may be large.

However, as pointed out in [12], the pre-break observations could be informative for forecasting even after the break. More specifically, it is appropriate to choose a high fraction of the pre-break observations especially when the break size is small, the variance parameter increases at the break point and the number of post break observations is small. Furthermore, the forecasting performance is sensitive to the choice of the observation window. A relatively long estimation window reduces the forecast error variance, but increases its bias; on the other hand, a short estimation window produces an increase in the forecast error variance although the bias decreases. Therefore, an optimal window size should balance the trade-off between an accurate estimate of the parameters and the possibility that the data come from different regimes. In this context, Pesaran and Timmermann [12] have proposed some methods to select the window size in the case of multiple discrete breaks when the errors of the model are serially uncorrelated and the regressors are strictly exogenous; Pesaran et al. [34] have derived optimal weights under continuous and discrete breaks in the case of independent errors and exogenous regressors; Giraitis et al. [35] have proposed to select a tuning parameter to downweight older data by using a cross-validation based method in the case of models without regressors; Inoue et al. [36] have suggested to choose the optimal window size that minimizes the conditional Mean Square Forecast Error (MSFE). However, in practice, the selection of a single best estimation window is not an easy task, and in many empirical studies, it is arbitrarily determined.

Alternatively, in order to deal with the uncertainty over the size of the estimation window, it is possible to combine forecasts generated from the same model, but over different estimation windows. This strategy is in the spirit of forecast combinations obtained by estimating a number of alternative models over the same sample period (for a review, see [37]). It has the advantage of avoiding the direct estimation of breakpoint parameters, and it is applicable to general dynamic models and for different estimation methods. In this context, Pesaran and Timmermann [12] have proposed forecast combinations formed by averaging across forecasts generated by using all possible window sizes subject to a minimum length requirement. Based on the same idea, more complex forecasting schemes have been proposed (see, for example, [34,38]).

The idea of forecast averaging over estimation windows has been fruitfully applied also in macroeconomic forecasting, in particular in the context of vector autoregressive models with weakly-exogenous regressors ([39,40] and in the context of GDP growth on the yield curve ([41]).

Pesaran and Pick [42] have discussed the theoretical advantages of using such combinations considering random walks with breaks in the drift and volatility and a linear regression model with a break in the slope parameter. They have shown that averaging forecasts over different estimation windows leads to a lower bias and root mean square forecast error than forecasts based on a single estimation window for all but the smallest breaks. Similar results are reported in [43]; they have highlighted that, in the presence of structural breaks, averaging forecasts obtained by using all the observations in the sample and forecasts obtained by using a window can be useful for forecasting. In this case, forecasts from only two different windows have been combined, and so, this procedure can be seen as a limited version of that proposed in [12].

In view of the above considerations, in this paper, attention has been focused on forecast schemes generated from the same model, but over different estimation windows. In particular, for each of the considered realized volatility models, different forecast combinations have been considered focusing on those that are feasible for financial time series and that do not explicitly incorporate the estimation of the break dates. Moreover, in the analysis, one-step ahead forecasts have been considered, and so, it is assumed that no structural breaks occur in the forecast period (for forecasting with structural breaks over the forecast period, see [44,45]).

4.1. Forecast Combination With Equal Weights

As previously pointed out, the forecast combination with equal weights is the simplest combination, but it is robust to structural breaks of unknown break dates and sizes. Moreover, it performs quite well especially when the break is of moderate magnitude and it is located close to the boundaries of the data sample [42].

Let ω be the minimum acceptable estimation window size. The forecast combination with equal weights is defined by:

$$\widehat{y}_{T+1} = \frac{1}{T-\omega} \sum_{\tau=1}^{T-\omega} \left(\mathbf{x}'_{T+1} \widehat{\beta}_{\tau+1:T} \right) \tag{26}$$

Many research works have highlighted the advantages of this scheme; it has good performance also when there is uncertainty about the presence of structural breaks in the data. This approach also avoids any estimation procedure for the weights.

4.2. Forecast Combination With Location Weights

By looking at Equation (26), it is evident that the weights in the equally-weighted combination can be converted into weights on the sample observations $\mathbf{x}_t$. As discussed in [38], the ω most recent observations are used in all of the forecasts, whereas the older observations are used less. Furthermore, the influence of each observation is inversely proportional to its distance from the forecasting origin: the most recent data are usually more relevant especially if the regression parameters have significant changes close to the end of the sample.

A way to place heavier weights on the forecasts that are based on more recent data much more than under the equally-weighted forecast combination is to use constant weights proportional to the location of τ in the sample.

More precisely, this combination, known as the forecast combination with location weights, is defined by:

$$\widehat{y}_{T+1} = \frac{1}{\sum_{\tau=1}^{T-\omega} \tau} \sum_{\tau=1}^{T-\omega} \tau \left(\mathbf{x}'_{T+1} \widehat{\beta}_{\tau+1:T} \right) \tag{27}$$

Also in this case, no estimation of the weights is needed.

4.3. Forecast Combination With MSFE Weights

This approach, proposed in [12], is based on the idea that the weights of the forecasters obtained with different estimation windows should be proportional to the inverse of the associated out-of-sample MSFE values. To this aim, a cross-validation approach is used.

To better understand, let m be the generic start point of the estimation window and assume that $\tilde{\omega}$ is the number of observations used in the cross-validation set, that is the observations used to measure pseudo out-of-sample forecasting performance. The recursive pseudo out-of-sample MSFE value is computed as:

$$MSFE(m|T, \tilde{\omega}) = \tilde{\omega}^{-1} \sum_{\tau=T-\tilde{\omega}}^{T-1} \left(y_{\tau+1} - \mathbf{x}'_{\tau} \widehat{\beta}_{m:\tau} \right)^2 \tag{28}$$

The forecast combination with MSFE weights is then defined as:

$$\widehat{y}_{T+1} = \frac{\sum_{m=1}^{T-\omega-\tilde{\omega}} \left(\mathbf{x}'_{T} \widehat{\beta}_{m:T} \right) \left(MSFE(m|T, \tilde{\omega}) \right)^{-1}}{\sum_{m=1}^{T-\omega-\tilde{\omega}} \left(MSFE(m|T, \tilde{\omega}) \right)^{-1}} \tag{29}$$

Together with the parameter ω, the length of the minimal estimation window, this method also requires the choice of the parameter $\tilde{\omega}$ and the length of the evaluation window. If this parameter is set too large, too much smoothing may result, and as a consequence, in the combination, the forecasting based on older data will be preferred. On the other hand, if $\tilde{\omega}$ is set too short, although a more precise estimation of the MSFE can be obtained, the ranking of the forecasting methods is more affected by noise. Of course, the selection of this parameter depends on the problem at hand and on the length of the series.

4.4. Forecast Combination With ROC Weights

This approach, proposed in [38], is based on by-products of the Reverse Ordered CUSUM (ROC) structural break test considered in [46].

It is a two-stage forecasting strategy. In the first step, a sequence of ROC test statistics, starting from the most recent observations and going backwards in time, is calculated. Each point in the sample is considered as a possible most recent break point.

This test is related to the classical CUSUM test, but in this case, the test sequence is made in reverse chronological order. In particular, the time series observations are placed in reverse order, and the standard CUSUM test is performed on the rearranged dataset.

In the paper [46], the test statistics are used to perform a formal structural break test and to estimate the last breakpoint in the sample.

In the second step, the ROC statistics are used to weight the associated post break forecast, developing a forecast combination. Moreover, the weights do not depend on finding and dating a structural break, but they are constructed in order to give more weights to observations subsequent to a potential structural break.

In the first step of the procedure, for $\tau = T - \omega + 1, T - \omega, \ldots, 2, 1$, let:

$$\mathbf{y}'_{T:\tau} = (y_T, y_{T-1}, \ldots, y_{\tau+1}, y_\tau) \tag{30}$$

$$\mathbf{X}'_{T:\tau} = (\mathbf{x}_T, \mathbf{x}_{T-1}, \ldots, \mathbf{x}_{\tau+1}, \mathbf{x}_\tau) \tag{31}$$

be the observation matrices, and let:

$$\widehat{\beta}^{(R)}_{T:\tau} = (\mathbf{X}'_{T:\tau}\mathbf{X}_{T:\tau})^{-1}\mathbf{X}'_{T:\tau}\mathbf{y}_{T:\tau} \tag{32}$$

be a sequence of least squares estimates of β associated with the reverse-ordered datasets.

The ROC test statistics s_t are defined as:

$$s_\tau = \frac{\sum_{t=\tau}^{T-\omega} \xi_t^2}{\sum_{t=1}^{T-\omega} \xi_t^2} \qquad \text{for } \tau = T - \omega, T - \omega - 1, \ldots, 2, 1 \tag{33}$$

where ξ_t are the standardized one-step-ahead recursive residuals defined as:

$$\xi_t = \frac{y_t - \mathbf{x}'_t \widehat{\beta}^{(R)}_{T:t+1}}{(1 + \mathbf{x}'_t(\mathbf{X}'_{T:t+1}\mathbf{X}_{T:t+1})^{-1}\mathbf{x}_t)^{1/2}} \tag{34}$$

In the second step of the procedure, all dates τ are considered as possible choices for the last breakpoint. The combination weight on each τ is constructed as:

$$cw_\tau = \frac{\left|s_\tau - \left(\frac{T-\omega-\tau+1}{T-\omega}\right)\right|}{\sum_{\tau=1}^{T-\omega}\left|s_t - \left(\frac{T-\omega-\tau+1}{T-\omega}\right)\right|} \qquad \tau = 1, 2, \ldots, T - \omega \tag{35}$$

Since, under the null hypothesis of no structural break in τ, it is:

$$E(s_\tau) = \frac{T - \omega - \tau + 1}{T - \omega} \tag{36}$$

the combination weights vary according to the absolute distances between s_τ and its expected value. As a consequence, cw_τ is larger if this distance is large, that is if the evidence of a structural break is stronger. On the contrary, if in τ, there is no evidence of substantial breakpoint, the associated weight is small.

Moreover, the weights do not depend on finding and dating a structural break. However, if the absolute values of the difference between the ROC statistics and their expectation, under the null hypothesis, start to grow (giving evidence of a potential structural break), the weights cw_τ increase giving more weights to the observations on data, subsequent to τ.

The one-step-ahead forecast based on ROC statistics is defined as:

$$\widehat{y}_{T+1} = \sum_{\tau=1}^{T-\omega} \left(cw_\tau(\mathbf{x}'_{T+1}\widehat{\beta}_{\tau+1:T})\right) \tag{37}$$

4.5. Forecast Combination With ROC Location Weights

In order to take into account a prior belief on the probability that a time τ could be the most recent break point, it is possible, in the definition of ROC weights, to incorporate an additional weight function l_τ.

Following [38], the new weights are defined as:

$$cw_\tau = \frac{\left|s_\tau - \left(\frac{T-\omega-\tau+1}{T-\omega}\right)\right| l_\tau}{\sum_{\tau=1}^{T-\omega}\left|s_t - \left(\frac{T-\omega-\tau+1}{T-\omega}\right)\right| l_\tau} \qquad \tau = 1,2,\ldots,T-\omega \tag{38}$$

For example, if a single break point seems to be equally likely at each time point, the natural choice is $l_\tau = 1$ for $\tau = 1,2,\ldots,T-\omega$. In this case, the weights depend only on the magnitude of the ROC statistics, and the combination defined in (37) is obtained.

However, in the forecasting context, where the identification of the most recent break is essential, the prior weight l_τ could be chosen as an increasing function of the location of time τ in the full sample. In the spirit of forecast combination with location weights, the most natural choice is $l_\tau = \tau$. Of course, different specifications are also allowed.

5. Empirical Application

The data were obtained from the Oxford-Man Institute's Realised library. It consists of 5-min realized volatility and daily returns of four U.S. stock market indices: S&P 500, Dow Jones Industrial Average, Russell 2000 and Nasdaq 100. The sample covers the period from 1 January 2012–4 February 2016; the plots of the RV series and the log-RV series are reported in Figure 1.

In order to investigate the constancy of the regression coefficients in all the considered models (HAR-RV, LHAR-RV and AHAR-RV), an analysis based on the recursive estimates test has been employed for all four series. The results are reported in Table 1. The rejection of the null hypothesis of the constancy of the regression parameters for all the models and for all the series is evident.

Table 1. Recursive estimates test. LHAR, Leverage Heterogeneous Autoregressive model; AHAR, Asymmetric Heterogeneous Autoregressive model; RV, Realized Volatility.

S&P 500			DJIA		
Model	statistic	*p*-value	Model	statistic	*p*-value
HAR-RV	2.369	0.0001	HAR-RV	2.167	0.0006
LHAR-RV	2.198	0.0010	LHAR-RV	1.943	0.0084
AHAR-RV	2.156	0.0011	AHAR-RV	2.089	0.0019
RUSSELL 2000			**NASDAQ 100**		
Model	statistic	*p*-value	Model	statistic	*p*-value
HAR-RV	2.216	0.0004	HAR-RV	1.848	0.0086
LHAR-RV	1.997	0.0048	LHAR-RV	1.659	0.0104
AHAR-RV	2.044	0.0028	AHAR-RV	1.782	0.0093

Moreover, in Figure 2, the fluctuation process, defined in Equation (21), is reported for each model specification and for each series, along with the boundaries obtained by its limiting process at level $\alpha = 0.05$.

The paths of the empirical fluctuation process confirm the parameters' instability: the boundaries are crossed for all the series and for all the models.

The above analysis seems to confirm the effectiveness of parameter instability in all three RV-models for all the considered indices. Moreover, it supports the use of forecasting methods that take into account the presence of structural breaks.

In order to evaluate and to compare the forecasting performance of the proposed forecast combinations for each class of model, an initial sub-sample, composed of the data from $t = 1$ to $t = R$, is used to estimate the model, and the one-step-ahead out-of-sample forecast is produced. The sample is then increased by one; the model is re-estimated using data from $t = 1$ to $t = R + 1$; and the one-step-ahead forecast is produced. The procedure continues until the end of the available out-of-sample period. In the following examples, R has been fixed such that the number of out of sample observations is 300. To generate the one-step-ahead forecasts, the five competing forecast combinations, defined in Section 4, have been considered together with a benchmark method.

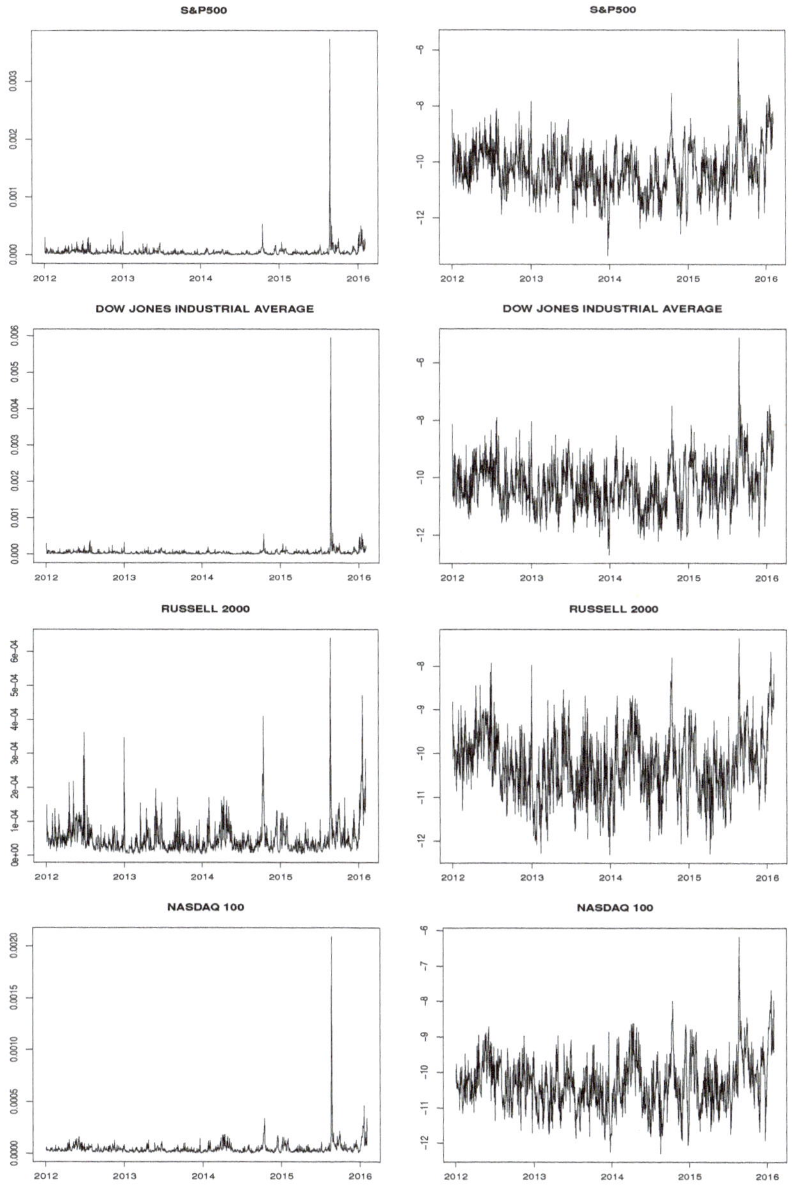

Figure 1. RV (**left**) and log-RV (**right**) for S&P 500, Dow Jones Industrial Average, Russell 2000 and Nasdaq 100.

As a natural benchmark, we refer to the expanding window method, which ignores the presence of structural breaks. In fact, it uses all the available observations. As pointed out in [12], this choice is optimal in situations with no breaks, and it is appropriate for forecasting when the data are generated by a stable model. For each class of models, this method produces out-of-sample forecasts using a recursive expanding estimation window.

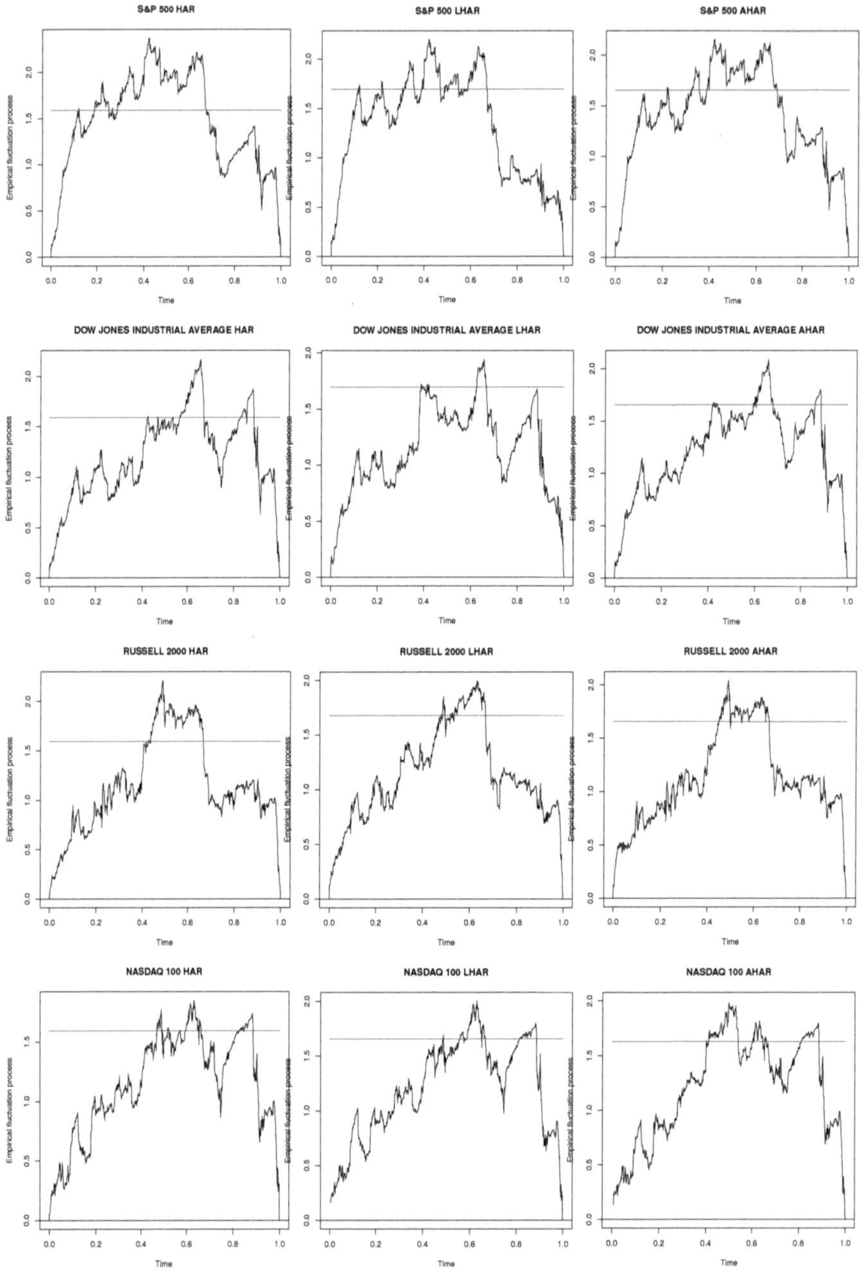

Figure 2. Empirical fluctuation process of the HAR-RV: (**left**) the LHAR-RV; (**middle**) the AHAR-RV; (**right**) model parameters for S&P 500, Dow Jones Industrial Average, Russell 2000 and Nasdaq 100. The red line refers to the boundary with significance level $\alpha = 0.05$.

Common to all the considered forecast combinations is the specification of the minimum acceptable estimation window size ω. It should not be smaller than the number of regressors plus one; however, as pointed out by [12] , to account for the very large effect of parameter estimation error, it should be at least 3 times the number of unknown parameters. For simplicity, the parameter ω has been held fixed at 40 for all the RV model specifications.

Moreover, for the MSFE weighted average combination, the length of the evaluation window has been held fixed at 100. This value allows a good estimation of the MSFE and, at the same time, a non-excessive loss of data at the end of the sample where a change point could be very influential for the forecasting.

In order to evaluate the quality of the volatility forecasts, the MSE and the QLIKE loss functions have been considered. They are defined as:

$$\text{MSE}_t = (y_t - \widehat{y}_t)^2 \tag{39}$$

$$\text{QLIKE}_t = \frac{y_t}{\widehat{y}_t} - \log\left(\frac{y_t}{\widehat{y}_t}\right) - 1 \tag{40}$$

where y_t is the actual value of the 5-min RV at time t and $\widehat{y}_t$ is the corresponding RV forecast. They are the most widely-used loss functions, and they provide robust ranking of the models in the context of volatility forecasts [13].

The QLIKE loss is a simple modification of the Gaussian log-likelihood in such a way that the minimum distance of zero is obtained when $y_t = \widehat{y}_{t+1}$. Moreover, according to [13], it is able to better discriminate among models and is less affected by the most extreme observations in the sample.

These loss functions have been used to rank the six competing forecasting methods for each RV model specification. To this aim, for every method, the average loss and the ratio between its value and the average loss of the benchmark method have been computed. Obviously, a value of the ratio below the unit indicates that the forecasting method "beats" the benchmark according to the loss function metric.

Moreover, to statistically assess if the differences in the performance are relevant, the model confidence set procedure has been used [14].

This procedure is able to construct a set of models from a specified collection, which consists of the best models in terms of a loss function with a given level of confidence, and it does not require the specification of a benchmark. Moreover, it is a stepwise method based on a sequence of significance tests in which the null hypothesis is that the two models under comparison have the same forecasting ability against the alternative that they are not equivalent. The test stops when the first hypothesis is not rejected, and therefore, the procedure does not accumulate Type I error ([14]). The critical values of the test, as well as the estimation of the variance useful to construct the test statistic are determined by using the block bootstrap. This re-sampling technique preserves the dependence structure of the series, and it works reasonably well under very weak conditions on the dependency structure of the data.

In the following, the results obtained for the three different model specifications for the realized volatility of the four considered series are reported and discussed.

Tables 2–4 provide the results of the analysis for the HAR-RV, LHAR-RV and AHAR-RV model specification respectively for the four considered series. In particular, they report the average MSE and the average QLIKE for each forecasting method, as well as their ratio for an individual forecasting method to the benchmark expanding window method and the ranking of the considered methods with respect to each loss function. Moreover, the value of the test statistic of the model confidence set approach and the associated p-value are also reported.

In the case of the HAR model, for all the considered series and for both loss functions, there is significant evidence that all the forecast combinations have better forecasting ability with respect to the expanding window procedure; the ratio values are all less than one. Moreover, for both loss functions, the best method is the forecast combination with ROC location weights for S&P 500 and Dow Jones

Industrial Average and with location weights for Russell 2000 and Nasdaq 100; this result confirms the importance of placing heavier weights on the forecast based on more recent data.

Table 2. Out of sample forecasting result for the HAR-RV model. ROC, Reverse Ordered Cumulative Sum.

	MSE	Ratio	rk	MCS	QLIKE	Ratio	rk	MCS
S&P 500								
Expanding window	0.5166	1.0000	6	2.154 (0.00)	0.4082	1.0000	6	2.035 (0.00)
MSFE weights	0.5016	0.9710	5	1.010 (1.00)	0.3879	0.9500	4	0.610 (1.00)
ROC weights	0.4986	0.9653	2	−0.956 (1.00)	0.3825	0.9370	2	−0.753 (1.00)
ROC location weights	0.4980	**0.9639**	1	−1.410 (1.00)	0.3794	**0.9294**	1	−1.534 (1.00)
Equal weights	0.5015	0.9708	4	0.917 (1.00)	0.3901	0.9557	5	1.191 (0.99)
Location weights	0.5007	0.9694	3	0.437 (1.00)	0.3874	0.9489	3	0.492 (1.00)
DJIA								
	MSE	Ratio	rk	MCS	QLIKE	Ratio	rk	MCS
Expanding window	0.6182	1.0000	6	2.112 (0.00)	0.5610	1.0000	6	2.007 (0.00)
MSFE weights	0.6088	0.9849	5	1.344 (0.97)	0.5410	0.9643	4	0.441 (1.00)
ROC weights	0.6066	0.9813	3	−0.213 (1.00)	0.5388	0.9603	2	−0.102 (1.00)
ROC location weights	0.6046	**0.9781**	1	−1.599 (1.00)	0.5319	**0.9480**	1	−1.789 (1.00)
Equal weights	0.6079	0.9834	4	0.690 (1.00)	0.5441	0.9699	5	1.208 (1.00)
Location weights	0.6066	0.9813	2	−0.220 (1.00)	0.5402	0.9629	3	0.252 (1.00)
RUSSELL 2000								
	MSE	Ratio	rk	MCS	QLIKE	Ratio	rk	MCS
Expanding window	0.3491	1.0000	6	1.904 (0.00)	0.2066	1.0000	6	1.649 (0.00)
MSFE weights	0.3450	0.9881	5	1.235 (0.98)	0.2039	0.9868	4	1.018 (1.00)
ROC weights	0.3445	0.9868	4	0.925 (1.00)	0.2040	0.9872	5	1.071 (0.65)
ROC location weights	0.3424	0.9808	2	−0.548 (1.00)	0.2011	0.9733	2	−0.702 (1.00)
Equal weights	0.3430	0.9826	3	−0.106 (1.00)	0.2025	0.9798	3	0.123 (1.00)
Location weights	0.3410	**0.9768**	1	−1.500 (1.00)	0.1998	**0.9670**	1	−1.507 (1.00)
NASDAQ 100								
	MSE	Ratio	rk	MCS	QLIKE	Ratio	rk	MCS
Expanding window	0.3509	1.0000	6	2.061 (0.00)	0.2397	1.0000	6	1.897 (0.00)
MSFE weights	0.3448	0.9825	4	1.007 (1.00)	0.2307	0.9697	5	1.106 (1.00)
ROC weights	0.3449	0.9829	5	1.130 (1.00)	0.2319	0.9676	4	0.907 (1.00)
ROC location weights	0.3433	0.9784	3	−0.229 (1.00)	0.2285	0.9535	2	−0.484 (1.00)
Equal weights	0.3432	0.9781	2	−0.319 (1.00)	0.2300	0.9595	3	0.108 (1.00)
Location weights	0.3417	**0.9738**	1	−1.578 (1.00)	0.2257	**0.9417**	1	−1.631 (1.00)

Note: For both the loss functions (MSE and QLIKE), the entries are: the values of the average loss; the ratio of the average loss to that of the expanding window method; the rank (rk) according to the average loss function; the statistic and the *p*-value of the Model Confidence Set (MCS) procedure with $\alpha = 0.10$. A bold entry denotes the value of the smallest average loss.

The model confidence set has the same structure for all four series and for both loss functions; it excludes only the forecast generated by the expanding window procedure and includes all the forecast combinations.

For the LHAR model, focusing on the results of S&P 500 and Dow Jones Industrial Average, which have very similar behavior, the forecast combination with MSFE weights offers the best improvement in forecasting accuracy according to the MSE metric, while the forecast combination with ROC weights according to the QLIKE metric.

For Russell 2000, the forecast combination with ROC location weights beats all the competing models according to MSE loss function, while, under the QLIKE, the best method is the forecast combination with location weights. This last method has better performance with respect to both of the loss function metrics for Nasdaq 100.

Table 3. Out of sample forecasting result for the LHAR-RV model.

	MSE	Ratio	rk	MCS	QLIKE	Ratio	rk	MCS
S&P 500								
Expanding window	0.4247	1.0000	5	1.776 (0.00)	0.2858	1.0000	6	2.143 (0.00)
MSFE weights	0.4176	**0.9834**	1	−1.258 (1.00)	0.2666	0.9328	3	1.316 (0.01)
ROC weights	0.4183	0.9851	2	−0.617 (1.00)	0.2657	**0.9298**	1	−0.632 (1.00)
ROC location weights	0.4204	0.9899	4	1.275 (0.13)	0.2658	0.9301	2	0.618 (0.55)
Equal weights	0.4197	0.9882	3	0.599 (1.00)	0.2678	0.9370	4	1.557 (0.00)
Location weights	0.4260	1.0031	6	1.536 (0.00)	0.2695	0.9430	5	1.673 (0.00)
DJIA								
	MSE	**Ratio**	**rk**	**MCS**	**QLIKE**	**Ratio**	**rk**	**MCS**
Expanding window	0.5291	1.0000	5	1.191 (0.09)	0.4236	1.0000	6	1.998 (0.00)
MSFE weights	0.5255	**0.9932**	1	−1.181 (1.00)	0.4012	0.9471	3	1.368 (0.00)
ROC weights	0.5260	0.9942	2	−0.844 (1.00)	0.3996	**0.9433**	1	−0.913 (1.00)
ROC location weights	0.5290	0.9998	4	1.092 (0.57)	0.3999	0.9440	2	0.912 (0.44)
Equal weights	0.5268	0.9957	3	−0.311 (1.00)	0.4063	0.9592	4	1.633 (0.00)
Location weights	0.5325	1.0063	6	1.783 (0.00)	0.4091	0.9658	5	1.522 (0.00)
RUSSELL 2000								
	MSE	**Ratio**	**rk**	**MCS**	**QLIKE**	**Ratio**	**rk**	**MCS**
Expanding window	0.3096	1.0000	6	2.082 (0.00)	0.1828	1.0000	6	1.564 (0.00)
MSFE weights	0.3058	0.9877	5	1.783 (0.00)	0.1821	0.9964	5	1.464 (0.00)
ROC weights	0.3044	0.9833	4	0.777 (1.00)	0.1809	0.9895	4	1.334 (0.00)
ROC location weights	0.3035	**0.9805**	1	−1.676 (1.00)	0.1794	0.9812	2	−0.989 (1.00)
Equal weights	0.3044	0.9832	3	0.686 (1.00)	0.1804	0.9867	3	1.268 (0.00)
Location weights	0.3042	0.9827	2	0.210 (1.00)	0.1789	**0.9786**	1	−0.993 (1.00)
NASDAQ 100								
	MSE	**Ratio**	**rk**	**MCS**	**QLIKE**	**Ratio**	**rk**	**MCS**
Expanding window	0.3180	1.0000	6	2.083 (0.00)	0.2070	1.0000	6	1.984 (0.00)
MSFE weights	0.3112	0.9787	4	1.058 (0.99)	0.1979	0.9562	5	1.066 (1.00)
ROC weights	0.3113	0.9788	5	1.086 (0.99)	0.1977	0.9551	4	0.970 (1.00)
ROC location weights	0.3099	0.9747	3	−0.034 (1.00)	0.1947	0.9406	2	−0.315 (1.00)
Equal weights	0.3093	0.9726	2	−0.586 (1.00)	0.1953	0.9436	3	−0.047 (1.00)
Location weights	0.3082	**0.9691**	1	−1.516 (1.00)	0.1915	**0.9252**	1	−1.665 (1.00)

Note: For both loss functions (MSE and QLIKE), the entries are: the values of the average loss; the ratio of the average loss to that of the expanding window method; the rank (rk) according to the average loss function; the statistic and the p-value of the Model Confidence Set procedure with $\alpha = 0.10$. A bold entry denotes the value of the smallest average loss.

By looking at the MSE ratios for S&P 500 and Dow Jones Industrial Average Index, it is evident that the forecast combination with location weights is unable to beat the expanding window procedure; for all the others, the combinations are able to outperform it consistently.

In the model confidence set, when the MSE loss function is used, the expanding window is eliminated from the model confidence set for all the series. However, the forecast combination with location weights for S&P 500 and Dow Jones Industrial Average and that with MSFE weights for Russell 2000 are also eliminated. A quite different situation arises when the loss function QLIKE is considered. In this case, the only surviving models in the model confidence set are the two combinations based on ROC statistics for S&P 500 and Dow Jones Industrial Average and those based on ROC location weights and on location weights for Russell 2000. For Nasdaq 100, all the combinations have the same forecasting accuracy. Excluding the last case, the QLIKE loss function, as pointed out previously, seems to better discriminate among forecasting methods.

Table 4. Out of sample forecasting result for the AHAR-RV model.

	MSE	Ratio	rk	MCS	QLIKE	Ratio	rk	MCS
				S&P 500				
Expanding window	0.4576	1.0000	6	2.186 (0.00)	0.3315	1.0000	6	2.151 (0.00)
MSFE weights	0.4397	0.9608	5	1.230 (0.94)	0.3072	0.9267	5	1.268 (0.00)
ROC weights	0.4371	0.9552	2	−0.860 (1.00)	0.3043	0.9179	3	1.393 (0.00)
ROC location weights	0.4364	**0.9537**	1	−1.416 (1.00)	0.3012	**0.9085**	1	−0.973 (1.00)
Equal weights	0.4391	0.9596	4	0.781 (1.00)	0.3063	0.9240	4	1.413 (0.00)
Location weights	0.4385	0.9582	3	0.252 (1.00)	0.3016	0.9097	2	−0.977 (0.38)
				DJIA				
	MSE	**Ratio**	**rk**	**MCS**	**QLIKE**	**Ratio**	**rk**	**MCS**
Expanding window	0.5669	1.0000	6	2.189 (0.00)	0.4718	1.0000	6	2.091 (0.00)
MSFE weights	0.5534	0.9762	5	1.015 (0.99)	0.4418	0.9365	5	0.986 (1.00)
ROC weights	0.5529	0.9754	3	0.291 (1.00)	0.4404	0.9335	3	0.688 (1.00)
ROC location weights	0.5516	**0.9731**	1	−1.744 (1.00)	0.4308	**0.9132**	1	−1.337 (1.00)
Equal weights	0.5532	0.9758	4	0.679 (0.99)	0.4407	0.9341	4	0.749 (1.00)
Location weights	0.5526	0.9748	2	−0.219 (1.00)	0.4320	0.9158	2	−1.083 (1.00)
				RUSSELL 2000				
	MSE	**Ratio**	**rk**	**MCS**	**QLIKE**	**Ratio**	**rk**	**MCS**
Expanding window	0.3193	1.0000	6	1.964 (0.00)	0.1898	1.0000	6	1.706 (0.00)
MSFE weights	0.3132	0.9811	5	1.498 (0.00)	0.1868	0.9842	5	1.348 (0.00)
ROC weights	0.3113	0.9750	4	1.183 (0.19)	0.1854	0.9769	4	1.385 (0.00)
ROC location weights	0.3086	0.9665	2	−0.735 (1.00)	0.1824	0.9610	2	0.995 (0.20)
Equal weights	0.3107	0.9731	4	0.760 (1.00)	0.1849	0.9741	3	1.385 (0.00)
Location weights	0.3079	**0.9644**	1	−1.204 (1.00)	0.1818	**0.9579**	1	−0.995 (1.00)
				NASDAQ 100				
	MSE	**Ratio**	**rk**	**MCS**	**QLIKE**	**Ratio**	**rk**	**MCS**
Expanding window	0.3299	1.0000	6	2.078 (0.00)	0.2223	1.0000	6	1.992 (0.00)
MSFE weights	0.3207	0.9719	4	0.825 (1.00)	0.2122	0.9545	4	0.936 (1.00)
ROC weights	0.3213	0.9738	5	1.212 (1.00)	0.2125	0.9560	5	1.073 (1.00)
ROC location weights	0.3192	0.9675	3	−0.092 (1.00)	0.2088	0.9394	2	−0.429 (1.00)
Equal weights	0.3189	0.9665	2	−0.298 (1.00)	0.2101	0.9451	3	0.085 (1.00)
Location weights	0.3167	**0.9599**	1	−1.643 (1.00)	0.2058	**0.9257**	1	−1.651 (1.00)

Note: For both loss functions (MSE and QLIKE), the entries are: the values of the average loss; the ratio of the average loss to that of the expanding window method; the rank (rk) according to the average loss function; the statistic and the p-value of the Model Confidence Set (MCS) procedure with $\alpha = 0.10$. A bold entry denotes the value of the smallest average loss.

Finally, in the case of the AHAR model, in line with the previous results, the expanding window appears to offer the worst forecasting performance overall.

Moreover, for both MSE and QLIKE loss functions, the method that offers the major improvement in forecasting accuracy is the forecast combination with ROC location weights for S&P 500 and Dow Jones Industrial Average Index and the the forecast combination with location weights for Russell 2000 and Nasdaq 100.

Focusing on the model confidence set, for the MSE loss function, the expanding window is always eliminated for all the series together with the forecast combination with MSFE weights for Russell 2000. With respect to the QLIKE loss function, for Dow Jones Industrial Average Index and Nasdaq 100, the only excluded method is the expanding window procedure. For the other series, the results confirm the better discriminative property of the QLIKE metric; the only surviving methods are forecast combination with ROC location weights and with location weights.

In conclusion, even if it is not clear which combination has the best forecasting performance, the forecast combination with ROC location weights and that with location weights seems to be always among the best methods. However, the forecast combination with ROC location weights always outperforms the expanding window method, and it is always in the top position with respect to the loss function ratio and is never excluded by the model confidence set.

6. Concluding Remarks

This paper has explored the relevance of taking into account the presence of structural breaks in forecasting realized volatility. The analysis has been based on 5-min realized volatility of four U.S. stock market indices: S&P 500, Dow Jones Industrial Average, Russell 2000 and Nasdaq 100. Three different model specifications of the log-realized volatility have been considered. For all the considered market indices, the instability in the parameters of the RV models has been verified through the recursive estimates test. In order to handle this problem, five forecast combinations, based on different estimation windows with alternative weighting schemes, have been introduced and compared with the expanding window method, a natural choice when the data are generated by a stable model. The forecasting performance has been evaluated, for each RV model specification, through two of the most relevant loss functions, the MSE and the QLIKE. Moreover, to this aim, the average loss function has been calculated, and in order to statistically assess if the differences in the performance are relevant, the model confidence set approach has been considered.

The analysis, repeated for each class of RV models separately, has highlighted the importance of taking into account structural breaks; in fact, the expanding window appears to offer the worst forecasting performance overall. In particular, in almost all the considered cases, the two combinations that make adjustments for accounting for the most recent possible break point (the forecast combination with location weights and with ROC location weights) are placed in first position and, as a consequence, have better forecasting performance. Nevertheless, the forecast combination with ROC location weights always outperforms the expanding window method; it is always in the top position with respect to the loss function ratio, and it is never excluded by the model confidence set.

Conflicts of Interest: The author declares no conflict of interest.

References

1. Liu, C.; Maheu, J.M. Are there structural breaks in Realized Volatility? *J. Financ. Econom.* **2008**, *6*, 326–360.
2. Kumar, D. Realized volatility transmission from crude oil to equity sectors: A study with economic significance analysis. *Int. Rev. Econ. Financ.* **2017**, *49*, 149–167.
3. Gong, X.; Lin, B. Structural breaks and volatility forecasting in the copper futures market. *J. Future Mark.* **2017**, doi:10.1002/fut.21867.
4. Ma, F.; Wahab, M.I.M.; Huang, D.; Xu, W. Forecasting the realized volatility of the oil futures market: A regime switching approach. *Energy Econ.* **2017**, *67*, 136–145.
5. Wang, Y.; Pan, Z.; Wu, C. Time-Varying Parameter Realized Volatility Models. *J. Forecast.* **2017**, *36*, 566–580.
6. Corsi, F. A Simple Long Memory Model of Realized Volatility. *Woking Paper*, University of Southern Switzerland, Manno, Switzerland, 2004.
7. Corsi, F.; Renó, R. *HAR Volatility Modelling with Heterogeneous Leverage and Jumps*; University of Siena: Siena, Italy, 2010.
8. Corsi, F. A Simple Approximate Long-Memory Model of Realized Volatility. *J. Financ. Econom.* **2009**, *7*, 174–196.
9. Sevi, B. Forecasting the volatility of crude oil futures using intraday data. *Eur. J. Oper. Res.* **2014**, *235*, 643–659.
10. Prokopczuk, M.; Symeonidis, L.; Wese Simen, C. Do jumps matter for volatility forecasting? evidence from energy markets. *J. Future Mark.* **2016**, *36*, 758–792.

11. Ploberger, W.; Krämer, W.; Kontrus, K. A new test for structural stability in the linear regression model. *J. Econom.* **1989**, *40*, 307–318.
12. Pesaran, M.H.; Timmermann, A. Selection of estimation windows in the presence of breaks. *J. Econom.* **2007**, *137*, 134–161.
13. Patton, A.J. Volatility forecast comparison using imperfect volatility proxies. *J. Econom.* **2011**, *160*, 246–256.
14. Hansen, P.R.; Lunde, A.; Nason, J.M. The model confidence set. *Econometrica* **2011**, *79*, 453–497.
15. Liu, L.Y.; Patton, A.J.; Sheppard, K. Does anything beat 5-min RV? A comparison of realized measures across multiple asset classes. *J. Econom.* **2011**, *187*, 293–311.
16. Andreou, E.; Ghysels, E.; Kourouyiannis, C. Robust Volatility Forecasts in the Presence of Structural Breaks. *Woking Paper*, University of Cyprus, Nicosia, Cyprus, 2012.
17. Andersen, T.G.; Bollerslev, T.; Diebold, F.X. Roughing It Up: Including Jump Components in the Measurement, Modeling anf Forecasting of Returns Volatility. *Rev. Econ. Stat.* **2007**, *89*, 701–720.
18. Asai, M.; McAleer, M.; Medeiros, M.C. Asymmetry and long memory in volatility modeling. *J. Financ. Econom.* **2012**, *10*, 495–512.
19. Chow, G.C. Tests of Equality Between Sets of Coefficients in Two Linear Regressions. *Econometrica* **1960**, *28*, 591–605.
20. Andrews, D.W.K.; Fair, R.C. Inference in Nonlinear Econometric Models with Structural Change. *Rev. Econ. Stud.* **1988**, *55*, 615–639.
21. Hansen, B.E. The new econometrics of structural change: Dating breaks in US labor productivity. *J. Econ. Perspect.* **2001**, *15*, 117–128.
22. Perron, P. Dealing with structural breaks. In *Palgrave Handbook of Econometrics, Vol. 1: Econometric Theory*; Patterson, K., Mills, T.C., Eds.; Palgrave Macmillan: Basingstote, UK, 2006; pp. 278–352.
23. Bai, J.; Perron, P. Estimating and testing linear models with multiple structural changes. *Econometrica* **1998**, *66*, 47–78.
24. Bai, J.; Perron, P. Computation and analysis of multiple structural change models. *J. Appl. Econom.* **2003**, *18*, 1–22.
25. Bai, J.; Perron, P. Critical values for multiple structural change tests. *Econom. J.* **2003**, *6*, 72–78.
26. Zeileis, A.; Kleiber, C.; Kramer, W.; Hornik, K. Testing and dating of structural changes in practice. *Comput. Stat. Data Anal.* **2003**, *44*, 109–123.
27. Kuan, C.M.; Hornik, K. The generalized fluctuation test: A unifying view. *Econom. Rev.* **1995**, *14*, 135–161.
28. Brown, R.L.; Durbin, J.; Evans, J.M. Techniques for Testing the Constancy of Regression Relationships over Time. *J. R. Stat. Soc.* **1975**, *37*, 149–163.
29. Ploberger, W.; Krämer, W. The CUSUM Test With OLS Residuals. *Econometrica* **1992**, *60*, 271–285.
30. Chu, C.S.J.; Hornik, K.; Kuan, C.M. MOSUM Tests for Parameter Constancy. *Biometrika* **1995**, *82*, 603–617.
31. Chu, C.S.J.; Hornik, K.; Kuan, C.M. The Moving-Estimates Test for Parameter Stability. *Econom. Theory* **1995**, *11*, 669–720.
32. Billingsley, P. *Convergence of Probability Measures*, 2nd ed.; John Wiley & Sons: New York, NY, USA, 2013.
33. Clements, M.P.; Hendry, D.F. Forecasting with breaks. In *Handbook of Economic Forecasting*; Elliot, G., Granger, C.W.J., Timmermann, A., Eds.; Elsevier: Amsterdam, The Netherlands **2006**, 605–657.
34. Pesaran, M.H.; Pick, A.; Pranovich, M. Optimal forecasts in the presence of structural breaks. *J. Econom.* **2013**, *177*, 134–152.
35. Giraitis, L.; Kapetanios, G.; Price, S. Adaptive forecasting in the presence of recent and ongoing structural change. *J. Econom.* **2013**, *177*, 153–170.
36. Inoue, A.; Jin, L.; Rossi, B. Rolling window selection for out-of-sample forecasting with time-varying parameters. *J. Econom.* **2017**, *196*, 55–67.
37. Timmermann, A. Forecast combinations. In *Handbook of Economic Forecasting*; Elliot, G., Granger, C.W.J., Timmermann, A., Eds.; Elsevier: Amsterdam, The Netherlands, 2006; pp. 135–196.
38. Tian, J.; Anderson, H.M. Forecast Combinations under Structural Break Uncertainty. *Int. J. Forecast.* **2014**, *30*, 161–175.
39. Assenmacher-Wesche, K.; Pesaran, M.H. Forecasting the Swiss economy using VECX models: An exercise in forecast combination across models and observation windows. *Natl. Inst. Econ. Rev.* **2008**, *203*, 91–108.
40. Pesaran, M.H.; Schuermann, T.; Smith, L.V. Forecasting economic and financial variables with global VARs. *Int. J. Forecast.* **2009**, *25*, 642–675.

41. Schrimpf, A.; Wang, Q. A reappraisal of the leading indicator properties of the yield curve under structural instability. *Int. J. Forecast.* **2010**, *26*, 836–857.
42. Pesaran, M.H.; Pick, A. Forecast combination across estimation windows. *J. Bus. Econ. Stat.* **2011**, *29*, 307–318.
43. Clark, T.E.; McCracken, M.W. Improving forecast accuracy by combining recursive and rolling forecasts. *Int. Econ. Rev.* **2009**, *50*, 363–395.
44. Pesaran, M.H.; Pettenuzzo, D.; Timmermann, A. Forecasting time series subject to multiple structural breaks. *Rev. Econ. Stud.* **2006**, *73*, 1057–1084.
45. Maheu, J.M.; Gordon, S. Learning, forecasting and structural breaks. *J. Appl. Econom.* **2008**, *23*, 553–583.
46. Pesaran, M.H.; Timmermann, A. Market Timing and Return Prediction under Model Instability. *J. Empir. Financ.* **2002**, *9*, 495–510.

mathematics

MDPI

Article

Stochastic Comparisons and Dynamic Information of Random Lifetimes in a Replacement Model

Antonio Di Crescenzo and Patrizia Di Gironimo *

Dipartimento di Matematica, Università degli Studi di Salerno, Via Giovanni Paolo II n. 132, 84084 Fisciano (SA), Italy; adicrescenzo@unisa.it

* Correspondence: pdigironimo@unisa.it

Received: 27 September 2018; Accepted: 14 October 2018; Published: 16 October 2018

Abstract: We consider a suitable replacement model for random lifetimes, in which at a fixed time an item is replaced by another one having the same age but different lifetime distribution. We focus first on stochastic comparisons between the involved random lifetimes, in order to assess conditions leading to an improvement of the system. Attention is also given to the relative ratio of improvement, which is proposed as a suitable index finalized to measure the goodness of the replacement procedure. Finally, we provide various results on the dynamic differential entropy of the lifetime of the improved system.

Keywords: reliability; stochastic orders; scale family of distributions; proportional hazard rates; differential entropy

MSC: 62N05 ; 60E15 ; 94A17

1. Introduction

In reliability theory, various stochastic models have been proposed in the past in order to describe replacement policies of system components. A classical model in this area is the relevation transform, which describes the overall lifetime of a component which is replaced at its (random) failure time by another component of the same age, whose lifetime distribution is possibly different. See the paper by Krakowski [1] that introduced this topic, and the further contributions by Baxter [2], Belzunce et al. [3], Chukova et al. [4], Shanthikumar and Baxter [5], Sordo and Psarrakos [6], for instance. A similar model, named reversed relevation transform, has been considered in Di Crescenzo and Toomaj [7] in order to describe the total lifetime of an item given that it is less than an independent random inspection time. Such transforms deserve a large interest since they can be employed in restoration models of failed units, or in the determination of optimal redundancy policy in coherent systems.

In both cases, the above models involve a random replacement (or inspection) time. In this paper, we aim to consider a different stochastic model dealing with replacement occurring at deterministic arbitrary instants. Specifically, we assume that an item having random lifetime X is planned to be replaced at time t by another item having the same age but possibly different random lifetime Y. The main aim is to investigate the effect of the replacement, with emphasis on criteria leading to attain better performance for the overall system.

The tools adopted in our investigation are based on stochastic orders and other typical notions of reliability theory. Specifically, we study the consequence of suitable assumptions by which the initial lifetime X is smaller than the lifetime Y of the replacing item according to some stochastic criteria. We also propose a suitable index finalized to assess the effective improvement gained by the system due to the replacement. In addition, we aim to propose the residual differential entropy as a dynamic measure of the information content of the replacement model.

This is the plan of the paper: in Section 2, we introduce the stochastic model and the main quantities of interest. Section 3 is devoted to establish some stochastic comparisons concerning the proposed model. We also deal with the case when the relevant random variables belong to the same scale family of distributions. In Section 4, we introduce the relative ratio of improvement for the considered model, and investigate its behavior in some examples. Section 5 is centered on results on two dynamic versions of the differential entropy for the proposed model, with reference to the residual and past entropy. Finally, some concluding remarks are given in Section 6.

Throughout the paper, as usual, we denote by $[X|B]$ a random variable having the same distribution of X conditional on B. The expectation of X is denoted by $\mathbb{E}[X]$. Moreover, $\mathbf{1}_A$ is the indicator function of A, i.e., $\mathbf{1}_A = 1$ if A is true, and $\mathbf{1}_A = 0$, otherwise. Furthermore, $a \stackrel{\text{sgn}}{=} b$ means that a and b have the same sign, and $\stackrel{\text{d}}{=}$ means equality in distribution.

2. The Stochastic Model

Let X be an absolutely continuous nonnegative random variable with cumulative distribution function (CDF) $F(t) = \mathbb{P}(X \leq t)$, probability density function (PDF) $f(t)$, and survival function $\overline{F}(t) = 1 - F(t)$. Bearing in mind possible applications to reliability theory and survival analysis, we assume that X describes the random lifetime of an item or a living organism. Let us now recall two functions of interest; as usual, we denote by

$$\lambda_X(t) = -\frac{\mathrm{d}}{\mathrm{d}t}\log\overline{F}(t) = \frac{f(t)}{\overline{F}(t)}, \qquad t \in \mathbb{R}_+, \ \overline{F}(t) > 0 \tag{1}$$

the hazard rate (or failure rate) of X, and by

$$\tau_X(t) = \frac{\mathrm{d}}{\mathrm{d}t}\log F(t) = \frac{f(t)}{F(t)}, \qquad t \in \mathbb{R}_+, \ F(t) > 0 \tag{2}$$

the reversed hazard rate function of X. See Barlow and Proschan [8] and Block et al. [9] for some illustrative results on these notions. Denote by Y another absolutely continuous nonnegative random variable with CDF $G(t)$, PDF $g(t)$, survival function $\overline{G}(t)$, hazard rate $\lambda_Y(t)$ and reversed hazard rate $\tau_Y(t)$.

We assume that X and Y are independent lifetimes of systems or items, both starting to work at time 0. A replacement of the first item by the second one (having the same age) is planned to occur at time t, provided that the first item is not failed before. Let us now define $I_t = \mathbf{1}_{\{0 \leq X \leq t\}}$, so that I_t is a Bernoulli random variable with parameter $P(I_t = 1) = F(t)$. Hence, denoting by X_t^Y the random duration of the (eventually replaced) system, it can be expressed as follows:

$$X_t^Y = [X|X \leq t]I_t + [Y|Y > t](1 - I_t) = \begin{cases} [X|X \leq t] & \text{if } 0 \leq X \leq t \\ [Y|Y > t] & \text{if } X > t \end{cases} \qquad (t \in \mathbb{R}_+). \tag{3}$$

In classical minimal repair models, an item, upon failure, is replaced by another item having the same failure distribution, and the same age of the previous item at the failure time. The present model also presumes that the item is replaced by another one having the same age at the failure time. The difference is that the replacement occurs at a preassigned deterministic time t, and that the new item possesses a possibly different failure distribution.

By (3), for any Borel set B, the following mixture holds:

$$\mathbb{P}(X_t^Y \in B) = \mathbb{P}(X \in B|X \leq t)\,F(t) + \mathbb{P}(Y \in B|Y > t)\,\overline{F}(t), \qquad t \in \mathbb{R}_+. \tag{4}$$

Then, the CDF and the PDF of X_t^Y are respectively

$$F_t^Y(x) = \mathbb{P}(X_t^Y \le x) = \begin{cases} F(x), & \text{if } 0 \le x \le t, \\ F(t) + \frac{\overline{F}(t)}{\overline{G}(t)}[G(x) - G(t)], & \text{if } x > t, \end{cases} \tag{5}$$

and

$$f_t^Y(x) = \frac{\mathrm{d}}{\mathrm{d}x} F_t^Y(x) = \begin{cases} f(x), & \text{if } 0 \le x \le t \\ \frac{\overline{F}(t)}{\overline{G}(t)} g(x), & \text{if } x > t, \end{cases} \tag{6}$$

so that the survival function of X_t^Y can be expressed as

$$\overline{F}_t^Y(x) = \begin{cases} \overline{F}(x), & \text{if } 0 \le x \le t, \\ \overline{F}(t) + \frac{\overline{F}(t)}{\overline{G}(t)}[\overline{G}(x) - \overline{G}(t)], & \text{if } x > t. \end{cases} \tag{7}$$

For instance, the special case when Y is uniformly distributed on $[0, 1]$ is treated in Example 4.18 of Santacroce et al. [10] concerning the analysis of some exponential models.

It should be pointed out that the replacement model given in (3) can be easily extended to integer-valued random variables. In this case, the model when Y is uniformly distributed on a set of integers is of interest in information theory (see, for instance, the operator considered in Equation (3) of Cicalese et al. [11]).

A relevant issue about model (3) is the following: to assess if the replacement planned at time t is beneficial to the system. This can be attained in several ways. As a first step, hereafter, we face the problem of establishing if the duration of the replaced system is larger than that of the originating lifetime (or smaller than that of the replaced lifetime) in some stochastic sense. To this aim, in the following section, we provide various useful comparisons based on stochastic orders.

3. Stochastic Comparisons

3.1. Definitions and Main Comparisons

In order to compare X and Y with X_t^Y, let us now recall some well-known definitions of partial stochastic orders, which involve the notions treated in Section 2. As a reference, see Shaked and Shanthikumar [12] or Belzunce et al. [13].

Definition 1. *Let X be an absolutely continuous random variable with support (l_X, u_X), CDF F, and PDF f. Similarly, let Y be an absolutely continuous random variable with support (l_Y, u_Y), CDF G, and PDF g. We say that X is smaller than Y in the*

(a) usual stochastic order ($X \le_{\mathrm{st}} Y$) if $\overline{F}(t) \le \overline{G}(t)\ \forall\, t \in \mathbb{R}$ or, equivalently, if $F(t) \ge G(t)\ \ \forall\, t \in \mathbb{R}$;

(b) hazard rate order ($X \le_{\mathrm{hr}} Y$) if $\overline{G}(t)/\overline{F}(t)$ increases in $t \in (-\infty, \max(u_X, u_Y))$ or, equivalently, if $\lambda_X(t) \ge \lambda_Y(t)$ for all $t \in \mathbb{R}$, where $\lambda_X(t) = f(t)/\overline{F}(t)$ and $\lambda_Y(t) = g(t)/\overline{G}(t)$ are respectively the hazard rates of X and Y, or equivalently if $f(x)\overline{G}(y) \ge g(x)\overline{F}(y)\ \ \forall x \le y$;

(c) likelihood ratio order ($X \le_{\mathrm{lr}} Y$) if $f(x)g(y) \ge f(y)g(x)$ for all $x \le y$, with $x, y \in (l_X, u_X) \cup (l_Y, u_Y)$ or, equivalently, $g(t)/f(t)$ increases in t over the union of supports of X and Y;

(d) reversed hazard rate order ($X \le_{\mathrm{rh}} Y$) if $G(t)/F(t)$ increases in $t \in (\min(l_X, l_Y), +\infty)$ or, equivalently, if $\tau_X(t) \le \tau_Y(t)$ for all $t \in \mathbb{R}$, where $\tau_X(t) = f(t)/F(t)$ and $\tau_Y(t) = g(t)/G(t)$ are respectively the reversed hazard rates of X and Y.

We recall the following relations among the above-mentioned stochastic orders:

$$X \le_{\mathrm{lr}} Y \Rightarrow X \le_{\mathrm{hr}} Y \Rightarrow X \le_{\mathrm{st}} Y, \qquad X \le_{\mathrm{lr}} Y \Rightarrow X \le_{\mathrm{rh}} Y \Rightarrow X \le_{\mathrm{st}} Y. \tag{8}$$

With reference to (3), we now investigate the effect of the replacement when the lifetime of the first item is stochastically smaller than the second in the sense of the criteria given in Definition 1.

Theorem 1. *Let X and Y be absolutely continuous nonnegative random variables. Then,*

(i) $X \leq_{\rm hr} Y \Rightarrow X \leq_{\rm hr} X_t^Y \leq_{\rm hr} Y \ \forall t > 0$;
(ii) $X \leq_{\rm lr} Y \Rightarrow X_t^Y \leq_{\rm lr} Y \ \forall t > 0$;
(iii) $X \leq_{\rm rh} Y \Rightarrow X_t^Y \leq_{\rm rh} Y \ \forall t > 0$;
(iv) $X \leq_{\rm st} Y \Leftrightarrow X_t^Y \leq_{\rm st} Y \ \forall t > 0$.

Proof. From (6) and (7), the hazard rate of X_t^Y is given by

$$\lambda_t^Y(x) = \frac{f_t^Y(x)}{\overline{F}_t^Y(x)} = \begin{cases} \lambda_X(x), & \text{if} \quad 0 \leq x \leq t, \\ \lambda_Y(x), & \text{if} \quad x > t. \end{cases} \tag{9}$$

We observe that, if $X \leq_{\rm hr} Y$, from (9), we immediately deduce $\lambda_X(x) \geq \lambda_t^Y(x) \geq \lambda_Y(x)$ for all $x \geq 0$, and so we obtain $X \leq_{\rm hr} X_t^Y \leq_{\rm hr} Y \ \forall t > 0$.

Now, by taking into account Equation (6), we get

$$\frac{g(x)}{f_t^Y(x)} = \begin{cases} \frac{g(x)}{f(x)}, & \text{if } 0 \leq x \leq t, \\ \frac{\overline{G}(t)}{\overline{F}(t)}, & \text{if } x > t. \end{cases}$$

Hence, since assumption $X \leq_{\rm lr} Y$ implies that $g(x)/f(x)$ is increasing in $x > 0$, and that $\frac{g(t)}{f(t)} \leq \frac{\overline{G}(t)}{\overline{F}(t)}$ for all $t > 0$ by the first of (8), we finally obtain $X_t^Y \leq_{\rm lr} Y$, this completing the proof of (ii).

Note that

$$\frac{F_t^Y(x)}{G(x)} = \begin{cases} \frac{F(x)}{G(x)}, & \text{if } 0 \leq x \leq t, \\ \frac{F(t)}{G(x)} + \frac{\overline{F}(t)}{\overline{G}(t)} \frac{[G(x)-G(t)]}{G(x)}, & \text{if } x > t. \end{cases}$$

Hence, for $0 \leq x \leq t$, we have that $F_t^Y(x)/G(x)$ is decreasing in x if and only if $X \leq_{\rm rh} Y$. Moreover $F_t^Y(x)/G(x)$ is continuous in $x = t$. Finally, it is not hard to see that the derivative of $F_t^Y(x)/G(x)$ is nonpositive if $G(t) \leq F(t)$ for all $t \geq 0$, i.e., $X \leq_{\rm st} Y$, this being ensured by assumption $X \leq_{\rm rh} Y$. The proof of (iii) is thus completed.

The proof of (iv) can be easily checked from (5), by seeing that $F_t^Y(x) \geq G(x)$ for all $x \geq 0$ and $t \geq 0$, if and only if assumption $X \leq_{\rm st} Y$ holds. □

Differently from case (i) of Theorem 1, condition $X \leq_{\rm lr} Y$ does not imply that $X \leq_{\rm lr} X_t^Y \ \forall t > 0$. This can be easily checked, for instance, when X and Y are exponentially distributed with rates λ_X and λ_Y, with $\lambda_X > \lambda_Y$. In this case, one has $X \leq_{\rm lr} Y$, whereas, recalling (6), the ratio $f_t^Y(x)/f(x)$ is not increasing for all $x > 0$, and thus $X \leq_{\rm lr} X_t^Y$ is not true. A similar conclusion holds for the cases (ii) and (iii). Indeed, in the following counterexample, we see that

$$X \leq_{\rm st} Y \not\Rightarrow X \leq_{\rm st} X_t^Y \quad \forall t > 0,$$

$$X \leq_{\rm rh} Y \not\Rightarrow X \leq_{\rm rh} X_t^Y \quad \forall t > 0.$$

Counterexample 1. Let X be exponentially distributed with parameter 1, and let $Y = \max\{X, Z\}$, where Z is Erlang distributed with parameters $(2, 2)$ and is independent from X. Hence, since $F(x) = 1 - e^{-x}, x \geq 0$, and $G(x) = F(x)H(x)$, with $H(x) = 1 - (1 + 2x)e^{-2x}, x \geq 0$, we immediately have that $X \leq_{\rm rh} Y$, and thus $X \leq_{\rm st} Y$. However, recalling (5), it is not hard to see that $F_t^Y(x)/F(x)$ is not monotonic and is not smaller than one for suitable choices of t, as shown in Figure 1. Hence, both the conditions $X \leq_{st} X_t^Y \; \forall t > 0$ and $X \leq_{\rm rh} X_t^Y \; \forall t > 0$ are not true.

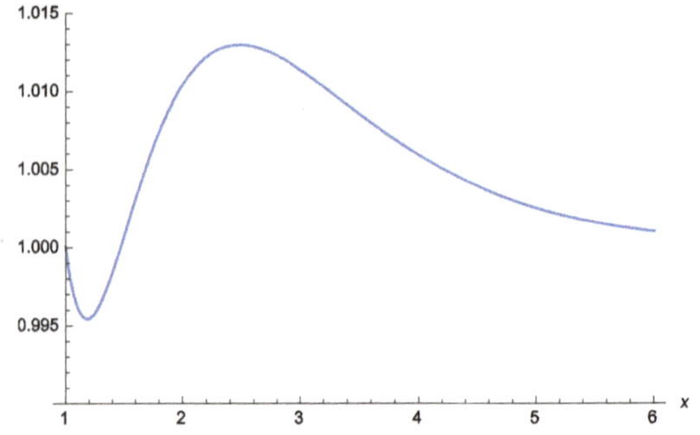

Figure 1. Plot of $F_t^Y(x)/F(x)$ for $t = 1$, with reference to Counterexample 1.

Let us now prove another result concerning stochastic orderings similar to those of Theorem 1.

Theorem 2. *Let X, Y be random lifetimes. If $X \leq_{\rm st} X_t^Y \;\; \forall t > 0$, then $X \leq_{\rm hr} Y$.*

Proof. From assumption $X \leq_{\rm st} X_t^Y \;\; \forall t > 0$, we have

$$F(x) \geq F(t) + \frac{\overline{F}(t)}{\overline{G}(t)}[G(x) - G(t)], \qquad x > t > 0,$$

so that

$$\frac{F(x) - F(t)}{x - t} \frac{1}{\overline{F}(t)} \geq \frac{G(x) - G(t)}{x - t} \frac{1}{\overline{G}(t)}.$$

In the limit as $x \downarrow t$, we have $\frac{f(t)}{\overline{F}(t)} \geq \frac{g(t)}{\overline{G}(t)}$ for all $t > 0$, thus $X \leq_{\rm hr} Y$. □

3.2. Scale Family of Distributions

Engineers in the manufacturing industries have used accelerated test experiments for many decades (see Arnold et al. [14], Escobar and Meeker [15], for instance). Various models for accelerated test involve time-transformations of suitable functions. The simplest case is based on linear transformations and on distribution functions. Then, let us now adapt the model (3) to the instance in which the distributions of X and Y belong to the same scale family.

Given the random lifetimes X and Y, having distribution functions $F(x)$ and $G(x)$ respectively, we assume that X and Y belong to the same scale family of distributions, i.e.,

$$G(x) = F(\alpha x) \quad \forall x \in \mathbb{R}, \quad 0 < \alpha < 1. \tag{10}$$

Hence, for $0 < \alpha < 1$, one has $X \leq_{\rm st} Y$. We recall that the quantile function of X is given by

$$Q_X(u) = \inf\{x \in \mathbb{R} | F(x) \geq u\}, \qquad 0 < u < 1.$$

Assumption (10) means that X and Y satisfy the proportional quantile functions model (see Section 4.1 of Di Crescenzo et al. [16]) expressed by $Q_X(u) = \alpha Q_Y(u) \;\; \forall u \in (0,1)$, where $Q_Y(u)$ is similarly defined as $Q_X(u)$. From (5), under the assumption (10) the distribution function of X_t^Y is

$$F_t^Y(x) = \begin{cases} F(x), & \text{if } 0 \leq x \leq t, \\ F(t) + \frac{\overline{F}(t)}{\overline{F}(\alpha t)}[F(\alpha x) - F(\alpha t)], & \text{if } x > t. \end{cases}$$

Among the quantities of interest in reliability theory, wide attention is devoted to the residual lifetime of a given random lifetime X, defined as

$$X_t := [X - t | X > t], \qquad t \geq 0. \tag{11}$$

The residual lifetime defined in (11) is involved in the following well-known notion of positive ageing.

Definition 2. *We say that X is IFR (increasing failure rate) if $X_t \leq_{\rm st} X_s$ for all $t \geq s \geq 0$, that is, if $\overline{F}(x)$ is logconcave, or equivalently the failure rate $\lambda_X(t)$ is increasing in $t \geq 0$.*

Remark 1. *If X and Y satisfy condition (10) and if X is IFR, then, for $0 < \alpha < 1$,*

$$\lambda_Y(t) = \alpha \lambda_X(\alpha t) < \lambda_X(\alpha t) \leq \lambda_X(t), \qquad t \geq 0,$$

so that $X \leq_{\rm hr} Y$, and hence $X \leq_{\rm hr} X_t^Y \leq_{\rm hr} Y$ due to point (i) of Theorem 1.

3.3. Further Results

From (7), the expected value of X_t^Y can be expressed as

$$\begin{aligned}\mathbb{E}[X_t^Y] &= \int_0^t \overline{F}(x)dx + \int_t^\infty \left[\overline{F}(t) + \frac{\overline{F}(t)}{\overline{G}(t)}[\overline{G}(x) - \overline{G}(t)]\right] \mathrm{d}x \\ &= \mathbb{E}[X] + \frac{1}{\overline{G}(t)} \int_t^\infty [\overline{F}(t)\overline{G}(x) - \overline{F}(x)\overline{G}(t)] \mathrm{d}x.\end{aligned}$$

Hence, recalling that the mean residual life of the random lifetime X is

$$m_X(t) = \mathbb{E}[X - t | X > t] = \frac{1}{\overline{F}(t)} \int_t^\infty \overline{F}(x)\mathrm{d}x, \qquad t \in \mathbb{R}_+, \ \overline{F}(t) > 0,$$

with $m_Y(t)$ similarly defined, we have

$$\mathbb{E}[X_t^Y] = \mathbb{E}[X] + \overline{F}(t)[m_Y(t) - m_X(t)]. \tag{12}$$

Let us now recall the mean residual life order (see Section 2.A of Shaked and Shanthikumar [12]).

Definition 3. *Let X and Y be absolutely continuous random variables with CDFs $F(t)$ and $G(t)$, and with finite mean residual lives $m_X(t)$ and $m_Y(t)$, respectively. We say that X is smaller than Y in the mean residual life order ($X \leq_{\rm mrl} Y$) if $m_X(t) \leq m_Y(t)$ for all t or, equivalently, if*

$$\frac{\int_t^\infty \overline{G}(x)\mathrm{d}x}{\int_t^\infty \overline{F}(x)\mathrm{d}x} \qquad \textit{is decreasing over } \{t : \int_t^\infty \overline{F}(x)\mathrm{d}x > 0\}.$$

Consequently, recalling (12), we immediately have the forthcoming result.

Proposition 1. *The relation $\mathbb{E}[X] \leq \mathbb{E}[X_t^Y]$ holds for all t if and only if $X \leq_{\rm mrl} Y$.*

We can now come to a probabilistic analogue of the mean value theorem.

Theorem 3. *Let X and Y be non-negative random variables satisfying $X \leq_{hr} Y$ and $\mathbb{E}[X] < \mathbb{E}[X_t^Y] < \infty$ and let $Z_t = \Psi(X, Y)$. Let g be a measurable and differentiable function such that $\mathbb{E}[g(X)]$ and $\mathbb{E}[g(Y)]$ are finite,* *and let its derivative g' be measurable and Riemann-integrable on the interval $[x, y]$ for all $y \geq x \geq 0$. Then, $\mathbb{E}[g'(Z)]$ is finite and*

$$\mathbb{E}[g(X_t^Y)] - \mathbb{E}[g(X)] = \mathbb{E}[g'(Z_t)]\{\mathbb{E}[X_t^Y] - \mathbb{E}[X]\}, \tag{13}$$

where Z_t is the absolutely continuous random variable having PDF

$$f_{Z_t}(x) = \frac{\overline{F}_t^Y(x) - \overline{F}(x)}{\mathbb{E}[X_t^Y] - \mathbb{E}[X]} = \frac{1}{m_Y(t) - m_X(t)} \left[\frac{\overline{G}(x)}{\overline{G}(t)} - \frac{\overline{F}(x)}{\overline{F}(t)} \right], \qquad x \geq t.$$

Proof. The proof follows from the Theorem 4.1 of Di Crescenzo [17]. □

It is interesting to point out that the relation (13) can be used in various applied contexts. For instance, if g is an utility function, then $\mathbb{E}[g(X)]$ can be viewed as the expected utility granted by an item having lifetime X. Accordingly, Equation (13) expresses the variation of the expected utility when such an item is subject to the replacement procedure described in Section 2. Clearly, it can be used to construct useful measures able to evaluate the goodness of the procedure. This specific task is not undertaken here, whereas in the following section we propose a different approach to assess the effectiveness of the replacement.

4. Relative Ratio of Improvement

Consider a system having random lifetime X, which is replaced by Y at time t. If X is smaller than Y according to some stochastic order, we expect that the reliability of the system at time $x > t > 0$ is improved. In order to measure the usefulness of replacing the lifetime X with Y at time t, let us now introduce the *relative ratio of improvement* evaluated at $x > t > 0$. It is defined in terms of (7) as

$$R_t(x) := \frac{\overline{F}_t^Y(x) - \overline{F}(x)}{\overline{F}(x)} = \frac{\overline{F}(t)}{\overline{F}(x)} - 1 + \frac{\overline{F}(t)}{\overline{G}(t)} \left[\frac{\overline{G}(x)}{\overline{F}(x)} - \frac{\overline{G}(t)}{\overline{F}(x)} \right]. \tag{14}$$

Clearly, if $X \leq_{\rm hr} Y$, then, from point (i) of Theorem 1, it follows that $X \leq_{\rm hr} X_t^Y$ and, in turn, $X \leq_{\rm st} X_t^Y$ so that $R_t(x) \geq 0$ for all $x > t > 0$.

Example 1. Let $\{Z(t), t \geq 0\}$ be an iterated Poisson process with parameters (μ, λ). In other terms, such process can be expressed as $Z(t) = M[N(t)]$, where $M(t)$ and $N(t)$ are independent Poisson processes with parameters $\mu, \lambda \in \mathbb{R}^+$, respectively (see Section 6 of Di Crescenzo et al. [18]). Denoting by

$$T_k = \inf\{t > 0 : Z(t) \geq k\} \tag{15}$$

the first crossing time (from below) of $Z(t)$ through the constant level $k \in \mathbb{N}$, the corresponding survival function is (cf. Section 7 of [18])

$$\mathbb{P}[T_k > t] = \exp\{-\lambda(1 - \mathrm{e}^{-\mu})t\} \sum_{j=0}^{k-1} \frac{\mu^j}{j!} B_j(\lambda \mathrm{e}^{-\mu} t), \qquad t \geq 0, \tag{16}$$

where $B_j(\cdot)$ is the j-th Bell polynomial. We consider a system subject to replacement policy as described in (3), where the relevant random lifetimes are given by the first-crossing times defined in (15), with $X \stackrel{\rm d}{=} T_1$ and $Y \stackrel{\rm d}{=} T_k$. The relative ratio of improvement of this system is then evaluated by means of (14). Figure 2 provides some plots of $R_t(x)$, showing that the relative ratio of improvement is increasing in x and k.

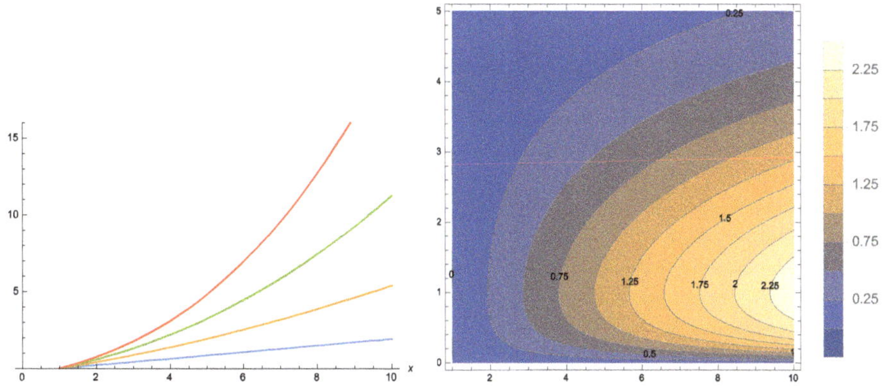

Figure 2. With reference to Example 1, **left**: plot of $R_t(x)$ for $1 < x < 10$, with $t = 1$ and $k = 2, 3, 4, 5$ (from bottom to top); **right**: contour plot of $R_t(x)$ for $1 < x < 10$ and $0 < \mu < 5$, with $k = 2$, $t = 1$ and $\lambda = 1$.

In the remaining part of this section, we restrict our attention to the special case in which X and Y satisfy the proportional hazard rates model (see Cox [19] or, for instance, the more recent contributions by Nanda and Das [20], and Ng et al. [21]). Hence, assuming that $\overline{G}(t) = [\overline{F}(t)]^{\theta}$, $\forall t \geq 0$, for $\theta > 0$, $\theta \neq 1$, the relative ratio defined in (14) becomes

$$R_t(x) = \frac{\overline{F}(t)}{\overline{F}(x)} - 1 + [\overline{F}(t)]^{1-\theta} \left\{ [\overline{F}(x)]^{\theta-1} - \frac{[\overline{F}(t)]^{\theta}}{\overline{F}(x)} \right\}, \qquad x > t > 0. \tag{17}$$

Here, the most interesting case is for $0 < \theta < 1$, since this assumption ensures that $X \leq_{\rm hr} Y$.

Example 2. Let X and Y be exponentially distributed with parameters 1 and θ, respectively, with $0 < \theta < 1$. Since $\overline{F}(t) = \mathrm{e}^{-t}$, $\overline{G}(t) = \mathrm{e}^{-\theta t}$, $t \geq 0$, from (17), we have

$$R_t(x) = \mathrm{e}^{-(t-x)(1-\theta)} - 1, \qquad x > t > 0.$$

Some plots of $R_t(x)$ are given in Figure 3, confirming that the relative ratio of improvement is increasing in $x - t > 0$, and is decreasing in $\theta \in (0, 1)$.

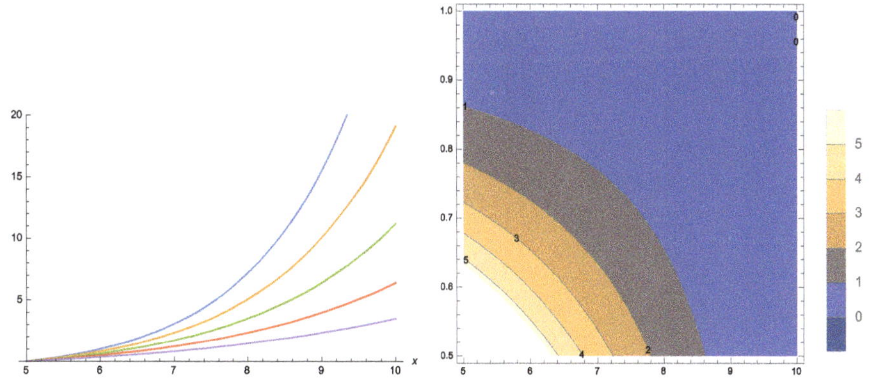

Figure 3. With reference to Example 2, **left**: plot of $R_t(x)$ for $5 < x < 10$, with $t = 5$ and $\theta = 0.3, 0.4, 0.5, 0.6, 0.7$ (from top to bottom); **right**: contour plot of $R_t(x)$ for $0.5 < \theta < 1$ and $5 < t < 10$, with $x = 10$.

5. Results on Dynamic Differential Entropies

In this section, we investigate some informational properties of the replacement model considered in Section 2.

A classical measure of uncertainty for an absolutely continuous random variable X is the differential entropy, defined as

$$H_X = -\int_{\mathbb{R}} f(x) \log f(x) \mathrm{d}x, \tag{18}$$

with $0 \log 0 = 0$ by convention. This measure has some analogies with the entropy of discrete random variables, even though the differential entropy lacks a number of properties that the Shannon discrete entropy possesses (see, for instance, Cover and Thomas [22] for details).

In the context of lifetimes truncated over intervals of the form $[0, t]$ or (t, ∞), specific forms of the differential entropy have been investigated in the recent decades (see the initial contributions by Muliere et al. [23]). Specifically, the following dynamic measure (named residual entropy of the lifetime X) has been extensively investigated:

$$H_X(t) = -\int_t^{\infty} \frac{f(x)}{\overline{F}(t)} \log \frac{f(x)}{\overline{F}(t)} \mathrm{d}x, \qquad t \in \mathbb{R}_+, \ \overline{F}(t) > 0. \tag{19}$$

This quantity is suitable for measuring the uncertainty in residual lifetimes defined as in (11). The residual entropy of X has been studied by Ebrahimi [24], Ebrahimi and Pellerey [25]; see also Asadi and Ebrahimi [26], Ebrahimi et al. [27] on this topic. A similar reasoning leads to the past entropy of X, defined as the differential entropy of the past lifetime $[X|X \le t]$, $t > 0$, i.e.,

$$\overline{H}_X(t) = -\int_0^{t} \frac{f(x)}{F(t)} \log \frac{f(x)}{F(t)} \mathrm{d}x, \qquad t \in \mathbb{R}_+, \ F(t) > 0. \tag{20}$$

This measure is also named 'past entropy' of X; it has been investigated in Di Crescenzo and Longobardi [28], Nanda and Paul [29], Kundu et al. [30]. Other results and applications of these dynamic information measures can be found in Sachlas and Papaioannou [31], Kundu and Nanda [32], and Ahmadi et al. [33].

Hereafter, we denote by

$$\mathbb{H}(t) = -F(t) \log F(t) - \overline{F}(t) \log \overline{F}(t), \qquad t > 0 \tag{21}$$

the partition entropy of X at time t (see Bowden [34]), which measures the information (in the sense of Shannon entropy) about the value of the random lifetime X derived from knowing whether $X \le t$ or $X > t$.

Under the conditions specified in Section 2, let us now provide a decomposition result for the differential entropy of (3).

Proposition 2. *For all $t > 0$, we have*

$$H_{X_t^Y} = \mathbb{H}(t) + F(t)\overline{H}_X(t) + \overline{F}(t)H_Y(t). \tag{22}$$

Proof. From (6) and (18), we have that, for $t > 0$,

$$H_{X_t^Y} = -\int_0^{\infty} f_t^Y(x) \log f_t^Y(x) \mathrm{d}x = -\int_0^t f(x) \log f(x) \mathrm{d}x - \int_t^{\infty} \frac{\overline{F}(t)}{\overline{G}(t)} g(x) \log \left[\frac{\overline{F}(t)}{\overline{G}(t)} g(x) \right] \mathrm{d}x.$$

Recalling the alternative expression of the residual entropy (19) given in (2.2) of [24], and the alternative expression of the past entropy (20) shown in (2.1) of [28], we have

$$-\int_0^t f(x)\log f(x)\mathrm{d}x = F(t)[\overline{H}_X(t) - \log F(t)],$$

and

$$-\int_t^\infty g(x)\log g(x)\mathrm{d}x = \overline{G}(t)[H_Y(t) - \log \overline{G}(t)].$$

Hence, we obtain

$$\begin{aligned} H_{X_t^Y} &= F(t)[\overline{H}_X(t) - \log F(t)] + \frac{\overline{F}(t)}{\overline{G}(t)}\left[\overline{G}(t)[H_Y(t) - \log \overline{G}(t)] - \overline{G}(t)\log\frac{\overline{F}(t)}{\overline{G}(t)}\right] \\ &= -F(t)\log F(t) - \overline{F}(t)\log \overline{F}(t) + F(t)\overline{H}_X(t) + \overline{F}(t)H_Y(t). \end{aligned}$$

This completes the proof of (22), due to (21). □

We note that Equation (22) can be interpreted as follows. The uncertainty about the failure time of an item having lifetime X_t^Y can be decomposed into three parts: (i) the uncertainty on whether the item has failed before or after time t, (ii) the uncertainty about the failure time in $(0, t)$ given that the item has failed before t, and (iii) the uncertainty about the failure time in $(t, +\infty)$ given that the item has failed after t, and thus the failure time is distributed as Y since the replacement occurred at time t.

Clearly, if X and Y are identically distributed, then Equation (22) becomes the identity given in Proposition 2.1 of [28], i.e.,

$$H_X = \mathbb{H}(t) + F(t)\overline{H}_X(t) + \overline{F}(t)H_X(t), \qquad t > 0. \tag{23}$$

The given results allow us to perform some comparisons involving the above entropies. To this aim, we recall a suitable stochastic order (see Definition 2.1 of Ebrahimi and Pellerey [25]):

Definition 4. *Let X and Y be random lifetimes; X is said to have less uncertainty than Y, and write $X \leq_{\text{LU}} Y$, if*

$$H_X(t) \leq H_Y(t) \qquad \textit{for all } t \geq 0.$$

From Proposition 2, we can now infer the following result.

Corollary 1. *If X and Y are random lifetimes such that $X \leq_{\text{LU}} Y$, then*

$$H_X \leq H_{X_t^Y} \qquad \forall t > 0.$$

Proof. By comparing Equations (22) and (23), we obtain

$$H_{X_t^Y} - H_X = \overline{F}(t)[H_Y(t) - H_X(t)], \qquad t > 0. \tag{24}$$

Thus, from Definition 4, we obtain the proof immediately. □

Let us now investigate some sufficient conditions leading to the monotonicity of the differential entropy of (3). To this aim, we recall the following notion, which was first considered in [24].

Definition 5. *Assume that the residual entropy of the random lifetime X is finite. If $H_X(t)$ is decreasing (increasing) in $t \geq 0$, we say that X has decreasing (increasing) uncertainty of residual life, i.e., X is DURL (IURL).*

Proposition 3.

(i) *Let $X \leq_{\rm LU} Y$. If X is IURL and Y is DURL, then X_t^Y is DURL.*
(ii) *Let $Y \leq_{\rm LU} X$. If X is DURL and Y is IURL, then X_t^Y is IURL.*
(iii) *If $\lambda_X(t) \leq {\rm e} \leq \lambda_Y(t)$ for all $t \geq 0$, and $H_Y(t) \geq 0$ for all $t \geq 0$, then X_t^Y is IURL.*

Proof.

(i) Under the given assumptions, we have that the right-hand-side of (24) is nonnegative and decreasing, so that $H_{X_t^Y}$ is decreasing.
(ii) Differentiating both sides of (24), we obtain:

$$\frac{\rm d}{{\rm d}t} H_{X_t^Y} = \overline{F}(t)\{\lambda_Y(t)[H_Y(t) - 1 + \log\lambda_Y(t)] - \lambda_X(t)[H_Y(t) - 1 + \log\lambda_X(t)]\}. \tag{25}$$

From Theorem 3.2 of Ebrahimi [24], we deduce that, if Y is IURL, then $H_Y(t) - 1 + \log\lambda_Y(t) \geq 0$. Moreover, since $Y \leq_{\rm LU} X$ and X is DURL, we have $H_Y(t) - 1 + \log\lambda_X(t) \leq 0$. Hence, from (25), it follows that $H_{X_t^Y}$ is increasing.
(iii) By expressing the derivative of $H_{X_t^Y}$ in an alternative form, we obtain

$$\frac{\rm d}{{\rm d}t} H_{X_t^Y} = \overline{F}(t)\{[\lambda_Y(t) - \lambda_X(t)]H_Y(t) + \lambda_Y(t)[-1 + \log\lambda_Y(t)] - \lambda_X(t)[-1 + \log\lambda_X(t)]\}.$$

Hence, thanks to the given hypothesis, the right-hand-side of the above identity is nonnegative.

□

We remark that the assumption $\lambda_X(t) \leq {\rm e} \leq \lambda_Y(t)$ for all $t \geq 0$, considered in point (iii) of Proposition 3, implies that Y is larger than X in the 'up hazard rate order', i.e., $Y \leq_{\rm hr\uparrow} X$ (see Theorem 6.21 of Lillo et al. [35]). This condition, in turn, implies that $Y \leq_{\rm hr} X$.

Classical studies in reliability theory show that the random lifetimes of items or systems follow suitable Weibull distributions. As an illustrative instance, in the forthcoming example, we investigate the effect of replacement for Weibull distributed lifetimes in terms of dynamic differential entropy.

Example 3. Assume that X and Y have Weibull distribution, with $F(t) = 1 - {\rm e}^{-(t/\lambda)^k}$, $t \geq 0$, and $G(t) = 1 - {\rm e}^{-(t/\mu)^h}$, $t \geq 0$, with $\lambda, k, \mu, h > 0$. The differential entropy of the lifetime (3) can be obtained by means of (22). However, we omit its expression since it is quite cumbersome. Some plots of the dynamic differential entropy $H_{X_t^Y}$ are given in Figure 4, in order to show the effect of the replacement at time t. Specifically, for the considered parameters, we have that $H_{X_t^Y}$ grows when μ increases and h decreases. Moreover, $H_{X_t^Y}$ has a reversed bathtub shape, with a single maximum attained for positive values of t. Clearly, such maximum provides useful information in order to chose optimal values (in the differential entropy sense) of the replacement instant.

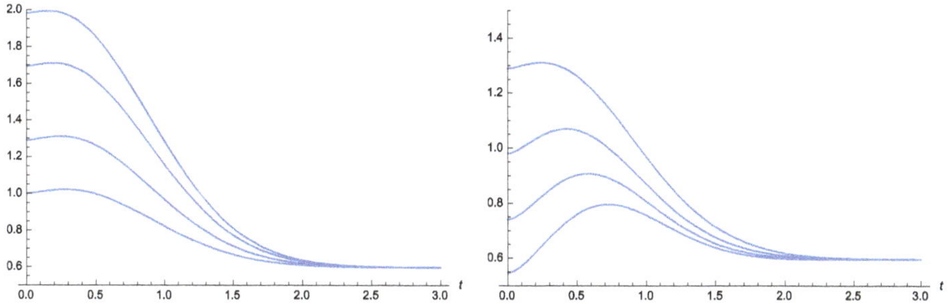

Figure 4. With reference to Example 3, plot of $H_{X_i^Y}$ for $\lambda = 1$ and $k = 2$. **Left:** $h = 2$ and $\mu = 1.5, 2, 3, 4$ (from bottom to top). **Right:** $\mu = 2$ and $h = 2, 3, 4, 5$ (from top to bottom).

6. Conclusions

Certain typical replacement models in reliability theory involve minimal repair instances, in which—upon failure—an item is replaced by another one having the same reliability of the failed item at the failure instant. The model considered in this paper deals with a different scenario, in which the replacement is planned in advance, the replaced item possessing a different failure distribution and having the same age of the replaced item.

The investigation has been centered first on the stochastic comparison of the resulting random lifetimes. We have proposed measuring the goodness of the replacement criteria by means of the relative ratio of improvement. The information amount provided by the dynamic version of the system lifetime differential entropy has also been considered as a relevant tool in this respect.

Possible future developments of the given results can be finalized to the extension of the model to more specific instances in which the replacement instant is constrained by operational guidelines, which can be implemented through suitable weight functions.

Author Contributions: All of the authors contributed equally to this work.

Funding: This research received no external funding.

Acknowledgments: A.D.C. and P.D.G. are members of the groups GNCS and GNAMPA of INdAM (Istituto Nazionale di Alta Matematica), respectively.

Conflicts of Interest: The authors declare no conflict of interest.

References

1. Krakowski, M. The relevation transform and a generalization of the gamma distribution function. *Rev. Fr. Autom. Inform. Rech. Opér.* **1973**, *7*, 107–120. [CrossRef]
2. Baxter, L.A. Reliability applications of the relevation transform. *J. Appl. Probab.* **1982**, *29*, 323–329. [CrossRef]
3. Belzunce, F.; Martínez-Riquelme, C.; Ruiz, J.M. Allocation of a relevation in redundancy problems. *Appl. Stoch. Model Bus. Ind.* **2018**, online first. [CrossRef]
4. Chukova, S.; Dimitrov, B.; Dion, J.P. On relevation transforms that characterize probability distributions. *J. Appl. Math. Stoch. Anal.* **1993**, *6*, 345–357. [CrossRef]
5. Shanthikumar, J.G.; Baxter, L.A. Closure properties of the relevation transform. *Nav. Res. Logist. Q.* **1985**, *32*, 185–189. [CrossRef]
6. Sordo, M.A.; Psarrakos, G. Stochastic comparisons of interfailure times under a relevation replacement policy. *J. Appl. Probab.* **2017**, *54*, 134–145. [CrossRef]
7. Di Crescenzo, A.; Toomaj, A. Extension of the past lifetime and its connection to the cumulative entropy. *J. Appl. Probab.* **2015**, *52*, 1156–1174. [CrossRef]
8. Barlow, R.; Proschan, F. *Mathematical Theory of Reliability;* Classics in Applied Mathematics (Book 17); With Contributions by Larry C. Hunter; SIAM: Philadelphia, PA, USA, 1996.

9. Block, H.W.; Savits, T.H.; Singh, H. The reversed hazard rate function. *Probab. Eng. Inf. Sci.* **1998**, *12*, 69–90. [CrossRef]
10. Santacroce, M.; Siri, P.; Trivellato, B. New results on mixture and exponential models by Orlicz spaces. *Bernoulli* **2016**, *22*, 1431–1447. [CrossRef]
11. Cicalese, F.; Gargano, L.; Vaccaro, U. Bounds on the entropy of a function of a random variable and their applications. *IEEE Trans. Inf. Theory* **2018**, *64*, 2220–2230. [CrossRef]
12. Shaked, M.; Shanthikumar, J.G. *Stochastic Orders*; Springer Series in Statistics; Springer: New York, NY, USA, 2007.
13. Belzunce, F.; Martínez-Riquelme, C.; Mulero, J. *An Introduction to Stochastic Orders*; Elsevier/Academic Press: Amsterdam, The Netherlands, 2016.
14. Arnold, B.C.; Castillo, E.; Sarabia, J.M. *Conditional Specification of Statistical Models*; Springer: New York, NY, USA, 1999.
15. Escobar, L.A.; Meeker, W.Q. A review of accelerated test models. *Stat. Sci.* **2006**, *21*, 552–577. [CrossRef]
16. Di Crescenzo, A.; Martinucci, B.; Mulero, J. A quantile-based probabilistic mean value theorem. *Probab. Eng. Inf. Sci.* **2016**, *30*, 261–280. [CrossRef]
17. Di Crescenzo, A. A probabilistic analogue of the mean value theorem and its applications to reliability theory. *J. Appl. Probab.* **1999**, *39*, 706–719. [CrossRef]
18. Di Crescenzo, A.; Martinucci, B.; Zacks, S. Compound Poisson process with a Poisson subordinator. *J. Appl. Probab.* **2015**, *52*, 360–374. [CrossRef]
19. Cox, D.R. Regression models and life tables (with Discussion). *J. R. Stat. Soc. Ser. B* **1972**, *34*, 187–220.
20. Nanda, A.K.; Das, S. Dynamic proportional hazard rate and reversed hazard rate models. *J. Stat. Plan. Inference* **2011**, *141*, 2108–2119. [CrossRef]
21. Ng, H.K.T.; Navarro, J.; Balakrishnan, N. Parametric inference from system lifetime data under a proportional hazard rate model. *Metrika* **2012**, *75*, 367–388. [CrossRef]
22. Cover, T.M.; Thomas, J.A. *Elements of Information Theory*, 2nd ed.; Wiley-Interscience: Hoboken, NJ, USA, 2006.
23. Muliere, P.; Parmigiani, G.; Polson, N.G. A note on the residual entropy function. *Probab. Eng. Inf. Sci.* **1993**, *7*, 413–420. [CrossRef]
24. Ebrahimi, N. How to measure uncertainty in the residual life time distribution. *Sankhyā Ser. A* **1996**, *58*, 48–56.
25. Ebrahimi, N.; Pellerey, F. New partial ordering of survival functions based on the notion of uncertainty. *J. Appl. Probab.* **1995**, *32*, 202–211. [CrossRef]
26. Asadi, M.; Ebrahimi, N. Residual entropy and its characterizations in terms of hazard function and mean residual life function. *Stat. Probab. Lett.* **2000**, *49*, 263–269. [CrossRef]
27. Ebrahimi, N.; Kirmani, S.N.U.A.; Soofi, E.S. Multivariate dynamic information. *J. Multivar. Anal.* **2007**, *98*, 328–349. [CrossRef]
28. Di Crescenzo, A.; Longobardi, M. Entropy-based measure of uncertainty in past lifetime distributions. *J. Appl. Probab.* **2002**, *39*, 434–440. [CrossRef]
29. Nanda, A.K.; Paul, P. Some properties of past entropy and their applications. *Metrika* **2006**, *64*, 47–61. [CrossRef]
30. Kundu, C.; Nanda, A.K.; Maiti, S.S. Some distributional results through past entropy. *J. Stat. Plan. Inference* **2010**, *140*, 1280–1291. [CrossRef]
31. Sachlas, A.; Papaioannou, T. Residual and past entropy in actuarial science and survival models. *Methodol. Comput. Appl. Probab.* **2014**, *16*, 79–99. [CrossRef]
32. Kundu, A.; Nanda, A.K. On study of dynamic survival and cumulative past entropies. *Commun. Stat. Theory Methods* **2016**, *45*, 104–122. [CrossRef]
33. Ahmadi, J.; Di Crescenzo, A.; Longobardi, M. On dynamic mutual information for bivariate lifetimes. *Adv. Appl. Probab.* **2015**, *47*, 1157–1174. [CrossRef]
34. Bowden, R.J. Information, measure shifts and distribution metrics. *Statistics* **2012**, *46*, 249–262. [CrossRef]
35. Lillo, R.E.; Nanda, A.K.; Shaked, M. Some shifted stochastic orders. In *Recent Advances in Reliability Theory*; Statistics for Industry and Technology; Birkhäuser: Boston, MA, USA, 2000; pp. 85–103.

mathematics

MDPI

Article

A Time-Non-Homogeneous Double-Ended Queue with Failures and Repairs and Its Continuous Approximation

Antonio Di Crescenzo [1], Virginia Giorno [2], Balasubramanian Krishna Kumar [3] and Amelia G. Nobile [2,*]

1 Dipartimento di Matematica, Università degli Studi di Salerno, Via Giovanni Paolo II n. 132, 84084 Fisciano (SA), Italy; adicrescenzo@unisa.it
2 Dipartimento di Informatica, Università degli Studi di Salerno, Via Giovanni Paolo II n. 132, 84084 Fisciano (SA), Italy; giorno@unisa.it
3 Department of Mathematics, Anna University, Chennai 600 025, India; drbkkumar@hotmail.com
* Correspondence: nobile@unisa.it

Received: 6 April 2018; Accepted: 9 May 2018; Published: 11 May 2018

Abstract: We consider a time-non-homogeneous double-ended queue subject to catastrophes and repairs. The catastrophes occur according to a non-homogeneous Poisson process and lead the system into a state of failure. Instantaneously, the system is put under repair, such that repair time is governed by a time-varying intensity function. We analyze the transient and the asymptotic behavior of the queueing system. Moreover, we derive a heavy-traffic approximation that allows approximating the state of the systems by a time-non-homogeneous Wiener process subject to jumps to a spurious state (due to catastrophes) and random returns to the zero state (due to repairs). Special attention is devoted to the case of periodic catastrophe and repair intensity functions. The first-passage-time problem through constant levels is also treated both for the queueing model and the approximating diffusion process. Finally, the goodness of the diffusive approximating procedure is discussed.

Keywords: double-ended queues; time-non-homogeneous birth-death processes; catastrophes; repairs; transient probabilities; periodic intensity functions; time-non-homogeneous jump-diffusion processes; transition densities; first-passage-time

MSC: 60J80; 60J25; 60J75; 60K25

1. Introduction

Double-ended queues are often adopted as stochastic models for queueing systems characterized by two flows of agents, i.e., customers and servers/resources. When there are a customer and a server in the system, the match between the request and service occurs immediately, and then, both agents leave the system. As a consequence, there cannot be simultaneously customers and servers in the system. Namely, denoting by $N(t)$ the state of the system at time t, it is assumed that $N(t) = n$, $n \in \mathbb{N}$, if there are n customers waiting for available servers, whereas $N(t) = -n$, $n \in \mathbb{N}$, if there are n servers waiting for new customers, and $N(t) = 0$, if the system is empty. Hence, typical models for $N(t)$ are bilateral continuous-time Markov chains or similar stochastic processes.

Double-ended queueing systems can be applied to model numerous situations in real-world scenarios. A classical example is provided by taxi-passenger systems, where the role of customers and servers is played by passengers and taxis, respectively. We recall the first papers on this topic by Kashyap [1,2] and the subsequent contributions by Sharma and Nair [3], Tarabia [4] and Conolly et al. [5]. Other examples are provided by the dynamical allocation of live organs (servers) to candidates (customers) needing transplantation (cf. Elalouf et al. [6] and the references therein). Double-ended queues are suitable also to describe different streams arriving at a system (see Takahashi et al. [7]).

In this area, the interest is typically in the determination of the transient distribution and the asymptotic distribution of the system state, the busy period density, the waiting time density and related indices such as means and variances. The difficulties related to the resolution of the birth-death processes describing the length of the queue, in some cases, can be overcome by means of suitable transformations as those presented in Di Crescenzo et al. [8]. Such a transformation-based approach has been successfully exploited also for diffusion processes (see Di Crescenzo et al. [9]), this being of interest for the analysis of customary diffusion approximations of queue-length processes.

Attention is given often also to variants of the relevant stochastic processes that are adopted to describe more complex situations, such as bulk arrivals, truncated queues, the occurrence of disasters and repairs, and so on. In this respect, we recall the recent paper by Di Crescenzo et al. [10], which is centered on the analysis (i) of a continuous-time stochastic process describing the state of a double-ended queue subject to disasters and repairs and (ii) of the Wiener process with jumps arising as a heavy-traffic approximation to the previous model.

In many queueing models of manufacturing systems, it is assumed that the times to failure and the times to repair of each machine are exponentially distributed. However, exponential distributions do not always accurately represent distributions encountered in real manufacturing systems. Some of these models adopt the phase-type distributions for failure and repair times (see, for instance, Altiok [11–13] and Dallery [14]).

In this paper, we propose and analyze an extension of the queueing model treated in [10] to a time-non-homogeneous setting in which the intensities of arrivals, services, disasters and repairs are suitably time dependent. Similarly, we investigate the related heavy-traffic jump-diffusion approximation, as well. The key features of our analysis and the motivations of the proposed study are based mainly on the following issues:

- Queueing systems subject to disasters are appropriate to model more realistic situations in which the number of customers is subject to an abrupt decrease by the effect of catastrophes occurring randomly in time and due to external causes. The literature on the area of stochastic systems evolving in the presence of catastrophes is very broad. We restrict ourselves to mentioning the papers by Economou and Fakinos [15,16], Kyriakidis and Dimitrakos [17], Krishna Kumar et al. [18], Di Crescenzo et al. [19], Zeifman and Korotysheva [20], Zeifman et al. [21] and Giorno et al. [22]. The analysis of some time-dependent queueing models with catastrophes has been performed in Di Crescenzo et al. [23] and, more recently, in Giorno et al. [24], with special attention to the $M(t)/M(t)/1$ and $M(t)/M(t)/\infty$ queues.
- We include a repair mechanism in the queueing system under investigation, since it is essential to model instances when the (random) repair times are not negligible. We remark that the interest in this feature is increasing in the recent literature on queueing theory (see, for instance, Dimou and Economou [25]).
- Heavy-traffic approximations are very often proposed in order to describe the queueing systems under proper limit conditions of the parameters involved. This allows one to come to more manageable formulas for the description of the queue content. Typically, a customary rescaling procedure allows one to approximate the queue length process by a diffusion process, as indicated in Giorno et al. [26]. Examples of diffusion models arising from heavy-traffic approximations of double-ended queues and of similar matching systems can be found in Liu et al. [27] and Büke and Chen [28], respectively. In the case of queueing systems subject to catastrophes, a customary approach leads to jump-diffusion approximating processes (see, for instance, Di Crescenzo et al. [29] and Dharmaraja et al. [30]).

Plan of the Paper

In Section 2, we consider a non-homogeneous double-ended queue, whose arrivals and departures occur with time-varying intensity functions $\lambda(t) > 0$ and $\mu(t) > 0$, respectively. We discuss various features of such a model, including the first-passage time through a constant level.

In Section 3, we consider the non-homogeneous double-ended queue subject to disasters and repairs, both occurring with time-varying rates. Specifically, we assume that catastrophes occur according to a non-homogeneous Poisson process with intensity function $\nu(t) > 0$. The effect of catastrophes moves the system into a spurious failure state, say F. The completion of a system's repair occurs with time-varying intensity function $\eta(t) > 0$. After any repair, the system starts afresh from the zero state. Our first aim is to determine the probability $q(t|t_0)$ that the system at time t is in the failure state and the probability $p_{0,n}(t|t_0)$ that the system at time t is in the state $n \in \mathbb{Z}$ (working state).

In Section 4, we study the asymptotic behavior of the state probabilities in two different cases: (i) when the rates $\lambda(t), \mu(t), \nu(t), \eta(t)$ admit finite positive limits as t tends to infinity and (ii) when the double-ended queue is time-homogeneous, the catastrophe intensity function $\nu(t)$ and the repair intensity function $\eta(t)$ being periodic functions with common period Q.

In Section 5, we consider a diffusion approximation, under a heavy-traffic regime, of the non-homogeneous double-ended queue discussed in Section 2. In this case, the approximating process is a time-non-homogeneous Wiener process. We discuss various results on this model, including a first-passage-time problem through a constant level.

In Section 6, we deal with the heavy-traffic jump-diffusion approximation for the discrete model with catastrophes and repairs. Various results shown for the basic diffusion process treated in the previous section are thus extended to the present case characterized by jumps. In both Sections 5 and 6, the goodness of the approximating procedure is discussed, as well.

In Section 7, we finally consider the asymptotic behavior of the densities in the same cases considered in Section 4. In conclusion, we perform some comparisons between the relevant quantities of the queueing system and of the approximating diffusion process under the heavy-traffic regime.

2. The Underlying Non-Homogeneous Double-Ended Queue

This section is devoted to the analysis of the basic time-non-homogeneous double-ended queue.

Let $\{\widetilde{N}(t),\ t \geq t_0\}$, with $t_0 \geq 0$, be a continuous-time Markov chain describing the number of customers in a time-non-homogeneous double-ended queueing system, with state-space $\mathbb{Z} = \{\ldots,-1,0,1,\ldots\}$. We assume that arrivals (upward jumps) and departures (downward jumps) at time t occur with intensity functions $\lambda(t) > 0$ and $\mu(t) > 0$, respectively, where $\lambda(t)$ and $\mu(t)$ are bounded and continuous functions for $t \geq t_0$, such that $\int_{t_0}^{+\infty} \lambda(t)\, dt = +\infty$ and $\int_{t_0}^{+\infty} \mu(t)\, dt = +\infty$. The given assumptions ensure that the eventual transitions from any state occur w.p. 1. The state diagram of $\widetilde{N}(t)$ is shown in Figure 1.

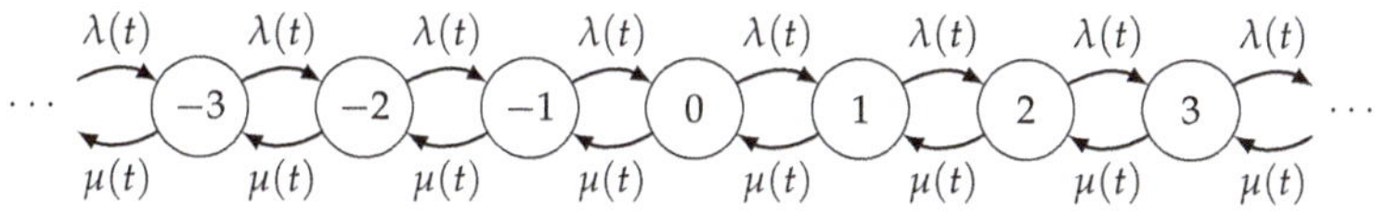

Figure 1. State diagram of the non-homogeneous double-ended queueing system.

For all $j, n \in \mathbb{Z}$ and $t > t_0 \geq 0$, the transition probabilities $\widetilde{p}_{j,n}(t|t_0) = P\{\widetilde{N}(t) = n|\widetilde{N}(t_0) = j\}$ are solutions of the system of Kolmogorov forward equations:

$$\frac{d\widetilde{p}_{j,n}(t|t_0)}{dt} = \lambda(t)\, \widetilde{p}_{j,n-1}(t|t_0) - [\lambda(t) + \mu(t)]\, \widetilde{p}_{j,n}(t|t_0) + \mu(t)\, \widetilde{p}_{j,n+1}(t|t_0), \qquad j, n \in \mathbb{Z}, \tag{1}$$

with the initial condition $\lim_{t\downarrow t_0} \widetilde{p}_{j,n}(t|t_0) = \delta_{j,n}$, where $\delta_{j,n}$ is the Kronecker delta function. For $t \geq t_0$ and $0 \leq z \leq 1$, let:

$$\widetilde{G}_j(z,t|t_0) = \mathrm{E}\left[z^{\widetilde{N}(t)}|\widetilde{N}(t_0)=j\right] = \sum_{n=-\infty}^{+\infty} z^n \widetilde{p}_{j,n}(t|t_0), \qquad j \in \mathbb{Z} \tag{2}$$

be the probability generating function of $\widetilde{N}(t)$. For any $t \geq t_0$, we denote the cumulative arrival and service intensity functions by:

$$\Lambda(t|t_0) = \int_{t_0}^{t} \lambda(\tau)\,d\tau, \qquad M(t|t_0) = \int_{t_0}^{t} \mu(\tau)\,d\tau. \tag{3}$$

Due to (1), for $t \geq t_0$, the probability generating Function (2) is the solution of the partial differential equation:

$$\frac{\partial}{\partial t}\widetilde{G}_j(z,t|t_0) = \left\{-[\lambda(t)+\mu(t)] + \lambda(t)\,z + \frac{\mu(t)}{z}\right\}\widetilde{G}_j(z,t|t_0), \qquad j \in \mathbb{Z}$$

to be solved with the initial condition $\lim_{t\downarrow t_0} \widetilde{G}_j(z,t|t_0) = z^j$. Hence, (2) can be expressed in terms of (3) as follows:

$$\widetilde{G}_j(z,t|t_0) = z^j \exp\left\{-[\Lambda(t|t_0)+M(t|t_0)] + \Lambda(t|t_0)\,z + \frac{M(t|t_0)}{z}\right\}, \qquad j \in \mathbb{Z}. \tag{4}$$

Recalling that (cf. Abramowitz [31], p. 376, n. 9.6.33):

$$\exp\left\{\frac{s}{2}\left(r+\frac{1}{r}\right)\right\} = \sum_{n=-\infty}^{+\infty} r^n\, I_n(s) \qquad (r \neq 0), \tag{5}$$

where:

$$I_\nu(z) = \sum_{m=0}^{\infty} \frac{(z/2)^{\nu+2m}}{m!\,\Gamma(\nu+m+1)} \qquad (\nu \in \mathbb{R})$$

denotes the modified Bessel function of first kind and by setting:

$$s = 2\sqrt{\Lambda(t|t_0)\,M(t|t_0)}, \qquad r = z\sqrt{\frac{\Lambda(t|t_0)}{M(t|t_0)}}$$

in (5), from (4), one has:

$$\widetilde{G}_j(z,t|t_0) = e^{-[\Lambda(t|t_0)+M(t|t_0)]} \sum_{k=-\infty}^{+\infty} z^{j+k}\left[\frac{\Lambda(t|t_0)}{M(t|t_0)}\right]^{k/2} I_k\left[2\sqrt{\Lambda(t|t_0)\,M(t|t_0)}\right], \qquad j \in \mathbb{Z}.$$

Hence, recalling (2), one obtains the transition probabilities:

$$\widetilde{p}_{j,n}(t|t_0) = e^{-[\Lambda(t|t_0)+M(t|t_0)]}\left[\frac{\Lambda(t|t_0)}{M(t|t_0)}\right]^{(n-j)/2} I_{n-j}\left[2\sqrt{\Lambda(t|t_0)\,M(t|t_0)}\right], \qquad j,n \in \mathbb{Z}. \tag{6}$$

We remark that, since $I_n(z) = I_{-n}(z)$ for $n \in \mathbb{N}$, the following symmetry relation holds:

$$\widetilde{p}_{j,n}(t|t_0) = \left[\frac{\Lambda(t|t_0)}{M(t|t_0)}\right]^{n-j} \widetilde{p}_{j,2j-n}(t|t_0) \qquad j,n \in \mathbb{Z}.$$

Moreover, from (6), we recover the conditional mean and variance of $\widetilde{N}(t)$, for $t \geq t_0$ and $j \in \mathbb{Z}$:

$$\mathrm{E}[\widetilde{N}(t)|\widetilde{N}(t_0)=j] = j + \Lambda(t|t_0) - M(t|t_0), \qquad \mathrm{Var}[\widetilde{N}(t)|\widetilde{N}(t_0)=j] = \Lambda(t|t_0) + M(t|t_0). \tag{7}$$

We point out that the transition probabilities given in (6) constitute the probability distribution of the difference of two independent non-homogeneous Poisson processes with intensities $\lambda(t)$ and $\mu(t)$, respectively, originated at zero (cf. Irwin [32] or Skellam [33] for the homogeneous case).

Let us now consider the first-passage-time (FPT) of $\widetilde{N}(t)$ through the state $n \in \mathbb{Z}$, starting from the initial state $j \in \mathbb{Z}$. Such a random variable will be denoted as:

$$\widetilde{T}_{j,n}(t_0) = \inf\{t \geq t_0 : \widetilde{N}(t) = n\}, \qquad \widetilde{N}(t_0) = j,\ j \neq n,$$

where $\widetilde{g}_{j,n}(t|t_0)$ is its probability density function (pdf). Special interest is given to $\widetilde{T}_{j,0}(t_0)$, which represents the busy period of the double-ended queue, with initial state $\widetilde{N}(t_0) = j$. As is well-known, due to the Markov property, $\widetilde{g}_{j,n}(t|t_0)$ satisfies the integral equation:

$$\widetilde{p}_{j,n}(t|t_0) = \int_{t_0}^{t} \widetilde{g}_{j,n}(\tau|t_0)\, \widetilde{p}_{n,n}(t|\tau)\, d\tau, \qquad j,n \in \mathbb{Z},\ j \neq n. \tag{8}$$

Hereafter, we consider the special case in which the arrival and departure intensity functions are proportional.

Remark 1. *Let $\lambda(t) = \lambda\varphi(t)$ and $\mu(t) = \mu\varphi(t)$, with λ, μ positive constants, where $\varphi(t)$ is a positive, bounded and continuous function of $t \geq t_0$, such that $\int_{t_0}^{\infty} \varphi(t)\, dt = +\infty$. By setting $\varrho = \lambda/\mu$ and:*

$$\Phi(t|t_0) = \int_{t_0}^{t} \varphi(\tau)\, d\tau, \qquad t \geq t_0, \tag{9}$$

then the transition probabilities (6) of the non-homogeneous double-ended queueing system $\widetilde{N}(t)$ can be expressed as:

$$\widetilde{p}_{j,n}(t|t_0) = e^{-(\lambda+\mu)\Phi(t|t_0)}\, \varrho^{(n-j)/2}\, I_{n-j}\big[2\lambda\mu\Phi(t|t_0)\big], \qquad j,n \in \mathbb{Z}. \tag{10}$$

Hence, from the results given in Section 5 of Giorno et al. [24], we have:

$$\widetilde{g}_{j,n}(t|t_0) = \frac{|n-j|\, \varphi(t)}{\Phi(t|t_0)}\, \widetilde{p}_{j,n}(t|t_0), \qquad j,n \in \mathbb{Z},\ j \neq n. \tag{11}$$

Furthermore, the FPT ultimate probability is given by:

$$P\{\widetilde{T}_{j,n}(t_0) < +\infty\} = \int_{t_0}^{+\infty} \widetilde{g}_{j,n}(t|t_0)\, dt = \begin{cases} 1, & (\lambda-\mu)(n-j) \geq 0, \\ \varrho^{n-j}, & (\lambda-\mu)(n-j) < 0. \end{cases}$$

3. The Queueing System with Catastrophes and Repairs

This section deals with the analysis of the queueing system with catastrophes and repairs.

Let $\{N(t),\ t \geq t_0\}$, with $t_0 \geq 0$, be a continuous-time Markov chain that describes the number of customers of a time-non-homogeneous double-ended queueing system subject to disasters and repairs. The state-space of $\{N(t),\ t \geq t_0\}$ is denoted by $S = \{F\} \cup \mathbb{Z} = \{F, 0, \pm 1, \pm 2, \ldots\}$, where F denotes the failure state. We assume that the catastrophes occur according to a non-homogeneous Poisson process with intensity function $\nu(t)$. If a catastrophe occurs, then the system goes instantaneously into the failure state F, and further, the completion of a repair occurs according to the intensity function $\eta(t)$ (cf. the diagram shown in Figure 2). We assume that the rates $\nu(t)$ and $\eta(t)$ are positive, bounded and continuous functions for $t \geq t_0$, such that $\int_{t_0}^{\infty} \nu(t)\, dt = +\infty$ and $\int_{t_0}^{\infty} \eta(t)\, dt = +\infty$. After every repair, the system starts again from the zero state.

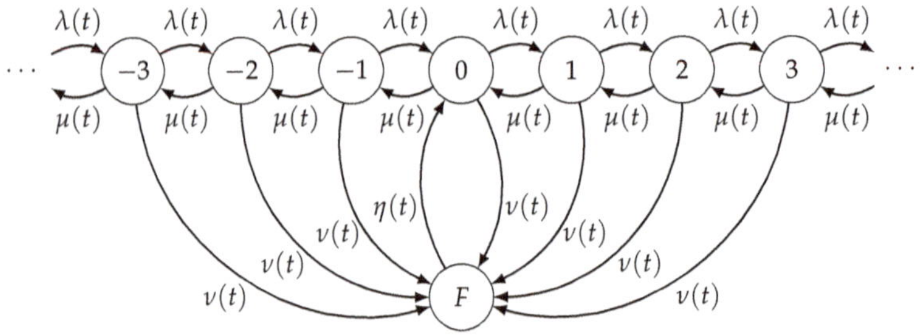

Figure 2. State diagram of the time-non-homogeneous double-ended queueing system with catastrophes and repairs.

For any $n \in \mathbb{Z}$ and $t > t_0 \geq 0$, we set:

$$p_{0,n}(t|t_0) = P\{N(t) = n|N(t_0) = 0\}, \qquad q(t|t_0) = P\{N(t) = F|N(t_0) = 0\}. \tag{12}$$

Hence, $p_{0,n}(t|t_0)$ is the transition probability from zero, at time t_0, to state n, at time t, when the system is active (in this case, we say that the system is in the "on" state), whereas $q(t|t_0)$ is the probability that the queueing system is in the state F (called the "failure" state) at time t starting from zero at time t_0. The probabilities given in (12) are the solution of the forward Kolmogorov system of differential equations:

$$\frac{dq(t|t_0)}{dt} = -\eta(t)\, q(t|t_0) + \nu(t)\, [1 - q(t|t_0)], \tag{13}$$

$$\frac{dp_{0,0}(t|t_0)}{dt} = -[\lambda(t) + \mu(t) + \nu(t)]\, p_{0,0}(t|t_0) + \lambda(t)\, p_{0,-1}(t|t_0) + \mu(t)\, p_{0,1}(t|t_0) + \eta(t)\, q(t|t_0), \tag{14}$$

$$\frac{dp_{0,n}(t|t_0)}{dt} = -[\lambda(t) + \mu(t) + \nu(t)]\, p_{0,n}(t|t_0) + \lambda(t)\, p_{0,n-1}(t|t_0) + \mu(t)\, p_{0,n+1}(t|t_0), \quad n \in \mathbb{Z} \setminus \{0\}, \tag{15}$$

to be solved with the following initial conditions, based on the Kronecker delta function:

$$\lim_{t \downarrow t_0} p_{n,0}(t|t_0) = \delta_{n,0}, \qquad \lim_{t \downarrow t_0} q(t|t_0) = 0. \tag{16}$$

Conditions (16) imply that at initial time t_0, the system is active and it starts from the zero state. In order to determine the transient probabilities of $N(t)$, similarly as in (3), in the following, we denote the cumulative catastrophe and repair intensity functions by:

$$V(t|t_0) = \int_{t_0}^{t} \nu(\tau)\, d\tau, \qquad H(t|t_0) = \int_{t_0}^{t} \eta(\tau)\, d\tau, \qquad t \geq t_0, \tag{17}$$

respectively.

Transient Probabilities

We first determine the probability that the system is under repair at time t. By solving Equation (13) with the second of the initial conditions (16), recalling (17), one obtains the probability that the process $N(t)$ is in the state F ("failure" state) at time t, starting from zero at time t_0:

$$q(t|t_0) = \int_{t_0}^{t} \nu(\tau)\, e^{-[V(t|\tau) + H(t|\tau)]}\, d\tau, \qquad t \geq t_0. \tag{18}$$

The transient analysis of the process $N(t)$ can be performed by relating the transient probabilities to those of the same process in the absence of catastrophes. Indeed, by conditioning on the time of the last catastrophe of $N(t)$ before t, the probabilities $p_{0,n}(t|t_0)$ can be expressed in terms of $\widetilde{p}_{0,n}(t|t_0)$ as follows, for $n \in \mathbb{Z}$ and $t \geq t_0$ (cf. [10,15,16]):

$$p_{0,n}(t|t_0) = e^{-V(t|t_0)}\, \widetilde{p}_{0,n}(t|t_0) + \int_{t_0}^{t} q(\tau|t_0)\, \eta(\tau)\, e^{-V(t|\tau)}\, \widetilde{p}_{0,n}(t|\tau)\, d\tau. \tag{19}$$

We note that the first term on the right-hand side of (19) expresses the probability that process $N(t)$ occupies state n at time t and that no catastrophes occurred in $[0,t]$. Similarly, the second term gives the probability that process $N(t)$ occupies state n at time t and that at least one catastrophe (with successive repair) occurred in $[0,t]$, i.e.,

- starting from zero at time t_0, at least a catastrophe and the subsequent repair occur before t; let $\tau \in (0,t)$ be the instant at which the last repair occurs, so that a transition entering in the zero state occurs at time τ with intensity $\eta(\tau)$;
- no catastrophe occurs in the interval (τ, t); then the system, starting from the zero state at time τ, reaches the state n at time t.

Note that Equation (19) is the suitable extension of (2.7) of [10], which refers to the case of constant rates. Furthermore, we remark that from (18) and (19), one obtains:

$$\sum_{n=-\infty}^{+\infty} p_{0,n}(t|t_0) + q(t|t_0) = 1, \qquad t \geq t_0. \tag{20}$$

Making use of (6) and (18) in (19), for $t \geq t_0$ and $n \in \mathbb{Z}$, one has the following expression for the transition probabilities of $N(t)$:

$$\begin{aligned} p_{0,n}(t|t_0) &= e^{-[\Lambda(t|t_0)+M(t|t_0)+V(t|t_0)]} \left[\frac{\Lambda(t|t_0)}{M(t|t_0)}\right]^{n/2} I_n\left[2\sqrt{\Lambda(t|t_0)\, M(t|t_0)}\right] \\ &+ \int_{t_0}^{t} d\tau\, \eta(\tau) e^{-[\Lambda(t|\tau)+M(t|\tau)+V(t|\tau)]} \left[\frac{\Lambda(t|\tau)}{M(t|\tau)}\right]^{n/2} I_n\left[2\sqrt{\Lambda(t|\tau)\, M(t|\tau)}\right] \int_{t_0}^{\tau} \nu(\vartheta) e^{-[V(\tau|\vartheta)+H(\tau|\vartheta)]}\, d\vartheta. \end{aligned}$$

Let us now introduce the r-th conditional moment of $N(t)$, for $r \in \mathbb{N}$:

$$\mathcal{M}_r(t|t_0) := \mathrm{E}[N^r(t)|N(t) \in \mathbb{Z}, N(t_0) = 0] = \frac{1}{1-q(t|t_0)} \sum_{n=-\infty}^{+\infty} n^r p_{0,n}(t|t_0). \tag{21}$$

From (19), it is not hard to see that the moments (21) can be expressed in terms of the conditional moments $\widetilde{\mathcal{M}}_r(t|t_0) := \mathrm{E}[\widetilde{N}^r(t)|\widetilde{N}(t_0) = 0]$ as follows, for $r \in \mathbb{N}$ and $t \geq t_0$:

$$\mathcal{M}_r(t|t_0) = \frac{1}{1-q(t|t_0)} \left\{ e^{-V(t|t_0)}\, \widetilde{\mathcal{M}}_r(t|t_0) + \int_{t_0}^{t} q(\tau|t_0)\, \eta(\tau) e^{-V(t|\tau)}\, \widetilde{\mathcal{M}}_r(t|\tau)\, d\tau \right\}. \tag{22}$$

Hence, by virtue of (7), from (22), the conditional mean and second order moment of $N(t)$ can be evaluated based on the knowledge of the relevant intensity functions.

Hereafter, we see that if the arrival and departure rates are constant, then some simplifications hold for the transition probabilities and conditional moments.

Theorem 1. *For the queueing system with catastrophes and repairs, having constant arrival rates $\lambda(t) = \lambda$ and departure rates $\mu(t) = \mu$, for $t \geq t_0$ and $n \in \mathbb{Z}$, one has:*

$$p_{0,n}(t|t_0) = e^{-V(t|t_0)}\, \widetilde{p}_{0,n}(t-t_0|0) + \int_0^{t-t_0} dx\, \nu(t-x)\, e^{-V(t|t-x)} \int_0^x \eta(t-u)\, e^{-H(t-u|t-x)}\, \widetilde{p}_{0,n}(u|0)\, du \tag{23}$$

and, for $r \in \mathbb{N}$,

$$\mathcal{M}_r(t|t_0) = \frac{1}{1-q(t|t_0)} \left\{ e^{-V(t|t_0)} \widetilde{\mathcal{M}}_r(t-t_0|0) + \int_0^{t-t_0} q(t-x|t_0)\, \eta(t-x) e^{-V(t|t-x)} \widetilde{\mathcal{M}}_r(x|0)\, dx \right\}. \quad (24)$$

Furthermore, it results:

$$p_{0,n}(t|t_0) = \int_{t_0}^{t} p_{0,0}(\tau|t_0)\, e^{-V(t|\tau)}\, \widetilde{g}_{0,n}(t|\tau)\, d\tau, \qquad n \in \mathbb{Z} \setminus \{0\},\ t \geq t_0. \quad (25)$$

Proof. Since $\lambda(t) = \lambda$ and $\mu(t) = \mu$, by virtue of (18), Relation (23) follows from (19), whereas Equation (24) derives from (22). Moreover, making use of (19) in the right-hand side of (25), one has:

$$\int_{t_0}^{t} p_{0,0}(u|t_0)\, e^{-V(t|u)}\, \widetilde{g}_{0,n}(t|u)\, du = e^{-V(t|t_0)} \int_{t_0}^{t} \widetilde{p}_{0,0}(u|t_0)\, \widetilde{g}_{0,n}(t|u)\, du$$
$$+ \int_{t_0}^{t} d\tau\, e^{-V(t|\tau)}\, \eta(\tau)\, q(\tau|t_0) \int_{\tau}^{t} \widetilde{p}_{0,0}(u|\tau) \widetilde{g}_{0,n}(t|u)\, du. \quad (26)$$

By virtue of (6), we note that $\widetilde{p}_{0,0}(t|t_0) = \widetilde{p}_{n,n}(t|t_0)$ for $n \in \mathbb{Z}$ and $t \geq t_0$. Moreover, since $\lambda(t) = \lambda$ and $\mu(t) = \mu$, we obtain:

$$\int_{t_0}^{t} \widetilde{p}_{0,0}(u|t_0)\, \widetilde{g}_{0,n}(t|u)\, du = \int_{t_0}^{t} \widetilde{p}_{n,n}(u|t_0)\, \widetilde{g}_{0,n}(t|u)\, du$$
$$= \int_0^{t-t_0} \widetilde{p}_{n,n}(t-t_0|\tau)\, \widetilde{g}_{0,n}(\tau|0)\, d\tau = \widetilde{p}_{0,n}(t-t_0|0) = \widetilde{p}_{0,n}(t|t_0). \quad (27)$$

Substituting (27) in (26), by virtue of (19), Relation (25) immediately follows. □

The integrand on the right-hand side of Equation (25) refers to the sample-paths of $N(t)$ that start from zero at time t_0, then reach the state zero at time $\tau \in (t_0, t)$ and, finally, go from zero at time τ to n at time t for the first time, without the occurrence of catastrophes in (τ, t).

4. Asymptotic Probabilities

In this section, we analyze the asymptotic behavior of the probabilities $q(t|t_0)$ and $p_{j,n}(t|t_0)$ of the process $N(t)$ in two different cases:

(i) the intensity functions $\lambda(t), \mu(t), \nu(t)$ and $\eta(t)$ admit finite positive limits as $t \to +\infty$,
(ii) the intensity functions $\lambda(t)$ and $\mu(t)$ are constant, whereas the rates $\nu(t)$ and $\eta(t)$ are periodic functions with common period Q.

4.1. Asymptotically-Constant Intensity Functions

In the following theorem, we determine the steady-state probabilities and the asymptotic failure probability of the process $N(t)$ when the intensity functions $\lambda(t), \mu(t), \nu(t)$ and $\eta(t)$ admit finite positive limits as t tends to $+\infty$.

Theorem 2. *If:*

$$\lim_{t\to+\infty} \lambda(t) = \lambda, \qquad \lim_{t\to+\infty} \mu(t) = \mu, \qquad \lim_{t\to+\infty} \nu(t) = \nu, \qquad \lim_{t\to+\infty} \eta(t) = \eta, \quad (28)$$

with λ, μ, ν, η positive constants, then the steady-state probabilities and the asymptotic failure probability of the process $N(t)$ are:

$$q^* := \lim_{t\to+\infty} q(t|t_0) = \frac{\nu}{\nu+\eta}, \tag{29}$$

$$p_0^* := \lim_{t\to+\infty} p_{0,0}(t|t_0) = \frac{\nu(1-q)}{\sqrt{(\lambda+\mu+\nu)^2-4\lambda\mu}},$$

$$p_n^* := \lim_{t\to+\infty} p_{0,n}(t|t_0) = \Big[\frac{\lambda+\mu+\nu-\sqrt{(\lambda+\mu+\nu)^2-4\lambda\mu}}{2\mu}\Big]^n p_0^*, \qquad n\in\mathbb{N}, \tag{30}$$

$$p_{-n}^* := \lim_{t\to+\infty} p_{0,-n}(t|t_0) = \Big[\frac{\lambda+\mu+\nu-\sqrt{(\lambda+\mu+\nu)^2-4\lambda\mu}}{2\lambda}\Big]^n p_0^*, \qquad n\in\mathbb{N}.$$

Furthermore, the asymptotic conditional mean, second order moment and variance are:

$$\mathcal{M}_1^* = \lim_{t\to+\infty} \mathcal{M}_1(t|t_0) = \frac{\lambda-\mu}{\nu}, \qquad \mathcal{M}_2^* = \lim_{t\to+\infty} \mathcal{M}_2(t|t_0) = \frac{2(\lambda-\mu)^2}{\nu^2} + \frac{(\lambda+\mu)}{\nu},$$
$$\mathcal{V}^* = \lim_{t\to+\infty} \mathrm{Var}(t|t_0) = \lim_{t\to+\infty} \{\mathcal{M}_2(t|t_0) - [\mathcal{M}_1(t|t_0)]^2\} = \frac{(\lambda-\mu)^2}{\nu^2} + \frac{\lambda+\mu}{\nu}. \tag{31}$$

Proof. The steady-state probabilities and the asymptotic failure probability of $N(t)$ can be obtained by taking the limit as $t\to+\infty$ in Equations (13)–(15), and then solving the corresponding balance equations. From (21), making use of (29) and (30), the asymptotic conditional mean and variance (31) immediately follow. □

4.2. Periodic Catastrophe and Repair Intensity Functions

Let us assume that the arrival and departure intensity functions are constant, whereas the catastrophe intensity function $\nu(t)$ and the repair intensity function $\eta(t)$ are periodic, such that $\nu(t+kQ) = \nu(t)$ and $\eta(t+kQ) = \eta(t)$ for all $k\in\mathbb{N}$, $t\ge t_0$, for a given constant period $Q>0$. We denote by:

$$\nu^* = \frac{1}{Q}\int_0^Q \nu(u)\,du, \qquad \eta^* = \frac{1}{Q}\int_0^Q \eta(u)\,du, \tag{32}$$

the average catastrophe and repair rates over the period Q. Since $\nu(t)$ and $\eta(t)$ are periodic functions, from (17), we have, for $t\ge t_0$:

$$V(t+kQ) = \int_t^{t+kQ} \nu(u)\,du = kQ\nu^*, \qquad H(t+kQ) = \int_t^{t+kQ} \eta(u)\,du = kQ\eta^*, \qquad k\in\mathbb{N}. \tag{33}$$

Let us now investigate the asymptotic distribution for the process $N(t)$, which can be defined as follows, for $t\ge t_0$:

$$q^*(t) = \lim_{k\to+\infty} q(t+kQ|t_0), \qquad p_{0,n}^*(t) = \lim_{k\to+\infty} p_{0,n}(t+kQ|t_0), \qquad n\in\mathbb{Z}. \tag{34}$$

Theorem 3. *For the queueing system with catastrophes and repairs, having constant arrival rates $\lambda(t)=\lambda>0$ and departure rates $\mu(t)=\mu>0$, with $\nu(t)$ and $\eta(t)$ continuous, positive and periodic functions, with period Q, for $t\ge t_0$, one has:*

$$p_{0,n}^*(t) = \int_0^{+\infty} dx\, \nu(t-x)e^{-V(t|t-x)} \int_0^x \eta(t-u)e^{-H(t-u|t-x)}\, \tilde{p}_{0,n}(u|0)\,du, \qquad n\in\mathbb{Z}. \tag{35}$$

$$q^*(t) = \int_0^{+\infty} \nu(t-x)e^{-[V(t|t-x)+H(t|t-x)]}\,dx. \tag{36}$$

Furthermore, an alternative expression for the failure asymptotic probability is:

$$q^*(t) = \frac{1}{e^{Q(\nu^*+\eta^*)}-1}\int_0^Q \nu(t+u)e^{[V(t+u|t)+H(t+u|t)]}\,du, \tag{37}$$

with ν^* and η^* given in (32).

Proof. Since $\lambda(t)=\lambda$ and $\mu(t)=\mu$, from (23), for $k\in\mathbb{N}_0$ and $t\geq t_0$, one has:

$$p_{0,n}(t+kQ|t_0) = e^{-V(t+kQ|t_0)}\,\widetilde{p}_{0,n}(t-t_0+kQ|0)+\int_0^{t-t_0+kQ} dx\,\nu(t-x)\,e^{-V(t+kQ|t-x+kQ)} \\ \times\int_0^x \eta(t-u)\,e^{-H(t-u+kQ|t-x+kQ)}\,\widetilde{p}_{0,n}(u|0)\,du. \tag{38}$$

Due to the periodicity of $\nu(t)$ and $\eta(t)$, the following equalities hold:

$$V(t+kQ|t-x+kQ)=V(t|t-x),\qquad H(t-u+kQ|t-x+kQ)=H(t-u|t-x).$$

Hence, from (38), it follows:

$$p_{0,n}(t+kQ|t_0) = e^{-V(t+kQ|t_0)}\,\widetilde{p}_{0,n}(t-t_0+kQ|0)+\int_0^{t-t_0+kQ} dx\,\nu(t-x)\,e^{-V(t+|t-x)} \\ \times\int_0^x \eta(t-u)\,e^{-H(t-u|t-x)}\,\widetilde{p}_{0,n}(u|0)\,du. \tag{39}$$

Then, taking the limit as $k\to+\infty$ in (39) and recalling the second of (34), one obtains (35). Hence, we note that:

$$\sum_{n=-\infty}^{+\infty} p^*_{0,n}(t) = \int_0^{+\infty} dx\,\nu(t-x)e^{-V(t|t-x)}\int_0^x \eta(t-u)e^{-H(t-u|t-x)}\,du \\ = 1-\int_0^{+\infty}\nu(t-x)\,e^{-[V(t|t-x)+H(t|t-x)]}\,dx. \tag{40}$$

Consequently, by virtue of (20), Equation (36) immediately follows. To prove Equation (37), we first consider (18), which implies:

$$q(t+kQ|t_0) = e^{-kQ(\nu^*+\eta^*)}\left[\int_{t_0}^t \nu(\tau)e^{-[V(t|\tau)+H(t|\tau)]}\,d\tau+\int_t^{t+kQ}\nu(\tau)e^{[V(t|\tau)+H(t|\tau)]}\,d\tau\right]. \tag{41}$$

Since $\nu(t)$ and $\eta(t)$ are periodic functions, one has:

$$\left.\int_t^{t+kQ}\nu(\tau)e^{[V(t|\tau)+H(t|\tau)]}\,d\tau\right] = \sum_{r=0}^{k-1}\int_0^Q \nu(t+x)e^{[V(t+rQ+x|t)+H(t+rQ+x|t)]}\,dx \\ = \left[\int_0^Q \nu(t+x)e^{[V(t+x|t)+H(t+x|t)]}\,dx\right]\sum_{r=0}^{k-1}e^{rQ(\nu^*+\eta^*)} \\ = \frac{e^{kQ(\nu^*+\eta^*)}-1}{e^{Q(\nu^*+\eta^*)}-1}\left[\int_0^Q \nu(t+x)e^{[V(t+x|t)+H(t+x|t)]}\,dx\right]. \tag{42}$$

Substituting (42) in (41) and taking the limit as $k\to+\infty$, one finally is led to (37). □

Under the assumptions of Theorem 3, by virtue of the periodicity of $\nu(t)$ and $\eta(t)$, from (35)–(37), one has that $p^*_{0,n}(t)$ and $q^*(t)$ are periodic functions with period Q. From (21), making use of (35), the asymptotic conditional moments are expressed as:

$$\mathcal{M}_r^*(t) := \lim_{k\to+\infty} \mathcal{M}_r(t+kQ|t_0)$$
$$= \frac{1}{1-q^*(t)} \int_0^{+\infty} dx\, \nu(t-x) e^{-V(t|t-x)} \int_0^x \eta(t-u) e^{-H(t-u|t-x)}\, \widetilde{\mathcal{M}}_r(u|0)\, du, \quad (43)$$

with $\widetilde{\mathcal{M}}_r(t|0) := \mathrm{E}[\widetilde{N}^r(t)|\widetilde{N}(0)=0]$ and $q^*(t)$ given in (36) or (37).

Example 1. *Assume that $N(t)$ has constant arrival rates $\lambda(t)=\lambda>0$ and departure rates $\mu(t)=\mu>0$. Furthermore, let the periodic catastrophe and repair intensity functions be given by:*

$$\nu(t) = \nu + \frac{\pi a}{Q}\sin\Big(\frac{2\pi t}{Q}\Big), \qquad \eta(t) = \eta + \frac{\pi b}{Q}\cos\Big(\frac{2\pi t}{Q}\Big), \qquad t\geq 0, \quad (44)$$

with $a>0$, $b>0$, $\nu>\pi a/Q$ and $\eta>\pi b/Q$. Clearly, from (32) and (44), we have that the averages of $\nu(t)$ and $\eta(t)$ in the period Q are equal to ν and η, respectively. In Figures 3–5, the relevant parameters are taken as:

$$\lambda = 0.2, \quad \mu = 0.1, \quad Q = 1, \quad \nu = 0.5, \quad a = 0.1, \quad \eta = 0.6, \quad b = 0.15.$$

On the left of Figure 3, the catastrophe intensity function $\nu(t)$ (black curve) is plotted with its average $\nu=0.5$ (black dashed line). The repair intensity function $\eta(t)$ (red curve) is plotted, as well, with its average $\eta=0.6$ (red dashed line). On the right of Figure 3, the failure probability $q(t|0)$, given in (18), is plotted and is compared with the asymptotic failure probability $q^*=\nu/(\nu+\eta)=0.454545$. The latter is obtained by considering constant intensity functions $\nu(t)=\nu$ and $\eta(t)=\eta$. As proved in Theorem 3, $q(t|0)$ admits an asymptotic periodic behavior, which is highlighted on the right of Figure 3. Instead, in Figure 4, we plot the probability $p_0(t|0)$ (magenta curve), on the left. Moreover, on the right of Figure 4, we show the probabilities $p_{-1}(t|0)$ (blue curve) and $p_1(t|0)$ (red curve), given in (23). The dashed lines indicate the steady-state probabilities $p_0^*=0.364447$ (magenta dashed line), $p_{-1}^*=0.0470761$ (blue dashed line) and $p_1^*=0.0941522$ (red dashed line), obtained by considering constant intensity functions $\nu(t)=\nu$ and $\eta(t)=\eta$. As shown in Figure 4, the probabilities admit an asymptotic periodic behavior, with period $Q=1$. Finally, in Figure 5, the mean $\mathcal{M}_1(t|0)$ and the variance $\mathrm{Var}(t|0)$ of the process $N(t)$, obtained via (24), are plotted and compared with the asymptotic values (dashed lines) $\mathcal{M}_1^*=0.2$ and $\mathcal{V}^*=0.64$, given in (31). Figures 3–5 show that the relevant quantities reflect the periodic nature of the rates and illustrate the limiting behavior as t grows.

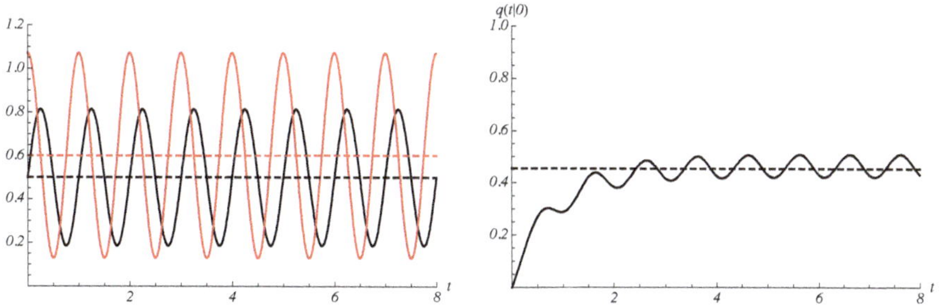

Figure 3. On the left: the periodic catastrophe intensity function (black curve) and repair intensity function (red curve), with their averages (dashed lines). On the right: the failure probability $q(t|t_0)$, given in (18), and the limit q^* (dashed line), given in (29). The parameters are specified in Example 1.

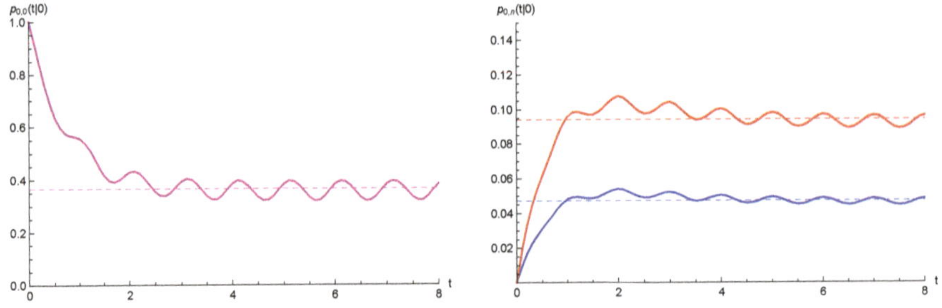

Figure 4. On the left: $p_0(t|0)$ and steady-state probability p_0^* (dashed line). On the right: $p_{-1}(t|0)$ (blue curve) and $p_1(t|0)$ (red curve), with steady-state probabilities p_{-1}^* and p_1^* (dashed lines), given in (30). The parameters are specified in Example 1.

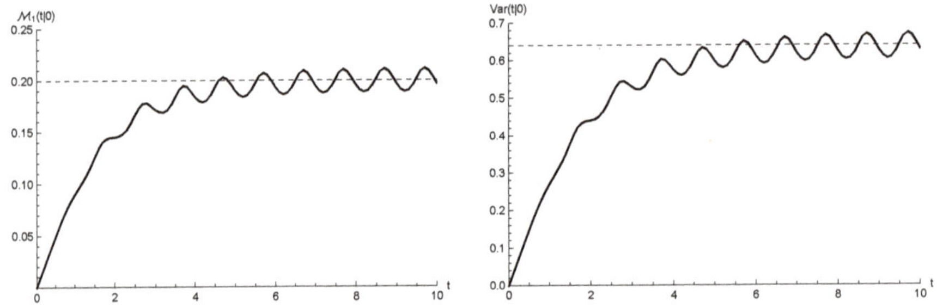

Figure 5. Plots of the mean $\mathcal{M}_1(t|0)$ (left) and the variance of $\text{Var}(t|0)$ (right) of the process $N(t)$, obtained by means of (24). The dashed lines indicate the asymptotic values $\mathcal{M}_1^*$ and $\mathcal{V}^*$. The parameters are specified in Example 1.

5. Diffusion Approximation of the Double-Ended Queueing System

With reference to the time-non-homogeneous double-ended queueing system discussed in Section 2, hereafter, we consider a heavy-traffic diffusion approximation of the queue-length process. This is finalized to obtain a more manageable description of the queueing system under a heavy-traffic regime. To this purpose, we shall adopt a suitable scaling procedure based on a scaling parameter ε. We first rename the intensity functions related to the double-ended process $\widetilde{N}(t)$, by setting:

$$\lambda(t) = \frac{\widehat{\lambda}(t)}{\varepsilon} + \frac{\omega^2(t)}{2\,\varepsilon^2}, \qquad \mu(t) = \frac{\widehat{\mu}(t)}{\varepsilon} + \frac{\omega^2(t)}{2\,\varepsilon^2}, \qquad n \in \mathbb{Z}. \tag{45}$$

Here, functions $\widehat{\lambda}(t)$, $\widehat{\mu}(t)$ and $\omega^2(t)$ are positive, bounded and continuous for $t \geq t_0$ and satisfy the conditions $\int_{t_0}^{+\infty} \widehat{\lambda}(t)\,dt = +\infty$, $\int_{t_0}^{+\infty} \widehat{\mu}(t)\,dt = +\infty$ and $\int_{t_0}^{+\infty} \omega^2(t)\,dt = +\infty$. Furthermore, the constant ε in the right-hand sides of (45) is positive and plays a relevant role in the following approximating procedure.

Let us consider the Markov process $\{\widetilde{N}_\varepsilon(t), t \geq t_0\}$, having state-space $\{0, \pm\varepsilon, \pm 2\varepsilon, \ldots\}$. Namely, it is defined as $\widetilde{N}_\varepsilon(t) = \epsilon\,\widetilde{N}(t)$, provided that the intensity functions are modified as in (45). By a customary scaling procedure similar to that adopted in [10,30], under suitable limit conditions and for $\varepsilon \downarrow 0$, the scaled process $\widetilde{N}_\varepsilon(t)$ converges weakly to a diffusion process $\{\widetilde{X}(t), t \geq t_0\}$ having state-space $\mathbb{R}$ and transition probability density function (pdf):

$$\widetilde{f}(x, t|x_0, t_0) = \frac{\partial}{\partial x} P\{\widetilde{X}(t) \leq x | \widetilde{X}(t_0) = x_0\}, \qquad x, x_0 \in \mathbb{R},\ t \geq t_0.$$

Indeed, with reference to System (1), substituting $\widetilde{p}_{j,n}(t|t_0)$ with $\widetilde{f}(n\varepsilon, t|j\varepsilon, t_0)\varepsilon$ in the Chapman–Kolmogorov forward differential-difference equation for $\widetilde{N}_\varepsilon(t)$, we have:

$$\frac{\partial \widetilde{f}(n\varepsilon, t|j\varepsilon, t_0)}{\partial t} = \Big[\frac{\widehat{\lambda}(t)}{\varepsilon} + \frac{\omega^2(t)}{2\,\varepsilon^2}\Big]\widetilde{f}[(n-1)\varepsilon, t|j\varepsilon, t_0] - \Big[\frac{\widehat{\lambda}(t)}{\varepsilon} + \frac{\widehat{\mu}(t)}{\varepsilon} + \frac{\omega^2(t)}{\varepsilon^2}\Big]\widetilde{f}(n\varepsilon, t|j\varepsilon, t_0)$$
$$+\Big[\frac{\widehat{\mu}(t)}{\varepsilon} + \frac{\omega^2(t)}{2\,\varepsilon^2}\Big]\widetilde{f}[(n+1)\varepsilon, t|j\varepsilon, t_0], \qquad j, n \in \mathbb{Z}.$$

After setting $x = n\varepsilon$ and $x_0 = j\varepsilon$, expanding $\widetilde{f}$ as Taylor series and taking the limit as $\varepsilon \downarrow 0$, we obtain the following partial differential equation:

$$\frac{\partial}{\partial t}\widetilde{f}(x, t|x_0, t_0) = -[\widehat{\lambda}(t) - \widehat{\mu}(t)]\,\frac{\partial}{\partial x}\widetilde{f}(x, t|x_0, t_0) + \frac{\omega^2(t)}{2}\,\frac{\partial^2}{\partial x^2}\widetilde{f}(x, t|x_0, t_0), \qquad x, x_0 \in \mathbb{R}. \tag{46}$$

The associated initial condition is $\lim_{t\downarrow t_0}\widetilde{f}(x, t|x_0, t_0) = \delta(x - x_0)$, where $\delta(x)$ is the Dirac delta-function. We remark that, due to (45), the limit $\varepsilon \downarrow 0$ leads to a heavy-traffic condition about the rates $\lambda(t)$ and $\mu(t)$ of process $\widetilde{N}(t)$. Hence, $\widetilde{X}(t)$ is a time-non-homogeneous Wiener process with drift $\widehat{\lambda}(t) - \widehat{\mu}(t)$ and infinitesimal variance $\omega^2(t)$, with initial state x_0 at time t_0. For $t \geq t_0$ and $s \in \mathbb{R}$, let:

$$H(s, t|x_0, t_0) = E[e^{is\widetilde{X}(t)}|\widetilde{X}(t_0) = x_0] = \int_{-\infty}^{+\infty} e^{isx}\,\widetilde{f}(x, t|x_0, t_0)\,dx, \qquad x_0 \in \mathbb{R} \tag{47}$$

be the characteristic function of $\widetilde{X}(t)$. Due to (46), the characteristic function (47) is the solution of the partial differential equation:

$$\frac{\partial}{\partial t}H(s, t|x_0, t_0) = \Big\{is\,[\lambda(t) - \mu(t)] - \frac{s^2}{2}\,\omega^2(t)\Big\}\,H(s, t|x_0, t_0), \qquad x_0 \in \mathbb{R},$$

to be solved with the initial condition $\lim_{t\downarrow t_0} H(s, t|x_0, t_0) = e^{isx_0}$. Hence, for $t \geq t_0$, one has:

$$H(s, t|x_0, t_0) = \exp\Big\{is\Big[x_0 + \widehat{\Lambda}(t|t_0) - \widehat{M}(t|t_0)\Big] - \frac{s^2}{2}\,\Omega(t|t_0)\Big\}, \qquad x_0 \in \mathbb{R}, \tag{48}$$

where we have set:

$$\widehat{\Lambda}(t|t_0) = \int_{t_0}^{t}\widehat{\lambda}(\tau)\,d\tau, \qquad \widehat{M}(t|t_0) = \int_{t_0}^{t}\widehat{\mu}(\tau)\,d\tau, \qquad \Omega(t|t_0) = \int_{t_0}^{t}\omega^2(\tau)\,d\tau, \qquad t \geq t_0. \tag{49}$$

Clearly, (48) is a normal characteristic function, so that the solution of (46) is the Gaussian pdf:

$$\widetilde{f}(x, t|x_0, t_0) = \frac{1}{\sqrt{2\,\pi\,\Omega(t|t_0)}}\,\exp\Big\{-\frac{[x - x_0 - \widehat{\Lambda}(t|t_0) + \widehat{M}(t|t_0)]^2}{2\,\Omega(t|t_0)}\Big\}, \qquad x, x_0 \in \mathbb{R},\ t \geq t_0. \tag{50}$$

Then, the conditional mean and variance are:

$$\mathrm{E}[\widetilde{X}(t)|\widetilde{X}(t_0) = x_0] = x_0 + \widehat{\Lambda}(t|t_0) - \widehat{M}(t|t_0), \qquad \mathrm{Var}[\widetilde{X}(t)|\widetilde{X}(t_0) = x_0] = \Omega(t|t_0), \qquad t \geq t_0. \tag{51}$$

Let us now consider a first-passage-time problem for $\widetilde{X}(t)$. We denote by $\widetilde{T}_{x_0,x}(t_0)$ the random variable describing the FPT of $\widetilde{X}(t)$ trough state $x \in \mathbb{R}$, starting from x_0 at time t_0, with $x_0 \neq x$. In analogy to (8), the Markov property yields:

$$\widetilde{f}(x, t|x_0, t_0) = \int_{t_0}^{t}\widetilde{g}(x, \tau|x_0, t_0)\,\widetilde{f}(x, t|x, \tau)\,d\tau, \qquad x_0, x \in \mathbb{R},\ x \neq x_0, \tag{52}$$

where $\widetilde{g}(x,t|x_0,t_0)$ is the pdf of $\widetilde{T}_{x_0,x}(t_0)$.

Hereafter, we deal with a special case, in which the functions $\widehat{\lambda}(t)$, $\widehat{\mu}(t)$ and $\omega^2(t)$ are proportional.

Remark 2. *Let $\widehat{\lambda}(t) = \widehat{\lambda}\varphi(t)$, $\widehat{\mu}(t) = \widehat{\mu}\varphi(t)$ and $\omega^2(t) = \omega^2\varphi(t)$, where $\widehat{\lambda}, \widehat{\mu}$, ω are positive constants and $\varphi(t)$ is a positive, bounded and continuous function for $t \geq t_0$, such that $\int_{t_0}^{\infty} \varphi(t)\,dt = +\infty$. Then, the transition pdf of $\widetilde{X}(t)$ becomes:*

$$\widetilde{f}(x,t|x_0,t_0) = \frac{1}{\sqrt{2\pi\omega^2\,\Phi(t|t_0)}} \exp\left\{-\frac{\left[x - x_0 - (\widehat{\lambda}-\widehat{\mu})\,\Phi(t|t_0)\right]^2}{2\,\omega^2\,\Phi(t|t_0)}\right\}, \qquad x, x_0 \in \mathbb{R},\ t \geq t_0,$$

where $\Phi(t|t_0)$ is defined in (9). Moreover, the FPT pdf of $\widetilde{T}_{x_0,x}(t_0)$ can be expressed as (see, for instance, [26]):

$$\widetilde{g}(x,t|x_0,t_0) = \frac{|x-x_0|\;\varphi(t)}{\Phi(t|t_0)}\,\widetilde{f}(x,t|x_0,t_0), \qquad x_0, x \in \mathbb{R},\ x \neq x_0.$$

The corresponding FPT ultimate probability is given by:

$$P\{\widetilde{T}_{x_0,x}(t_0) < +\infty\} = \int_{t_0}^{+\infty} \widetilde{g}(x,t|x_0,t_0)\,dt = \begin{cases} 1, & (\widehat{\lambda}-\widehat{\mu})(x-x_0) \geq 0, \\ e^{2(\widehat{\lambda}-\widehat{\mu})(x-x_0)/\omega^2}, & (\widehat{\lambda}-\widehat{\mu})(x-x_0) < 0. \end{cases}$$

Clearly, $\widetilde{T}_{x_0,0}(t_0)$ is a suitable approximation of the busy period $\widetilde{T}_{j,0}(t_0)$ considered in Section 2.

Goodness of the Approximating Procedure

Thanks to the above heavy-traffic approximation, the state of the time-non-homogeneous double-ended queue $\widetilde{N}(t)$ has been approximated by the non-homogeneous Wiener process $\widetilde{X}(t)$, with the transition pdf given in (50).

A first confirmation of the goodness of the approximating procedure can be obtained by the comparing mean and variance of $\widetilde{N}(t)$ with those of $\widetilde{X}(t)/\varepsilon$, for $\lambda(t)$ and $\mu(t)$ chosen as in (45). Recalling (7) and (51), the means satisfy the following identity, for all $\varepsilon > 0$:

$$\mathrm{E}[\widetilde{X}(t)|\widetilde{X}(t_0) = j\varepsilon] = \varepsilon\mathrm{E}[\widetilde{N}(t)|\widetilde{N}(t_0) = j]. \tag{53}$$

Moreover, for the variances, we have:

$$\lim_{\varepsilon\downarrow 0} \frac{\mathrm{Var}[\widetilde{N}(t)|\widetilde{N}(t_0) = j]}{\mathrm{Var}\left[\frac{\widetilde{X}(t)}{\varepsilon}\Big|\frac{\widetilde{X}(t_0)}{\varepsilon} = j\right]} = \lim_{\varepsilon\downarrow 0} \frac{\varepsilon^2\mathrm{Var}[\widetilde{N}(t)|\widetilde{N}(t_0) = j]}{\mathrm{Var}[\widetilde{X}(t)|\widetilde{X}(t_0) = j\varepsilon]} = \lim_{\varepsilon\downarrow 0} \frac{\varepsilon^2\,[\Lambda(t|t_0) + M(t|t_0)]}{\Omega(t|t_0)} = 1,$$

so that for ε close to zero, one has:

$$\mathrm{Var}[\widetilde{X}(t)|\widetilde{X}(t_0) = j\varepsilon] \simeq \varepsilon^2\mathrm{Var}[\widetilde{N}(t)|\widetilde{N}(t_0) = j]. \tag{54}$$

The discussion of the goodness of the heavy-traffic approximation involves also the probability distributions. Let us denote by $\widetilde{p}_{j,n}^{(\varepsilon)}(t)$ the transition probabilities of the process $\widetilde{N}(t)$, for $\lambda(t)$ and $\mu(t)$ given in (45), and for $n = x/\varepsilon$, $j = x_0/\varepsilon$. The following theorem holds.

Theorem 4. *For $t \geq t_0$, one has:*

$$\lim_{\substack{\varepsilon\downarrow 0,\\ n\varepsilon = x, j\varepsilon = x_0}} \frac{\widetilde{p}_{j,n}^{(\varepsilon)}(t|t_0)}{\varepsilon} = \widetilde{f}(x,t|x_0,t_0), \tag{55}$$

with $\widetilde{f}(x,t|x_0,t_0)$ given in (50).

Proof. To prove (55), we consider separately the following cases: (i) $n = j$ and (ii) $n \neq j$, with $j, n \in \mathbb{Z}$.

(i) For $n = j$, from (6), one has:

$$\widetilde{p}_{n,n}^{(\varepsilon)}(t|t_0) = \exp\Big\{-\frac{\widehat{\Lambda}(t|t_0)+\widehat{M}(t|t_0)}{\varepsilon} - \frac{\Omega(t|t_0)}{\varepsilon^2}\Big\}\, I_0\Bigg[2\sqrt{\Big[\frac{\widehat{\Lambda}(t|t_0)}{\varepsilon}+\frac{\Omega(t|t_0)}{2\varepsilon^2}\Big]\Big[\frac{\widehat{M}(t|t_0)}{\varepsilon}+\frac{\Omega(t|t_0)}{2\varepsilon^2}\Big]}\Bigg]. \tag{56}$$

We recall that $I_\nu(z) \simeq e^z/\sqrt{2\pi z}$ (cf. [31], p. 377, n. 9.71) when $|z|$ is large, for ν fixed. Hence, from (56) as ε is close to zero, one has:

$$\begin{aligned}\frac{\widetilde{p}_{n,n}^{(\varepsilon)}(t|t_0)}{\varepsilon} &\simeq \frac{1}{2\varepsilon\sqrt{\pi}}\Big[\frac{\widehat{\Lambda}(t|t_0)}{\varepsilon}+\frac{\Omega(t|t_0)}{2\varepsilon^2}\Big]^{-1/4}\Big[\frac{\widehat{M}(t|t_0)}{\varepsilon}+\frac{\Omega(t|t_0)}{2\varepsilon^2}\Big]^{-1/4}\\ &\quad\times\exp\Bigg\{-\Bigg[\sqrt{\frac{\widehat{\Lambda}(t|t_0)}{\varepsilon}+\frac{\Omega(t|t_0)}{2\varepsilon^2}}-\sqrt{\frac{\widehat{M}(t|t_0)}{\varepsilon}+\frac{\Omega(t|t_0)}{2\varepsilon^2}}\Bigg]^2\Bigg\}, \qquad n\in\mathbb{Z}.\end{aligned} \tag{57}$$

We note that:

$$\begin{aligned}&\lim_{\varepsilon\downarrow 0}\frac{1}{2\varepsilon\sqrt{\pi}}\Big[\frac{\widehat{\Lambda}(t|t_0)}{\varepsilon}+\frac{\Omega(t|t_0)}{2\varepsilon^2}\Big]^{-1/4}\Big[\frac{\widehat{M}(t|t_0)}{\varepsilon}+\frac{\Omega(t|t_0)}{2\varepsilon^2}\Big]^{-1/4}\\ &\quad=\frac{1}{2\sqrt{\pi}}\lim_{\varepsilon\downarrow 0}\Big[\varepsilon\widehat{\Lambda}(t|t_0)+\frac{\Omega(t|t_0)}{2}\Big]^{-1/4}\Big[\varepsilon\widehat{M}(t|t_0)+\frac{\Omega(t|t_0)}{2}\Big]^{-1/4}=\frac{1}{\sqrt{2\pi\,\Omega(t|t_0)}}\end{aligned}$$

and:

$$\begin{aligned}&\lim_{\varepsilon\downarrow 0}\exp\Bigg\{-\Bigg[\sqrt{\frac{\widehat{\Lambda}(t|t_0)}{\varepsilon}+\frac{\Omega(t|t_0)}{2\varepsilon^2}}-\sqrt{\frac{\widehat{M}(t|t_0)}{\varepsilon}+\frac{\Omega(t|t_0)}{2\varepsilon^2}}\Bigg]^2\Bigg\}\\ &\quad=\lim_{\varepsilon\downarrow 0}\exp\Bigg\{-\big[\widehat{\Lambda}(t|t_0)-\widehat{M}(t|t_0)\big]^2\Bigg[\sqrt{\varepsilon\widehat{\Lambda}(t|t_0)+\frac{\Omega(t|t_0)}{2}}+\sqrt{\varepsilon\widehat{M}(t|t_0)+\frac{\Omega(t|t_0)}{2}}\Bigg]^{-2}\Bigg\}\\ &\quad=\exp\Bigg\{-\frac{\big[\widehat{\Lambda}(t|t_0)-\widehat{M}(t|t_0)\big]^2}{2\,\Omega(t|t_0)}\Bigg\},\end{aligned}$$

so that, taking the limit as $\varepsilon \downarrow 0$ in (57), Equation (55) follows for $n = j$ and $x = x_0$.

(ii) For $n \neq j$, recalling that $I_n(z) = I_{-n}(z)$, from (6), one has:

$$\begin{aligned}\widetilde{p}_{j,n}^{(\varepsilon)}(t|t_0) &= \exp\Big\{-\frac{\widehat{\Lambda}(t|t_0)+\widehat{M}(t|t_0)}{\varepsilon}-\frac{\Omega(t|t_0)}{\varepsilon^2}\Big\}\Big[\frac{2\varepsilon\widehat{\Lambda}(t|t_0)+\Omega(t|t_0)}{2\varepsilon\widehat{M}(t|t_0)+\Omega(t|t_0)}\Big]^{(x-x_0)/(2\varepsilon)}\\ &\quad\times I_{|x-x_0|/\varepsilon}\Bigg[2\sqrt{\Big[\frac{\widehat{\Lambda}(t|t_0)}{\varepsilon}+\frac{\Omega(t|t_0)}{2\varepsilon^2}\Big]\Big[\frac{\widehat{M}(t|t_0)}{\varepsilon}+\frac{\Omega(t|t_0)}{2\varepsilon^2}\Big]}\Bigg].\end{aligned} \tag{58}$$

Making use of the asymptotic result (cf. [31], p. 378, n. 9.7.7):

$$I_\nu(\nu z) \simeq \frac{1}{\sqrt{2\pi\nu}\,(1+z^2)^{1/4}}\exp\Big\{\nu\Big[\sqrt{1+z^2}+\ln\frac{z}{1+\sqrt{1+z^2}}\Big]\Big\}, \qquad \nu\to+\infty,\ 0<z<+\infty,$$

from (58), we obtain:

$$\frac{\widetilde{p}_{j,n}^{(\varepsilon)}(t|t_0)}{\varepsilon} \simeq \prod_{j=1}^{4} A_j^{(\varepsilon)}(x,t|x_0,t_0), \tag{59}$$

where:

$$
\begin{aligned}
A_1^{(\varepsilon)}(x,t|x_0,t_0) &= \left[\frac{2\varepsilon\widehat{\Lambda}(t|t_0)+\Omega(t|t_0)}{2\varepsilon\widehat{M}(t|t_0)+\Omega(t|t_0)}\right]^{(x-x_0)/(2\varepsilon)} \\
A_2^{(\varepsilon)}(x,t|x_0,t_0) &= \frac{1}{\sqrt{2\pi}}\Big\{\varepsilon^2(x-x_0)^2+\Big[2\varepsilon\widehat{\Lambda}(t|t_0)+\Omega(t|t_0)\Big]\Big[2\varepsilon\widehat{M}(t|t_0)+\Omega(t|t_0)\Big]\Big\}^{-1/4} \\
A_3^{(\varepsilon)}(x,t|x_0,t_0) &= \left\{\frac{\sqrt{[2\varepsilon\widehat{\Lambda}(t|t_0)+\Omega(t|t_0)]\,[2\varepsilon\widehat{M}(t|t_0)+\Omega(t|t_0)]}}{\varepsilon\,|x-x_0|+\sqrt{\varepsilon^2\,(x-x_0)^2+[2\varepsilon\widehat{\Lambda}(t|t_0)+\Omega(t|t_0)]\,[2\varepsilon\widehat{M}(t|t_0)+\Omega(t|t_0)]}}\right\}^{\frac{|x-x_0|}{\varepsilon}} \\
A_4^{(\varepsilon)}(x,t|x_0,t_0) &= \exp\Big\{-\frac{\widehat{\Lambda}(t|t_0)+\widehat{M}(t|t_0)}{\varepsilon}-\frac{\Omega(t|t_0)}{\varepsilon^2}\Big\} \\
&\quad\times\exp\Big\{\frac{1}{\varepsilon^2}\sqrt{\varepsilon^2(x-x_0)^2+[2\varepsilon\widehat{\Lambda}(t|t_0)+\Omega(t|t_0)]\,[2\varepsilon\widehat{M}(t|t_0)+\Omega(t|t_0)]}\Big\}.
\end{aligned}
$$

Since:

$$
\begin{aligned}
\lim_{\varepsilon\downarrow 0} A_1^{(\varepsilon)}(x,t|x_0,t_0) &= \exp\Big\{(x-x_0)\,\frac{\widehat{\Lambda}(t|t_0)-\widehat{M}(t|t_0)}{\Omega(t|t_0)}\Big\}, \\
\lim_{\varepsilon\downarrow 0} A_2^{(\varepsilon)}(x,t|x_0,t_0) &= \frac{1}{\sqrt{2\,\pi\,\Omega(t|t_0)}}, \\
\lim_{\varepsilon\downarrow 0} A_3^{(\varepsilon)}(x,t|x_0,t_0) &= \exp\Big\{-\frac{(x-x_0)^2}{\Omega(t|t_0)}\Big\}, \\
\lim_{\varepsilon\downarrow 0} A_4^{(\varepsilon)}(x,t|x_0,t_0) &= \exp\Big\{\frac{(x-x_0)^2}{2\Omega(t|t_0)}\Big\}\,\exp\Big\{-\frac{[\widehat{\Lambda}(t|t_0)-\widehat{M}(t|t_0)]^2}{2\Omega(t|t_0)}\Big\},
\end{aligned}
$$

by taking the limit as $\varepsilon\downarrow 0$ in (59), Equation (55) follows for $n\neq j$ and $x\neq x_0$. □

Finally, the goodness of the heavy-traffic approximation is confirmed by the approximation:

$$\widetilde{p}_{j,n}^{(\varepsilon)}(t|t_0)\simeq\varepsilon\widetilde{f}(x,t|x_0,t_0),$$

which is a consequence of Equation (55) and is valid for ε close to zero.

6. Diffusion Approximation of the Double-Ended Queueing System with Catastrophes and Repairs

In this section, we consider a heavy-traffic approximation of the time-non-homogeneous double-ended queueing system subject to disasters and repairs, discussed in Section 3. The continuous approximation of the discrete model leads to a jump-diffusion process and is similar to the scaling procedure employed in Section 5. The relevant difference is that the state-space of the process $N(t)$ presents also a spurious state F.

Let us now consider the continuous-time Markov process $\{N_\varepsilon(t), t\geq t_0\}$, having state-space $\{F,0,\pm\varepsilon,\pm 2\varepsilon,\ldots\}$. Under suitable limit conditions, as $\varepsilon\downarrow 0$, the scaled process $N_\varepsilon(t)$ converges weakly to a jump-diffusion process $\{X(t), t\geq t_0\}$ having state-space $\{F\}\cup\mathbb{R}$. The limiting procedure is analogous to that used in Buonocore et al. [34], which involves spurious states, as well. As in the previous section, for the approximating procedure, we first assume that the rates $\lambda(t)$ and $\mu(t)$ are modified as in (45). Hence, the limit $\varepsilon\downarrow 0$ leads to a heavy-traffic condition about such intensity functions. Instead, the catastrophe rate $\nu(t)$ and the repair rate $\eta(t)$ are not affected by the scaling procedure.

We note that $X(t)$ describes the motion of a particle, which starts at the origin at time t_0 and then behaves as a non-homogeneous Wiener process, with drift $\widehat{\lambda}(t)-\widehat{\mu}(t)$ and infinitesimal variance $\omega^2(t)$, until a catastrophe occurs. We remark that the catastrophes arrive according to a time-non-homogeneous Poisson process with intensity function $\nu(t)$. As soon as a catastrophe occurs, the process enters into the failure state F and remains therein for a random time (corresponding to the repair time) that ends according to the time-dependent intensity function $\eta(t)$. Clearly, catastrophes are not allowed during a repair period. The effect of a repair is the instantaneous transition of the process $X(t)$ to the state zero. After that, the motion is subject again to diffusion and proceeds as before. We recall that $\nu(t)$ and $\eta(t)$ are positive, bounded and continuous functions for $t \geq t_0$, such that $\int_{t_0}^{+\infty} \nu(t)\,dt = +\infty$ and $\int_{t_0}^{+\infty} \eta(t)\,dt = +\infty$. We denote by:

$$f(x,t|0,t_0) = \frac{\partial}{\partial x} P\{X(t) \leq x | X(t_0) = 0\}, \qquad x \in \mathbb{R},\ t \geq 0 \tag{60}$$

the transition density of $X(t)$ and by $q(t|t_0) = P\{X(t) = F|X(t_0) = 0\}$ the probability that the process is in the failure-state at time t starting from zero at time t_0. We point out that the adopted scaling procedure does not affect the spurious state, so that $q(t|t_0)$ is identical to the analogous probability of the process $N(t)$ and is given in (18). From (15), proceeding similarly as for (46), one obtains that (60) is the solution of the following partial differential equation, for $t > t_0$:

$$\frac{\partial}{\partial t} f(x,t|0,t_0) = -\nu(t)\, f(x,t|0,t_0) - [\widehat{\lambda}(t) - \widehat{\mu}(t)]\, \frac{\partial}{\partial x} f(x,t|0,t_0) + \frac{\omega^2(t)}{2}\frac{\partial^2}{\partial x^2} f(x,t|0,t_0), \quad x \in \mathbb{R} \setminus \{0\}, \tag{61}$$

to be solved with the initial condition $\lim_{t\downarrow t_0} f(x,t|0,t_0) = \delta(x)$ and, in analogy to (20), with the compatibility condition:

$$\int_{-\infty}^{+\infty} f(x,t|0,t_0)\,dx + q(t|t_0) = 1, \qquad t \geq t_0. \tag{62}$$

6.1. Transient Distribution

Similarly to the discrete model discussed in Section 3, the pdf (60) can be expressed as follows, in terms of the transition pdf of the time-non-homogeneous Wiener process $\widetilde{X}(t)$ treated in Section 5:

$$f(x,t|0,t_0) = e^{-V(t|t_0)}\, \widetilde{f}(x,t|0,t_0) + \int_{t_0}^{t} q(\tau|t_0)\,\eta(\tau)\,e^{-V(t|\tau)}\, \widetilde{f}(x,t|0,\tau)\,d\tau, \qquad x \in \mathbb{R},\ t \geq 0, \tag{63}$$

with $q(t|t_0)$ and $\widetilde{f}(x,t|x_0,t_0)$ given in (18) and (50), respectively. Making use of (18) and (50) in (63), for $t \geq t_0$ and $x \in \mathbb{R}$, one has:

$$\begin{aligned} f(x,t|0,t_0) &= \frac{e^{-V(t|t_0)}}{\sqrt{2\pi\,\Omega(t|t_0)}} \exp\left\{-\frac{\left[x - \widehat{\Lambda}(t|t_0) + \widehat{M}(t|t_0)\right]^2}{2\,\Omega(t|t_0)}\right\} \\ &+ \int_{t_0}^{t} d\tau\, \eta(\tau)\, \frac{e^{-V(t|\tau)}}{\sqrt{2\pi\,\Omega(t|\tau)}} \exp\left\{-\frac{\left[x - \widehat{\Lambda}(t|\tau) + \widehat{M}(t|\tau)\right]^2}{2\,\Omega(t|\tau)}\right\} \int_{t_0}^{\tau} \nu(\vartheta) e^{-[V(\tau|\vartheta)+H(\tau|\vartheta)]}\,d\vartheta. \end{aligned} \tag{64}$$

For $r \in \mathbb{N}$, let us now consider the r-th conditional moment of $X(t)$:

$$\mathfrak{M}_r(t|t_0) := \mathrm{E}[X^r(t)|X(t) \in \mathbb{R}, X(t_0) = 0] = \frac{1}{1 - q(t|t_0)} \int_{-\infty}^{+\infty} x^r\, f(x,t|0,t_0)\,dx. \tag{65}$$

From (63), for $r \in \mathbb{N}$, it results:

$$\mathfrak{M}_r(t|t_0) = \frac{1}{1 - q(t|t_0)} \left\{ e^{-V(t|t_0)}\, \widetilde{\mathfrak{M}}_r(t|t_0) + \int_{t_0}^{t} q(\tau|t_0)\,\eta(\tau) e^{-V(t|\tau)}\, \widetilde{\mathfrak{M}}_r(t|\tau)\,d\tau \right\}, \tag{66}$$

where $\widetilde{\mathfrak{M}}_r(t|t_0) := \mathrm{E}[\widetilde{X}^r(t)|\widetilde{X}(t_0) = 0]$ is the r-th conditional moment of $\widetilde{X}(t)$. Hence, by virtue of (51), from (66), we obtain the conditional moments $\mathfrak{M}_r(t|t_0)$.

In the following theorem, we discuss the special case when the functions $\widehat{\lambda}(t) - \widehat{\mu}(t)$ and $\omega^2(t)$ are constant.

Theorem 5. *Consider the process $X(t)$ such that $\widehat{\lambda}(t) - \widehat{\mu}(t) = \widehat{\lambda} - \widehat{\mu}$ and $\omega^2(t) = \omega^2$ for all $t \geq t_0$. Then, for $t \geq t_0$ and $x \in \mathbb{R}$, one has:*

$$f(x,t|0,t_0) = e^{-V(t|t_0)}\widetilde{f}(x,t-t_0|0,0) + \int_0^{t-t_0} dx\, \nu(t-x)e^{-V(t|t-x)} \int_0^x \eta(t-u)e^{-H(t-u|t-x)}\widetilde{f}(x,u|0,0)\, du \quad (67)$$

and, for $r \in \mathbb{N}$,

$$\mathfrak{M}_r(t|t_0) = \frac{1}{1-q(t|t_0)} \left\{ e^{-V(t|t_0)}\widetilde{\mathfrak{M}}_r(t-t_0|0) + \int_0^{t-t_0} q(t-x|t_0)\,\eta(t-x)e^{-V(t|t-x)}\widetilde{\mathfrak{M}}_r(x|0)\, dx \right\}. \quad (68)$$

Furthermore, it results:

$$f(x,t|0,t_0) = \int_{t_0}^t f(0,\tau|0,t_0)\, e^{-V(t|\tau)}\, \widetilde{g}(x,t|0,\tau)\, d\tau, \qquad x \in \mathbb{R}\setminus\{0\},\ t \geq t_0, \quad (69)$$

where $\widetilde{g}(x,t|0,\tau)$ is the FPT pdf of $\widetilde{T}_{0,x}(\tau)$, introduced in Section 5.

Proof. It proceeds similarly to the proof of Theorem 1. □

6.2. Goodness of the Approximating Procedure

Let us now analyze the goodness of the heavy-traffic approximation considered above. The time-non-homogeneous process describing the state of the double-ended queueing system with catastrophes and repairs has been approximated by the diffusion process $X(t)$, whose transition pdf is given in (63).

First of all, we compare the mean, second order moment and variance of $N(t)$ with those of $X(t)/\varepsilon$, when $\lambda(t)$ and $\mu(t)$ are chosen as in (45). By virtue of (53) and (54), one has:

$$\widetilde{\mathfrak{M}}_1(t|t_0) = \mathrm{E}[\widetilde{X}(t)|\widetilde{X}(t_0)=0] = \varepsilon\mathrm{E}[\widetilde{N}(t)|\widetilde{N}(t_0)=0] = \varepsilon\mathcal{M}_1(t|t_0),$$
$$\widetilde{\mathfrak{M}}_2(t|t_0) = \mathrm{E}[\widetilde{X}^2(t)|\widetilde{X}(t_0)=0] \simeq \varepsilon^2\mathrm{E}[\widetilde{N}^2(t)|\widetilde{N}(t_0)=0] = \mathcal{M}_2(t|t_0) \qquad \text{as } \varepsilon \downarrow 0.$$

Hence, recalling (22) and (66), one has:

$$\mathfrak{M}_1(t|t_0) \equiv \mathrm{E}[X(t)|X(t_0)=0] = \varepsilon\mathrm{E}[N(t)|N(t_0)=0] \equiv \varepsilon\mathcal{M}_1(t|t_0).$$

Moreover,

$$\lim_{\varepsilon\downarrow 0} \frac{\mathrm{E}[N^2(t)|N(t_0)=0]}{\mathrm{E}\left[\frac{X^2(t)}{\varepsilon^2}\middle|\frac{X(t_0)}{\varepsilon}=0\right]} = \lim_{\varepsilon\downarrow 0} \frac{\varepsilon^2\left[e^{-V(t|t_0)}\widetilde{\mathcal{M}}_2(t|t_0) + \int_{t_0}^t q(\tau|t_0)\,\eta(\tau)e^{-V(t|\tau)}\widetilde{\mathcal{M}}_2(t|\tau)\,d\tau\right]}{e^{-V(t|t_0)}\widetilde{\mathfrak{M}}_2(t|t_0) + \int_{t_0}^t q(\tau|t_0)\,\eta(\tau)e^{-V(t|\tau)}\widetilde{\mathfrak{M}}_2(t|\tau)\,d\tau} = 1,$$

so that the variances satisfy the following relation, for ε close to zero:

$$\mathrm{Var}[X(t)|X(t_0)=0] \simeq \varepsilon^2\mathrm{Var}[N(t)|N(t_0)=0].$$

Furthermore, we denote by $p_{j,n}^{(\varepsilon)}(t)$ the transition probabilities of the process $N(t)$, when $n = x/\varepsilon$ and the intensity functions $\lambda(t)$ and $\mu(t)$ are given in (45). The following theorem holds.

Theorem 6. *For $t \geq t_0$, one has:*

$$\lim_{\varepsilon \downarrow 0,\, n\varepsilon = x} \frac{p_{0,n}^{(\varepsilon)}(t|t_0)}{\varepsilon} = f(x,t|0,t_0), \tag{70}$$

with $f(x,t|,0,t_0)$ given in (63).

Proof. From (19), one obtains:

$$\frac{p_{0,n}^{(\varepsilon)}(t|t_0)}{\varepsilon} = e^{-V(t|t_0)}\, \frac{\widetilde{p}_{0,n}^{(\varepsilon)}(t|t_0)}{\varepsilon} + \int_{t_0}^{t} q(\tau|t_0)\, \eta(\tau)\, e^{-V(t|\tau)}\, \frac{\widetilde{p}_{0,n}^{(\varepsilon)}(t|\tau)}{\varepsilon}\, d\tau. \tag{71}$$

Taking the limit as $\varepsilon \downarrow 0$ on both sides of (71) and recalling (55), for $t \geq t_0$, one has:

$$\lim_{\varepsilon \downarrow 0,\, n\varepsilon = x} \frac{p_{0,n}^{(\varepsilon)}(t|t_0)}{\varepsilon} = e^{-V(t|t_0)}\, \widetilde{f}(x,t|0,t_0) + \int_{t_0}^{t} q(\tau|t_0)\, \eta(\tau)\, e^{-V(t|\tau)}\, \widetilde{f}(x,t|0,\tau)\, d\tau,$$

so that (70) immediately follows by using (63). □

As a consequence of Theorem 6, for $\lambda(t)$ and $\mu(t)$ chosen as in (45) and under heavy-traffic conditions, the probability $p_{0,n}^{(\varepsilon)}(t|t_0)$ of the discrete process $N(t)$ is close to $\varepsilon\, f(n\varepsilon,t|0,t_0)$ for ε near to zero.

7. Asymptotic Distributions

Similar to the analysis performed in Section 4, in this section, we consider the asymptotic behavior of the density $f(x,t|0,t_0)$ of the process $X(t)$ in two different cases:

(i) the functions $\widehat{\lambda}(t)$, $\widehat{\mu}(t)$, $\omega^2(t)$, $\nu(t)$ and $\eta(t)$ admit finite positive limits as $t \to +\infty$,
(ii) the functions $\widehat{\lambda}(t)$, $\widehat{\mu}(t)$ and $\omega^2(t)$ are constant, and the rates $\nu(t)$ and $\eta(t)$ are periodic functions with common period Q.

7.1. Asymptotically-Constant Intensity Functions

We assume that the functions $\widehat{\lambda}(t)$, $\widehat{\mu}(t)$, $\omega^2(t)$, $\nu(t)$ and $\eta(t)$ admit finite positive limits as t tends to $+\infty$. In this case, the failure asymptotic probability $q^* = \lim_{t\to+\infty} q(t|t_0)$ of the process $X(t)$ is provided in (29). Moreover, the steady-state density of the process $X(t)$ is an asymmetric bilateral exponential density, as given in the following theorem.

Theorem 7. *Assuming that:*

$$\lim_{t\to+\infty} \lambda(t) = \lambda, \quad \lim_{t\to+\infty} \mu(t) = \mu, \quad \lim_{t\to+\infty} \omega^2(t) = \omega^2, \quad \lim_{t\to+\infty} \nu(t) = \nu, \quad \lim_{t\to+\infty} \eta(t) = \eta, \tag{72}$$

with $\widehat{\lambda}$, $\widehat{\mu}$, ω^2, ν, η positive constants, then the steady-state pdf of the process $X(t)$ is, for $x \in \mathbb{R}$,

$$f^*(x) := \lim_{t\to+\infty} f(x,t|0,t_0) = \frac{\eta\nu}{\eta+\nu} \frac{1}{\sqrt{(\widehat{\lambda}-\widehat{\mu})^2 + 2\omega^2\nu}} \exp\Big\{ \frac{(\widehat{\lambda}-\widehat{\mu})}{\omega^2}\, x - \frac{\sqrt{(\widehat{\lambda}-\widehat{\mu})^2 + 2\omega^2\nu}}{\omega^2}\, |x| \Big\}. \tag{73}$$

Furthermore, the asymptotic conditional mean, second order moment and variance are:

$$\begin{aligned} &\lim_{t\to+\infty} \mathfrak{M}_1(t|t_0) = \frac{\widehat{\lambda}-\widehat{\mu}}{\nu}, \qquad \lim_{t\to+\infty} \mathfrak{M}_2(t|t_0) = \frac{2(\widehat{\lambda}-\widehat{\mu})^2}{\nu^2} + \frac{\omega^2}{\nu}, \\ &\lim_{t\to+\infty} \mathrm{Var}(t|t_0) = \lim_{t\to+\infty} \{\mathfrak{M}_2(t|t_0) - [\mathfrak{M}_1(t|t_0)]^2\} = \frac{(\widehat{\lambda}-\widehat{\mu})^2}{\nu^2} + \frac{\omega^2}{\nu}. \end{aligned} \tag{74}$$

Proof. The steady-state density can be obtained by taking the limit as $t \to +\infty$ in Equations (61) and (62) and recalling (29). Moreover, the asymptotic conditional mean and variance (74) follow from (65), making use of (29) and (73). $\square$

7.2. Periodic Intensity Functions

Let us assume that the functions $\widehat{\lambda}(t)$, $\widehat{\mu}(t)$ and $\omega^2(t)$ are constant and that the catastrophe intensity function $\nu(t)$ and the repair intensity function $\eta(t)$ are periodic, so that $\nu(t+kQ) = \nu(t)$ and $\eta(t+kQ) = \eta(t)$ for $k \in \mathbb{N}$ and $t \geq t_0$. The average catastrophe and repair rates in the period Q are defined in (32). The asymptotic distribution of the process $X(t)$ is described by the following functions, for $t \geq t_0$,

$$q^*(t) := \lim_{k\to+\infty} q(t+kQ|t_0), \qquad f^*(x,t) := \lim_{k\to+\infty} f(x,t+kQ|0,t_0), \quad x \in \mathbb{R}. \tag{75}$$

Note that the asymptotic failure probability $q^*(t)$ is given in (36) or, alternatively, in (37). Moreover, the asymptotic density $f^*(x,t)$ is determined in the following theorem.

Theorem 8. *Consider the stochastic process $X(t)$ and assume $\widehat{\lambda}(t) - \widehat{\mu}(t) = \widehat{\lambda} - \widehat{\mu}$ and $\omega^2(t) = \omega^2$, and that the intensities $\nu(t)$ and $\eta(t)$ are continuous, positive and periodic functions with period Q. Then, one has:*

$$f^*(x,t) = \int_0^{+\infty} dx\, \nu(t-x) e^{-V(t|t-x)} \int_0^x \eta(t-u) e^{-H(t-u|t-x)}\, \widetilde{f}(x,u|0,0)\, du, \qquad x \in \mathbb{R}. \tag{76}$$

Proof. It proceeds similarly to the proof of Theorem 3, by starting from Equation (67). $\square$

By virtue of the periodicity of $\nu(t)$ and $\eta(t)$, from (76), one has that $f^*(x,t)$ is a periodic function with period Q. From (65), making use of (76), the asymptotic conditional moments are:

$$\begin{aligned}\mathfrak{M}_r^*(t) &:= \lim_{k\to+\infty} \mathfrak{M}_r(t+kQ|t_0) \\ &= \frac{1}{1-q^*(t)} \int_0^{+\infty} dx\, \nu(t-x) e^{-V(t|t-x)} \int_0^x \eta(t-u) e^{-H(t-u|t-x)}\, \widetilde{\mathfrak{M}}_r(u|0)\, du,\end{aligned} \tag{77}$$

where $\widetilde{\mathfrak{M}}_r(t|0) = E[\widetilde{X}(t)^r | \widetilde{X}(0) = 0]$ and where $q^*(t)$ is given in (36) or in (37).

The following illustrative example concludes the section.

Example 2. *Let $X(t)$ be the approximating jump-diffusion process, subject to disasters and repairs, with drift $\widehat{\lambda} - \widehat{\mu}$ and infinitesimal variance ω^2, where $\widehat{\lambda} = 2.0$, $\widehat{\mu} = 1.0$ and $\omega^2 = 0.2$ and with periodic catastrophe intensity function $\nu(t)$ and repair intensity function $\eta(t)$ given by (44). The parameters ν, a, η, b, Q are taken as in Example 1. For these choices, probability $q(t|0)$ is identical as for the discrete model. It is plotted in Figure 3, on the right.*

We now consider the two choices $\varepsilon = 0.05$ and $\varepsilon = 0.025$. Then, the parameters λ and μ are determined according to (45), so that for $\varepsilon = 0.05$, we have $\lambda = 80$ and $\mu = 60$, whereas for $\varepsilon = 0.025$, we have $\lambda = 240$ and $\mu = 200$. To show the validity of the approximating procedure, we compare the quantity $\varepsilon f(\varepsilon n, t|0,0)$ with the probability $p_{0,n}(t|0)$, for $n = 0, -1, 1$, in Figures 6–8, respectively. The case $\varepsilon = 0.05$ is shown on the left, and $\varepsilon = 0.025$ is on the right. Recall that $f(x,t|0,0)$ is given in (67), whereas $p_{0,n}(t|0)$ is given in (23). We note that the goodness of the continuous approximation for $p_{0,n}(t|0)$ improves as ε decreases, this corresponding to an increase of traffic in the double-ended queue with catastrophes and repairs, due to (45).

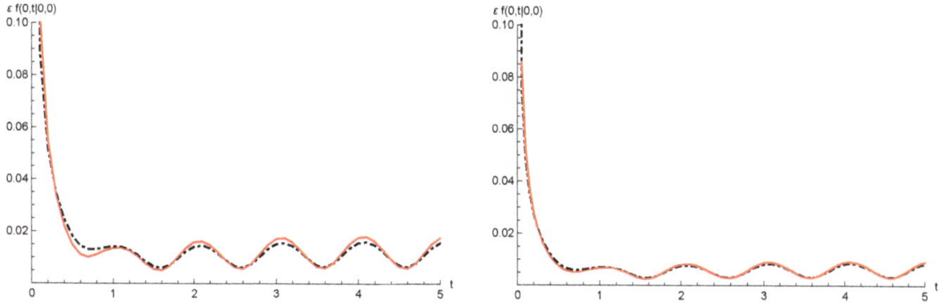

Figure 6. For $\widehat{\lambda} = 2.0$, $\widehat{\mu} = 1.0$, $\omega^2 = 0.2$, the function $\varepsilon f(0, t|0, 0)$ (red curve) is shown with the probability $p_{0,0}(t|0)$ (black dashed curve) for $\varepsilon = 0.05$ (left) and $\varepsilon = 0.025$ (right). The parameters λ and μ are shown in Example 2, according to (45).

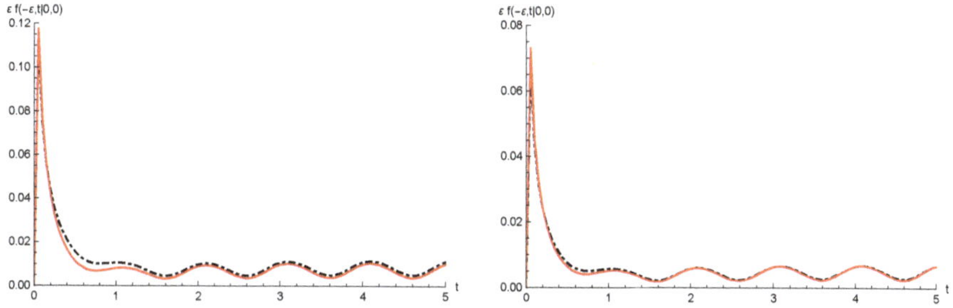

Figure 7. For the same choices of parameters of Figure 6, the function $\varepsilon f(-\varepsilon, t|0, 0)$ (red curve) is compared with the probability $p_{0,-1}(t|0)$ (black dashed curve) for $\varepsilon = 0.05$ (left) and $\varepsilon = 0.025$ (right).

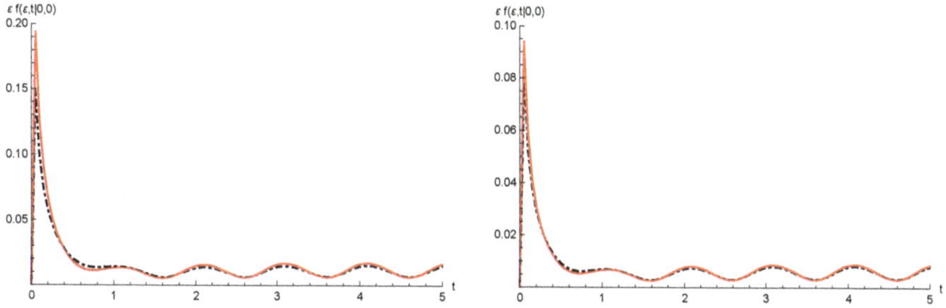

Figure 8. For the same choices of parameters of Figure 6, the function $\varepsilon f(\varepsilon, t|0, 0)$ (red curve) is compared with the probability $p_{0,1}(t|0)$ (black dashed curve) for $\varepsilon = 0.05$ (left) and $\varepsilon = 0.025$ (right).

8. Conclusions

We analyzed a continuous-time stochastic process that describes the state of a double-ended queue subject to disasters and repairs. The system is time-non-homogeneous, since arrivals, services, disasters and repairs are governed by time-varying intensity functions. This model is a suitable generalization of the queueing system investigated in [10]. Indeed, the previous model is characterized by constant rates of arrivals, services, catastrophes and repairs. However, motivated by the need to describe more realistic situations in which the system evolution reflects daily or seasonal fluctuations, in this paper, we investigated the case where all such rates are time-dependent. Whereas in the previous model, the approach involved the Laplace transforms, in the present case, the analysis cannot be based on such a method, but rather on a direct analysis of the relevant equations. Our analysis involved also the heavy-traffic approximation of the system, which leads to a time-non-homogeneous diffusion process useful to describe the queue-length dynamics via more manageable formulas. Future developments of the present investigation will be centered on the inclusion of multiple types of customers and more general forms of catastrophe/repair mechanisms.

Author Contributions: All the authors contributed equally to this work.

Acknowledgments: A.D.C., V.G. and A.G.N. are members of the group GNCS of INdAM.

Conflicts of Interest: The authors declare no conflict of interest.

References

1. Kashyap, B.R.K. A double-ended queueing system with limited waiting space. *Proc. Natl. Inst. Sci. India A* **1965**, *31*, 559–570. doi:10.1007/BF02613588. [CrossRef]
2. Kashyap, B.R.K. The double-ended queue with bulk service and limited waiting space. *Oper. Res.* **1966**, *14*, 822–834. doi:10.1287/opre.14.5.822. [CrossRef]
3. Sharma, O.P.; Nair, N.S.K. Transient behaviour of a double ended Markovian queue. *Stoch. Anal. Appl.* **1991**, **9**, 71–83. doi:10.1080/07362999108809226. [CrossRef]
4. Tarabia, A.M.K. On the transient behaviour of a double ended Markovian queue. *J. Combin. Inf. Syst. Sci.* **2001**, *26*, 125–134.
5. Conolly, B.W.; Parthasarathy, P.R.; Selvaraju, N. Doubled-ended queues with impatience. *Comput. Oper. Res.* **2002**, *29*, 2053–2072. doi:10.1016/S0305-0548(01)00075-2. [CrossRef]
6. Elalouf, A.; Perlman, Y.; Yechiali, U. A double-ended queueing model for dynamic allocation of live organs based on a best-fit criterion. *Appl. Math. Model.* **2018**, in press. doi:10.1016/j.apm.2018.03.022. [CrossRef]
7. Takahashi, M.; Ōsawa, H.; Fujisawa, T. On a synchronization queue with two finite buffers. *Queueing Syst.* **2000**, *36*, 107–123. doi:10.1023/A:1019127002333. [CrossRef]
8. Di Crescenzo, A.; Giorno, V.; Nobile, A.G. Constructing transient birth-death processes by means of suitable transformations. *Appl. Math. Comput.* **2016**, *281*, 152–171. doi:10.1016/j.amc.2016.01.058. [CrossRef]
9. Di Crescenzo, A.; Giorno, V.; Nobile, A.G. Analysis of reflected diffusions via an exponential time-based transformation. *J. Stat. Phys.* **2016**, *163*, 1425–1453. doi:10.1007/s10955-016-1525-9. [CrossRef]
10. Di Crescenzo A.; Giorno V.; Krishna Kumar B.; Nobile, A.G. A double-ended queue with catastrophes and repairs, and a jump-diffusion approximation. *Methodol. Comput. Appl. Probab.* **2012**, *14*, 937–954. doi:10.1007/s11009-011-9214-2. [CrossRef]
11. Altiok, T. On the phase-type approximations of general distributions. *IIE Trans.* **1985**, *17*, 110–116. [CrossRef]
12. Altiok, T. Queueing modeling of a single processor with failures. *Perform. Eval.* **1989**, *9*, 93–102. [CrossRef]
13. Altiok, T. *Performance analysis of manufacturing systems. Springer Series in Operations Research*; Springer: New York, NY, USA, 1997.
14. Dallery, Y. On modeling failure and repair times in stochastic models of manufacturing systems using generalized exponential distributions. *Queueing Syst.* **1994**, *15*, 199–209. [CrossRef]
15. Economou, A.; Fakinos, D. A continuous-time Markov chain under the influence of a regulating point process and applications in stochastic models with catastrophes. *Eur. J. Oper. Res.* **2003**, *149*, 625–640. doi:10.1016/S0377-2217(02)00465-4. [CrossRef]
16. Economou, A.; Fakinos, D. Alternative approaches for the transient analysis of Markov chains with catastrophes. *J. Stat. Theory Pract.* **2008**, *2*, 183–197. doi:10.1080/15598608.2008.10411870. [CrossRef]
17. Kyriakidis, E.; Dimitrakos, T. Computation of the optimal policy for the control of a compound immigration process through total catastrophes. *Methodol. Comput. Appl. Probab.* **2005**, *7*, 97–118. doi:10.1007/s11009-005-6657-3. [CrossRef]
18. Krishna Kumar, B.; Vijayakumar, A.; Sophia, S. Transient analysis for state-dependent queues with catastrophes. *Stoch. Anal. Appl.* **2008**, *26*, 1201–1217. doi:10.1080/07362990802405786. [CrossRef]
19. Di Crescenzo, A.; Giorno, V.; Nobile, A.G.; Ricciardi, L.M. A note on birth-death processes with catastrophes. *Stat. Probab. Lett.* **2008**, *78*, 2248–2257. doi:10.1016/j.spl.2008.01.093. [CrossRef]
20. Zeifman, A.; Korotysheva, A. Perturbation bounds for $M_t/M_t/N$ queue with catastrophes. *Stoch. Models* **2012**, *28*, 49–62. doi:10.1080/15326349.2011.614900. [CrossRef]
21. Zeifman, A.; Satin, Y.; Panfilova, T. Limiting characteristics for finite birth-death-catastrophe processes. *Math. Biosci.* **2013**, *245*, 96–102. doi:10.1016/j.mbs.2013.02.009. [CrossRef] [PubMed]

22. Giorno, V.; Nobile, A.G.; Pirozzi, E. A state-dependent queueing system with asymptotic logarithmic distribution. *J. Math. Anal. Appl.* **2018**, *458*, 949–966. doi:10.1016/j.jmaa.2017.10.004. [CrossRef]
23. Di Crescenzo, A.; Giorno, V.; Nobile, A.G.; Ricciardi, L.M. On time non-homogeneous stochastic processes with catastrophes. In *Cybernetics and Systems 2010, Proceedings of the Austrian Society for Cybernetics Studies (EMCSR 2010), Vienna, Austria, 6–9 April 2010*; Trappl, R., Ed.; Austrian Society for Cybernetic Studies: Vienna, Austria, 2010; pp. 169–174.
24. Giorno, V.; Nobile, A.G.; Spina, S. On some time non-homogeneous queueing systems with catastrophes. *Appl. Math. Comput.* **2014**, *245*, 220–234. doi:10.1016/j.amc.2014.07.076. [CrossRef]
25. Dimou, S.; Economou, A. The single server queue with catastrophes and geometric reneging. *Methodol. Comput. Appl. Probab.* **2013**, *15*, 595–621. doi:10.1007/s11009-011-9271-6. [CrossRef]
26. Giorno, V.; Nobile, A.G.; Ricciardi, L.M. On some time-non-homogeneous diffusion approximations to queueing systems. *Adv. Appl. Probab.* **1987**, *19*, 974–994. doi:10.2307/1427111. [CrossRef]
27. Liu, X.; Gong, Q.; Kulkarni, V.G. Diffusion models for double-ended queues with renewal arrival processes. *Stoch. Syst.* **2015**, *5*, 1–61. doi:10.1214/13-SSY113. [CrossRef]
28. Büke, B.; Chen, H. Fluid and diffusion approximations of probabilistic matching systems. *Queueing Syst.* **2017** *86*, 1–33. doi:10.1007/s11134-017-9516-3. [CrossRef]
29. Di Crescenzo, A.; Giorno, V.; Nobile, A.G.; Ricciardi, L.M. On the $M/M/1$ queue with catastrophes and its continuous approximation. *Queueing Syst.* **2003**, *43*, 329–347. doi:10.1023/A:1023261830362. [CrossRef]
30. Dharmaraja, S.; Di Crescenzo, A.; Giorno, V.; Nobile, A.G. A continuous-time Ehrenfest model with catastrophes and its jump-diffusion approximation. *J. Stat. Phys.* **2015**, *161*, 326–345. doi:10.1007/s10955-015-1336-4. [CrossRef]
31. Abramowitz, M.; Stegun, I. *Handbook of Mathematical Functions with Formulas, Graphs and Mathematical Tables*; Dover Publications, Inc.: New York, NY, USA, 1972.
32. Irwin, J.O. The frequency-distribution of the difference between two independent variates following the same Poisson distribution. *J. R. Stat. Soc.* **1937**, *100*, 415–416. [CrossRef]
33. Skellam, J.G. The frequency-distribution of the difference between two Poisson variates belonging to different populations distribution. *J. R. Stat. Soc.* **1946**, *109*, 296. [CrossRef]
34. Buonocore, A.; Di Crescenzo, A.; Giorno, V.; Nobile, A.G.; Ricciardi, L.M. A Markov chain-based model for actomyosin dynamics. *Sci. Math. Jpn.* **2009**, *70*, 159–174.

mathematics

MDPI

Article

On the Number of Periodic Inspections During Outbreaks of Discrete-Time Stochastic SIS Epidemic Models

Maria Gamboa and Maria Jesus Lopez-Herrero *

Department of Statistics and Data Science, Faculty of Statistical Studies, Complutense University of Madrid, 28040 Madrid, Spain; mgamboa@ucm.es

* Correspondence: lherrero@ucm.es

Received: 4 July 2018; Accepted: 20 July 2018; Published: 24 July 2018

Abstract: This paper deals with an infective process of type SIS, taking place in a closed population of moderate size that is inspected periodically. Our aim is to study the number of inspections that find the epidemic process still in progress. As the underlying mathematical model involves a discrete time Markov chain (DTMC) with a single absorbing state, the number of inspections in an outbreak is a first-passage time into this absorbing state. Cumulative probabilities are numerically determined from a recursive algorithm and expected values came from explicit expressions.

Keywords: discrete time stochastic model; first-passage time; time between inspections

1. Introduction

The spread of infectious diseases is a major concern for human populations. Disease control by any therapy measure lies on the understanding of the disease itself. Epidemic modeling is an interdisciplinary subject that can be addressed from deterministic applied mathematical models and stochastic processes theory [1,2] to quantitative social and biological science and empirical analysis of data [3]. Epidemic models are also used to describe spreading processes connected with technology, business, marketing or sociology, where the interest is related to dissemination of news, rumors or ideas among different groups [4–8]. Mathematical models provide an essential tool for understanding and forecasting the spread of infectious diseases and also to suggest control policies.

There is a large variety of models for describing the spread of an infective process [1,2]. An essential distinction is done between deterministic and stochastic models. Deterministic models constitute a vast majority of the existing literature and are formulated in terms of ordinary differential equations (ODE); consequently they predict the same dynamic for an infective process given the same initial conditions. However this is not what it is expected to happen in real world diseases, outbreaks do not involve the same people becoming infected at the same time and uncertainty should be included when modeling diseases. The stochastic models [2], analogous to those defined by ODE, take into account the random nature of the events and, hence, they are mainly used to measure probabilities of major outbreaks, disease extinction and, in general, to make statistical analysis of some relevant epidemic descriptors.

In any case, both deterministic and stochastic frameworks are important but stochastic models seems to be more appropriate to describe the evolution of an infective process evolving in a small community, rather than their deterministic counterparts (e.g., [2,9]). In deterministic models, persistence or extinction of the epidemic process is determined by the basic reproduction number, R_0 (see for instance [10,11]). Usually, stochastic models inherited the basic reproduction number from their deterministic counterparts however, in stochastic epidemic where Markov chains model disease spread (see [12,13]), two alternative random variables (namely, the exact reproduction number and the population transmission number) provide a real measurement of the spread of the disease at the initial time or at any time during the epidemic process.

Most of the research works, both deterministic and stochastic, deal with continuous-time models. However, one of the earliest works is the model studied by Reed and Frost in 1928, who formulate an SIR model using a discrete-time Markov chain (DTMC) [1]. In recent years, literature shows an increasing interest in using discrete-time models. Emmert and Allen, in [14], consider a discrete-time model to investigate the spread of a disease within a structured host population. Allen and van den Driessche [15] applied the next generation approach for calculating the basic reproduction number to several models related to hantavirus and chytridiomycosis wildlife diseases. In [16], authors introduce probabilities to formulate death, recovery and incidence rates of epidemic diseases on discrete-time SI and SIS epidemic models with vital dynamics. Bistability of discrete-time SI and SIS epidemic models with vertical transmission is the subject matter of [17]. An SIS model with a temporary vaccination program is studied in [18]. D'Onofrio et al. [19] analyze a discrete-time SIR model where vaccination coverage depends on the risk of infection. In [20,21], the Reed–Frost model is generalized by Markovian models of SIR and SEIR types, where transition probabilities depend on the total number of diseased, the number of daily encounters and the probability of transmission per contact. Under demographic population dynamics, van den Driessche and Yakubu [22] use the next generation method to compute R_0 and to investigate disease extinction and persistence. An approximation of the deterministic multiple infectious model by a Branching process is employed in [23] to extract information about disease extinction. Accuracy of discrete-time approaches for studying continuous time contagion dynamics is the topic developed in [24], who show potential limitations of this approach depending on the time-step considered.

In this paper we deal with a stochastic SIS epidemic model, that describes diseases such as tuberculosis, meningitis or some sexually transmitted diseases, in which infected individuals do not present an exposed period and are recovered with no immunity. Hence, individuals have reoccurring infections. The host population is divided into two groups: susceptible (S) or infected (I). We assume that disease transmission depends on the number of infective individuals and also on a contact rate, α; in addition individual recoveries depend on the recovery rate γ. For any event (in our model, either contact or recovery), the event rate or intensity provides the mean number of events per unit time. Hence, in case of recoveries $1/\gamma$ denotes the mean infectious time. To control the epidemic spread, the population is observed at a fixed time interval.

The aim of this paper is to analyze, for a discrete-time SIS model, the number of inspections that find an active epidemic. We remark that this quantity is the discrete-time analogous of the extinction time that describes the length of the epidemic process.

The extinction time has been the subject of interest of many papers. Many of them focus on the determination of the moments and a few also on whole distribution. In that sense, assuming a finite birth-death process, Norden [25] first obtained an explicit expression for the mean time to extinction and established that the extinction time, when the initial distribution equals the quasi-stationary distribution, follows a simple exponential distribution. Allen and Burgin [26], for SIS and SIR models in discrete-time, investigated numerically the expected duration. Stone et al. [27], for an SIS model with external infection, determine also expressions for higher order moments. Artalejo et al. [28], for general birth-death and SIR models, develop algorithmic schemes to analyze Laplace transforms and moments of the extinction time and other continuous measures.

We model the evolution of the epidemics in terms of an absorbing DTMC providing the amount of infective individuals at each stage or inspection point, that introduces in the model individual variations coming from chance circumstances. As the extinction of the epidemic process is certain, we will investigate both the distribution and expected values of the number of inspections taking place prior the epidemic end, conditioned to the initial number of infected.

The rest of the paper is organized as follows. In Section 2, we introduce the discrete-time SIS model and the DTMC describing the evolution of the epidemic. In Section 3, we present recursive results and develop algorithmic schemes for the distribution of the random variable representing the number of inspections that find an active epidemic process. For any number of initial infected individuals, the expected number of inspections will be explicitly determined. Finally, numerical results regarding the effect of the model parameters are displayed in Section 4. That also includes an application to evaluate cost and benefit per outbreak in such epidemics.

2. Model Description

Let us consider a closed population of N individuals that is affected by a communicable disease transmitted by direct contact. In addition, it is assumed that there are no births or deaths during disease outbreaks, therefore population size remains constant. Individuals in the population are classified as susceptible, S, or infective, I, according to their health state regarding the disease. Transitions from susceptible to infective depend on the contact rate between individuals and also on the quantity of infective present in the population. Once recovered, individuals become again susceptible to the disease. Consequently, individuals can be infected several times during an epidemic process but the epidemic stops as soon as there are no infective individuals to transmit the contagious disease.

In a discrete-time study, time is discretized into time steps Δt and transitions from states occur during this time interval with certain probabilities. In chain-Binomial models [1,3] the time step corresponds to the length of the incubation period, contact process depend on the Binomial distribution and during a fixed time interval zero, one or even more infections may happen. In [16] time step is one and probabilities depend on effective transmission through the time. In [23] a branching process describes transitions and survival probabilities during any stage.

In our model, the population is observed periodically at time points $n * \Delta t$, with $n \geq 0$, where interval's length Δt is chosen as to guarantee that between consecutive inspections at most one change—either an infection or a recovery—occurs. The underlying mathematical model is the discrete-time SIS model described, for instance, in [26] assuming zero births or deaths per individual in the time interval Δt.

Due to the constant population hypothesis, the evolution of the disease can be represented by a one-dimensional Markov chain, $\{I_n; n \geq 0\}$ where I_n is a random variable giving the number of infective individuals in the population at the $n-$th inspection. State space is finite and contains a single absorbing state, the state zero.

Non-negative transition probabilities depend on the time interval and have the following form, for $0 \leq i \leq N$:

$$\begin{aligned} P\{ I_{n+1} = i-1 | \, I_n = i\} &= p_{i,i-1} = \gamma i \Delta t, \\ P\{ I_{n+1} = i | \, I_n = i\} &= p_{i,i} = 1 - \gamma i \Delta t - \frac{\alpha}{N} i(N-i)\Delta t, \\ P\{ I_{n+1} = i+1 | \, I_n = i\} &= p_{i,i+1} = \frac{\alpha}{N} i(N-i)\Delta t, \end{aligned} \tag{1}$$

where α represents the contact rate and γ represents the recovery rate.

We need to fix an interval length, Δt, providing that probabilities in (1) are well defined and therefore that the chosen time step guarantees that at most one change occurs between successive inspections. In particular, for any choice on the model parameters it is required that,

$$1 - \gamma i \Delta t - \frac{\alpha}{N} i(N-i)\Delta t \geq 0, \text{ for } 0 \leq i \leq N.$$

Which, after some algebra, provides a bound for time step. Hence, in what follows Δt will be chosen small enough as to satisfy

$$\Delta t \leq \begin{cases} \frac{1}{\gamma N}, & \text{if } \alpha \leq \gamma, \\ \frac{4\alpha}{(\alpha+\gamma)^2 N}, & \text{if } \alpha > \gamma. \end{cases} \tag{2}$$

Notice that the bound given in (2) can be written in terms of the basic reproduction number, $R_0 = \alpha/\gamma$, as in [26].

As $\{I_n : n \geq 0\}$ is a reducible aperiodic DTMC, with a single absorbing state, the standard theory of Markov chains (see for instance [29]) gives that

$$\lim_{n\to\infty} P\{I_n = j \,|\, I_0 = i\} = \delta_{j0}, \text{ for every } 1 \leq i, j \leq N, \tag{3}$$

where δ_{ab} is the Kronecker's delta, defined as one for $a = b$ and 0, otherwise.

The limiting behavior result (3) states that in the long term, for any choice on model parameters, there will be no infective individuals in the population. Hence, the end of any outbreak of the disease occurs almost surely, but it may take a long number of inspections until the disease disappears from the population; as it was observed in [27,28] for continuous-time models. Thus, a theoretical study of this random variable is well supported.

3. Analysis of the Number of Inspections

We consider the random variable T that counts the number of inspections of the population that find an active epidemic process; i.e., T is the number of steps that it takes, to the DTMC $\{I_n : n \geq 0\}$, to reach the state zero. Thus, T can be seen as a first-passage time and we define it as

$$T = \min\{n \geq 0 : I_n = 0\}.$$

In this section we describe its probabilistic behavior in terms of distribution functions and expected values. Theoretical discussion is based on the conditional first-passage times T_i, for $1 \leq i \leq N$, defined as the number of inspections that take place during an outbreak, given that at present population contains $I_0 = i$ infected. Notice that, even for a finite population, $T_i = (T|I_0 = i)$ is a discrete random variable with countable infinite mass points.

Next we introduced some notation for point and cumulative probabilities, and expected values regarding random variables T_i, for $1 \leq i \leq N$.

$$\begin{aligned} \alpha_i(n) &= P\{T_i = n\} = P\{T = n \,|\, I_0 = i\}, \text{ for } n \geq 0, \\ u_i(n) &= P\{T_i \leq n\} = P\{T \leq n \,|\, I_0 = i\}, \text{ for } n \geq 0, \\ m_i &= E[T_i] = E[T \,|\, I_0 = i]. \end{aligned}$$

We want to point out some trivial facts. Notice that

$$\alpha_i(0) = u_i(0) = 0, \text{ for } 1 \leq i \leq N, \tag{4}$$

$$u_i(n) = 0, \text{ whenever } 0 \leq n < i, \tag{5}$$

(4) is trivially true due to the definition of T_i as a first-passage time. On the other hand, condition (5) follows from the fact that, starting with i infective individuals and as at each inspection we observe at most one event, we need at least i inspections in order to observe that all initial infected have been recovered. Because point probabilities, $\alpha_i(n)$, can be determined from cumulative probabilities, $u_i(n)$, with the help of the well-known relationship

$$\alpha_i(n) = u_i(n) - u_i(n-1), \text{ for } n \geq 1 \text{ and } 1 \leq i \leq N,$$

we present results in order to deal with the cumulative ones.

Next theorem provides a recursive scheme for determining cumulative probabilities associated to an initial number of infective individuals.

Theorem 1. *For any initial number of infective individuals, $I_0 = i$ with $1 \leq i \leq N$, the set of cumulative probabilities satisfies the following recursive conditions, for $n \geq 1$:*

$$\begin{aligned} u_i(n) &= p_{i,0} + (1-\delta_{i1})p_{i,i-1}u_{i-1}(n-1) \\ &\quad + p_{i,i}u_i(n-1) + (1-\delta_{iN})p_{i,i+1}u_{i+1}(n-1). \end{aligned} \tag{6}$$

Proof. The proof is an application of a first-step analysis by conditioning on the first transition out of the current state. □

Remark 1. *Notice that $P\{T_i < \infty\} = 1$, for $1 \leq i \leq N$, because states $\{1, 2, ..., N\}$ are a non-decomposable set of states. Consequently, T_i is a non-defective random variable and $\lim_{n \to \infty} u_i(n) = 1$, for $1 \leq i \leq N$.*

Remark 2. *The use of the iterative Equation (6) produces a sequence of increasing probabilities converging to one but, for computational purposes, a stopping criteria should be provided in order to avoid longer computation runs.*

For each number of inspections, $n \geq 0$, Equation (6) is solved recursively, with the help of the boundary conditions (4) and the trivial result (5). Numerical results appearing in the following section have been obtained with the help of a recursive algorithm, that stops as soon as a certain percentile value of the distribution is accumulated. For each initial number of infective individuals, $I_0 = i_0$, cumulative probabilities are computed up to the q-th percentile, using the following pseudo-code.

Remark 3. *For any appropriate time interval Δt satisfying (2), point and cumulative probabilities are determined numerically from Algorithm 1. Moreover, for $n = 1, 2$ inspections they present the following explicit forms:*

$$P\{T_i = 1\} = \delta_{i1}\gamma\Delta t, \text{ for } 1 \leq i \leq N,$$

$$P\{T_i \leq 2\} = \begin{cases} \gamma\Delta t\left(2 - \frac{2\alpha(N-2)}{N}\Delta t - \gamma\Delta t\right), & \text{for } i = 1, \\ 2(\gamma\Delta t)^2, & \text{for } i = 2, \\ 0, & \text{for } 3 \leq i \leq N. \end{cases}$$

Explicit values displayed in Remark 3 indicate that distribution of the random variable T_i depends on rates α and γ not only through its ratio. Consequently, models sharing the same basic reproductive number, R_0, present different probabilistic characteristics.

Algorithm 1:

The sequence of cumulative probabilities conditioned to the current number of infective individuals, $\{u_i(n) : n \geq 0\}$, for $1 \leq i \leq N$, are determined as follows:

Step 0: Set $q \in (0, 1)$ and $i_0 \geq 1$.
Step 1: Set $n = 0$ and $u_i(n) = 0$, for $1 \leq i \leq N$.
Step 2: Set $n = n + 1$ and $i = 0$.
Step 2a: Set $i = i + 1$ and $u_i(n) = 0$.
Step 2b: If $i \leq n$ compute $u_i(n)$ using Equation (6).
Step 2c: If $i = i_0$ and $u_i(n) \geq q$ then Stop.
Step 2d: If $i < N$ go to step 2a.
Step 3: Go to step 2.

Expected values m_i, for $1 \leq i \leq N$, provide the long-run average value of inspections prior to the epidemic end, given that the outbreak started with i infected. Typically, expected values can be computed from mass distribution functions but, in our case, the use of the recursive procedure given in

Algorithm 1 produces a lower approximation of the true value. Instead of that, next theorem provides a closed form expression for m_i, given any initial number of infective individuals.

Theorem 2. *Expected values $m_i = E[T_i]$, present the following form:*

$$m_i = \frac{1}{\gamma\Delta t}\sum_{k=1}^{i} A_k + \frac{1}{N\gamma\Delta t}\sum_{k=1}^{i}(N-k)!\left(\frac{\alpha}{N\gamma}\right)^{N-k}, \text{ for } 1 \le i \le N-1, \tag{7}$$

$$m_N = \frac{1}{\gamma\Delta t}\sum_{k=1}^{N-1} A_k + \frac{1}{N\gamma\Delta t}\sum_{k=0}^{N-1} k!\left(\frac{\alpha}{N\gamma}\right)^{k}. \tag{8}$$

where $A_k = \frac{1}{k} + \sum_{j=k+1}^{N-1} \frac{(\alpha/N\gamma)^{j-k}}{j} \prod_{s=k}^{j-1}(N-s)$, *for* $1 \le k \le N-1$.

Proof. The proof is based on a first-step argument. By conditioning on the state visited by the underlying Markov chain after first transition, we get the following set of equations that involve expected values m_i, for $1 \le i \le N$ initial infective.

$$\begin{aligned} m_i &= E[T\,|I_0 = i] = E[T\,|I_0 = i, I_1 = i-1]p_{i,i-1} \\ &\quad + E[T\,|I_0 = i, I_1 = i]p_{i,i} + (1-\delta_{iN})E[T\,|I_0 = i, I_1 = i+1]p_{i,i+1}. \end{aligned} \tag{9}$$

But $E[T\,|I_0 = 1, I_1 = 0] = 1$ and $E[T\,|I_0 = i, I_1 = j] = 1 + E[T\,|I_0 = j] = 1 + m_j$, when $j \neq 0$. Substituting in (9) yields

$$\begin{aligned} m_1 &= p_{1,0} + (1+m_1)p_{1,1} + (1+m_2)p_{1,2}, \\ m_i &= (1+m_{i-1})p_{i,i-1} + (1+m_i)p_{i,i} + (1-\delta_{iN})(1+m_{i,i+1})p_{i,i+1}, \text{ for } 1 < i \le N, \end{aligned}$$

that is in accordance with results appearing in [26] (see Section 2.2.5).

Using the normalization condition $p_{i,i-1} + p_{i,i} + (1-\delta_{iN})p_{i.i+1} = 1$, we get that conditioned moments satisfy the following tridiagonal system:

$$(1-p_{i,i})m_i = 1 + (1-\delta_{i0})p_{i,i-1}m_{i-1} + (1-\delta_{iN})p_{i,i+1}m_{i+1}, \tag{10}$$

that can be solved explicitly. Note that Equation (10) can be rewritten as

$$p_{1,0}m_1 = 1 + p_{1,2}(m_2 - m_1), \tag{11}$$

$$p_{i,i-1}(m_i - m_{i-1}) = 1 + p_{i,i+1}(m_{i+1} - m_i), \text{ for } 1 < i < N, \tag{12}$$

$$p_{N,N-1}(m_N - m_{N-1}) = 1. \tag{13}$$

Now we use a method of finite difference equations. First, we introduce differences d_i defined as follows

$$d_i = m_i - (1-\delta_{i1})m_{i-1}, \text{for } 1 \le i \le N, \tag{14}$$

and then substitute (14) in Equations (11)–(13). The tridiagonal system is reduced to a bidiagonal one

$$p_{i,i-1}d_i = 1 + p_{i,i+1}d_i, \text{ for } 1 \le i < N, \tag{15}$$

$$p_{N,N-1}d_N = 1. \tag{16}$$

Moreover, Equation (16) provides a closed expression for d_N :

$$d_N = \frac{1}{p_{N,N-1}} = \frac{1}{N\gamma\Delta t} \tag{17}$$

and remainder differences can be expressed in terms of (17) by noticing that

$$d_i = \frac{1}{p_{i,i-1}} + \frac{p_{i,i+1}}{p_{i,i-1}} d_{i+1}, \text{for } 1 \leq i < N. \tag{18}$$

Using backward substitution and the expressions for transition probabilities (1), we get that

$$d_i = \frac{1}{\gamma \Delta t} A_i + B_i d_N, \text{ for } 1 \leq i < N, \tag{19}$$

where $A_i = \frac{1}{i} + \sum_{j=i+1}^{N-1} \frac{(\alpha/N\gamma)^{j-i}}{j} \prod_{s=i}^{j-1} (N-s)$ and $B_i = (N-i)!(\alpha/N\gamma)^{N-i}$.

On the other hand, from definition (14) and by noticing that $m_i = d_i + (1 - \delta_{i1}) m_{i-1}$, we can write

$$m_i = \sum_{k=1}^{i} d_k, \text{ for } 1 \leq i \leq N. \tag{20}$$

Finally, using (19) repeatedly in combination with (20) gives the explicit expressions (7)–(8). □

Remark 4. *Notice that, from (7) and (8), expected values m_i depend on contact and recovery rates not only through its ratio R_0. Hence, SIS models sharing the basic reproduction number can present different long run average values for the number of inspections prior to the end of the infective process.*

4. Numerical Results

The objective of this section is to reveal the main insights of the mathematical characteristic that is the subject matter of this paper. In the previous section we have derived theoretical and algorithmic results regarding the probabilistic behavior of the random variables T_i, for $i \leq i \leq N$. Probability distribution, conditioned on the initial number of infected, is obtained as the solution of a system of linear equations. But unfortunately, we have not reached a well-known, or even a closed form, distribution for T_i and, in addition, the model relies on a group of parameters that varies over a fairly broad range. Hence, we are going to examine and quantify the effect of changing one or more of the parameter's value in the possible outcomes of the number of inspections. In more details, numerical results come from the application of Theorem 1 and Algorithm 1, when we focus on probabilities of different outcomes, and from the explicit Equations (7)–(8) when the interest are expected values m_i.

Our aim is two-fold: to investigate the influence of the model parameters in the probabilistic behavior of these random variables and show a possible application in evaluating benefits associated to the quantity of inspections conducted over an outbreak of a discrete-time SIS epidemic model.

4.1. Influence of Model Parameters

First we assume that we are able to detect the epidemics as soon as first infection appears, that is $I_0 = 1$. The objective is to characterize the random variable T_1 that counts, from the very beginning of the epidemic, the total number of periodic inspections taking place prior to the epidemics end. We choose a contact rate $\alpha = 2.0$ and a recovery rate $\gamma = 1.0$.

Figure 1 is a bar chart of $P\{T_1 = n\}$, for $n \geq 1$. We consider a population of $N = 100$ individuals and time interval between inspections is $\Delta t = 0.01$ time units. Mass function presents a decreasing shape, with a single mode for $n = 1$. From numerical results, we get that 1% of the outbreaks end by the time of the first inspection, but also 1% of the outbreaks will last more than 30,000 inspections.

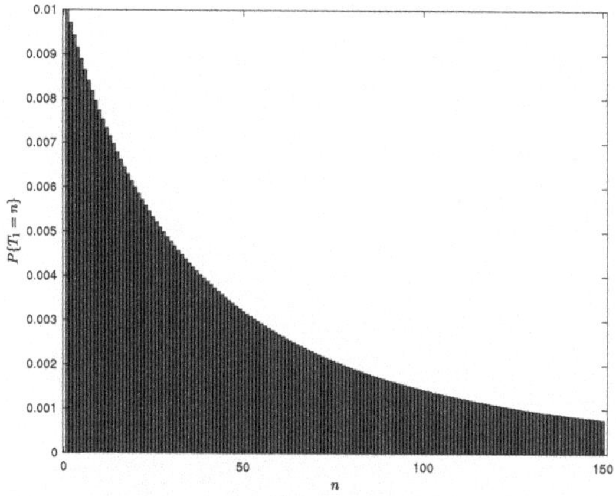

Figure 1. Mass function for $N = 50$.

Table 1 displays several cumulative probabilities up to 150 inspections and the expected value m_1, for a population of $N = 5$, 25, 50 and 75 individuals. We keep rates $\alpha = 2.0$, $\gamma = 1.0$ and time interval length as $\Delta t = 0.01$. For each population, only 1% of the outbreaks are inspected once before extinction. For a fixed number of inspections, cumulative probabilities are smaller when population size is larger. During outbreaks, large populations are inspected more times, in average, than smaller ones. Even for a small population of 5 individuals, we observe that the 48% of the outbreaks are still in progress at the 150-th inspection and an average of around 295 inspections take place while the epidemic is active.

Table 1. Cumulative probabilities for different population sizes.

	$N = 5$	$N = 25$	$N = 50$	$N = 75$
$P\{T_1 \leq 1\}$	0.01	0.01	0.01	0.01
$P\{T_1 \leq 10\}$	0.0894	0.0882	0.0880	0.0880
$P\{T_1 \leq 20\}$	0.1599	0.1558	0.1553	0.1552
$P\{T_1 \leq 30\}$	0.2172	0.2093	0.2083	0.2080
$P\{T_1 \leq 50\}$	0.3047	0.2879	0.2858	0.2852
$P\{T_1 \leq 75\}$	0.3827	0.3536	0.3501	0.3489
$P\{T_1 \leq 100\}$	0.4400	0.3983	0.3932	0.3916
$P\{T_1 \leq 125\}$	0.4856	0.4302	0.4236	0.4215
$P\{T_1 \leq 150\}$	0.5226	0.4539	0.4458	0.4431
m_1	294.66	7094.24	582627.67	58366793.08

Next, we describe the distribution of T_1 using a Box-Whiskers plot diagram. The objective is to compare the patterns of the number of inspections when we vary the contact rate. The box encloses the middle central part of the distribution, lower and upper edges of the box correspond to the lower and the upper quartile, respectively, and the line drawn across the box shows the median of the distribution. Finally, whiskers below start at 1 and whiskers above the boxes reach up to the 99% of the distribution.

Figure 2 shows the distribution of T_1 for $\alpha = 0.5, 1.0$ and 2.0. We consider a recovery rate $\gamma = 2.0$, a population of $N = 20$ individuals and a time interval of $\Delta t = 0.01$ units length between successive inspections. The distribution of the random variable is skewed to the right and longer right tails are observed for larger contact rates. Additional numerical results for $\gamma = 5.0$ show a similar shape for box plot diagrams, with 99th quantile under 200 inspections. This fact is according to the intuition, because large recovery rates give more chance to recoveries than to new infections and, consequently, outbreaks will involve lesser infective individuals and present shorter extinction times in comparison with Figure 2's scenario.

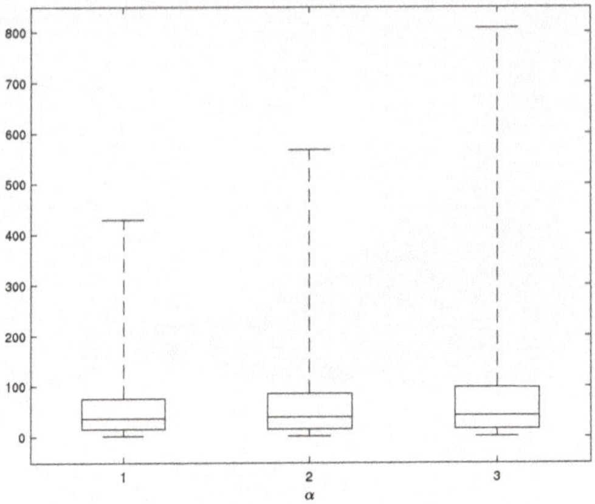

Figure 2. Box plot for T_1.

In Table 2, we display the expected number of inspections for outbreaks starting from a single infective case, we consider several values for contact and recovery rates. Population contains 20 individuals and time interval between inspections is $\Delta t = 0.01$, for every pair of rates. As was stated on Remark 4, results show that expected values are not a function of the basic reproductive number $R_0 = \alpha/\gamma$. The expected number of inspections prior the end of the infective outbreak increases as a function of contact rate α and decreases as a function of recovery rate. That remark is according to the intuition too, because the epidemic length enlarges when contacts between individuals occur more often and/or when individuals need longer times to be recovered.

Table 2. Expected inspections before extinction time.

	$\alpha = 0.5$	$\alpha = 1.0$	$\alpha = 2.0$	$\alpha = 5.0$
$\gamma = 0.5$	444.73	6242.14	1.294×10^7	1.917×10^{13}
$\gamma = 1.0$	134.19	222.36	3121.07	1.536×10^8
$\gamma = 2.0$	56.99	67.09	111.18	11,325.60
$\gamma = 5.0$	21.01	22.16	25.02	44.47

Next we focus on outbreaks that are first observed when the epidemic process involves i, not necessarily one, infected. Our aim is to describe the expected values m_i when varying the initial number of infective individuals. Notice that Equation (20) guarantees that $m_i \geq m_{i-1}$, for $1 < i \leq N$. Thus, the expected number of inspections taking place prior to the epidemics end is a non-decreasing function regarding the amount of initial infective.

Figure 3 provides a numerical illustration for a population of $N = 20$, contact rate $\alpha = 1.8$, recovery rate $\gamma = 0.8$ and time interval between periodic inspections $\Delta t = 0.01$. Pictured graph agrees with theoretical results, it quantifies the growth of the number of inspections when infective rises and it shows the importance of an early detection of such a epidemic process. More specifically, outbreaks detected at the first infection will be active, in mean terms, about 4000 inspections while if the outbreak is first checked when two infected are present in the population then, the expected inspections will rise up to 6000 times or up to 8000 inspections when first checking shows five infected.

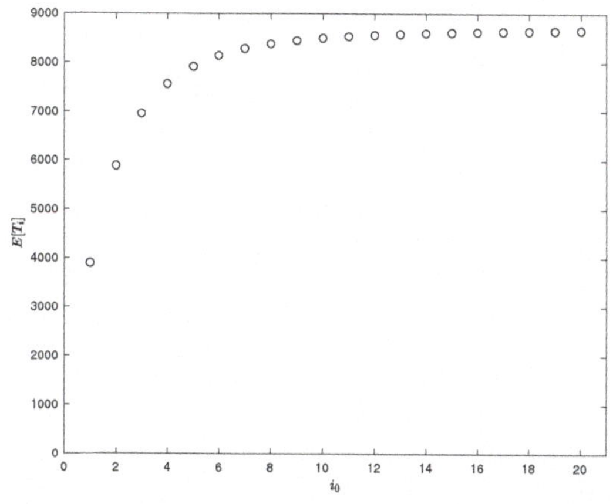

Figure 3. Expected number of inspections versus initial infective.

Additional numerical results, not included here, report that when we choose intervals between inspections with decreasing length Δt, the mass distribution function of any T_i, for $1 \leq i \leq N$, provides an aproximation to the density function of the extinction time of an epidemic process starting with i infected [28], that is the continuous counterpart of the number of periodic inspections taking place while the epidemic process is active.

In the following section we present a possible use of the probabilistic behavior of $T_i = (T|I_0 = i)$.

4.2. *Application to Evaluate Outbreak Benefits*

Let us assume that every inspection has a travel or approaching cost c_1 and, whenever there is a change in the population regarding the immediate previous inspection, we get a profit in terms of information that depends on the type of event. Let g_R and g_I represent recovery and infection detection's gain, respectively.

Associated to every outbreak, the random variable T_i provides the total number of inspections conducted during an outbreak that starts from $I_0 = i$ infective individuals. On the other hand, for outbreaks starting with $I_0 = i$ infective individuals, Artalejo et al. introduced in [30] the random variables N_i^R and N_i^I defined as the number of recoveries and infections per outbreak, respectively. By noticing that the number of recoveries in an outbreak agrees with the total number of infections in the same outbreak, we get

$$N_i^R = i + N_i^I, \text{ for } i \geq 1. \tag{21}$$

With the help of the above random variables and its relationship (21), we can determine outbreak's benefit, for instance, just by defining a benefit function conditioned to the initial number of infective, as follows

$$B(T, R|I_0 = i) = (g_R + g_I)N_i^R - ig_I - c_1 T_i. \tag{22}$$

The expected benefit per outbreak will depend on the mean values of N_i^R and T_i, but also on the choice of travel cost and information profits.

Figure 4 represents expected benefit when the initial number of infective varies in $1 \leq i \leq 11$, for a population of $N = 25$, with contact rate $\alpha = 2.5$, recovery rate of $\gamma = 1.0$ and a time interval of $\Delta t = 0.01$ units. We fixed a unitary travel cost per the inspection; i.e. $c_1 = 1$, and gain values per recovery or infection have been chosen as $g_R = g_I = c_2$. Several graphs are drawn by fixing $c_2 > c_1$. We notice first that, for a fixed number of infective and in order to obtain a positive benefit, recovery gain c_2 must satisfy $c_2 > c_1 E[T_i]/(2E[N_i^R] - i)$. Hence, the trivial restriction $c_2 > c_1$ does not guarantee a positive expected benefit per outbreak. In any case, the expected benefit is a linear increasing function of the gain value c_2.

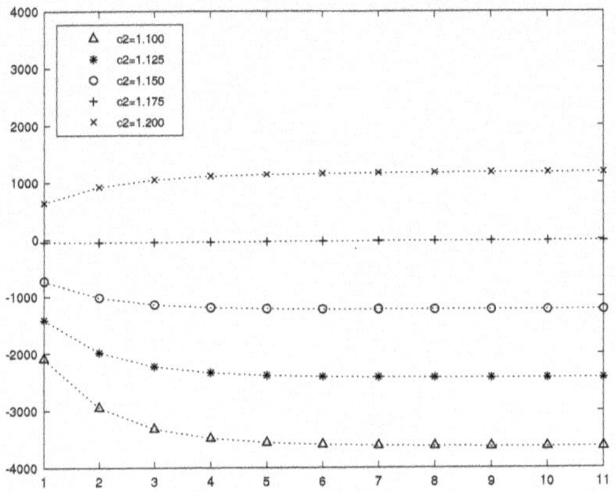

Figure 4. Expected Benefit versus initial infective.

As we can see in Figure 4, depending on the choice for c_2, expected benefit functions present different shape as a function of the initial number of infective individuals. Numerical results show that, for $c_2 \leq 1.125$, expected benefit decreases as the initial infected increases. For $c_2 = 1.15$ we obtain a minimum expected benefit at $I_0 = 9$ infective. When we set $c_2 = 1.175$ the minimum corresponds to the value $I_0 = 2$ but we obtain expected values close to 0. For $c_2 \geq 1.2$, expected benefit remains almost constant for $I_0 \geq 5$. These facts illustrate that a deep knowledge on $E[N_i^R]$ or $E[T_i]$ will help in decision making process regarding travel cost and gain values.

5. Discussion and Conclusions

Literature in mathematical modelization of epidemics includes both continuous and discrete-time models. Continuous-time models are more accurate but more difficult to implement than discrete-time ones. On the other hand, discrete-time models fit better with real information; data related to real-world epidemic processes are often given by unit time, so it is natural to preserve dynamic features by modeling a dynamical system from observations at discrete times which are adapted to time step.

This paper focuses on the discrete-time SIS model, where transition probabilities for event occurring during time-steps are described in terms of an absorbing DTMC. The population is observed at periodic time points assuming that at most one event takes in a time-step. The discrete-time stochastic epidemic SIS models are formulated as DTMC which may be considered approximations to the continuous-time Markov jump processes. The size of the time step must be controlled to assure that the model gives genuine probability distributions.

Our purpose is to study the extinction time counterpart in discrete-time, that is the random variable that counts the total number of inspections that find an active epidemic process.

Subject to the initial number of infective individuals, mass probability function of the number of inspections, T_i, is numerically determined through a recursive scheme; complementing the probabilistic knowledge of this variable provided by its expected value, that comes directly from an explicit expression.

A really interesting extension of this work arises when considering equidistant time inspections relaxing the requirement about the maximum number of events observed during inspections. This problem, that appears to be analytically intractable, is the aim of the paper [31], where authors tackle this difficulty via the total area between the sample paths of the numbers of infective individuals in the continuous-time process and its discrete-time counterpart.

Author Contributions: These authors contributed equally to this work.

Funding: This work was supported by the Government of Spain (Department of Science, Technology and Innovation) and the European Commission through project MTM 2014-58091.

Conflicts of Interest: The authors declare no conflict of interest.

References

1. Bailey, N.T.J. *The Mathematical Theory of Infectious Diseases*; Griffin & Co.: New York, NY, USA, 1975; ISBN 0852642318, 9780852642313.
2. Andersson, H.; Britton, T. *Stochastic Epidemic Models And Their Statistical Analysis*; Lectures Notes in Statistics; Springer: New York, NY, USA, 2000; Volume 151, ISBN 978-1-4612-1158-7.
3. Becker, N.G. *Analysis of Infectious Disease Data*; Chapman & Hall/CRC: Boca Raton, FL, USA, 2000; ISBN 0-412-30990-4.
4. Kawachi, K. Deterministic models for rumor transmission. *Nonlinear Anal. Real World Appl.* **2008**, *9*, 1989–2028. [CrossRef]
5. Isham, V.; Harden, S.; Nekovee, M. Stochastic epidemics and rumours on finite random networks. *Phys. A Stat. Mech. Appl.* **2010**, *389*, 561–576. [CrossRef]
6. Amador, J.; Artalejo, J.R. Stochastic modeling of computer virus spreading with warning signals. *J. Frankl. Inst.* **2013**, *350*, 1112–1138. [CrossRef]
7. De Martino, G.; Spina, S. Exploiting the time-dynamics of news diffusion on the Internet through a generalized Susceptible-Infected model. *Phys. A Stat. Mech. Appl.* **2015**, *438*, 634–644. [CrossRef]
8. Giorno, V.; Spina, S. Rumor spreading models with random denials. *Phys. A Stat. Mech. Appl.* **2016**, *461*, 569–576. [CrossRef]
9. Britton, T. Stochastic epidemic models: A survey. *Math. Biosci.* **2010**, *225*, 24–35. [CrossRef] [PubMed]
10. Diekman, O.; Heesterbeek, H.; Britton, T. *Mathematical for Understanding Infectious Disease Dynamics*; Princeton University Press: Princeton, NJ, USA, 2012; ISBN 9781400845620.
11. Van den Driessche, P. Reproduction numbers of infectious disease models. *Infect. Dis. Model* **2017**, *2*, 288–303. [CrossRef] [PubMed]
12. Artalejo, J.R.; Lopez-Herrero, M.J. On the exact measure of disease spread in stochastic epidemic models. *Bull. Math. Biol.* **2013**, *75*, 1031–1050. [CrossRef] [PubMed]
13. Lopez-Herrero, M.J. Epidemic transmission on SEIR stochastic models with nonlinear incidence rate. *Math. Methods Appl. Sci.* **2016**, *40*, 2532–2541. [CrossRef]
14. Emmert, K.E.; Allen, L.J.S. Population persistence and extinction in a discrete-time, stage-structured epidemic model. *J. Differ. Equ. Appl.* **2004**, *10*, 1177–1199. DOI 10.1080/10236190410001654151. [CrossRef]
15. Allen, L.J.S.; van den Driessche, P. The basic reproduction number in some discrete-time epidemic models. *J. Differ. Equ. Appl.* **2008**, *14*, 1127–1147. [CrossRef]
16. Li, J.Q.; Lou, J.; Lou, M.Z. Some discrete SI and SIS epidemic models. *Appl. Math. Mech.* **2008**, *29*, 113–119. [CrossRef]
17. Zhang, J.; Jin, Z. Discrete time SI and SIS epidemic models with vertical transmission. *J. Biol. Syst.* **2009**, *17*, 201–212. [CrossRef]
18. Farnoosh, R.; Parsamanesh, M. Disease extinction and persistence in a discrete-time SIS epidemic model with vaccination and varying population size. *Filomat* **2017**, *31*, 4735–4747. [CrossRef]
19. d'Onofrio, A.; Manfredi, P.; Salinelli, E. Dynamic behaviour of a discrete-time SIR model with information dependent vaccine uptake. *J. Differ. Equ. Appl.* **2016**, *22*, 485–512. [CrossRef]
20. Tuckell, H.C.; Williams, R.J. Some properties of a simple stochastic epidemic model of SIR type. *Math. Biosci.* **2007**, *208*, 76–97. [CrossRef] [PubMed]
21. Ferrante, M.; Ferraris, E.; Rovira, C. On a stochastic epidemic SEIHR model and its diffusion approximation. *TEST* **2016**, *25*, 482–502. [CrossRef]
22. Van den Driessche, P.; Yakubu, A. Disease extinction versus persistence in discrete-time epidemic models. *Bull. Math. Biol.* **2018**. [CrossRef] [PubMed]

23. Allen, L.J.S.; van den Driessche, P. Relations between deterministic and stochastic thresholds for disease extinction in continuous- and discrete-time infectious disease models. *Math. Biosci.* **2013**, *243*, 99–108. [CrossRef] [PubMed]
24. Fennell, P.G.; Melnik, S.; Gleeson, J.P. Limitations of discrete-time approaches to continuous-time contagion dynamics. *Phys. Rev. E* **2016**, *94*, 052125. [CrossRef] [PubMed]
25. Norden, R.H. On the distribution of the time to extinction in the stochastic logistic population model. *Adv. Appl. Probab.* **1982**, *14*, 687–708. [CrossRef]
26. Allen, L.J.S.; Burgin, A.M. Comparison of deterministic and stochastic SIS and SIR models in discrete time. *Math. Biosci.* **2000**, *163*, 1–33. [CrossRef]
27. Stone, P.; Wilkinson-Herbots, H.; Isham, V. A stochastic model for head lice infections. *J. Math. Biol.* **2008**, *56*, 743–763. [CrossRef] [PubMed]
28. Artalejo, J.R.; Economou, A.; Lopez-Herrero, M.J. Stochastic epidemic models revisited: Analysis of some continuous performance measures. *J. Biol. Dyn.* **2012**, *6*, 189–211. [CrossRef] [PubMed]
29. Kulkarni, V.G. *Modeling and Analysis of Stochastic Systems*; Chapman & Hall: London, UK, 1995; ISBN 0-412-04991-0.
30. Artalejo, J.R.; Economou, A.; Lopez-Herrero, M.J. On the number of recovered individuals in the SIS and SIR stochastic epidemic models. *Math. Biosci.* **2010**, *228*, 45–55. [CrossRef] [PubMed]
31. Gómez-Corral, A.; López-García, M.; Rodríguez-Bernal, M.T. On time-discretized versions of SIS epidemic models. 2018, in press.

mathematics

MDPI

Article

Large Deviation Results and Applications to the Generalized Cramér Model †

Rita Giuliano [1] and Claudio Macci [2,*]

1 Dipartimento di Matematica, Università di Pisa, Largo Bruno Pontecorvo 5, I-56127 Pisa, Italy; rita.giuliano@unipi.it

2 Dipartimento di Matematica, Università di Roma Tor Vergata, Via della Ricerca Scientifica, I-00133 Rome, Italy

* Correspondence: macci@mat.uniroma2.it

† The support of INdAM (Fondi GNAMPA) and Università di Pisa (Fondi di Ateneo) is acknowledged. The first version of this paper was written during the staying of the first author at the University Jean Monnet (St. Etienne).

Received: 2 March 2018; Accepted: 27 March 2018; Published: 2 April 2018

Abstract: In this paper, we prove large deviation results for some sequences of weighted sums of random variables. These sequences have applications to the probabilistic generalized Cramér model for products of primes in arithmetic progressions; they could lead to new conjectures concerning the (non-random) set of products of primes in arithmetic progressions, a relevant topic in number theory.

Keywords: arithmetic progressions; first Chebyshev function; products of primes; regularly varying functions; slowly varying functions

1. Introduction

The aim of this paper is to prove asymptotic results for a class of sequences of random variables, i.e.,

$$\left\{\frac{\sum_{k=1}^{n} L_k X_k}{b_n} : n \geq 1\right\} \tag{1}$$

for suitable sequences of real numbers $\{b_n : n \geq 1\}$ and $\{L_n : n \geq 1\}$ (see Condition 1 in Section 3) and suitable random independent variables $\{X_n : n \geq 1\}$ defined on the same probability space $(\Omega, \mathcal{F}, P)$. We also present analogue results for the slightly different sequence

$$\left\{\frac{L_n \sum_{k=1}^{n} X_k}{b_n} : n \geq 1\right\}. \tag{2}$$

More precisely we refer to the theory of large deviations, which gives an asymptotic computation of small probabilities on an exponential scale (see, e.g., [1] as a reference on this topic). We recall [2] as a recent reference on large deviations for models of interest in number theory.

The origin and the motivation of our research rely on the study of some random models similar in nature to the celebrated *Cramér model for prime numbers*: i.e., what we have called *the generalized model (for products of prime numbers in arithmetic progressions)*. We are not aware of any work where these probabilistic models are studied. Details on these structures will be given in Section 2. Here we only point out that, as the classical probabilistic model invented by Cramér has been used to formulate conjectures on the (non-random) set of primes (see [3] for details), in a similar way we can draw out conjectures also for the non-random sets of products of primes or products of primes in arithmetic progressions. The large deviation results for the sequences concerning these structures will be given in Corollary 1.

We also remark that the particular form of the sequence (1) is motivated by analogy with the first Chebyshev function, as will be explained in Section 2.

It is worth noting that also some moderate deviation properties can be proved (in terms of suitable bounds on cumulants and central moments) for the centered sequences

$$\left\{\frac{\sum_{k=1}^{n} L_k(X_k - \mathbb{E}[X_k])}{b_n} : n \geq 1\right\} \text{ and } \left\{\frac{L_n \sum_{k=1}^{n}(X_k - \mathbb{E}[X_k])}{b_n} : n \geq 1\right\}.$$

Such propositions will not be dealt with in the sequel since, though some specific assumptions must be made in the present setting, these results are in the same direction as those of the paper [4], where moderate deviations from the point of view of cumulants and central moments are fully investigated.

It should be noted that our results are a contribution to the recent literature on limit theorems of interest in probability and number theory; here, we recall [5], where the results are formulated in terms of the mod-φ convergence (see also [6] where the simpler mod-Gaussian convergence is studied).

We here introduce some terminology and notation. We always set $0 \log 0 = 0$, $\frac{c}{\infty} = 0$ for $c \neq 0$, and $\lfloor x \rfloor := \max\{k \in \mathbb{Z} : k \leq x < k+1\}$ for all $x \in \mathbb{R}$. Moreover, we write

- $a_n \sim b_n$ to mean that $\lim_{n\to\infty} \frac{a_n}{b_n} = 1$;
- $Z \overset{\text{law}}{\sim} \mathcal{B}(p)$, for $p \in [0,1]$, to mean that $P(Z=1) = p = 1 - P(Z=0)$;
- $Z \overset{\text{law}}{\sim} \mathcal{P}(\lambda)$, for $\lambda > 0$, to mean that $P(Z=k) = \frac{\lambda^k}{k!}e^{-\lambda}$ for all integers $k \geq 0$.

The outline of this paper is as follows: We start with some preliminaries in Section 2, and we present the results in Section 3. The results for the generalized Cramér model (for products of primes in arithmetic progressions) are presented in Corollary 1.

2. Preliminaries

On large deviations.

We refer to [1] (pages 4–5). Let $\mathcal{Z}$ be a topological space equipped with its completed Borel σ-field. A sequence of $\mathcal{Z}$-valued random variables $\{Z_n : n \geq 1\}$ satisfies the large deviation principle (LDP) with speed function v_n and rate function I if the following is true: $\lim_{n\to\infty} v_n = \infty$, and the function $I : \mathcal{Z} \to [0,\infty]$ is lower semi-continuous.

$$\limsup_{n\to\infty} \frac{1}{v_n} \log P(Z_n \in F) \leq -\inf_{z\in F} I(z) \text{ for all closed sets } F$$

$$\liminf_{n\to\infty} \frac{1}{v_n} \log P(Z_n \in G) \geq -\inf_{z\in G} I(z) \text{ for all open sets } G.$$

A rate function I is said to be *good* if its level sets $\{\{z \in \mathcal{Z} : I(z) \leq \eta\} : \eta \geq 0\}$ are compact.

Throughout this paper, we prove LDPs with $\mathcal{Z} = \mathbb{R}$. We recall the following known result for future use.

Theorem 1 (Gärtner–Ellis Theorem). *Let $\{Z_n : n \geq 1\}$ be a sequence of real valued random variables. Assume that the function $\Lambda : \mathbb{R} \to (-\infty, \infty]$ defined by*

$$\Lambda(\theta) := \lim_{n\to\infty} \frac{1}{v_n} \log \mathbb{E}\left[e^{v_n \theta Z_n}\right] \quad (\textit{for all } \theta \in \mathbb{R}) \tag{3}$$

exists; assume, moreover, that Λ is essentially smooth (see e.g., Definition 2.3.5 in [1]) and lower semi-continuous. Then $\{Z_n : n \geq 1\}$ satisfies the LDP with speed function v_n and good rate function $\Lambda^* : \mathbb{R} \to [0,\infty]$ defined by

$$\Lambda^*(z) := \sup_{\theta\in\mathbb{R}} \{\theta z - \Lambda(\theta)\}.$$

Proof. See, e.g., Theorem 2.3.6 in [1]. □

The main application of Theorem 1 in this paper concerns Theorem 2, where we have

$$\Lambda(\theta) = e^{\theta} - 1, \text{ which yields } \Lambda^*(x) = \begin{cases} x\log x - x + 1 & \text{if } x \geq 0 \\ \infty & \text{if } x < 0. \end{cases} \tag{4}$$

The LDP in Theorem 3 will instead be proved by combining Theorem 4.2.13 in [1] with Theorem 2, i.e., by checking the exponential equivalence (see, e.g., Definition 4.2.10 in [1]) of the involved sequences.

On the generalized Cramér model (for products of primes in arithmetic progressions).

The Cramér model for prime numbers consists in a sequence of independent random variables $\{X_n : n \geq 1\}$ such that, for every $n \geq 2$,

$$X_n \overset{\text{law}}{\sim} \mathcal{B}(1/\log n). \tag{5}$$

This model can be justified by the prime numbers theorem (PNT), which roughly asserts that the expected density of primes around x is $\frac{1}{\log x}$: the cardinality of prime numbers $\leq n$ is

$$\pi(n) := \sum_{p \leq n} 1 \sim \text{li}(n) := \int_2^n \frac{1}{\log t} dt,$$

and, with the words of [7] (see footnote on p. 6), "the quantity $\frac{1}{\log n}$ appears here naturally as the derivative of $\text{li}(x)$ evaluated at $x = n$". Since $\int_2^n \frac{1}{\log t} dt \sim \frac{n}{\log n}$, another way of stating the PNT is

$$\frac{\pi(n)}{n} \sim \frac{1}{\log n}. \tag{6}$$

A first extension of this formula concerns the case of integers n which are products of exactly r prime factors ($r \geq 2$). More precisely, we consider the sets

$$A_r(n) := \{k \leq n : \Omega(k) = r\} \text{ and } B_r(n) := \{k \leq n : \omega(k) = r\}$$

where $\omega(n)$ is the number of distinct prime factors of n, and $\Omega(n)$ counts the number of prime factors of n (with multiplicity); this means that, letting (by the canonical prime factorization of n) $n = \prod_{i=1}^{\omega(n)} p_i^{\alpha_i}$, where $p_1, \ldots, p_n$ are the distinct prime factors of n, we have

$$\Omega(n) := \sum_{i=1}^{\omega(n)} \alpha_i.$$

A result proved by Landau in 1909 (see, e.g., [8]) states that the cardinalities $\tau_r(n)$ and $\pi_r(n)$ of $A_r(n)$ and $B_r(n)$ respectively verify

$$\tau_r(n) := \sum_{k \in A_r(n)} 1 \sim \frac{n(\log\log n)^{r-1}}{(r-1)!\log n} \text{ and } \pi_r(n) := \sum_{k \in B_r(n)} 1 \sim \frac{n(\log\log n)^{r-1}}{(r-1)!\log n};$$

see also, e.g., Theorem 437 in [9] (Section 22.18, page 368) or [10] (II.6, Theorems 4 and 5). Note that this formula for $\pi_r(n)$ reduces to Equation (6) when $r = 1$.

Going a little further, for fixed integers a and q, we can consider the sets of products of primes in arithmetic progressions

$$A_r^{(q)}(n) =: \{k \leq n : \Omega(k) = r,\ k \equiv a \bmod q\} \text{ and } B_r^{(q)}(n) =: \{k \leq n : \omega(k) = r,\ k \equiv a \bmod q\}.$$

One can prove (by similar methods as in [10,11]) that, for any a and q with $(a,q)=1$, the cardinalities $\tau_r^{(q)}(n)$ and $\pi_r^{(q)}(n)$ of $A_r^{(q)}(n)$ and $B_r^{(q)}(n)$ respectively verify

$$\tau_r^{(q)}(n) := \sum_{k\in A_r^{(q)}(n)} 1 \sim \frac{1}{\phi(q)} \cdot \frac{n(\log\log n)^{r-1}}{(r-1)!\log n} \text{ and } \pi_r^{(q)}(n) := \sum_{k\in B_r^{(q)}(n)} 1 \sim \frac{1}{\phi(q)} \cdot \frac{n(\log\log n)^{r-1}}{(r-1)!\log n},$$

where ϕ is Euler's totient function. Notice that, for $r=1$, we recover the sets of primes in arithmetic progressions, considered for instance in [8,10] II.8, or [11]; the case $r=2$ is studied in [12]; the general case $r\geq 1$ is considered in the recent preprint [13]; for $q=1$, we recover the sets and the formulas for the model described above.

Therefore, following Cramér's heuristic, Equation (5), we can define the generalized Cramér model for products of r prime numbers (or products of r prime numbers in arithmetic progression) as a sequence of independent random variables $\{X_n : n\geq 1\}$ such that

$$X_n \overset{\text{law}}{\sim} \mathcal{B}(\lambda_n), \text{ where } \lambda_n := \frac{\ell_n}{\log n} \text{ and } \ell_n := \frac{1}{\phi(q)} \cdot \frac{(\log\log n)^{r-1}}{(r-1)!}. \tag{7}$$

Obviously in Equation (7) we take $n\geq n_0$, where n_0 is an integer, such that $\lambda_n\in(0,1]$ for $n\geq n_0$; the definition of λ_n for $n<n_0$ is arbitrary.

Large deviation results for this model will be presented in Corollary 1 as a consequence of Theorem 3 and Remark 2, with

$$L_n := \log n \text{ and } b_n := n\ell_n; \tag{8}$$

thus, the sequences in Equations (1) and (2) become

$$\frac{\sum_{k=1}^n (\log k)X_k}{n\ell_n} \text{ and } \frac{(\log n)\sum_{k=1}^n X_k}{n\ell_n} \tag{9}$$

respectively. Moreover, by taking into account Remark 3 presented below, the sequences in Equation (9) converge almost surely to 1 (as $n\to\infty$).

On the first Chebyshev function.

The first Chebyshev function is defined by

$$\theta(x) := \sum_{p\leq x} \log p,$$

where the sum is extended over all prime numbers $p\leq x$.

Therefore, when considering the classical Cramér model, this function is naturally modeled with $\sum_{k=1}^n (\log k)X_k$ (and we obtain the numerator of the first fraction in Equation (9)).

It must be noted that T. Tao, in his blog (see [14]), considers the same random variable $\sum_{k\leq x}(\log k)X_k$ and proves that almost surely one has

$$\sum_{k\leq x}(\log k)X_k = x + O_\varepsilon(x^{1/2+\varepsilon})$$

for all $\varepsilon>0$ (where the implied constant in the $O_\varepsilon(\cdot)$ notation is allowed to be random). In particular, almost surely one has

$$\lim_{n\to\infty} \frac{\sum_{k\leq n}(\log k)X_k}{n} = 1.$$

It appears clearly that in this setting we have a sequence of the form of Equation (1), with the particular choices $L_n=\log n$ and $b_n=n$. What we are going to investigate in the sequel is how the sequence of random variables $\{X_n : n \geq 1\}$ and the two sequences of numbers $\{L_n : n \geq 1\}$ and $\{b_n : n \geq 1\}$ must be connected in order to obtain large deviations and convergence results (see also Equations (8) and (9) above).

On slowly and regularly varying functions (at infinity).

Here we recall the following basic definitions. A positive measurable function H defined on some neighborhood of $[x_0, \infty)$ of infinity is said to be *slowly varying* at infinity (see, e.g., [15], page 6) if

$$\lim_{t\to\infty} \frac{H(tx)}{H(t)} = 1 \text{ for all } x > 0.$$

Similarly, a positive measurable function M defined on some neighborhood of $[x_0, \infty)$ of infinity is said to be *regularly varying* at infinity of index ρ (see, e.g., [15], page 18) if

$$\lim_{t\to\infty} \frac{M(tx)}{M(t)} = x^\rho \text{ for all } x > 0.$$

Obviously, we recover the slowly varying case if $\rho = 0$. Recall the following well-known result for slowly varying functions.

Lemma 1 (Karamata's representation of slowly varying functions). *A function H is slowly varying at infinity if and only if*

$$H(x) = c(x) \exp\left(\int_{x_0}^{x} \frac{\phi(t)}{t} dt\right)$$

where $\phi(x) \to 0$ and $c(x) \to c_\infty$ for some $c_\infty > 0$ (as $x \to \infty$).

Proof. See, e.g., Theorem 1.3.1 in [15]. □

In view of what follows we also present the following results. They are more or less known; but we prefer to give detailed proofs in order to ensure that the paper is self-contained.

Lemma 2. *Let M be a regularly varying function (at infinity) of index $\rho \geq 0$. Then,*

$$\lim_{t\to\infty} \frac{M(\lfloor tx \rfloor)}{M(t)} = x^\rho \textit{ for all } x > 0.$$

Proof. It is well-known (see, e.g., Theorem 1.4.1 in [15]) that we have $M(x) = x^\rho H(x)$ for a suitable slowly varying function H. Thus, it is easy to check that it suffices to prove the result for the case $\rho = 0$ (namely for a slowly varying function H), i.e.,

$$\lim_{t\to\infty} \frac{H(\lfloor tx \rfloor)}{H(t)} = 1 \text{ for all } x > 0. \tag{10}$$

By Lemma 1, for all $x > 0$, we have

$$\frac{H(\lfloor tx \rfloor)}{H(t)} = \frac{c(\lfloor tx \rfloor)}{c(t)} \exp\left(\int_t^{\lfloor tx \rfloor} \frac{\phi(v)}{v} dv\right)$$

for $t > 0$. Obviously, $\frac{c(\lfloor tx \rfloor)}{c(t)} \to 1$ (as $t \to \infty$). Moreover, for all $\varepsilon > 0$, we have

$$\left|\int_t^{\lfloor tx \rfloor} \frac{\phi(v)}{v} dv\right| \leq \varepsilon |\log(\lfloor tx \rfloor / t)|$$

for $t > 0$, and $\log(\lfloor tx \rfloor / t) \to \log x$ (as $t \to \infty$); thus,

$$\int_t^{\lfloor tx \rfloor} \frac{\phi(v)}{v} dv \to 0 \text{ (as } t \to \infty)$$

by the arbitrariness of $\varepsilon > 0$. Thus, Equation (10) holds, and the proof is complete. □

Lemma 3. *Let H be a slowly varying function (at infinity). Then,*

$$\lim_{x\to\infty} \frac{xH(x)}{\sum_{k=1}^{\lfloor x \rfloor} H(k)} = 1.$$

Proof. By the representation of H in Lemma 1, for all $\varepsilon > 0$ there is an integer $n_0 \geq 1$ such that, for all $x > n_0$, we have $c_\infty - \varepsilon < c(x) < c_\infty + \varepsilon$ and $-\varepsilon < \phi(x) < \varepsilon$. Then, we take $x \geq n_0 + 1$, and

$$\frac{\sum_{k=1}^{\lfloor x \rfloor} H(k)}{xH(x)} = \frac{\sum_{k=1}^{n_0} H(k)}{xH(x)} + \frac{\sum_{k=n_0+1}^{\lfloor x \rfloor} H(k)}{xH(x)}.$$

The first summand in the right hand side can be ignored since, if we take $\varepsilon \in (0,1)$, for a sufficient high x, we have

$$H(x) > \frac{c_\infty}{2} \exp\left(-\varepsilon \int_{x_0}^x \frac{1}{t} dt\right) = \frac{c_\infty}{2} \left(\frac{x}{x_0}\right)^{-\varepsilon},$$

which yields $xH(x) > c_1 x^{1-\varepsilon}$ for a suitable constant $c_1 > 0$ (and $x^{1-\varepsilon} \to \infty$ as $x \to \infty$). Therefore, we concentrate our attention on the second summand and, by taking into account again the representation of H in Lemma 1, for a sufficiently high x, we have

$$\frac{\sum_{k=n_0+1}^{\lfloor x \rfloor} H(k)}{xH(x)} = \frac{\sum_{k=n_0+1}^{\lfloor x \rfloor} c(k) \exp\left(\int_{x_0}^k \frac{\phi(t)}{t} dt\right)}{xc(x) \exp\left(\int_{x_0}^x \frac{\phi(t)}{t} dt\right)} = \frac{\sum_{k=n_0+1}^{\lfloor x \rfloor} \frac{c(k)}{c(x)} \exp\left(-\int_k^x \frac{\phi(t)}{t} dt\right)}{x}.$$

Moreover,

$$\frac{\sum_{k=n_0+1}^{\lfloor x \rfloor} \frac{c(k)}{c(x)} \exp\left(-\int_k^x \frac{\phi(t)}{t} dt\right)}{x} \leq \frac{c_\infty + \varepsilon}{c_\infty - \varepsilon} \frac{\sum_{k=n_0+1}^{\lfloor x \rfloor} k^{-\varepsilon}}{x^{1-\varepsilon}} \to \frac{c_\infty + \varepsilon}{c_\infty - \varepsilon} \frac{1}{1-\varepsilon} \text{ (as } x \to \infty)$$

and

$$\frac{\sum_{k=n_0+1}^{\lfloor x \rfloor} \frac{c(k)}{c(x)} \exp\left(-\int_k^x \frac{\phi(t)}{t} dt\right)}{x} \geq \frac{c_\infty - \varepsilon}{c_\infty + \varepsilon} \frac{\sum_{k=n_0+1}^{\lfloor x \rfloor} k^{\varepsilon}}{x^{1+\varepsilon}} \to \frac{c_\infty - \varepsilon}{c_\infty + \varepsilon} \frac{1}{1+\varepsilon} \text{ (as } x \to \infty),$$

and the proof is complete by the arbitrariness of ε. □

3. Results

In this section we present large deviation results for Equations (1) and (2). We start with the case of Poisson distributed random variables (see Theorem 2 and Remark 1), and later we consider the case of Bernoulli distributed random variables (see Theorem 3 and Remark 2). Our large deviation results yield the almost sure convergence to 1 (as $n \to \infty$) of the involved random variables (see Remark 3 for details). In particular, the results for Bernoulli distributed random variables can be applied to the sequences of the generalized Cramér model in Equation (9) (see Corollary 1).

In all our results, we assume the following condition.

Condition 1. *The sequence $\{b_n : n \geq 1\}$ is eventually positive; $\{L_n : n \geq 1\}$ is eventually positive and non-decreasing.*

In general, we can ignore the definition of $\{b_n : n \geq 1\}$ and $\{L_n : n \geq 1\}$ for a finite number of indices; therefore, in order to simplify the proofs, we assume that $\{b_n : n \geq 1\}$ and $\{L_n : n \geq 1\}$ are positive sequences and that $\{L_n : n \geq 1\}$ is non-decreasing.

We start with the case where $\{X_n : n \geq 1\}$ are (independent) Poisson distributed random variables.

Theorem 2 (the Poisson case; the sequence in Equation (1)). *Let $\{b_n : n \geq 1\}$ and $\{L_n : n \geq 1\}$ be two sequences as in Condition 1. Assume that*

$$\{L_n : n \geq 1\} \text{ is the restriction (on } \mathbb{N}) \text{ of a slowly varying function (at infinity).} \tag{11}$$

$$\text{For all } c \in (0,1),\ \alpha(c) := \lim_{n\to\infty} \frac{b_{\lfloor cn \rfloor}}{b_n} \text{ exists, and } \lim_{c\downarrow 0} \alpha(c) = 0. \tag{12}$$

$$\lim_{n\to\infty} \frac{L_n}{b_n} = 0. \tag{13}$$

Moreover, assume that $\{X_n : n \geq 1\}$ are independent and $X_n \overset{\text{law}}{\sim} \mathcal{P}(\lambda_n)$ for all $n \geq 1$, where $\{\lambda_n : n \geq 1\}$ are positive numbers such that

$$\sum_{k=1}^{n} \lambda_k \sim \frac{b_n}{L_n}. \tag{14}$$

The sequence in Equation (1) then satisfies the LDP with speed function $v_n = \frac{b_n}{L_n}$ and good rate function Λ^* defined by Equation (4).

We point out that Equation (12) is satisfied if the sequence $\{b_n : n \geq 1\}$ is nondecreasing and is the restriction (on $\mathbb{N}$) of a regularly varying function with positive index (at infinity); this is a consequence of Lemma 2.

Proof. We apply Theorem 1, i.e., we check that Equation (3) holds with $Z_n = \frac{\sum_{k=1}^{n} L_k X_k}{b_n}$, $v_n = \frac{b_n}{L_n}$, and Λ as in Equation (4) (in fact, Equation (3) holds even without assuming (13); however, Equation (13) must be required in order that $v_n = \frac{b_n}{L_n}$ be a speed function). We remark that

$$\frac{L_n}{b_n} \log \mathbb{E}\left[e^{\frac{b_n}{L_n}\theta \frac{\sum_{k=1}^{n} L_k X_k}{b_n}}\right] = \frac{L_n}{b_n} \log \mathbb{E}\left[e^{\theta \frac{\sum_{k=1}^{n} L_k X_k}{L_n}}\right] = \frac{L_n}{b_n} \sum_{k=1}^{n} \log \mathbb{E}\left[e^{(\theta L_k/L_n) X_k}\right]$$
$$= \frac{L_n}{b_n} \sum_{k=1}^{n} \log(e^{\lambda_k (e^{\theta L_k/L_n} - 1)}) = \frac{L_n}{b_n} \sum_{k=1}^{n} \lambda_k (e^{\theta L_k/L_n} - 1) \text{ for all } \theta \in \mathbb{R}.$$

Equation (3) trivially holds for $\theta = 0$. The proof is divided in two parts: the proof of the upper bound,

$$\limsup_{n\to\infty} \frac{L_n}{b_n} \log \mathbb{E}\left[e^{\frac{b_n}{L_n}\theta \frac{\sum_{k=1}^{n} L_k X_k}{b_n}}\right] \leq e^{\theta} - 1 \text{ for all } \theta \in \mathbb{R}, \tag{15}$$

and that of the lower bound,

$$\liminf_{n\to\infty} \frac{L_n}{b_n} \log \mathbb{E}\left[e^{\frac{b_n}{L_n}\theta \frac{\sum_{k=1}^{n} L_k X_k}{b_n}}\right] \geq e^{\theta} - 1 \text{ for all } \theta \in \mathbb{R}. \tag{16}$$

We start with the proof of Equation (15). For $\theta > 0$, we have

$$\frac{L_n}{b_n} \log \mathbb{E}\left[e^{\frac{b_n}{L_n}\theta \frac{\sum_{k=1}^{n} L_k X_k}{b_n}}\right] = \frac{L_n}{b_n} \sum_{k=1}^{n} \lambda_k (e^{\theta L_k/L_n} - 1) \leq \frac{L_n}{b_n} \sum_{k=1}^{n} \lambda_k (e^{\theta} - 1)$$

since $\{L_n : n \geq 1\}$ is nondecreasing, and we obtain Equation (15) by letting n go to infinity and by taking into account Equation (14). For $\theta < 0$, we take $c \in (0,1)$ and

$$\gamma := \sup\{L_n : n \geq 1\}$$

(possibly infinite). Recalling that $\{L_n : n \geq 1\}$ is nondecreasing and that $\frac{L_{\lfloor cn \rfloor}}{L_n} \to 1$ (it is a consequence of Lemma 2), we have

$$\begin{aligned}
\frac{L_n}{b_n} \log \mathbb{E}\left[e^{\frac{b_n}{L_n}\theta \frac{\sum_{k=1}^n L_k X_k}{b_n}}\right] &= \frac{L_n}{b_n} \sum_{k=1}^n \lambda_k (e^{\theta L_k / L_n} - 1) \\
&\leq \frac{L_n}{b_n} \sum_{k=1}^{\lfloor cn \rfloor} \lambda_k (e^{\theta L_1/\gamma} - 1) + \frac{L_n}{b_n} \sum_{k=\lfloor cn \rfloor + 1}^{n} \lambda_k (e^{\theta L_{\lfloor cn \rfloor}/L_n} - 1) \\
&= \frac{L_n}{L_{\lfloor cn \rfloor}} \frac{b_{\lfloor cn \rfloor}}{b_n} \left\{ \frac{L_{\lfloor cn \rfloor}}{b_{\lfloor cn \rfloor}} \sum_{k=1}^{\lfloor cn \rfloor} \lambda_k \right\} (e^{\theta L_1/\gamma} - 1) \\
&\quad + \left(\left\{ \frac{L_n}{b_n} \sum_{k=1}^{n} \lambda_k \right\} - \frac{L_n}{L_{\lfloor cn \rfloor}} \frac{b_{\lfloor cn \rfloor}}{b_n} \left\{ \frac{L_{\lfloor cn \rfloor}}{b_{\lfloor cn \rfloor}} \sum_{k=1}^{\lfloor cn \rfloor} \lambda_k \right\} \right) (e^{\theta L_{\lfloor cn \rfloor}/L_n} - 1).
\end{aligned}$$

Then, by Equation (11) (and Lemma 2 with $\rho = 0$), (12) and (14), we obtain

$$\limsup_{n\to\infty} \frac{L_n}{b_n} \log \mathbb{E}\left[e^{\frac{b_n}{L_n}\theta \frac{\sum_{k=1}^n L_k X_k}{b_n}}\right] \leq \alpha(c)(e^{\theta L_1/\gamma} - 1) + (1 - \alpha(c))(e^{\theta} - 1).$$

Using Equation (12), we conclude by letting $c \downarrow 0$.

The proof of Equation (16) is similar with reversed inequalities; hence, we only sketch it here. For $\theta < 0$, we have

$$\frac{L_n}{b_n} \log \mathbb{E}\left[e^{\frac{b_n}{L_n}\theta \frac{\sum_{k=1}^n L_k X_k}{b_n}}\right] = \frac{L_n}{b_n} \sum_{k=1}^n \lambda_k (e^{\theta L_k/L_n} - 1) \geq \frac{L_n}{b_n} \sum_{k=1}^n \lambda_k (e^{\theta} - 1),$$

and we obtain Equation (16) by letting n go to infinity and by taking into account (14). For $\theta \geq 0$, we take $c \in (0,1)$ and, for γ defined as above, after some manipulations, we obtain

$$\liminf_{n\to\infty} \frac{L_n}{b_n} \log \mathbb{E}\left[e^{\frac{b_n}{L_n}\theta \frac{\sum_{k=1}^n L_k X_k}{b_n}}\right] \geq \alpha(c)(e^{\theta L_1/\gamma} - 1) + (1 - \alpha(c))(e^{\theta} - 1).$$

We conclude by letting $c \downarrow 0$ (by Equation (12)). □

Remark 1 (The Poisson case; the sequence in Equation (2)). *The LDP in Theorem 2 holds also for the sequence in Equation (2) in place of the sequence in Equation (1). In this case we only need to use Condition 1 and to assume Equations (13) and (14), whereas Equations (11) and (12) (which were required in the proof of Theorem 2) can be ignored. For the proof, we still apply Theorem 1, so we have to check that Equation (3) holds with* $Z_n = \frac{L_n \sum_{k=1}^n X_k}{b_n}$, $v_n = \frac{b_n}{L_n}$, *and* Λ *as in Equation (4). This can be easily checked noting that*

$$\begin{aligned}
\frac{L_n}{b_n} \log \mathbb{E}\left[e^{\frac{b_n}{L_n}\theta \frac{L_n \sum_{k=1}^n X_k}{b_n}}\right] &= \frac{L_n}{b_n} \log \mathbb{E}\left[e^{\theta \sum_{k=1}^n X_k}\right] = \frac{L_n}{b_n} \sum_{k=1}^n \log \mathbb{E}\left[e^{\theta X_k}\right] \\
&= \frac{L_n}{b_n} \sum_{k=1}^n \log(e^{\lambda_k (e^{\theta} - 1)}) = \frac{L_n}{b_n} \sum_{k=1}^n \lambda_k (e^{\theta} - 1) \to e^{\theta} - 1 \text{ for all } \theta \in \mathbb{R}
\end{aligned}$$

where the limit relation holds by Equation (14).

The next result is for Bernoulli distributed random variables $\{X_n : n \geq 1\}$. Here we shall use the concept of exponential equivalence (see, e.g., Definition 4.2.10 in [1]). The proof is similar to the one of Proposition 3.5 in [16] (see also Remark 3.6 in the same reference). We point out that it is not unusual to prove a convergence result for Bernoulli random variables $\{X_n : n \geq 1\}$ starting from a similar one for Poisson random variables $\{Y_n : n \geq 1\}$ and by setting $X_n := Y_n \wedge 1$ for all $n \geq 1$; see, for instance, Lemmas 1 and 2 in [17].

Theorem 3 (The Bernoulli case; the sequence in Equation (1)). *Let $\{b_n : n \geq 1\}$ and $\{L_n : n \geq 1\}$ be as in Theorem 2 (thus, Condition 1 together with Equations (11)–(13) hold). Moreover, assume that $\{X_n : n \geq 1\}$ are independent and $X_n \overset{\text{law}}{\sim} \mathcal{B}(\lambda_n)$ for all $n \geq 1$ and that Equation (14) and $\lim_{n\to\infty} \lambda_n = 0$ hold. The sequence in Equation (1) satisfies the LDP with speed function $v_n = \frac{b_n}{L_n}$ and the good rate function Λ^* defined by Equation (4).*

Proof. Let n_0 such that $\lambda_n \in [0,1)$ for all $n \geq n_0$ (recall that $\lambda_n \to 0$ as $n \to \infty$), and let $\{X_n^* : n \geq 1\}$ be independent random variables such that $X_n^* \overset{\text{law}}{\sim} \mathcal{P}(\hat{\lambda}_n)$ (for all $n \geq 1$), where $\hat{\lambda}_n := \log \frac{1}{1-\lambda_n}$ for $n \geq n_0$ (the definition of $\hat{\lambda}_n$ for $n < n_0$ is arbitrary). Notice that

$$\sum_{k=1}^{n} \hat{\lambda}_k \sim \sum_{k=1}^{n} \lambda_k$$

because $\sum_{k=1}^{n} \lambda_k \to \infty$ (as $n \to \infty$) by Equations (13) and (14) and, by the Cesaro theorem,

$$\lim_{n\to\infty} \frac{\sum_{k=1}^{n} \hat{\lambda}_k}{\sum_{k=1}^{n} \lambda_k} = \lim_{n\to\infty} \frac{\hat{\lambda}_n}{\lambda_n} = \lim_{n\to\infty} \frac{\log \frac{1}{1-\lambda_n}}{\lambda_n} = 1.$$

Hence, the assumption of Equation (14) and Theorem 2 are in force for the sequence $\{X_n^* : n \geq 1\}$ (in fact, we have Equation (14) with $\{\hat{\lambda}_n : n \geq 1\}$ in place of $\{\lambda_n : n \geq 1\}$) and, if we set $X_n := X_n^* \wedge 1$ (for all $n \geq 1$), the sequence $\{X_n : n \geq 1\}$ is indeed an instance of the sequence appearing in the statement of the present theorem since, by construction, $X_n \overset{\text{law}}{\sim} \mathcal{B}(1 - e^{-\hat{\lambda}_n})$ and $1 - e^{-\hat{\lambda}_n} = \lambda_n$.

The statement will be proved by combining Theorem 4.2.13 in [1] and Theorem 2 (for the sequence $\{X_n^* : n \geq 1\}$). This means that we have to check the exponential equivalence condition

$$\limsup_{n\to\infty} \frac{L_n}{b_n} \log P(\Delta_n > \delta) = -\infty \text{ (for all } \delta > 0) \tag{17}$$

where

$$\Delta_n := \left| \frac{1}{b_n} \sum_{k=1}^{n} L_k X_k - \frac{1}{b_n} \sum_{k=1}^{n} L_k X_k^* \right|. \tag{18}$$

We remark that

$$\Delta_n \leq \frac{L_n}{b_n} \sum_{k=1}^{n} |X_k - X_k^*| \tag{19}$$

by the monotonicity and the nonnegativeness of $\{L_n : n \geq 1\}$; therefore, if we combine Equation (19) and the Chernoff bound, for each arbitrarily fixed $\theta \geq 0$, we obtain

$$P(\Delta_n > \delta) \leq P\left(\frac{L_n}{b_n} \sum_{k=1}^{n} |X_k - X_k^*| > \delta \right) \leq \frac{\mathbb{E}\left[e^{\theta \sum_{k=1}^{n} |X_j - X_j^*|} \right]}{e^{\theta \delta b_n / L_n}}.$$

Therefore,

$$\frac{L_n}{b_n} \log P(\Delta_n > \delta) \leq \frac{L_n}{b_n} \sum_{k=1}^{n} \log \mathbb{E}\left[e^{\theta |X_k - X_k^*|} \right] - \theta \delta.$$

Moreover, if we set

$$\rho_k^{(\theta)} := \frac{e^{\lambda_k e^\theta} - 1}{\lambda_k e^\theta},$$

we have

$$\begin{aligned}\mathbb{E}\left[e^{\theta|X_k - X_k^*|}\right] &= P(X_k^* = 0) + P(X_k^* = 1) + \sum_{h=2}^{\infty} e^{\theta|1-h|} P(X_k^* = h) \\ &= e^{-\lambda_k} + \lambda_k e^{-\lambda_k} + \sum_{h=2}^{\infty} e^{\theta(h-1)} \frac{\lambda_k^h}{h!} e^{-\lambda_k} = e^{-\lambda_k} + \lambda_k e^{-\lambda_k} + e^{-\theta} e^{-\lambda_k}\left(e^{\lambda_k e^\theta} - 1 - \lambda_k e^\theta\right) \\ &= e^{-\lambda_k} + e^{-\theta} e^{-\lambda_k}\left(e^{\lambda_k e^\theta} - 1\right) = e^{-\lambda_k}\left(1 + e^{-\theta}\left(e^{\lambda_k e^\theta} - 1\right)\right) = e^{-\lambda_k}\left(1 + \lambda_k \rho_k^{(\theta)}\right).\end{aligned}$$

Therefore,

$$\frac{L_n}{b_n} \log P(\Delta_n > \delta) \le -\frac{L_n}{b_n} \sum_{k=1}^{n} \lambda_k + \frac{L_n}{b_n} \sum_{k=1}^{n} \log\left(1 + \lambda_k \rho_k^{(\theta)}\right) - \theta\delta. \tag{20}$$

The proof will be complete if we show that, for all $\theta > 0$,

$$\limsup_{n\to\infty} \frac{L_n}{b_n} \sum_{k=1}^{n} \log\left(1 + \lambda_k \rho_k^{(\theta)}\right) \le 1. \tag{21}$$

In fact, by Equations (14) and (21), we deduce from Equation (20) that

$$\limsup_{n\to\infty} \frac{L_n}{b_n} \log P(\Delta_n > \delta) \le -\theta\delta,$$

and we obtain Equation (17) by letting θ go to infinity.

Thus, we prove Equation (21). We remark that $\rho_n^{(\theta)} \to 1$ because $\lambda_n \to 0$ (as $n \to \infty$). Hence, for all $\varepsilon \in (0,1)$, there exists n_0 such that, for all $n > n_0$, we have $\rho_n^{(\theta)} < 1 + \varepsilon$ and

$$\begin{aligned}\frac{L_n}{b_n} \sum_{k=1}^{n} \log\left(1 + \lambda_k \rho_k^{(\theta)}\right) &= \frac{L_n}{b_n} \sum_{k=1}^{n_0} \log\left(1 + \lambda_k \rho_k^{(\theta)}\right) + \frac{L_n}{b_n} \sum_{k=n_0+1}^{n} \log\left(1 + \lambda_k \rho_k^{(\theta)}\right) \\ &\le \frac{L_n}{b_n} \sum_{k=1}^{n_0} \log\left(1 + \lambda_k \rho_k^{(\theta)}\right) + \frac{L_n}{b_n} \sum_{k=n_0+1}^{n} \log\left(1 + \lambda_k(1+\varepsilon)\right).\end{aligned}$$

Moreover, $\frac{L_n}{b_n} \sum_{k=1}^{n_0} \log\left(1 + \lambda_k \rho_k^{(\theta)}\right) \to 0$ (as $n \to \infty$) by Equation (13) and

$$\frac{L_n}{b_n} \sum_{k=n_0+1}^{n} \log\left(1 + \lambda_k(1+\varepsilon)\right) \le (1+\varepsilon) \frac{L_n}{b_n} \sum_{k=n_0+1}^{n} \lambda_k = (1+\varepsilon)\left(\frac{L_n}{b_n} \sum_{k=1}^{n} \lambda_k - \frac{L_n}{b_n} \sum_{k=1}^{n_0} \lambda_k\right).$$

Hence, Equation (21) follows from Equations (13) and (14), and the arbitrariness of ε. □

Remark 2 (The Bernoulli case; the sequence in Equation (2)). *The LDP in Theorem 3 holds also for the sequence in Equation (2) in place of the sequence in Equation (1). The proof is almost identical to the one of Theorem 3: in this case, we have*

$$\Delta_n := \left|\frac{L_n}{b_n} \sum_{k=1}^{n} X_k - \frac{L_n}{b_n} \sum_{k=1}^{n} X_k^*\right|$$

in place of Equation (18), and Inequality (19) still holds (even without the monotonicity of $\{L_n : n \ge 1\}$).

Remark 3 (Almost sure convergence to 1 of the sequences in Theorems 2 and 3). *Let $\{Z_n : n \ge 1\}$ be either the sequence in Equation (1) or the sequence in Equation (2), where $\{X_n : n \ge 1\}$ is as in Theorem 2 or as in Theorem 3* *(so we also consider Remarks 1 and 2). Then, by a straightforward consequence of the Borel–Cantelli lemma, the sequence $\{Z_n : n \geq 1\}$ converges to 1 almost surely (as $n \to \infty$) if*

$$\textstyle\sum_{n\geq 1} P(Z_n \in C) < \infty \text{ for closed set } C \text{ such that } 1 \notin C.$$

Obviously this condition holds if $C \subset (-\infty, 0)$ because $\{Z_n : n \geq 1\}$ are nonnegative random variables. On the other hand, if $C \cap [0, \infty)$ is not empty, $\Lambda^*(C) := \inf_{x\in C} \Lambda^*(x)$ is finite; moreover, $\Lambda^*(C) \in (0, \infty)$ because $1 \notin C$. Then, by the upper bound of the closed set, for all $\delta > 0$, there exists n_δ such that, for all $n > n_\delta$, we have

$$P(Z_n \in C) \leq e^{-(\Lambda^*(C)-\delta)b_n/L_n}.$$

Thus, again by the Borel–Cantelli lemma, $\{Z_n : n \geq 1\}$ converges almost surely to 1 (as $n \to \infty$) if, for all $\kappa > 0$, we have

$$\sum_{n\geq 1} e^{-\kappa b_n/L_n} < \infty. \tag{22}$$

Then, by the Cauchy condensation test, Equation (22) holds if and only if $\sum_{n\geq 1} 2^n e^{-\kappa b_{2^n}/L_{2^n}} < \infty$ and, as we see below, the convergence of the condensed series is a consequence of the ratio test and of some hypotheses of Theorems 2 and 3. In fact,

$$\frac{2^{n+1}e^{-\kappa b_{2^{n+1}}/L_{2^{n+1}}}}{2^n e^{-\kappa b_{2^n}/L_{2^n}}} = 2\exp\left(-\kappa \frac{b_{2^{n+1}}}{L_{2^{n+1}}}\left(1 - \frac{b_{2^n}}{b_{2^{n+1}}} \cdot \frac{L_{2^{n+1}}}{L_{2^n}}\right)\right) \to 0 \text{ (as } n \to \infty)$$

because $\frac{b_{2^n}}{b_{2^{n+1}}} \to \alpha(1/2)$ by Equation (12), $\frac{L_{2^{n+1}}}{L_{2^n}} \to 1$ by Equation (11) and $\frac{b_{2^{n+1}}}{L_{2^{n+1}}} \to +\infty$ by Equation (13).

We conclude with the results for the generalized Cramér model (the sequences in Equation (9)).

Corollary 1 (Application to the sequences in Equation (9))**.** *Let $\{X_n : n \geq 1\}$ be the random variables in Equation (7), and let $\{b_n : n \geq 1\}$ and $\{L_n : n \geq 1\}$ be defined by Equation (8). Then, the sequences $\left\{\frac{\sum_{k=1}^n (\log k)X_k}{n\ell_n} : n \geq 1\right\}$ and $\left\{\frac{(\log n)\sum_{k=1}^n X_k}{n\ell_n} : n \geq 1\right\}$ in Equation (9) satisfy the LDP with speed function $v_n = \frac{b_n}{L_n} = \frac{n\ell_n}{\log n}$ and the good rate function Λ^* defined by Equation (4).*

Proof. In this proof, the sequences in Equation (9) play the roles of the sequences in Equations (1) and (2) in Theorem 3 and Remark 2, respectively. Therefore, we have to check that the hypotheses of Theorem 3 are satisfied. Condition 1 and Equations (11) and (13) and $\lim_{n\to\infty} \lambda_n = 0$ can be easily checked. Moreover, one can also check Equation (12) with $\alpha(c) = c$; note that in this case, we have a regularly varying function with index $\rho = 1$ (as $n \to \infty$), and $\{b_n : n \geq 1\}$ is eventually nondecreasing. Finally, Equation (14), which is

$$\lim_{n\to\infty} \frac{(\log n)\sum_{k=1}^n \frac{\ell_k}{\log k}}{n\ell_n} = 1,$$

can be obtained as a consequence of Lemma 3; in fact, $\{\ell_n : n \geq 1\}$ and $\{\ell_n/(\log n) : n \geq 1\}$ are restrictions (on $\mathbb{N}$) of slowly varying functions at infinity. □

In conclusion, we can say that, roughly speaking, for any Borel set A such that $1 \notin \bar{A}$ (where $\bar{A}$ is the closure of A), the probabilities $P\left(\frac{\sum_{k=1}^n (\log k)X_k}{n\ell_n}\right)$ and $P\left(\frac{(\log n)\sum_{k=1}^n X_k}{n\ell_n}\right)$ decay exponentially as $e^{-\frac{n\ell_n}{\log n}\inf_{x\in A}\Lambda^*(x)}$ (as $n \to \infty$). Thus, in the spirit of Tao's remark, we are able to suggest estimations concerning a sort of "generalized" Chebychev function defined by $\frac{\sum_{p_1\cdots p_r \leq x} \log(p_1\cdots p_r)}{x\ell_x}$ or by $\frac{(\log x)\sum_{p_1\cdots p_r\leq x} 1}{x\ell_x}$. To our knowledge, such estimations are not available for $r > 1$.

Author Contributions: Rita Giuliano and Claudio Macci equally contributed to the proofs of the results. The paper was also written and reviewed cooperatively.

Conflicts of Interest: The authors declare no conflict of interest.

References

1. Dembo, A.; Zeitouni, O. *Large Deviations Techniques and Applications*, 2nd ed.; Springer: New York, NY, USA, 1998.
2. Fang, L. Large and moderate deviation principles for alternating Engel expansions. *J. Number. Theory* **2015**, *156*, 263–276.
3. Granville, A. Harald Cramér and the distribution of prime numbers. *Scand. Actuar. J.* **1995**, *1995*, 12–28.
4. Döring, H.; Eichelsbacher, P. Moderate deviations via cumulants. *J. Theor. Probab.* **2013**, *26*, 360–385.
5. Féray, V.; Méliot, P.L.; Nikeghbali, A. Mod-φ Convergence, I: Normality Zones and Precise Deviations. Unpublished Manuscript, 2015. Available online: http://arxiv.org/pdf/1304.2934.pdf (accessed on 23 November 2015).
6. Jacod, J.; Kowalski, E.; Nikeghbali, A. Mod-Gaussian convergence: new limit theorems in probability and number theory. *Forum Math.* **2011**, *23*, 835–873.
7. Tenenbaum, G.; Mendès France, M. *The Prime Numbers and Their Distribution*; (Translated from the 1997 French original by P.G. Spain); American Mathematical Society: Providence, RI, USA, 2000.
8. Landau, E. *Handbuch der Lehre von der Verteilung der Primzahlen (2 Volumes)*, 3rd ed.; Chelsea Publishing: New York, NY, USA, 1974.
9. Hardy, G.H.; Wright, E.M. *An Introduction to the Theory of Numbers*, 4th ed.; Oxford University Press: London, UK, 1975.
10. Tenenbaum, G. *Introduction to Analytic and Probabilistic Number Theory*, 3rd ed.; (Translated from the 2008 French Edition by P.D.F. Ion); American Mathematical Society: Providence, RI, USA, 2015.
11. Davenport, H. *Multiplicative Number Theory*, 3rd ed.; Springer: New York, NY, USA; Berlin, Germany, 2000.
12. Ford, K.; Sneed, J. Chebyshev's bias for products of two primes. *Exp. Math.* **2010**, *19*, 385–398.
13. Meng, X. Chebyshev's Bias for Products of k Primes. Unpublished Manuscript, 2016. Available online: http://arxiv.org/pdf/1606.04877v2.pdf (accessed on 16 August 2016).
14. Tao, T. Probabilistic Models and Heuristics for the Primes (Optional). In Terence Tao Blog. 2015. Available online: https://terrytao.wordpress.com/2015/01/04/254a-supplement-4-probabilistic-models-and-heuristics-for-the-primes-optional/ (accessed on 4 January 2015).
15. Bingham, N.H.; Goldie, C.M.; Teugels, J.L. Regular variation. In *Encyclopedia of Mathematics and its Applications*; Cambridge University Press: Cambridge, UK, 1987; Volume 27.
16. Giuliano, R.; Macci, C. Asymptotic results for weighted means of random variables which converge to a Dickman distribution, and some number theoretical applications. *ESAIM Probab. Stat.* **2015**, *19*, 395–413.
17. Arratia, R.; Tavaré, S. Independent processes approximations for random combinatorial structures. *Adv. Math.* **1994**, *104*, 90–154.

mathematics

MDPI

Article

A Within-Host Stochastic Model for Nematode Infection

Antonio Gómez-Corral [1,*] and Martín López-García [2]

1 Instituto de Ciencias Matemáticas CSIC-UAM-UC3M-UCM, Calle Nicolás Cabrera 13-15, 28049 Madrid, Spain
2 Department of Applied Mathematics, School of Mathematics, University of Leeds, Leeds LS2 9JT, UK; m.lopezgarcia@leeds.ac.uk
* Correspondence: antonio.gomez@icmat.es

Received: 25 June 2018; Accepted: 11 August 2018; Published: 21 August 2018

Abstract: We propose a stochastic model for the development of gastrointestinal nematode infection in growing lambs under the assumption that nonhomogeneous Poisson processes govern the acquisition of parasites, the parasite-induced host mortality, the natural (no parasite-induced) host mortality and the death of parasites within the host. By means of considering a number of age-dependent birth and death processes with killing, we analyse the impact of grazing strategies that are defined in terms of an intervention instant t_0, which might imply a move of the host to safe pasture and/or anthelmintic treatment. The efficacy and cost of each grazing strategy are defined in terms of the transient probabilities of the underlying stochastic processes, which are computed by means of Strang–Marchuk splitting techniques. Our model, calibrated with empirical data from Uriarte et al and Nasreen et al., regarding the seasonal presence of nematodes on pasture in temperate zones and anthelmintic efficacy, supports the use of dose-and-move strategies in temperate zones during summer and provides stochastic criteria for selecting the exact optimum time instant t_0 when these strategies should be applied.

Keywords: host-parasite interaction; nematode infection; nonhomogeneous Poisson process; seasonal environment; Strang–Marchuk splitting approach

1. Introduction

Gastrointestinal (GI) nematodes are arguably (see [1,2]) the major cause of ill health and poor productivity in grazing sheep worldwide, especially in young stock. Productivity losses result from both parasite challenge and parasitism, while regular treatment of the infections is costly in terms of chemicals and labour. The relative cost of GI parasitism has become greater in recent decades as the availability of effective broad-spectrum anthelmintics (see Chapter 5 of [1]) has enabled the intensification of pastoral agriculture. To an extent, it appears the success of the various anthelmintic products developed since the 1960s has created a rod for our own backs, particularly as resistance has arisen to each active family in turn (see, for example, [3,4]). Options for the control of GI nematode infections (which do not rely uniquely on the use of anthelmintics) include management procedures (involving intervention with anthelmintics, grazing management, level of nutrition and bioactive forages), biological control (with nematophagous fungi), selection for genetic resistance in sheep (within breed/use of selected breeds) and vaccination. The article by Stear et al. [5] gives an overview of alternatives to anthelmintics for the control of nematodes in livestock, and it complements and extends other review articles by Hein et al. [6], Knox [7], Sayers and Sweeney [8] and Waller and Thamsborg [9]. Moreover, we refer the reader to the article by Smith et al. [10] for stochastic and deterministic models of anthelmintic resistance emergence and spread.

The aim of this paper is to present a stochastic model for quantitatively comparing among various grazing strategies involving isolation, movement or treatment of the host, but without incorporating the risk of selecting for resistance. This amounts to the assumption that the nematodes in our model have not been previously exposed to the anthelmintic treatments under consideration; see, for example, [11]. We point out that the effect of the resistance in the dynamics is usually limited by the rotation of different anthelmintic classes on an annual basis (see [12,13]).

A wide range of mathematical models can be found in the literature for modelling the infection dynamics of nematodes in ruminants. Originally, simple deterministic models were proposed in terms of systems of ordinary differential equations describing the population dynamics of infected ruminants and nematodes on pasture. By describing these dynamics in a deterministic way, the resulting models were tractable from a mathematical and analytical point of view [14]. However, efforts were soon redirected towards stochastic approaches given the importance of stochastic effects in these systems [15]. These stochastic effects are related to, among others, spatial dynamics, clumped infection events or individual heterogeneities related to the host's immune response to infection [16]. Without any aim of an exhaustive enumeration, we refer the reader to [15,17,18] for deterministic and stochastic models of nematode infection in ruminants for a population of hosts maintaining a fixed density.

In this paper, we develop a mathematical model for the within-host GI nematode infection dynamics, to compare the effectiveness and cost of various worm control strategies, which are related to pasture management practices and/or strategic treatments based on the use of a single anthelmintic drug. Control criteria are applied to the development of GI nematode parasitism in growing lambs. Specifically, the interest is in the parasite *Nematodirus* spp. with *Nematodirus battus*, *Nematodirus filicollis* and *Nematodirus spathiger* as the main species. The resulting grazing management strategies are specified in terms of an intervention instant t_0 that, under certain specifications, implies moving animals to safe pastures and/or anthelmintic treatment. For a suitable selection of t_0, we present two control criteria that provide a suitable balance between the efficacy and cost of intervention. Our methodology is based on simple stochastic principles and time-dependent continuous-time Markov chains; see the book by Allen [19] for a review of the main results for deterministic and stochastic models of interacting biological populations.

Our work in this paper is directly related to that in [20], where we examine stochastic models for the parasite load of a single host and where the interest is in analysing the efficacy of various grazing management strategies. In [20], we defined control strategies based on isolation and treatment of the host at a certain age t_0. This means that the host is free living in a seasonal environment, and it is transferred to an uninfected area at age t_0. In the uninfected area, the host does not acquire new parasites, undergoes an anthelmintic treatment to decrease the parasite load and varies in its susceptibility to parasite-induced mortality and natural (no parasite-induced) mortality. From a mathematical point of view, an important feature of the analysis in [20] is that the underlying processes, recording the number of parasites infesting the host at an arbitrary time t, can be thought of as age-dependent versions of a pure birth process with killing and a pure death process with killing, which are both defined on a finite state space.

Here, we complement the treatment of control strategies applied to GI nematode burden that we started in [20] by focusing on strategies that are not based on isolation of the host; to be concrete, our interest is in three grazing strategies that reflect the use of a paddock with safe pasture and/or the efficacy of an anthelmintic drug. Seasonal fluctuations in the acquisition of parasites, the death of parasites within the host and the natural and parasite-induced host mortality are incorporated into our model by using nonhomogeneous Poisson processes. Contrary to [20], grazing management strategies considered in this work lead to, instead of pure birth/death processes with killing, the analysis of several age-dependent birth and death processes with killing. The efficacy and cost of each grazing strategy are defined in terms of the transient probabilities of each of the underlying stochastic processes; that is, the probability that the parasite load of the infected host is at any given level at each time instant, given that a particular control strategy has been applied at the intervention instant t_0. In order to compute these probabilities, we apply Strang–Marchuk splitting techniques for solving the corresponding systems of differential equations.

The paper is organized as follows. In Section 2, we define the mathematical model used in various grazing management strategies, derive the analytical solution of the underlying time-dependent systems of linear differential equations and present two criteria allowing us to find the instant t_0 that appropriately balances effectiveness and cost of intervention in these grazing strategies. In Section 3, we examine seasonal changes of GI nematode burden in growing lambs. Finally, concluding remarks are given in Section 4.

2. Stochastic Within-Host Model and Control Criteria

In this section, we first propose a mathematical stochastic model for the within-host infection dynamics by GI nematodes in growing lambs, define grazing management strategies and set down a set of equations governing the dynamics of the underlying processes. We then present control criteria based on stochastic principles. For the sake of brevity, we refer the reader to Appendix A where we comment on the equivalence used in Table 1 of [20] in the identification of the degree of infestation, level of infection, eggs per gram (EPG) value, number of L_3 infective larvae in the small intestine and the points system. Further details on the taxonomy and morphology of the parasite *Nematodirus* spp. and the treatment and control of parasite gastroenteritis in sheep can be found in [1,2,21].

2.1. Grazing Management Strategies: A Stochastic Within-Host Model

We define the mathematical model in terms of levels of infection and let the random variable $M(t)$ record the infection level of the host at time t. From Table 1 of [20], this means that the degree of infestation is null if $M(t) = 0$, light if $M(t) = m$ with $m \in \{1,2,3\}$, moderate if $M(t) = m$ with $m \in \{4,5,6,7\}$, high if $M(t) = m$ with $m \in \{8,9,10,11\}$ and heavy if $M(t) = -1$. In the setting of GI nematode parasitism, the value $M_0 = 11$ amounts to a critical level that does not permit the host to develop immunity to the nematode infection, in such a way that an eventual intervention is assumed to be ineffective as the degree of infestation is heavy. Therefore, we let $M(t) = -1$ be equivalent to the degree heavy of infestation (i.e., the number of L_3 infective larvae in the small intestine is greater than 12,000 worms) or the death of the host. Let $\mathcal{S}$ denote the set $\{0,1,\dots,M_0\}$ of infection levels, with $M_0 = 11$.

We consider individual-based grazing strategies, which are related to a single lamb (host) that is born, parasite-free, at time $t = 0$ and, over its lifetime, is exposed to parasites at times that form a nonhomogeneous Poisson process of rate $\lambda(t)$. At every exposure instant, the host acquires parasites, independently of one exposure to another. It is assumed that the number of acquired parasites does not allow the level $M(t)$ of infection to increase more than one unit at any acquisition instant, which is a plausible assumption in our examples where increments in the number of L_3 infective larvae in the small intestine are registered at daily intervals. Let $\delta(t)$ be the death rate of the host at age t in the absence of any parasite burden, and assume that this rate is increased by an amount $\gamma_m(t)$, which is related to the parasite-induced host mortality as the infection level equals m at age t. For later use, we define the functions $\lambda_m(t) = (1 - 1_{-1,m})\lambda(t)$ and $\delta_m(t) = (1 - 1_{-1,m})(\delta(t) + \gamma_m(t))$ for levels $m \in \{-1\} \cup \mathcal{S}$, where $1_{k,m}$ denotes Kronecker's delta.

At age τ, the interest is in the level $M(\tau)$ of infection as, under certain grazing assumptions, intervention is prescribed at a certain age $t_0 < \tau$. Note that the host at age t_0 can be dead or its degree of infestation can be heavy ($M(t_0) = -1$), and it can be alive and parasite-free ($M(t_0) = 0$), or it can be alive and infected ($M(t_0) = m$ with $m \in \{1,2,\dots,M_0\}$).

In analysing the process $\mathcal{Z} = \{M(t) : 0 \le t \le \tau\}$, we distinguish between the free-living interval $[0, t_0)$ and the post-intervention interval $[t_0, \tau]$; for ease of presentation, we first digress to briefly recall the analytic solution for ages $t \in [0, t_0)$ given in [20]. For a host that has survived at age t with $t < t_0$, the infection dynamics within the host are analysed in terms of transient probabilities

$\pi_m(t) = P(M(t) = m | M(0) = 0)$, for levels $m \in \{-1\} \cup \mathcal{S}$. That is, $\pi_m(t)$ represents the probability of the host being at infection level m at time t. These dynamics lead us to a pure birth process with killing on the state space $\{-1\} \cup \mathcal{S}$ (see Figure 1 in [20]), the age-dependent birth and killing rates of which are given by $\lambda_m(t) = \lambda(t)$ and $\delta_m(t) = \delta(t) + \gamma_m(t)$, respectively, for $m \in \mathcal{S}$, and where -1 is an absorbing state. Expressions for $\pi_m(t)$ can be then evaluated following our arguments in Section 2.2 of [20].

Next, we focus on three grazing strategies that are defined in terms of the intervention instant t_0. This implies that, at post-intervention ages $t \in (t_0, \tau]$, the rates $\lambda(t)$, $\delta(t)$ and $\gamma_m(t)$ are replaced by functions $\lambda'(t)$, $\delta'(t)$ and $\gamma'_m(t)$, respectively, allowing us to show concrete effects of an intervention on the lamb and its environmental conditions. To be concrete, the functions $\lambda'(t)$, $\delta'(t)$ and $\gamma'_m(t)$ appropriately reflect the use of a paddock with safe pasture and/or the efficacy of an anthelmintic treatment, in accordance with the following grazing strategies:

Strategy UM: The host is left untreated, but moved to a paddock with safe pasture at age t_0. The resulting process $\mathcal{Z}$ can be thought of as an age-dependent pure birth process with killing, the birth rates of which are given by $\lambda_m(t) = \lambda(t)$ if $t \in [0, t_0)$, and $\lambda'(t)$ if $t \in [t_0, \tau]$, and killing rates are defined by $\delta_m(t) = \delta(t) + \gamma_m(t)$ if $t \in [0, t_0)$ and $\delta'(t) + \gamma'_m(t)$ if $t \in [t_0, \tau]$, for $m \in \mathcal{S}$.

Strategy TS: The host is treated with anthelmintics and set-stocked at age t_0. Let $\eta'_m(t)$ be the death rate of parasites when the infection level of the host is $m \in \mathcal{S}$ at time t with $t > t_0$. In this case, $\mathcal{Z}$ can be seen as an age-dependent birth and death process with killing. The birth and death rates are defined by $\lambda_m(t) = \lambda(t)$ if $t \in [0, \tau]$, $\eta_m(t) = 0$ if $t \in [0, t_0)$ and $\eta'_m(t)$ if $t \in [t_0, \tau]$, for $m \in \mathcal{S}$, respectively. Killing rates are defined identically to the rates $\delta_m(t)$ in strategy UM.

Strategy TM: The host is treated with anthelmintics and moved to safe pasture at age t_0. In a similar manner to strategy TS, the process $\mathcal{Z}$ may be formulated as an age-dependent birth and death process with killing. Birth, death and killing rates are identical to those in strategy TS with the exception of $\lambda_m(t)$ for time instants $t \in [t_0, \tau]$, which has the form $\lambda_m(t) = \lambda'(t)$.

In strategies TS and TM, a single anthelmintic drug is used. In accordance with the empirical data in [22], resistance is not incorporated into modelling aspects since $\tau = 1$ year is assumed in Section 3. The resulting models are thus related to the rotation of different anthelmintic classes on an annual basis, which has been widely promoted and adopted as a strategy to delay the development of anthelmintic resistance in nematode parasites; see, e.g., [12,13].

For the sake of completeness, we introduce the term **scenario US** to reflect no intervention, that is the host is left untreated and set-stocked. Note that, in such a case, the process $\mathcal{Z}$ is an age-dependent pure birth process with killing, and its birth and killing rates are specified by $\lambda_m(t) = \lambda(t)$ and $\delta_m(t) = \delta(t) + \gamma_m(t)$ if $t \in [0, \tau]$, for $m \in \mathcal{S}$. It follows then that the transient distribution of $\mathcal{Z}$ is readily derived from [20] for time instants $t \in (0, \tau]$.

A slight modification of our arguments in solving Equations (1) and (2) of [20] allows us to derive explicit expressions for the transient solution at post-intervention instants $t \in (t_0, \tau]$ in grazing strategy UM. For time instants $t \in [t_0, \tau]$, we introduce the probability $\pi_m^{UM}(t_0; t) = P(M(t) = m)$ of the process being at infection level m at time t, given that strategy UM is implemented at the intervention instant t_0, and initial conditions $\pi_m^{UM}(t_0; t_0) = \pi_m(t_0)$, for $m \in \{-1\} \cup \mathcal{S}$. Then, the transient solution at time instants $t \in (t_0, \tau]$ can be readily expressed as:

$$\pi_m^{UM}(t_0; t) = e^{-\Lambda(t_0;t) - \Delta_m(t_0;t)} \left(\pi_m^{UM}(t_0; t_0) + (1 - 1_{0,m}) \sum_{j=0}^{m-1} \pi_j^{UM}(t_0; t_0) K_{m-1}^{UM,j}(t_0; t) \right), \quad (1)$$

where $\Lambda(t_0; t) = \int_{t_0}^{t} \lambda'(u) du$ and $\Delta_m(t_0; t) = \int_{t_0}^{t} (\delta'(u) + \gamma'_m(u)) du$. Starting from:

$$K_{m-1}^{UM,m-1}(t_0;t) = \int_{t_0}^{t} \lambda'(u) e^{\tilde{\Delta}_{m-1}(t_0;u)} du,$$

the functions $K_{m-1}^{UM,j}(t_0;t)$, for values $0 \le j \le m-2$, are iteratively computed as:

$$K_{m-1}^{UM,j}(t_0;t) = \int_{t_0}^{t} \lambda'(u) e^{\tilde{\Delta}_{m-1}(t_0;u)} K_{m-2}^{UM,j}(t_0;u) du,$$

with $\tilde{\Delta}_{m-1}(t_0;t) = \Delta_m(t_0;t) - \Delta_{m-1}(t_0;t)$.

2.2. Splitting Techniques

For grazing strategies TS and TM, the transient solution at time instants $t \in (t_0, \tau]$ can be numerically derived by using splitting techniques; see [23]. In a unifying manner, we may observe that, for a host that has survived at age t with $t_0 < t < \tau$ and $M(t) = m \in \mathcal{S}$, the possible transitions (in both strategies TS and TM) are as follows (Figure 1):

(i) $m \to m+1$ at rate $\lambda_m(t)$, for levels $m \in \{0, 1, \dots, M_0 - 1\}$;
(ii) $m \to m-1$ at rate $\eta_m(t)$, for levels $m \in \{1, 2, \dots, M_0\}$;
(iii) $m \to -1$ at rate $\delta_m(t)$, for levels $m \in \{0, 1, \dots, M_0 - 1\}$;
(iv) $M_0 \to -1$ at rate $\delta_{M_0}(t) + \lambda_{M_0}(t)$.

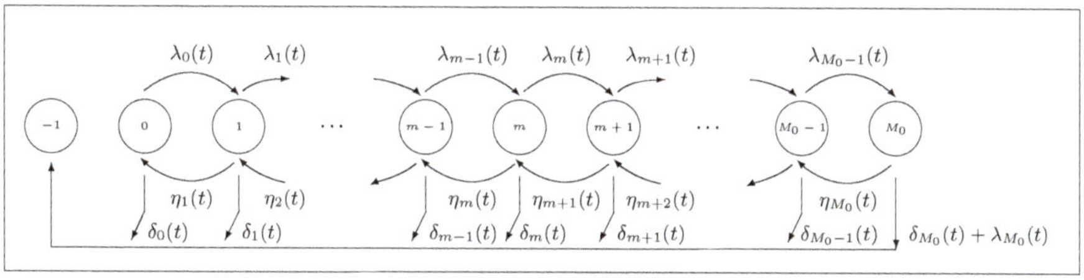

Figure 1. State space and transitions at post-intervention instants $t \in [t_0, \tau]$. Grazing strategies TS and TM.

Then, if we select a certain grazing strategy s with $s \in \{TS, TM\}$, the resulting probabilities $\pi_m^s(t_0;t) = P(M(t) = m)$, for $m \in \{-1\} \cup \mathcal{S}$ and time instants $t \in [t_0, \tau]$, satisfy the equality:

$$\pi_{-1}^s(t_0;t) = 1 - \sum_{m=0}^{M_0} \pi_m^s(t_0;t),$$

and the time-dependent linear system of differential equations:

$$\frac{d}{dt}\mathbf{\Pi}^s(t_0;t) = \mathbf{B}(t)\mathbf{\Pi}^s(t_0;t), \tag{2}$$

where $\mathbf{\Pi}^s(t_0;t) = (\pi_0^s(t_0;t), \pi_1^s(t_0;t), \dots, \pi_{M_0}^s(t_0;t))^T$, and $\mathbf{B}(t)$ is a tridiagonal matrix with entries:

$$(\mathbf{B}(t))_{i,j} = \begin{cases} -(\lambda_i(t) + \delta_i(t) + (1 - 1_{0,i})\eta_i(t)), & \text{if } 0 \le i \le M_0, j = i, \\ \eta_{i+1}(t), & \text{if } 0 \le i \le M_0 - 1, j = i+1, \\ \lambda_{i-1}(t), & \text{if } 1 \le i \le M_0, j = i-1, \\ 0, & \text{otherwise.} \end{cases}$$

Needless to say, initial conditions in Equation (2) are given by $\mathbf{\Pi}^s(t_0;t_0) = (\pi_0(t_0), \pi_1(t_0), \ldots, \pi_{M_0}(t_0))^T$ where the values for $\pi_m(t_0)$ with $m \in \mathcal{S}$ do not depend on the grazing strategy under consideration.

In principle, the system (2) of differential equations could be solved in many ways, but Strang–Marchuk splitting techniques are concretely used in Section 3 to derive its solution. Following the approach in [23], the original problem given by Equation (2) is first split into several subsystems that are then solved cyclically one after the other. This procedure is particularly advisable when tailor-made numerical methods can be applied for each split subsystem or when, as occurs in our case, explicit solutions for the subsystems can be derived.

The approach in Section 1.3 of [23] is of particular interest when, for a certain splitting $\mathbf{B}(t) = \mathbf{U}(t) + \mathbf{V}(t)$, the time-dependent linear systems of differential equations:

$$\begin{aligned}
\frac{d}{dt}\mathbf{\Pi}^s(t_0;t) &= \mathbf{U}(t)\mathbf{\Pi}^s(t_0;t), \quad t_0 \le t \le \tau, \\
\frac{d}{dt}\mathbf{\Pi}^s(t_0;t) &= \mathbf{V}(t)\mathbf{\Pi}^s(t_0;t), \quad t_0 \le t \le \tau,
\end{aligned}$$

can be accurately and efficiently solved, which is our case here. In our examples in Section 3, we consider the splitting $\mathbf{B}(t) = \mathbf{U}(t) + \mathbf{V}(t)$ with:

$$\begin{aligned}
\mathbf{U}(t) &= \begin{pmatrix} -(\lambda_0(t)+\delta_0(t)) & & & \\ \lambda_0(t) & -(\lambda_1(t)+\delta_1(t)) & & \\ & \ddots & \ddots & \\ & & \lambda_{M_0-1}(t) & -(\lambda_{M_0}(t)+\delta_{M_0}(t)) \end{pmatrix}, \\
\mathbf{V}(t) &= \begin{pmatrix} 0 & \eta_1(t) & & & \\ & -\eta_1(t) & \eta_2(t) & & \\ & & \ddots & \ddots & \\ & & & -\eta_{M_0-1}(t) & \eta_{M_0}(t) \\ & & & & -\eta_{M_0}(t) \end{pmatrix},
\end{aligned}$$

and we evaluate numerically the transient solution $\pi^s_m(t_0;t)$ at time instants $t \in \{t_0, t_0+1, \ldots, \tau\}$ by solving a sequence of four time-dependent linear subsystems of differential equations.

In order to determine the probabilities $\pi^s_m(t_0;t_0+1)$ for levels $m \in \mathcal{S}$ and a certain grazing strategy s with $s \in \{TS, TM\}$, we first select the splitting time-step as $\Delta t = N^{-1}$ with $N = 10^3$, and introduce the notation:

$$a_n = t_0 + (n-1)\Delta t, \quad n \in \{1, 2, \ldots, N+1\}, \tag{3}$$

$$b_n = t_0 + (n-0.5)\Delta t, \quad n \in \{1, 2, \ldots, N\}. \tag{4}$$

At step n with $n \in \{1, 2, \ldots, N\}$, we solve the subsystems $(S_1)_n$, $(S_2)_n$, $(S_3)_n$ and $(S_4)_n$ cyclically on successive intervals of length Δt, using the solution of one subsystem as the initial condition of the other one as follows:

$$\text{Subsystem } (S_1)_n \equiv \begin{cases} \frac{d}{dt}\mathbf{\Pi}^s_1(a_n;t) = \mathbf{U}(t)\mathbf{\Pi}^s_1(a_n;t), \; a_n \le t \le b_n, \\ \mathbf{\Pi}^s_1(a_n;a_n) = \begin{cases} \mathbf{\Pi}^s(t_0;t_0), & \text{if } n = 1, \\ \mathbf{\Pi}^s_4(b_{n-1};a_n), & \text{if } 2 \le n \le N. \end{cases} \end{cases}$$

$$\text{Subsystem } (S_2)_n \equiv \begin{cases} \frac{d}{dt}\mathbf{\Pi}_2^s(a_n;t) = \mathbf{V}(t)\mathbf{\Pi}_2^s(a_n;t), \ a_n \leq t \leq b_n, \\ \mathbf{\Pi}_2^s(a_n;a_n) = \mathbf{\Pi}_1^s(a_n;b_n). \end{cases}$$

$$\text{Subsystem } (S_3)_n \equiv \begin{cases} \frac{d}{dt}\mathbf{\Pi}_3^s(b_n;t) = \mathbf{V}(t)\mathbf{\Pi}_3^s(b_n;t), \ b_n \leq t \leq a_{n+1}, \\ \mathbf{\Pi}_3^s(b_n;b_n) = \mathbf{\Pi}_2^s(a_n;b_n). \end{cases}$$

$$\text{Subsystem } (S_4)_n \equiv \begin{cases} \frac{d}{dt}\mathbf{\Pi}_4^s(b_n;t) = \mathbf{U}(t)\mathbf{\Pi}_4^s(b_n;t), \ b_n \leq t \leq a_{n+1}, \\ \mathbf{\Pi}_4^s(b_n;b_n) = \mathbf{\Pi}_3^s(b_n;a_{n+1}). \end{cases}$$

This procedure results in the solution at $t = t_0 + 1$, which is given by $\mathbf{\Pi}^s(t_0;t) = \mathbf{\Pi}_4^s(b_N;a_{N+1})$ since $a_{N+1} = t_0 + 1$. Then, we may proceed similarly in the numerical evaluation of the transient solution at subsequent time instants $t = t_0 + k$ with $k \geq 2$ and $t_0 + k \leq \tau$, by replacing t_0 by $t_0 + k$ in (3) and (4), so that the solution of the previous subsystems at time instant $t = t_0 + k - 1$ is now used as the initial condition in the subsystem $(S_1)_n$ at step $n = 1$. We refer the reader to [23] for qualitative properties of the operator splitting approach and convergence order.

For grazing strategy $s \in \{TS, TM\}$, the entries $\pi_m^s(a_n;t)$, for levels $m \in \mathcal{S}$, of the vector $\mathbf{\Pi}_1^s(a_n;t)$ are given by Equation (1) for time instants $t \in [a_n, b_n]$, with t_0 replaced by a_n, and the function $\lambda'(t)$ replaced by $\lambda(t)$ in the case TS.

The solution $\mathbf{\Pi}_2^s(a_n;t)$ at time instants $t \in [a_n, b_n]$ has entries:

$$\pi_m^s(a_n;t) = e^{-H_m(a_n;t)} \left(\pi_m^s(a_n;a_n) + (1 - 1_{m,M_0}) \sum_{j=m+1}^{M_0} \pi_j^s(a_n;a_n) K_{m+1}^{s,M_0-j}(a_n;t) \right), \tag{5}$$

where $H_m(a_n;t) = (1 - 1_{0,m}) \int_{a_n}^t \eta'_m(u)du$ and, starting from:

$$K_{m+1}^{s,M_0-(m+1)}(a_n;t) = \int_{a_n}^t \eta'_{m+1}(u) e^{\tilde{H}_{m+1}(a_n;u)} du,$$

the functions $K_{m+1}^{s,M_0-j}(a_n;t)$, for values $m + 2 \leq j \leq M_0$, can be iteratively evaluated as:

$$K_{m+1}^{s,M_0-j}(a_n;t) = \int_{a_n}^t \eta'_{m+1}(u) e^{\tilde{H}_{m+1}(a_n;u)} K_{m+2}^{s,M_0-j}(a_n;u)du,$$

with $\tilde{H}_m(a_n;t) = H_{m-1}(a_n;t) - H_m(a_n;t)$.

In a similar manner, the solution $\mathbf{\Pi}_3^s(b_n;t)$ at time instants $t \in [b_n, a_{n+1}]$ has the form (5), with a_n replaced by b_n. The entries $\pi_m^s(b_n;t)$, for levels $m \in \mathcal{S}$, of the solution $\mathbf{\Pi}_4^s(b_n;t)$ are given by Equation (1) for time instants $t \in [b_n, a_{n+1}]$, with t_0 replaced by b_n and $\lambda'(t)$ replaced by $\lambda(t)$ in the case TS.

2.3. Control Criteria Based on Stochastic Principles

For grazing strategies UM, TS and TM, we define a control strategy by means of an age t_0 and an intervention rule, which is related to a concrete infection level $m' \in \{1, 2, \ldots, M_0\}$ and the resulting probability:

$$P_{\geq m'}(t) = \sum_{m=m'}^{M_0} \pi_m(t).$$

This age-dependent probability allows us to determine a set $I_{\geq m'}$ of potential intervention instants $t \in (0, \tau)$ satisfying the inequality $P_{\geq m'}(t) \geq p$ for a predetermined index $p \in (0,1)$; note that $I_{\geq m'} = (0, \tau)$ in the case $p = 0$ regardless of the threshold m'.

It should be pointed out that, in grazing strategies UM and TM, maintaining safe-pasture conditions in a paddock for the whole year does not seem feasible in practice. Moreover, treating the host with anthelmintic drugs (cases TS and TM) within early days of the year will not yield optimal results, since profits of treatment cannot be obtained before host exposure to infection; see Figure A1 in Appendix A. Thus, we focus on values $p > 0$ in such a way that, for a fixed pair (m', p) with $p > 0$, those instants $t \notin I_{\geq m'}$ can be seen as either low-risk (states $m \in \{0, 1, \ldots, m' - 1\}$) or extreme-risk ($m = -1$) intervention instants, and consequently, they are not taken into account in our next arguments.

In carrying out our examples, we select the threshold $m' = 4$ yielding a moderate degree of infestation (according to Table 1 of [20]) and the index $p \in \{0.1, 0.2, \ldots, 0.7\}$. Then, for each resulting set $I_{\geq m'}$ of potential intervention instants, the problem is to find a single instant $t_0 \in I_{\geq m'}$ that appropriately balances the effectiveness and cost of intervention in the grazing strategy under consideration. In our approach, the effectiveness and cost functions can be seen as alternative measures of the efficacy of an intervention, with a negative significance in the case of the cost function. To be concrete, in an attempt to reflect the effect of the parasite burden on the lamb weight at age τ, effectiveness is measured in terms of:

$$eff^s(t_0; \tau) = \sum_{m=0}^{3} \pi_m^s(t_0; \tau),$$

which corresponds to the probability that the degree of infestation at age τ is null or light as the intervention is prescribed at age t_0 in accordance with the grazing strategy s with $s \in \{UM, TS, TM\}$. In contrast, we make the cost of intervention depend on the probability:

$$cost^s(t_0; \tau) = \sum_{m=8}^{11} \pi_m^s(t_0; \tau) + \pi_{-1}^s(t_0; \tau)$$

that either the host does not survive or its degree of infestation is high at age τ. It is worth noting that operational (financial) costs are not considered within the modelling framework, which will allow us to derive a single intervention instant t_0 regardless of concrete specifications for the cost of maintaining safe-pasture conditions, or the cost of purchasing the anthelmintic drugs. Then, the proposed cost function $cost^s(t_0; \tau)$, which can be seen as a negative measure of efficacy, is related to productivity losses corresponding to high levels of infection, and it may be advisable when the financial fluctuations (in comparing various drugs, how prices of anthelmintics change) over time are not known in advance.

For a suitable choice of t_0, the following control criteria are suggested:

Criterion 1: We select the intervention instant t_0 verifying $cost^s(t_0; \tau) = \inf\left\{cost^s(t; \tau) : t \in J^1_{\geq m'}\right\}$, where the subset $J^1_{\geq m'}$ consists of those potential intervention instants $t \in I_{\geq m'}$ satisfying the inequality $eff^s(t; \tau) \geq p_1$, for a certain probability $p_1 \in (0, 1)$.

Criterion 2: We select the intervention instant t_0 such that $eff^s(t_0; \tau) = \sup\left\{eff^s(t; \tau) : t \in J^2_{\geq m'}\right\}$, where the subset $J^2_{\geq m'}$ is defined by those time instants $t \in I_{\geq m'}$ verifying $cost^s(t; \tau) \leq p_2$, for a certain probability $p_2 \in (0, 1)$.

Our objective in Criterion 1 is thus to minimize the cost of intervention and to maintain a minimum level of effectiveness, which is translated into the probability $p_1 \in (0, 1)$. In Criterion 2, the objective is to maximize the effectiveness and to set an upper bound $p_2 \in (0, 1)$ to the cost of intervention. An alternative manner to measure the effectiveness and cost of intervention at a certain age $t_0 < \tau$ is given by the respective values:

$$\tau^{-1}E^s(t_0;\tau) = \tau^{-1}\int_0^{\tau}\sum_{m=0}^{3}\tilde{\pi}_m^s(t_0;u)du,$$
$$\tau^{-1}C^s(t_0;\tau) = \tau^{-1}\int_0^{\tau}\left(\sum_{m=8}^{11}\tilde{\pi}_m^s(t_0;u)+\tilde{\pi}_{-1}^s(t_0;u)\right)du,$$

where $\tilde{\pi}_m^s(t_0;u) = \pi_m^{US}(u)$ if $u \in (0,\tau)$ in scenario US and $\tilde{\pi}_m^s(t_0;u) = \pi_m^s(u)$ if $u \in (0,t_0)$, and $\pi_m^s(t_0;u)$ if $u \in [t_0,\tau)$ in grazing strategy s with $s \in \{UM, TS, TM\}$; then, values for $\tau^{-1}E^s(t_0;\tau)$ and $\tau^{-1}C^s(t_0;\tau)$ are related to the expected proportions of time that the degree of infestation is either null or light, and either high or heavy, respectively.

3. Empirical Data, Age-Dependent Rates and Results

Age-dependent patterns are from now on specified to reflect that the parasite-induced death of the host is negligible, and death rates in the absence of any parasite burden at free-living instants and post-intervention instants are identical, that is $\delta_m(t) = \delta(t) = \delta'(t)$ for levels $m \in \mathcal{S}$. Nevertheless, we point out that, in a general setting, the analytical solution in Equations (1) and (2) allows $\delta'(t)$ and $\gamma_m'(t)$ to be potentially different from $\delta(t)$ and $\gamma_m(t)$, respectively, and it can be therefore applied when, among other circumstances, maintaining identical environmental conditions at free-living and post-intervention instants is not possible (i.e., different rates $\delta(t)$ and $\delta'(t)$) and/or anthelmintic resistance must be considered within the modelling framework (i.e., different functions $\gamma_m(t)$ and $\gamma_m'(t)$). In our examples, we select $\delta(t) = \delta'(t)$=$e^{-10.0t}$, from which it follows that the probability that, in absence of any parasite burden, the host dies in the interval $[0,\tau]$ with $\tau = 1$ year equals 9.5162%, and the conditional probability that the host death occurs within the first 24 hours, given that it dies in the interval $[0,\tau]$, equals 99.9995%.

In Section 3.1, the age-dependent rates $\lambda(t)$ and $\eta_m(t)$ defining grazing strategies UM, TS and TM are inherently connected to the empirical data in [24] and Figure 2 of [22]. To be concrete, we first use the results in Section 3.2 of [20] to specify the function $\lambda(t)$ for time instants $t \in [0,\tau]$ in scenario US and for time instants $t \in [0,t_0]$ in grazing strategies UM, TS and TM. Concrete specifications for age-dependent patterns at time instants $t \in (t_0,\tau]$ are then derived by suitably modifying these functions under the distributional assumptions in the cases UM, TS and TM. Results yielding scenario US are related to the study conducted by Uriate et al. [22], which was designed to describe monthly fluctuations of nematode burden in sheep (Rasa Aragonesa female lambs) raised under irrigated conditions in Ebro Valley, Spain, by using worm-free tracer lambs and monitoring the faecal excretion of eggs by ewes. Specifically, the age-dependent rate $\lambda(t)$ for ages $t \in [0,\tau]$ in scenario US and grazing strategy TS and ages $t \in [0,t_0]$ in the cases UM and TM is obtained by following our arguments in Section 3.2 of [20]. Therefore, the function $\lambda(t)$ is related to increments in the number of L_3 infective larvae in the small intestine of the lamb (Figures 2–4, shaded area), and it is computed as a function of the infection levels in Table 1 of [20]. To reflect the use of safe pasture in grazing strategies UM and TM, it is assumed that $\lambda'(t) = 0.25\lambda(t)$ for ages $t \in (t_0,\tau]$, where $\lambda(t)$ denotes the previously specified function, which is linked to the original paddock. In grazing strategies TS and TM, the empirical data in [22] are appropriately combined with those data in [24] on the clinical efficacy assessment of ivermectin, fenbendazole and albendazole in lambs parasited with nematode infective larvae; similarly to Section 3.2 in [20], the death rates of parasites in the cases TS and TM are then given by $\eta_m'(t) = m\eta(t)$, for levels $m \in \mathcal{S}$, where $\eta(t)$ reflects the chemotherapeutic efficacy of a concrete anthelmintic over time. More details on the specific form of $\lambda(t)$ and $\eta(t)$ can be found in Appendix A.

3.1. Preliminary Analysis

Because of seasonal conditions, a preliminary analysis of the probabilities $eff^s(t_0;\tau)$ and $cost^s(t_0;\tau)$ in the cases UM, TS and TM is usually required to determine values p_1 and p_2 in such a way that Criteria 1 and 2 lead us to non-empty subsets $J^1_{\geq m'}$ and $J^2_{\geq m'}$ of potential intervention instants

$t_0 \in I_{\geq m'}$, for a predetermined threshold m'. A graphical representation of $eff^s(t_0;\tau)$ and $cost^s(t_0;\tau)$ can help in measuring allowable values for the minimum value of effectiveness and the maximum cost of intervention in terms of concrete values for p_1 and p_2, respectively. Figures 2 and 3 show how $eff^s(t_0;\tau)$ and $cost^s(t_0;\tau)$ behave in terms of t_0 for grazing strategies UM, TS and TM. We remark here that, in scenario US, the effectiveness (respectively, cost of intervention) is given by $\sum_{m=0}^{3} \pi_m^{US}(\tau)$ (respectively, $\sum_{m=8}^{11} \pi_m^{US}(\tau) + \pi_{-1}^{US}(\tau)$), which is a constant as a function of t_0. It is worth noting that the value $\sum_{m=0}^{3} \pi_m^{US}(\tau)$ (respectively, $\sum_{m=8}^{11} \pi_m^{US}(\tau) + \pi_{-1}^{US}(\tau)$) results in a lower bound (respectively, upper bound) to the corresponding values of effectiveness (respectively, cost of intervention) in grazing strategies UM, TS and TM.

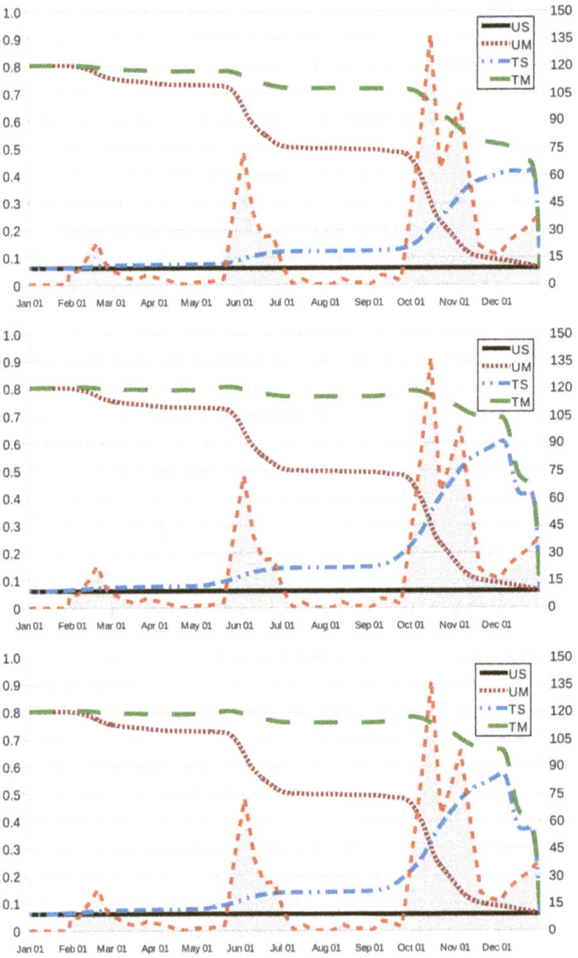

Figure 2. Effectiveness $eff^s(t_0;\tau)$ as a function of the intervention age t_0 for $\tau = 1$ year and increments in the number of L_3 infective larvae in the small intestine (shaded area, right vertical axis). Scenario US, and grazing strategies UM, TS and TM with the anthelmintics ivermectin, fenbendazole and albendazole (from top to bottom).

The effectiveness and cost functions for strategies UM and TM are monotonic in one direction, while the corresponding curves for strategy TS are largely in the opposite direction. This corroborates that an early movement of the host to safe pasture results in a more effective (Figure 2) and less expensive (Figure 3) solution regardless of other actions on the use of anthelmintic drugs, which is related to the safe-pasture conditions having 75% less free-living L_3 than the original paddock. On the contrary, set-stocking conditions made an early intervention seem inadvisable and, due to the effect of the therapeutic period (28 days), intervention should be prescribed by the end of November in the case TS.

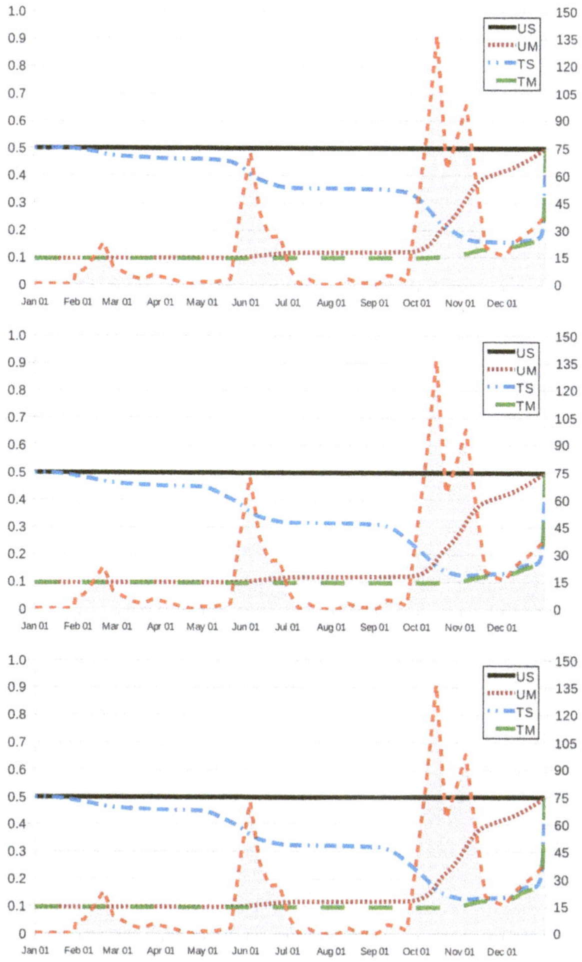

Figure 3. Cost $cost^s(t_0; \tau)$ of intervention as a function of the intervention age t_0 for $\tau = 1$ year and increments in the number of L_3 infective larvae in the small intestine (shaded area, right vertical axis). Scenario US, and grazing strategies UM, TS and TM with the anthelmintics ivermectin, fenbendazole and albendazole (from top to bottom).

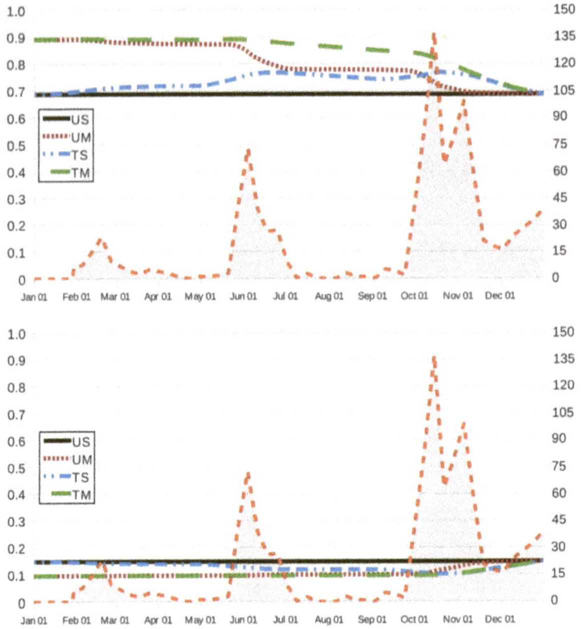

Figure 4. Expected proportions $\tau^{-1}E^s(t_0;\tau)$ (**top**) and $\tau^{-1}C^s(t_0;\tau)$ (**bottom**) versus the intervention age t_0 for $\tau = 1$ year and increments in the number of L_3 infective larvae in the small intestine (shaded area, right vertical axis). Scenario US, and grazing strategies UM, TS and TM with the anthelmintic fenbendazole.

As intuition tells us, grazing strategy TM results in the most effective procedure for every time instant t_0, regardless of the anthelmintic treatment. In Figures 2 and 3, it is also seen that grazing strategy UM is preferred to grazing strategy TS when intervention is prescribed at ages $t_0 < 293$ (21 October), 285 (13 October) and 286 (14 October) as the respective anthelmintics ivermectin, fenbendazole and albendazole are used in the case TS; on the contrary, the latter is preferred to the former at intervention instants $t_0 > 293$, 285 and 286. This behaviour is also noted in Figure 4, where we make the effectiveness and cost of intervention amount to $\tau^{-1}E^s(t_0;\tau)$ and $\tau^{-1}C^s(t_0;\tau)$, respectively; in such a case, grazing strategy UM is preferred to grazing strategy TS for intervention instants $t_0 < 278$ (6 October) as the host is treated with fenbendazole, and the latter is preferred to the former in the case of intervention instants $t_0 > 281$ (9 October). For grazing strategy TS, it is seen in Figure 2 (respectively, Figure 3) that the effectiveness function $eff^{TS}(t_0;\tau)$ (respectively, the cost function $cost^{TS}(t_0;\tau)$) appears to behave as an increasing (respectively, decreasing) function of the intervention instant t_0 as $t_0 < 346$ (13 December) and 338 (5 December) if the anthelmintic ivermectin and the anthelmintics fenbendazole and albendazole are administered to the host (respectively, $t_0 < 309$ (6 November), 308 (5 November) and 339 (6 December) if anthelmintics ivermectin, fenbendazole and albendazole are used); moreover, its variation over time seems to be more apparent, in agreement with three periods of maximum pasture contamination, with 42.0 L_3 kg^{-1} DM (by mid-February), 68.0 L_3 kg^{-1} DM (by 2 June) and 80.0 L_3 kg^{-1} DM (between October and November) as maximum values of infective larvae on herbage. Figure 2 (respectively, Figure 3) allows us to remark that, in comparison with the case TS, these periods of maximum pasture contamination influence in an opposite manner the effectiveness (respectively, cost of intervention) in grazing strategies UM and TM.

3.2. *Intervention Instants t_0*

In Table 1, we list the value of the effectiveness $eff^s(t_0;\tau)$ and the cost $cost^s(t_0;\tau)$ of intervention for certain intervention instants t_0 derived by applying Criteria 1 and 2 in grazing strategies UM, TS and TM, for probabilities $p_1 \in \{0.50, 0.60, 0.70\}$ and $p_2 \in \{0.15, 0.20, 0.25\}$ and a variety of values of the index p; in scenario US, effectiveness and cost are replaced by the probabilities $\sum_{m=0}^{3} \pi_m^{US}(\tau)$ and $\sum_{m=8}^{11} \pi_m^{US}(\tau) + \pi_{-1}^{US}(\tau)$, respectively. A detailed discussion on the instants t_0 in Table 1 and some related consequences can be found in Appendix A.2. It can be noticed that the selection $t_0 = 273$ (1 October), which is related to the index $p = 0.1$ in the case TM with the anthelmintic fenbendazole, results in the minimum cost of intervention (0.09589, instead of 0.49951 in scenario US) and the maximum effectiveness (0.79086, instead of 0.06072 in scenario US), and it can be thus taken as optimal for our purposes. Moreover, the anthelmintic fenbendazole is found the most effective drug since the highest values of $eff^s(t_0;\tau)$ and the smallest values of $cost^s(t_0;\tau)$ are observed in Table 1 for every grazing strategy $s \in \{TS, TM\}$ and fixed intervention instant t_0.

Table 1. Effectiveness and cost of intervention. Scenario US and grazing strategies UM, TS and TM with the anthelmintics ivermectin, fenbendazole and albendazole.

Strategy (s)	Anthelmintic	t_0	Criteria	$eff^s(t_0;\tau)$	$cost^s(t_0;\tau)$	$\tau^{-1}E^s(t_0;\tau)$	$\tau^{-1}C^s(t_0;\tau)$
US	—	—	—	0.06072	0.49951	0.68645	0.14746
UM		170	1 & 2	0.54431	0.11049	0.79996	0.09726
		274	2	0.45540	0.12524	0.76629	0.09983
		281	2	0.38981	0.14216	0.74973	0.10267
		286	2	0.32115	0.16811	0.73306	0.10715
		290	2	0.26634	0.19763	0.72023	0.11233
		298	2	0.20886	0.24130	0.70769	0.11984
TS	ivermectin	358	2	0.41766	0.16608	0.69160	0.14217
	fenbendazole	308	1	0.50340	0.12350	0.75871	0.10433
		336	1	0.60161	0.13144	0.71941	0.12421
		338	2	0.60604	0.13209	0.71613	0.12578
	albendazole	313	1	0.50240	0.12842	0.74908	0.10793
		338	2	0.57385	0.13407	0.71312	0.12626
TM	ivermectin	170	1 & 2	0.73224	0.09721	0.86987	0.09525
		274	1 & 2	0.71025	0.09797	0.82480	0.09580
		281	1 & 2	0.69119	0.09877	0.81634	0.09602
		286	1 & 2	0.66653	0.10011	0.80686	0.09644
		290	1 & 2	0.64110	0.10197	0.79743	0.09713
		298	1 & 2	0.61142	0.10528	0.78209	0.09911
		308	1 & 2	0.56977	0.11374	0.76202	0.10372
	fenbendazole	273	1 & 2	0.79086	0.09589	0.83891	0.09557
		274	1 & 2	0.79080	0.09589	0.83820	0.09558
		281	1 & 2	0.78559	0.09601	0.83107	0.09573
		286	1 & 2	0.77604	0.09636	0.82304	0.09605
		290	1 & 2	0.76467	0.09707	0.81476	0.09662
		298	1 & 2	0.75182	0.09895	0.79922	0.09852
		308	1 & 2	0.72721	0.10573	0.77734	0.10310
	albendazole	272	1 & 2	0.78128	0.09605	0.83749	0.09558
		274	1 & 2	0.78102	0.09606	0.83605	0.09560
		281	1 & 2	0.77361	0.09623	0.82838	0.09576
		286	1 & 2	0.76132	0.09666	0.81971	0.09610
		290	1 & 2	0.74737	0.09747	0.81089	0.09671
		298	1 & 2	0.73134	0.09945	0.79492	0.09867
		308	1 & 2	0.70211	0.10641	0.77271	0.10336

Values for $\tau^{-1}E^s(t_0;\tau)$ and $\tau^{-1}C^s(t_0;\tau)$ in Table 1 correspond to the expected proportions of time that the host infection level $M(t)$ remains in the subsets of levels $\{0,1,2,3\}$ and $\{8,9,10,11\} \cup \{-1\}$, respectively. It is remarkable to note that the maximum effectiveness $\tau^{-1}E^s(t_0;\tau) = 0.86987$ (instead of 0.68645 in scenario US) and the minimum cost of intervention $\tau^{-1}C^s(t_0;\tau) = 0.09525$ (instead of 0.14746 in scenario US) are both related to the selection $t_0 = 170$ (19 June) in grazing strategy TM with the anthelmintic ivermectin. It should be noted that $t_0 = 170$ results in the longest post-intervention interval $[t_0, \tau]$ in our examples; similarly to the case of control strategies based on isolation and anthelmintic treatment of the host (see Section 3.3 in [20]), the maintenance of stable safe-pasture conditions for a long period of time may often be difficult and highly expensive, so that the choice $t_0 = 170$ might be unsustainable for practical use.

An interesting question concerns the comparative analysis between the mass functions $\{\pi_m^s(t_0;\tau) : m \in \{-1\} \cup \mathcal{S}\}$ of the parasite burden at age $\tau = 1$ year in grazing strategies UM, TS and TM and the corresponding mass function $\{\pi_m^{US}(\tau) : m \in \{-1\} \cup \mathcal{S}\}$ in the case of no intervention. In Figure 5, we first focus on this question as intervention is prescribed at age $t_0 = 170$ in grazing strategies UM, TS and TM, with the anthelmintic drug ivermectin in the cases TS and TM. The movement of the host to safe pasture (strategies UM and TM) at day $t_0 = 170$ yields a significant decrease in the probability that the host does not survive at age $\tau = 1$ year (0.09528 and 0.09516 in the cases UM and TM, respectively, instead of 0.15708 in scenario US), as well as an important decrease in the expected degree of infestation in the case of survival; more particularly, the degree of infestation is expected to be light as either anthelmintic drugs are used ($E^{TM,t_0}[M(\tau)|M(\tau) \neq -1] = 2.25085$) or the host is transferred to a paddock with safe pasture ($E^{UM,t_0}[M(\tau)|M(\tau) \neq -1] = 3.20312$), instead of moderate and nearly high in the case US ($E^{US}[M(\tau)|M(\tau) \neq -1] = 6.88878$). Set-stocking conditions are not as effective as the movement of the host to safe pasture since the expected degree of infestation amounts to a moderate degree in the case of survival ($E^{TS,t_0}[M(\tau)|M(\tau) \neq -1] = 6.13509$); moreover, for grazing strategy TS, the decrease in the probability of no-survival is apparent, but it is not as notable as for strategies UM and TM. In Figure 6, we plot the mass function of the parasite burden $M(\tau)$ at age $\tau = 1$ year in scenario US versus its counterpart in grazing strategy TM, when animals are treated with ivermectin, fenbendazole and albendazole at ages $t_0 = 170, 273$ and 272, respectively. By Tables A2 and A3 in Appendix A, ages $t_0 = 170, 273$ and 272 are all feasible intervention instants, which leads us to mass functions that are essentially comparable in magnitude. On the contrary, the shape and magnitudes of the mass function in grazing strategy TM are dramatically different from the shape and magnitudes in scenario US, where no intervention is prescribed, irrespective of the anthelmintic product.

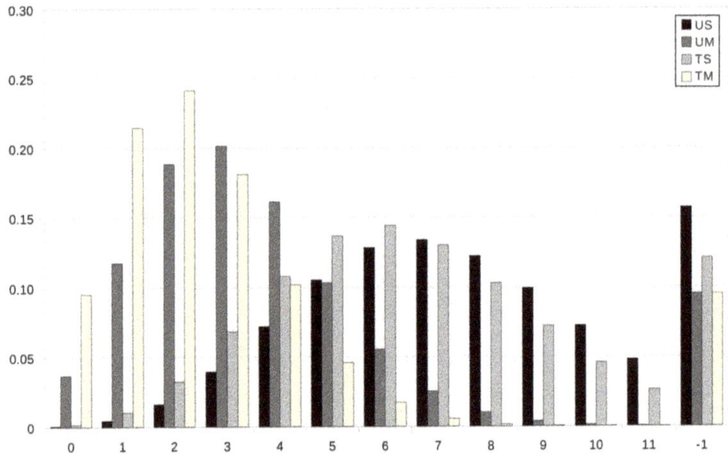

Figure 5. The mass function of the parasite burden $M(\tau)$ at age $\tau = 1$ year. Scenario US and grazing strategies UM, TS and TM (from left to right) with the anthelmintic ivermectin as the intervention prescribed at age $t_0 = 170$.

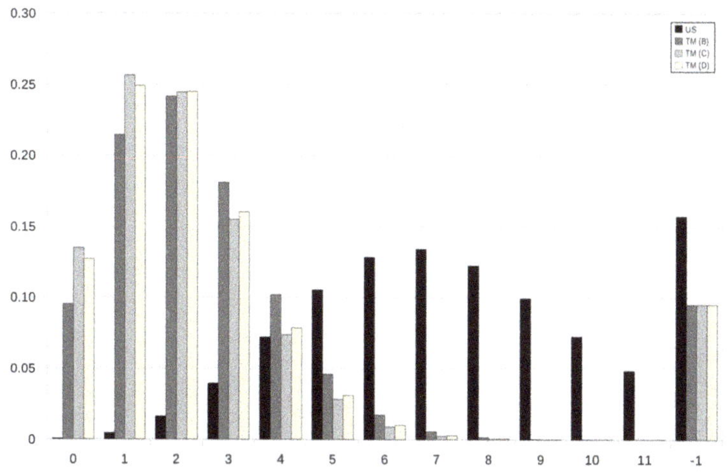

Figure 6. The mass function of the parasite burden $M(\tau)$ at age $\tau = 1$ year. Scenario US and grazing strategy TM with the anthelmintics ivermectin, fenbendazole and albendazole (from left to right) as the intervention prescribed at ages $t_0 = 170$, 273 and 272, respectively.

4. Conclusions

It is of fundamental importance in the development of GI nematode infection in sheep to understand the role of grazing management in reducing anthelmintic use and improving helminth control. With empirical data of [22,24], we present a valuable modelling framework for better understanding the host-parasite interaction under fluctuations in time, which arguably represents the most realistic setting for assessing the impact of seasonal changes in the parasite burden of a growing lamb. Grazing strategies UM, TS and TM in Section 2.1 are defined in terms of an eventual movement to safe pasture and/or chemotherapeutic treatment of the host at a certain age $t_0 \in (0, \tau)$. For a suitable choice of t_0, we suggest to use two control criteria that adequately balance the effectiveness and cost of intervention at age t_0 by using simple stochastic principles. Specifically, each intervention instant t_0 in Table 1 yields an individual-based grazing strategy for a lamb that is born, parasite-free, at time $t = 0$ (1 January, in our examples). The individual-based grazing strategies UM, TS and TM can be also thought of as group-based grazing strategies in the case of a flock consisting of young lambs, essentially homogeneous in age. In such a case, intervention at age t_0 is prescribed (in accordance with a predetermined grazing strategy) by applying our methodology to a typical lamb that is assumed to be born, parasite-free, at a certain average day t'. Then, results may be routinely derived by handling the set of empirical data in Figure 2 in [22] starting from day t', instead of Day 0, since intervention at time instant $t' + t_0$ amounts to age t_0 of the typical lamb in the paddock. From an applied perspective, the descriptive model in Section 2 becomes a prescriptive model as the set of empirical data in Figure 2 in [22] is appropriately replaced by a set of data derived by taking the average of annual empirical data from historical records.

For practical use, the profits of applying Criteria 1 and 2 in grazing strategies UM, TS and TM should be appropriately compared with experimental results. To that end, we first comment on general guidelines (see Part II of [25]) for control of GI nematode infection. From an experimental perspective, the dose-and-move strategy (termed TM) is usually recommended in mid-July, this recommendation being applicable in temperate zones where the maximum numbers of L_3 infective larvae do not occur before midsummer, which is our case (Figures 2–4, shaded area). As stated in [25], midsummer movement to safe pasture without deworming (strategy UM) is thought of as a low cost control measure; it can even be effective at moderate levels of pasture infectivity, and it has the advantage of creating no anthelmintic resistance by drug selection. We may translate these specifications into an intervention at day $t_0 = 195$ (15 July) in strategies TM and UM. Guidelines for the application of an anthelmintic drug without movement (strategy TS) are not so clearly available, and various alternatives (based mainly on the several-dose approach) are applied in a variety of circumstances that strongly depend on the geographical region, climate and farming system. For comparative purposes in strategy TS, we compare our results (derived by applying Criteria 1 and 2) with an intervention at day $t_0 = 287$ (15 October), when the maximum number of infective larvae L_3 on herbage is observed (Figures 2–4, shaded area).

We present in Table 2 a sample of our results when the anthelmintic fenbendazole is used in grazing strategies TS and TM. Table 2 lists values of the reduction in the mean infection level (RMIL) at age $\tau = 1$ year, under the taboo that the host survives at age $\tau = 1$ year, and the reduction in the total lost probability (RTLP) when, instead of scenario US, intervention is prescribed at day $t_0 < \tau$ by a certain grazing strategy s with $s \in \{UM, TS, TM\}$. The indexes RMIL and RTLP for grazing strategy s, with $s \in \{UM, TS, TM\}$, are defined by:

$$\begin{aligned} RMIL^{s,t_0} &= 100 \times \left(1 - \frac{E^{s,t_0}[M(\tau)|M(\tau) \neq -1]}{E^{US}[M(\tau)|M(\tau) \neq -1]}\right)\%, \\ RTLP^{s,t_0} &= 100 \times \left(1 - \frac{P^{s,t_0}(M(\tau) = -1)}{P^{US}(M(\tau) = -1)}\right)\%, \end{aligned}$$

where $E^{s,t_0}[M(\tau)|M(\tau) \neq -1]$ denotes the conditional expected infection level of the host at age $\tau = 1$ year, given that it survives at age $\tau = 1$ year, and $P^{s,t_0}(M(\tau) = -1)$ is the probability that the host does not survive at age $\tau = 1$ year, when intervention is prescribed at day t_0 according to grazing strategy s. The values $E^{US}[M(\tau)|M(\tau) \neq -1]$ and $P^{US}(M(\tau) = -1)$ are related to scenario US, and they reflect no intervention.

Table 2. Indexes reduction in the mean infection level (RMIL) and reduction in the total lost probability (RTLP) for strategies UM, TS and TM with the anthelmintic fenbendazole.

Strategy (s)	Criteria	t_0	$RMIL^{s,t_0}$	$RTLP^{s,t_0}$
UM	1 & 2	170	53.50%	39.34%
	2	274	46.98%	39.17%
	2	281	41.85%	38.88%
	2	286	36.00%	38.21%
	2	290	30.78%	37.12%
	2	298	24.54%	34.86%
	Midsummer	195	50.28%	39.28%
TS	1	308	51.01%	33.27%
	1	336	59.19%	19.77%
	2	338	59.56%	19.10%
	Maximum pasture contamination	287	37.78%	38.17%
TM	1 & 2	273	72.36%	39.41%
	1 & 2	274	72.36%	39.41%
	1 & 2	281	71.88%	39.39%
	1 & 2	286	71.03%	39.26%
	1 & 2	290	70.07%	38.92%
	1 & 2	298	69.09%	37.82%
	1 & 2	308	67.47%	33.63%
	Midsummer	195	70.47%	39.42%

In grazing strategies UM and TM, the experimental selection $t_0 = 195$ (midsummer) is found to be near an optimal solution, and Table 2 permits us to analyse the effects of the stochastic control criteria in a more detailed manner. Based on the decreasing monotonic behaviour of RMIL and RTLP with respect to the intervention instant t_0, it is noticed that the later we apply grazing strategies UM and TM, the worse the results we obtain. This is closely related to the important role played in the cases UM and TM by the use of safe pasture, which is reflected in the 75% contamination reduction with respect to the original paddock. Therefore, the movement of the host to safe pasture appears to be dominant in the use of anthelmintics, so that the sooner the host is moved, the safer it is for the host. The maintenance of stable safe-pasture conditions for a long period of time may be difficult and/or highly expensive, whence additional considerations should be taken into account when selecting the time instant t_0 for moving the host. In grazing strategy UM, the intervention instant $t_0 = 170$ should be considered as optimal for our purposes, and it yields a reduction of 53.50% in the mean infection level at the end of the year, as well as a reduction of 39.34% in the probability of no-survival. However, an intervention at day $t_0 = 274$ (i.e., moving the host to safe pasture more than one hundred days later) would result in significantly lower operational costs, but predicted reductions are still around high levels ($RMIL^{UM,274} = 46.98\%$, $RTLP^{UM,274} = 39.17\%$). It is clear that a balance between operational costs and the magnitudes of the indexes RMIL and RTLP should be made. It is seen that the experimental selection $t_0 = 195$ seems to implicitly incorporate this balance, delaying the movement of the host almost a month with respect to $t_0 = 170$, at the expense of losing 3.22% and 0.06% of efficiency in the indexes RMIL and RTLP, respectively. Although the selection of t_0 may depend on external factors, the movement of the host to safe pasture before day $t_0 = 287$ (maximum pasture contamination) is highly recommendable, and intervention instants $t_0 = 290$ and 298 should be discarded in the light of these results.

Similar comments can be made for grazing strategy TM. In this case, the experimental selection $t_0 = 195$ allows us to achieve a good index RMIL in comparison with those time instants t_0 obtained by applying Criteria 1 and 2, while obtaining the highest index RTLP. The intervention at day $t_0 = 195$ is more than two months advanced with respect to the day $t_0 = 273$, which is derived by applying Criteria 1 and 2. The experimental selection $t_0 = 195$ results in higher operational costs due to an early movement, and it amounts to a minor improvement of 0.01% in the index RTLP; it is also seen that the option $t_0 = 273$ yields the value $RMIL^{TM,273} = 72.36\%$, which is higher than the corresponding value for the experimental choice. Thus, when comparing grazing strategy TM with strategy UM, the use of an anthelmintic drug seems to permit delaying the movement of the host to safe pasture, while maintaining good indexes RMIL and RTLP; note that it is still possible to have values of RMIL and RTPL above 70% and 39%, respectively, if the intervention is delayed at day $t_0 = 286$.

Under set-stocking conditions, the use of an anthelmintic drug at day $t_0 = 287$ (15 October) may be seen as optimal in terms of the index RTLP, but at the expense of an unacceptable value 37.78% of RMIL. Note that an application of Criteria 1 and 2 leads us to intervention instants $t_0 = 308, 336$ and 338, with values $RMIL^{TS,t_0}$ varying between 50% and 60%. In particular, the time instant $t_0 = 308$ permits us to achieve a significant improvement of the index RMIL (51.01% instead of 37.78%) and maintain the index RTLP above 33%, which is comparable with the value 38.17% resulting from an intervention when maximum values of L_3 on herbage are observed.

One of the simplifying assumptions in Section 3 (see also Appendix A) is related to the effect that the infestation degree of the lambs might have on the pasture infection level itself. We deal with a non-infectious assumption, and specifically, the empirical data in Figure 2 in [22] allow us to partially incorporate this effect into the age-dependent patterns in terms of the infection level of a standard paddock during the year. The analytical solution in Section 2.1 can be however used to examine the infectious nature of the parasite in a more explicit manner. Although it is an additional topic for further study, we stress that the infectious nature of the parasite appears to be a relevant feature in grazing strategy UM, where the force-of-infection in a field seeded with untreated lambs would likely increase back up to a similar level to the original paddock. In an attempt to address this question, various variants of the age-dependent rate $\lambda'(t)$ can be conjectured, such as the function $\lambda'(t) = (0.25 + 0.75h^{-1}(t - t_0))\lambda(t)$ at post-intervention instants, with $h > 0$. Then, under proper data availability, the selection $\lambda'(t)$ reflects the use of a paddock with safe pasture at initial post-intervention instants ($\lambda'(t_0) = 0.25\lambda(t_0)$) and how the pasture infection level reaches the pre-intervention level, represented by $\lambda'(t_0 + h) = \lambda(t_0 + h)$, after a period consisting of h days.

Author Contributions: A.G.-C. and M.L.-G. conceived of the model, carried out the stochastic analysis and designed the numerical results. M.L.-G. programmed the computer codes and obtained results for all figures and tables. A.G.C. wrote the first version of the draft. A.G.-C. and M.L.-G. finalised the paper and revised the literature.

Funding: This research was funded by the Ministry of Economy, Industry and Competitiveness (Government of Spain), project MTM2014-58091-P and grant BES-2009-018747.

Acknowledgments: The authors thank two anonymous referees whose comments and suggestions led to improvements in the manuscript.

Conflicts of Interest: The authors declare no conflict of interest.

Appendix A. GI Nematode Infection in Growing Lambs

In its complete life cycle, the parasitic phase of *Nematodirus* spp. commences when worms in the larval stage L_3 encounter the host, which is a largely passive process with the grazing animal inadvertently ingesting larvae with herbage as it feeds. As a result, infection occurs by ingestion of the free-living L_3, with an establishment proportion (i.e., the proportion of ingested free-living L_3 that become established in the small intestine of the host) ranging between 45% and 60%; see, e.g., [26–28]. Various external factors (moisture levels, temperature and the availability of oxygen) are key drivers that affect how quickly eggs hatch and larvae develop and how long larvae and eggs survive on pasture. Therefore, the occurrence of nematode infections in sheep is inherently linked to seasonal conditions, and it is therefore connected to a diversity of physiographic and climatic conditions; see [22,29,30], among others. The adverse effects of GI nematode parasites on productivity are diverse, and reductions of live weight gain in growing stock have been recorded as being as high as 60–100%. Anthelmintics, such as ivermectin, fenbendazole and albendazole, are drugs that are effective in removing existing burdens or that prevent establishment of ingested L_3.

Faecal examination for the presence of worm eggs or larvae is the most common routine aid to diagnosis employed. In the faecal egg count (FEC) reduction test, animals are allocated to groups of ten based on pre-treatment FEC, with one group of ten for each anthelmintic treatment tested and a further untreated control group. For instance, this requires the use of forty animals in [24], where the efficacy of three anthelmintics (ivermectin, fenbendazole and albendazole) against GI nematodes is investigated. A full FEC reduction test is understandably expensive and takes a significant length of time before farmers are presented with the results; in addition, accurate larval differentiation also demands a high degree of skill. As an alternative test, a points system (see [21]) may serve as a crude guide to interpreting worm counts, which is based on the fact that one point is equivalent to the presence of 4000 worms, a total of two points in a young sheep is likely to be causing measurable losses of productivity and clinical signs and deaths are unlikely unless the total exceeds three points.

Based on the above comments, Table 1 in [20] presents an equivalence in the identification of the degree of infestation, level of infection, eggs per gram (EPG) value, number of L_3 infective larvae in the small intestine and the points system, which can be used to study the parasite load of a lamb in a unified manner. We refer the reader to [1,2,21] for further details on nematode taxonomy and morphology and the treatment and control of parasite gastroenteritis in sheep.

Appendix A.1. Empirical Data and Age-Dependent Rates

In this section, we first use the results in Section 3.2 of [20] to specify the functions $\lambda(t)$ and $\eta_m(t)$ for time instants $t \in [0, \tau]$ in scenario US and for time instants $t \in [0, t_0]$ in grazing strategies UM, TS and TM. Concrete specifications for age-dependent patterns at time instants $t \in (t_0, \tau]$ are then derived by suitably modifying these functions under the distributional assumptions in the cases UM, TS and TM. Results yielding scenario US are related to the study conducted by Uriate et al. [22], which is designed to describe monthly fluctuations of nematode burden in sheep (Rasa Aragonesa female lambs) raised under irrigated conditions in Ebro Valley, Spain, by using worm-free tracer lambs and monitoring the faecal excretion of eggs by ewes. Specifically, we use the set of empirical data in Figure 2 of [22] recording the number of L_3 infective larvae on herbage samples at weekly intervals from a fixed paddock of the farm. In grazing strategies TS and TM, the empirical data in [22] are appropriately combined with those data in [24] on the clinical efficacy assessment of ivermectin, fenbendazole and albendazole in lambs parasited with nematode infective larvae.

In Figure 2 of [22], results are expressed as infective larvae per kilogram of dry matter (L_3 kg^{-1} DM) after drying the herbage overnight at 60° C, and the numbers of L_3 infective larvae on herbage samples correspond to *Chabertia ovina* and *Haemonchus* spp. (9.6%), *Nematodirus* spp. (4.0%), *Ostertagia* spp. (71.4%) and *Trichostrongylus* spp. (15.0%). In our work, the increments in the number of L_3 infective larvae in the small intestine (Figure 1, shaded area) are estimated by fixing the value 55% as the establishment proportion and incorporating concrete specifications for the lamb growth pre- and post-weaning. To be concrete, it is assumed that, for a host that is born on 1 January (Day 0), the lamb birth weight equals 5 kg, the pre-weaning period consists of four weeks and the lamb growth rate from birth to weaning is given by 0.3 kg per day. The lamb growth rate on pasture post-weaning is assumed to be equal to 0.15 kg per day, and the daily DM intake is given by the 6% of body weight (BW); see [31] for details on lamb growth rates on pasture.

These specifications determine the age-dependent rate $\lambda_m(t) = \lambda(t)$ for ages $t \in [0, \tau]$ in scenario US and grazing strategy TS, and for ages $t \in [0, t_0]$ in grazing strategies UM and TM, with $\tau = 1$ year. More concretely, the function $\lambda(t)$ is defined to be the piecewise linear function formed by connecting the points $(n, \lambda(n))$ in order by segments, where the value $\lambda(n)$ at the n-th day is determined in [20] as a function of the number of L_3 infective larvae of *Nematodirus* spp. on pasture, from Figure 2 of [22], the DM intake at the n-th day, the establishment proportion and the interval length $l = 10^3$ used in Table 1 of [20] to define infection levels $m \in \mathcal{S}$ in terms of numbers of infective larvae in the small intestine. To reflect the use of safe pasture in grazing strategies UM and TM, it is assumed that $\lambda'(t) = 0.25\lambda(t)$ for ages $t \in (t_0, \tau]$ where $\lambda(t)$ denotes the previously specified function, which is related to the original paddock.

Similarly to Section 3.2 in [20], the death rates of parasites in grazing strategies TS and TM are given by $\eta'_m(t) = m\eta(t)$ for levels $m \in \mathcal{S}$, where $\eta(t)$ reflects the chemotherapeutic efficacy of a concrete anthelmintic over time. We use the empirical data of [24], where the efficacy of three anthelmintic products against GI nematodes is investigated. In the FEC reduction test of [24], animals were allocated to four groups termed A, B, C and D. Animals of Group A served as the control, whereas animals of Groups B, C and D were orally administered ivermectin (0.2 mg·kg^{-1}·BW), fenbendazole (5.0 mg·kg^{-1}·BW) and albendazole (7.5 mg·kg^{-1}·BW), respectively. Animals were sampled for FEC at Day 0 immediately before administering the drug and thereafter on Days 3, 7, 14, 21 and 28. Then, the function $\eta(t)$ associated with each anthelmintic is defined as the polyline connecting the points $(t_n, \eta(t_n))$, where the instants t_n are given by t_0, $t_1 = t_0 + 3$, $t_2 = t_0 + 7$, $t_3 = t_0 + 14$, $t_4 = t_0 + 21$ and $t_5 = t_0 + 28$. The length of the therapeutic period is assumed to be equal to 28 days, so that $\eta(t) = 0$ for instants $t > t_5$. Values $\eta(t_n)$ are determined in Table 1 of [20] from the EPG value and the infection level at time t_n, as well as the length $l' = 50$ used to define levels of infection in terms of EPG values.

Appendix A.2. Intervention Instants t_0

Values of t_0 are listed in Table A1 for grazing strategy UM and denoted by t_0^1 and t_0^2 as they are derived by applying Criteria 1 and 2, respectively. In Tables A2 and A3, values of t_0 are listed for grazing strategies TS and TM and the anthelmintics ivermectin (Group B), fenbendazole (Group C) and albendazole (Group D), which are denoted by t_0^B, t_0^C and t_0^D, respectively. The selection $m' = 4$ in Tables A1–A3 amounts to a degree of infestation that is moderate (Figure A1), and consequently, a measurable presence of worms is observed.

Table A1. Intervention instants t_0 versus the index p and the lower bound p_1 for effectiveness (Criterion 1) and the upper bound p_2 for the cost of intervention (Criterion 2) for $m' = 4$. Grazing strategy UM.

p	$I_{\geq 4}$	p_1	$J^1_{\geq 4}$	t^1_0	p_2	$J^2_{\geq 4}$	t^2_0
0.1	[170, 365)	0.70	—	—	0.25	[170, 299]	170
		0.60	—	—	0.20	[170, 290]	170
		0.50	[170, 194]	170	0.15	[170, 282]	170
0.2	[274, 365)	0.70	—	—	0.25	[274, 299]	274
		0.60	—	—	0.20	[274, 290]	274
		0.50	—	—	0.15	[274, 282]	274
0.3	[281, 365)	0.70	—	—	0.25	[281, 299]	281
		0.60	—	—	0.20	[281, 290]	281
		0.50	—	—	0.15	[281, 282]	281
0.4	[286, 365)	0.70	—	—	0.25	[286, 299]	286
		0.60	—	—	0.20	[286, 290]	286
		0.50	—	—	0.15	—	—
0.5	[290, 365)	0.70	—	—	0.25	[290, 299]	290
		0.60	—	—	0.20	[290, 290]	290
		0.50	—	—	0.15	—	—
0.6	[298, 365)	0.70	—	—	0.25	[298, 299]	298
		0.60	—	—	0.20	—	—
		0.50	—	—	0.15	—	—
0.7	[308, 365)	0.70	—	—	0.25	—	—
		0.60	—	—	0.20	—	—
		0.50	—	—	0.15	—	—

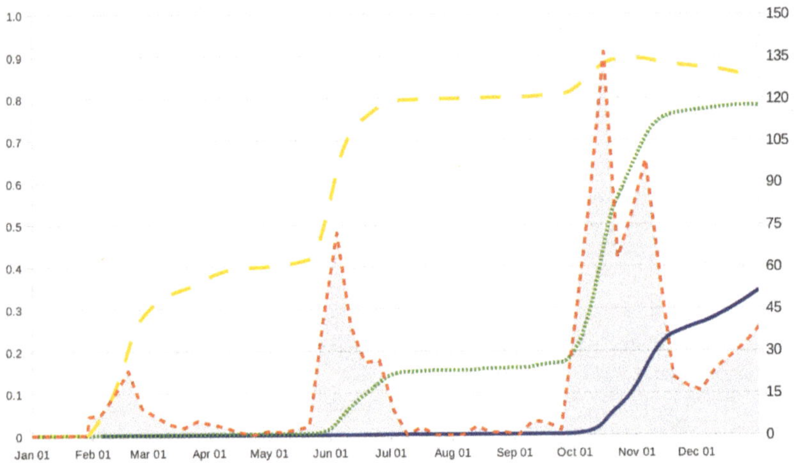

Figure A1. The age-dependent probability $P_{\geq m'}(t)$ as a function of the age $t \in (0, \tau)$ with $\tau = 1$ year, for $m' = 1$ (**broken** line), 4 (**dotted** line) and 8 (**solid** line), and increments in the number of L_3 infective larvae on the small intestine (**shaded** area, **right** vertical axis).

Table A2. Intervention instants t_0 versus the index p and the lower bound p_1 for effectiveness (Criterion 1) for $m' = 4$. Grazing strategies TS and TM with the anthelmintics *ivermectin* (B), *fenbendazole* (C) and *albendazole* (D).

p	$I_{\geq 4}$	p_1		$J^{1,B}_{\geq 4}$	t^B_0	$J^{1,C}_{\geq 4}$	t^C_0	$J^{1,D}_{\geq 4}$	t^D_0
0.1	[170, 365)	0.70	TS	—	—	—	—	—	—
			TM	[170, 278]	170	[170, 319]	273	[170, 308]	272
		0.60	TS	—	—	[336, 339]	336	—	—
			TM	[170 ,301]	170	[170, 344]	273	[170, 342]	272
		0.50	TS	—	—	[308, 344]	308	[313, 343]	313
			TM	[170, 343]	170	[170, 348]	273	[170, 346]	272
0.2	[274, 365)	0.70	TS	—	—	—	—	—	—
			TM	[274, 278]	274	[274, 319]	274	[274, 308]	274
		0.60	TS	—	—	[336, 339]	336	—	—
			TM	[274, 301]	274	[274, 344]	274	[274, 342]	274
		0.50	TS	—	—	[308, 344]	308	[313, 343]	313
			TM	[274, 343]	274	[274, 348]	274	[274, 346]	274
0.3	[281, 365)	0.70	TS	—	—	—	—	—	—
			TM	—	—	[281, 319]	281	[281, 308]	281
		0.60	TS	—	—	[336, 339]	336	—	—
			TM	[281, 301]	281	[281, 344]	281	[281, 342]	281
		0.50	TS	—	—	[308, 344]	308	[313, 343]	313
			TM	[281, 343]	281	[281, 348]	281	[281, 346]	281
0.4	[286, 365)	0.70	TS	—	—	—	—	—	—
			TM	—	—	[286, 319]	286	[286, 308]	286
		0.60	TS	—	—	[336, 339]	336	—	—
			TM	[286, 301]	286	[286, 344]	286	[286, 342]	286
		0.50	TS	—	—	[308, 344]	308	[313, 343]	313
			TM	[286, 343]	286	[286, 348]	286	[286, 346]	286
0.5	[290, 365)	0.70	TS	—	—	—	—	—	—
			TM	—	—	[290, 319]	290	[290, 308]	290
		0.60	TS	—	—	[336, 339]	336	—	—
			TM	[290, 301]	290	[290, 344]	290	[290, 342]	290
		0.50	TS	—	—	[308, 344]	308	[313, 343]	313
			TM	[290, 343]	290	[290, 348]	290	[290, 346]	290
0.6	[298, 365)	0.70	TS	—	—	—	—	—	—
			TM	—	—	[298, 319]	298	[298, 308]	298
		0.60	TS	—	—	[336, 339]	336	—	—
			TM	[298, 301]	298	[298, 344]	298	[298, 342]	298
		0.50	TS	—	—	[308, 344]	308	[313, 343]	313
			TM	[298, 343]	298	[298, 348]	298	[298, 346]	298
0.7	[308, 365)	0.70	TS	—	—	—	—	—	—
			TM	—	—	[308, 319]	308	[308, 308]	308
		0.60	TS	—	—	[336, 339]	336	—	—
			TM	—	—	[308, 344]	308	[308, 342]	308
		0.50	TS	—	—	[308, 344]	308	[313, 343]	313
			TM	[308, 343]	308	[308, 348]	308	[308, 346]	308

Table A3. Intervention instants t_0 versus the index p and the upper bound p_2 for the cost of intervention (Criterion 2) for $m' = 4$. Grazing strategies TS and TM with the anthelmintics *ivermectin* (B), *fenbendazole* (C) and *albendazole* (D).

p	$I_{\geq 4}$	p_2		$J^{2,B}_{\geq 4}$	t^B_0	$J^{2,C}_{\geq 4}$	t^C_0	$J^{2,D}_{\geq 4}$	t^D_0
0.1	[170, 365)	0.25	TS	[286, 363]	358	[268, 363]	338	[270, 363]	338
			TM	[170, 363]	170	[170, 363]	273	[170, 363]	272
		0.20	TS	[299, 362]	358	[279, 362]	338	[281, 361]	338
			TM	[170, 362]	170	[170, 362]	273	[170, 362]	272
		0.15	TS	—	—	[290, 346]	338	[292, 344]	338
			TM	[170, 350]	170	[170, 350]	273	[170, 348]	272
0.2	[274, 365)	0.25	TS	[286, 363]	358	[274, 363]	338	[274, 363]	338
			TM	[274, 363]	274	[274, 363]	274	[274, 363]	274
		0.20	TS	[299, 362]	358	[279, 362]	338	[281, 361]	338
			TM	[274, 362]	274	[274, 362]	274	[274, 362]	274
		0.15	TS	—	—	[290, 346]	338	[292, 344]	338
			TM	[274, 350]	274	[274, 350]	274	[274, 348]	274
0.3	[281, 365)	0.25	TS	[286, 363]	358	[281, 363]	338	[281, 363]	338
			TM	[281, 363]	281	[281, 363]	281	[281, 363]	281
		0.20	TS	[299, 362]	358	[281, 362]	338	[281, 361]	338
			TM	[281, 362]	281	[281, 362]	281	[281, 362]	281
		0.15	TS	—	—	[290, 346]	338	[292, 344]	338
			TM	[281, 350]	281	[281, 350]	281	[281, 348]	281
0.4	[286, 365)	0.25	TS	[286, 363]	358	[286, 363]	338	[286, 363]	338
			TM	[286, 363]	286	[286, 363]	286	[286, 363]	286
		0.20	TS	[299, 362]	358	[286, 362]	338	[286, 361]	338
			TM	[286, 362]	286	[286, 362]	286	[286, 362]	286
		0.15	TS	—	—	[290, 346]	338	[292, 344]	338
			TM	[286, 350]	286	[286, 350]	286	[286, 348]	286
0.5	[290, 365)	0.25	TS	[290, 363]	358	[290, 363]	338	[290, 363]	338
			TM	[290, 363]	290	[290, 363]	290	[290, 363]	290
		0.20	TS	[299, 362]	358	[290, 362]	338	[290, 361]	338
			TM	[290, 362]	290	[290, 362]	290	[290, 362]	290
		0.15	TS	—	—	[290, 346]	338	[292, 344]	338
			TM	[290, 350]	290	[290, 350]	290	[290, 348]	290
0.6	[298, 365)	0.25	TS	[298, 363]	358	[298, 363]	338	[298, 363]	338
			TM	[298, 363]	298	[298, 363]	298	[298, 363]	298
		0.20	TS	[299, 362]	358	[298, 362]	338	[298, 361]	338
			TM	[298, 362]	298	[298, 362]	298	[298, 362]	298
		0.15	TS	—	—	[298, 346]	338	[298, 344]	338
			TM	[298, 350]	298	[298, 350]	298	[298, 348]	298
0.7	[308, 365)	0.25	TS	[308, 363]	358	[308, 363]	338	[308, 363]	338
			TM	[308, 363]	308	[308, 363]	308	[308, 363]	308
		0.20	TS	[308, 362]	358	[308, 362]	338	[308, 361]	338
			TM	[308, 362]	308	[308, 362]	308	[308, 362]	308
		0.15	TS	—	—	[308, 346]	338	[308, 344]	338
			TM	[308, 350]	308	[308, 350]	308	[308, 348]	308

An examination of the resulting instants t_0 in Tables A1–A3 reveals the following important consequences:

(i) In applying Criterion 1 (respectively, Criterion 2) to grazing strategy TM, values of the lower bound $p_1 \in \{0.5, 0.6, 0.7\}$ for effectiveness (respectively, the upper bound $p_2 \in \{0.15, 0.2, 0.25\}$ for the cost of intervention) result in identical intervention instants t_0, irrespective of the anthelmintic drug, with the exception of the case $p = 0.1$. More concretely, we observe that, in the case $p = 0.1$, identical intervention instants t_0 are derived for each fixed anthelmintic drug, but a replacement of the predetermined drug by another anthelmintic results in different intervention instants.

(ii) For every anthelmintic drug and fixed index p, Criteria 1 and 2 applied to grazing strategy TM yield identical intervention instants t_0, with the exception of those pairs (p, p_1) for the anthelmintic ivermectin leading us to empty subsets $J^{1,B}_{\geq 4}$. In order to maintain high values of the minimum level of effectiveness (Criterion 1), we have therefore to handle smaller values of the index p (0.1 and 0.2 in Table A2) for grazing strategy TM, which means that low-risk intervention instants should become potential intervention instants.

(iii) For every anthelmintic, the intervention instant t_0 derived in grazing strategy TM behaves as an increasing function of the index p, regardless of the control criterion.

(iv) For every anthelmintic and fixed value p_1, the intervention instant t_0 in grazing strategy TS appears to be constant as a function of the index p. This is in agreement with the fact that the maximum levels of effectiveness (Figure 2) and the minimum costs of intervention (Figure 3) are observed at the end of the year (November–December), in such a way that this period of time always consists of potential intervention instants (Figure A1) for the index p ranging between 0.1 and 0.7.

(v) In contrast to grazing strategies TS and TM, the values $p_1 \in \{0.5, 0.6, 0.7\}$ for grazing strategy UM lead us to empty subsets $J^1_{\geq 4}$ of potential intervention instants, with the exception of the pair $(p, p_1) = (0.1, 0.5)$. This observation is closely related to the monotonic behaviour of the effectiveness (Figure 2) and cost (Figure 3) functions, which links the first months of the year to the highest effectiveness and the minimum cost of intervention.

(vi) The upper limit of the set $I_{\geq 4}$ in Tables A1–A3 is always at Day 365, which can be readily explained from the monotone behaviour (Figure A1) of the age-dependent probability $P_{\geq m'}(t)$ in the case $m' = 4$. It is clear that other thresholds m' will not necessarily yield Day 365; for example, $I_{\geq m'} = (280, 360)$ in the case $m' = 1$ with $p = 0.85$.

(vii) For strategies UM (Table A1) and TM (Tables A2 and A3), the lower limits of the resulting sets $J^1_{\geq 4}$ and $J^2_{\geq 4}$ always coincide with the lower limit of the set $I_{\geq 4}$ of potential intervention instants t_0, but this is not the case for strategy TS. This means that an early movement of the host to safe pasture should lead to feasible intervention instants.

The values of the effectiveness $eff^s(t_0; \tau)$ and the cost $cost^s(t_0; \tau)$ of intervention for instants t_0 in Tables A1–A3 are listed in Table 1 and analysed in more detail in Section 3.2.

References

1. Sutherland, I.; Scott, I. *Gastrointestinal Nematodes of Sheep and Cattle. Biology and Control*; Wiley-Blackwell: Chichester, UK, 2010.
2. Taylor, M.A.; Coop, R.L.; Wall, R.L. *Veterinary Parasitology*, 3rd ed.; Blackwell: Oxford, UK, 2007.
3. Bjørn, H.; Monrad, J.; Nansen, P. Anthelmintic resistance in nematode parasites of sheep in Denmark with special emphasis on levamisole resistance in Ostertagia circumcincta. *Acta Vet. Scand.* **1991**, *32*, 145–154. [PubMed]
4. Entrocasso, C.; Alvarez, L.; Manazza, J.; Lifschitz, A.; Borda, B.; Virkel, G.; Mottier, L.; Lanusse, C. Clinical efficacy assessment of the albendazole-ivermectin combination in lambs parasited with resistant nematodes. *Vet. Parasitol.* **2008**, *155*, 249–256. [CrossRef] [PubMed]
5. Stear, M.J.; Doligalska, M.; Donskow-Schmelter, K. Alternatives to anthelmintics for the control of nematodes in livestock. *Parasitology* **2007**, *134*, 139–151. [CrossRef] [PubMed]
6. Hein, W.R.; Shoemaker, C.B.; Heath, A.C.G. Future technologies for control of nematodes of sheep. *N. Z. Vet. J.* **2001**, *49*, 247–251. [CrossRef] [PubMed]
7. Knox, D.P. Technological advances and genomics in metazoan parasites. *Int. J. Parasitol.* **2004**, *34*, 139–152. [CrossRef] [PubMed]
8. Sayers, G.; Sweeney, T. Gastrointestinal nematode infection in sheep—A review of the alternatives to anthelmintics in parasite control. *Anim. Health Res. Rev.* **2005**, *6*, 159–171. [CrossRef] [PubMed]
9. Waller, P.J.; Thamsborg, S.M. Nematode control in 'green' ruminant production systems. *Trends Parasitol.* **2004**, *20*, 493–497. [CrossRef] [PubMed]

10. Smith, G.; Grenfell, B.T.; Isham, V.; Cornell, S. Anthelmintic resistance revisited: Under-dosing, chemoprophylactic strategies, and mating probabilities. *Int. J. Parasitol.* **1999**, *29*, 77–91. [CrossRef]
11. Praslička, J.; Bjørn, H.; Várady, M.; Nansen, P.; Hennessy, D.R.; Talvik, H. An in vivo dose-response study of fenbendazole against *Oesophagostomum dentatum* and *Oesophagostomum quadrispinulatum* in pigs. *Int. J. Parasitol.* **1997**, *27*, 403–409. [CrossRef]
12. Coles, G.C.; Roush, R.T. Slowing the spread of anthelmintic resistant nematodes of sheep and goats in the United Kingdom. *Vet. Res.* **1992**, *130*, 505–510. [CrossRef]
13. Prichard, R.K.; Hall, C.A.; Kelly, J.D.; Martin, I.C.A.; Donald, A.D. The problem of anthelmintic resistance in nematodes. *Aust. Vet. J.* **1980**, *56*, 239–250. [CrossRef] [PubMed]
14. Anderson, R.M.; May, R.M. *Infectious Diseases of Humans: Dynamics and Control*; Oxford University Press: Oxford, UK, 1992.
15. Marion, G.; Renshaw, E.; Gibson, G. Stochastic effects in a model of nematode infection in ruminants. *IMA J. Math. Appl. Med. Biol.* **1998**, *15*, 97–116. [CrossRef] [PubMed]
16. Cornell, S.J.; Isham, V.S.; Grenfell, B.T. Stochastic and spatial dynamics of nematode parasites in farmed ruminants. *Proc. R. Soc. B Biol. Sci.* **2004**, *271*, 1243–1250. [CrossRef] [PubMed]
17. Roberts, M.G.; Grenfell, B.T. The population dynamics of nematode infections of ruminants: Periodic perturbations as a model for management. *IMA J. Math. Appl. Med. Biol.* **1991**, *8*, 83–93. [CrossRef] [PubMed]
18. Roberts, M.G.; Grenfell, B.T. The population dynamics of nematode infections of ruminants: The effect of seasonality in the free-living stages. *IMA J. Math. Appl. Med. Biol.* **1992**, *9*, 29–41. [CrossRef] [PubMed]
19. Allen, L.J.S. *An Introduction to Stochastic Processes with Applications to Biology*; Pearson Education: Hoboken, NJ, USA, 2003.
20. Gómez-Corral, A.; López García, M. Control strategies for a stochastic model of host-parasite interaction in a seasonal environment. *J. Theor. Biol.* **2014**, *354*, 1–11. [CrossRef] [PubMed]
21. Abbott, K.A.; Taylor, M.; Stubbings, L.A. *Sustainable Worm Control Strategies for Sheep*, 4th ed.; A Technical Manual for Veterinary Surgeons and Advisers; SCOPS: Worcestershire, UK, 2012. Available online: http://www.scops.org.uk/workspace/pdfs/scops-technical-manual-4th-edition-updated-september-2013.pdf (accessed on 1 June 2018).
22. Uriarte, J.; Llorente, M.M.; Valderrábano, J. Seasonal changes of gastrointestinal nematode burden in sheep under an intensive grazing system. *Vet. Parasitol.* **2003**, *118*, 79–92. [CrossRef] [PubMed]
23. Faragó, I.; Havasi, A.; Horváth, R. On the order of operator splitting methods for time-dependent linear systems of differential equations. *Int. J. Numer. Anal. Model. Ser. B* **2011**, *2*, 142–154.
24. Nasreen, S.; Jeelani, G.; Sheikh, F.D. Efficacy of different anthelmintics against gastro-intestinal nematodes of sheep in Kashmir Valley. *VetScan* **2007**, *2*, 1.
25. Kassai, T. *Veterinary Helminthology*; Butterworth-Heinemann: Oxford, UK, 1999.
26. Barger, I.A. Genetic resistance of hosts and its influence on epidemiology. *Vet. Parasitol.* **1989**, *32*, 21–35. [CrossRef]
27. Barger, I.A.; Le Jambre, L.F.; Georgi, J.R.; Davies, H.I. Regulation of *Haemonchus contortus* populations in sheep exposed to continuous infection. *Int. J. Parasitol.* **1985**, *15*, 529–533. [CrossRef]
28. Dobson, R.J.; Waller, P.J.; Donald, A.D. Population dynamics of *Trichostrongylus colubriformis* in sheep: The effect of infection rate on the establishment of infective larvae and parasite fecundity. *Int. J. Parasitol.* **1990**, *20*, 347–352. [CrossRef]
29. Bailey, J.N.; Kahn, L.P.; Walkden-Brown, S.W. Availability of gastro-intestinal nematode larvae to sheep following winter contamination of pasture with six nematode species on the Northern Tablelands of New South Wales. *Vet. Parasitol.* **2009**, *160*, 89–99. [CrossRef] [PubMed]
30. Valderrábano, J.; Delfa, R.; Uriarte, J. Effect of level of feed intake on the development of gastrointestinal parasitism in growing lambs. *Vet. Parasitol.* **2002**, *104*, 327–338. [CrossRef]
31. Grennan, E.J. *Lamb Growth Rate on Pasture: Effect of Grazing Management, Sward Type and Supplementation*; Teagasc Research Centre: Athenry, Ireland, 1999.

mathematics

MDPI

Article

On Small Deviation Asymptotics In L_2 of Some Mixed Gaussian Processes

Alexander I. Nazarov [1,2] and Yakov Yu. Nikitin [2,3,*]

1 St. Petersburg Department of the Steklov Mathematical Institute, Fontanka 27, 191023 St. Petersburg, Russia; al.il.nazarov@gmail.com
2 Saint-Petersburg State University, Universitetskaya nab. 7/9, 199034 St. Petersburg, Russia
3 National Research University, Higher School of Economics, Souza Pechatnikov 16, 190008 St. Petersburg, Russia
* Correspondence: y.nikitin@spbu.ru; Tel.: +7-921-5832100

Received: 23 March 2018; Accepted: 3 April 2018; Published: 5 April 2018

Abstract: We study the exact small deviation asymptotics with respect to the Hilbert norm for some mixed Gaussian processes. The simplest example here is the linear combination of the Wiener process and the Brownian bridge. We get the precise final result in this case and in some examples of more complicated processes of similar structure. The proof is based on Karhunen–Loève expansion together with spectral asymptotics of differential operators and complex analysis methods.

Keywords: mixed Gaussian process; small deviations; exact asymptotics

MSC: 60G15, 60J65, 62J05

1. Introduction

The problem of small deviation asymptotics for Gaussian processes was intensively studied in last years. Such a development was stimulated by numerous links between the small deviation theory and such mathematical problems as the accuracy of discrete approximation for random processes, the calculation of the metric entropy for functional sets, and the law of the iterated logarithm in the Chung form. It is also known that the small deviation theory is related to the functional data analysis and nonparametric Bayesian estimation.

The history of the question is described in the surveys [1,2], see also [3] for recent results. The most explored is the case of L_2-norm. For an arbitrary square-integrable random process X on $[0,1]$ put

$$||X||_2 = \left(\int_0^1 X^2(t)dt\right)^{\frac{1}{2}}.$$

We are interested in the exact asymptotics as $\varepsilon \to 0$ of the probability $\mathbb{P}\{||X||_2 \le \varepsilon\}$.

Usually one studies the *logarithmic* asymptotics while the *exact* asymptotics was found only for several special processes. Most of them are so-called *Green Gaussian processes*. This means that the covariance function G_X is the Green function for the ordinary differential operator (ODO)

$$\mathcal{L}u \equiv (-1)^\ell \left(p_\ell(t)u^{(\ell)}\right)^{(\ell)} + \left(p_{\ell-1}(t)u^{(\ell-1)}\right)^{(\ell-1)} + \cdots + p_0(t)u, \qquad (1)$$

$(p_\ell(t) > 0)$ subject to proper homogeneous boundary conditions. This class of processes contains, e.g., the integrated Brownian motion, the Slepian process and the Ornstein–Uhlenbeck process, see [4–10]. Notice that some strong and interesting results were obtained recently for non-Green processes by Kleptsyna et al., see [11] and references therein.

In the present paper, we are interested in small deviations of so-called *mixed* Gaussian processes which are the sum (or the linear combination) of two independent Gaussian processes, usually with zero mean values. Mixed random processes arise quite naturally in the mathematical theory of finances and engineering applications and are known long ago.

Cheredito [12] considered the linear combination of the standard Wiener process W and the fractional Brownian motion (fBm) W^H with the Hurst index H, namely the process

$$Y^H_{(\beta)}(t) = W(t) + \beta W^H(t),$$

where $\beta \neq 0$ is a real constant. It is assumed that the processes W and W^H are independent. The covariance function of this process is $\min(s,t) + \beta^2 G_{W^H}(s,t)$, where the covariance function of the fBm is given by the well-known formula

$$G_{W^H}(s,t) = \frac{1}{2}(s^{2H} + t^{2H} - |s-t|^{2H}),$$

and $H \in (0,1)$ is the so-called Hurst index. For $H = 1/2$ the fBm process turns into the usual Wiener process.

This paper strongly stimulated the probabilistic study of such process and its generalizations concerning the regularity of its trajectories, its martingale properties, the innovation representations, etc. The papers [13–15] are the typical examples.

The small deviations of the process $Y^H_{(\beta)}$ were studied at the logarithmical level in [16], where the following result was obtained. We cite it in the simplified form (without the weight function).

Proposition 1. *As $\varepsilon \to 0$ the following asymptotics holds*

$$\ln \mathbb{P}\{||Y^H_{(\beta)}||_2 \le \varepsilon\} \sim \begin{cases} \ln \mathbb{P}\{||W||_2 \le \varepsilon\}, & \text{if } \quad H > 1/2; \\ \beta^{1/H} \ln \mathbb{P}\{||W^H||_2 \le \varepsilon\}, & \text{if } \quad H < 1/2. \end{cases}$$

From [17] we know that as $\varepsilon \to 0$

$$\ln \mathbb{P}\{||W^H|| \le \varepsilon\} \sim -\frac{H}{(2H+1)^{\frac{2H+1}{2H}}} \left(\frac{\Gamma(2H+1)\sin(\pi H)}{\left(\sin\left(\frac{\pi}{2H+1}\right)\right)^{2H+1}} \right)^{\frac{1}{2H}} \varepsilon^{-1/H},$$

and the exact small deviation asymptotics of W is given below, see (3).

However, the exact small deviations of mixed processes have not been explored. In general case it looks like a very complicated problem. First steps were made in a special case when a Gaussian process is mixed with some finite-dimensional "perturbation". The general theory was built in [18], later some refined results were obtained in the case of Durbin processes (limiting processes for empirical processes with estimated parameters), see [19] as a typical example.

We can give the solution in two cases. In Section 2 we consider the linear combination of two processes whose covariance functions are Green functions for two different boundary value problems to the same differential equation. The simplest example here is given by the standard Wiener process $W(t)$ and the Brownian bridge $B(t)$. Also we provide the exact small deviation asymptotics for more complicated mixtures containing the Ornstein–Uhlenbeck processes.

In Section 3 we deal with pairs of processes whose covariance functions are kernels of integral operators which are powers (or, more general, polynomials) of the same integral operator. The basic example here is the Brownian bridge and the integrated centered Wiener process

$$\overline{W}(t) = \int_0^t \left(W(s) - \int_0^1 W(u)du \right) ds. \qquad (2)$$

Another series of examples is given by the Wiener process and the so-called Euler integrated Wiener process.

2. Mixed Green Processes Related to the Same Ordinary Differential Operator (ODO)

Let X_1 and X_2 be independent zero mean Gaussian processes on $[0,1]$. We assume that their covariance functions $G_1(s,t)$ and $G_2(s,t)$ are the Green functions for the same ODO (1) with different boundary conditions. This means they satisfy the equation

$$\mathcal{L}G_i(s,t) = \delta(s-t), \qquad i = 1,2.$$

in the sense of distributions and satisfy corresponding boundary conditions.

We consider the mixed process

$$Z^\beta(t) = X_1(t) + \beta X_2(t), \qquad t \in [0,1].$$

Since X_1 and X_2 are independent, it is easy to see that its covariance function equals

$$G_{Z^\beta}(s,t) = G_1(s,t) + \beta^2 G_2(s,t)$$

and satisfies the equation

$$\mathcal{L}G_{Z^\beta}(s,t) = (1+\beta^2)\delta(s-t)$$

in the sense of distributions. Therefore, it is the Green function for the ODO $\frac{1}{1+\beta^2}\mathcal{L}$ subject to some (in general, more complicated) boundary conditions. This allows us to apply general results of [6,8] on the small ball behavior of the Green Gaussian processes and to obtain the asymptotics of $\mathbb{P}\{||Z^\beta||_2 \le \varepsilon\}$ as $\varepsilon \to 0$ up to a constant. Then the sharp constant can be found by the complex variable method as shown in [7], see also [20].

To illustrate this algorithm we begin with the simplest mixed process

$$Z_1^\beta(t) = B(t) + \beta W(t), \qquad t \in [0,1].$$

The covariance function $G_{Z_1^\beta}$ is given by $(1+\beta^2)\min(s,t) - st$, and the integral equation for eigenvalues is equivalent to the boundary value problem

$$-f''(t) = \frac{1+\beta^2}{\lambda} f(t), \qquad f(0) = 0, \quad f(1) + \beta^2 f'(1) = 0.$$

It is easy to see that the process $\frac{1}{\sqrt{1+\beta^2}} Z_1^\beta(t)$ coincides in distribution with the process $W^{(\beta)}$, so-called *"elongated" Brownian bridge from zero to zero with length* $1+\beta^2$, see ([21], Section 4.4.20). Therefore, we obtain, as $\varepsilon \to 0$,

$$\mathbb{P}\{||Z_1^\beta||_2 \le \varepsilon\} = \mathbb{P}\Big\{||W^{(\beta)}||_2 \le \frac{\varepsilon}{\sqrt{1+\beta^2}}\Big\} \overset{(*)}{\sim} \frac{\sqrt{1+\beta^2}}{|\beta|} \cdot \mathbb{P}\Big\{||W||_2 \le \frac{\varepsilon}{\sqrt{1+\beta^2}}\Big\}$$

(the relation $(*)$ was derived in ([7], Proposition 1.9), see also ([18], Example 6)).

The last asymptotics was obtained long ago:

$$\mathbb{P}\{||W||_2 \le \varepsilon\} \sim \frac{4}{\sqrt{\pi}} \cdot \varepsilon \cdot \exp\Big(-\frac{1}{8}\varepsilon^{-2}\Big), \qquad \varepsilon \to 0, \tag{3}$$

and we arrive at the following Theorem:

Theorem 1. *The following asymptotic relation holds as* $\varepsilon \to 0$*:*

$$\mathbb{P}\{||B+\beta W||_2 \le \varepsilon\} \sim \frac{4}{\sqrt{\pi}} \cdot \frac{\varepsilon}{|\beta|} \cdot \exp\Big(-\frac{1+\beta^2}{8}\varepsilon^{-2}\Big).$$

The next process we consider is

$$Z_2^{\beta}(t) = \mathring{U}_{(\alpha)}(t) + \beta U_{(\alpha)}(t), \qquad t \in [0,1].$$

Here $\mathring{U}_{(\alpha)}(t)$ is the Ornstein–Uhlenbeck process starting at the origin and $U_{(\alpha)}(t)$ is the stationary Ornstein–Uhlenbeck process. Both them are Gaussian processes with zero mean-value. Their covariance functions are, respectively,

$$G_{\mathring{U}_{(\alpha)}}(s,t) = (e^{-\alpha|t-s|} - e^{-\alpha(t+s)})/(2\alpha); \qquad G_{U_{(\alpha)}}(s,t) = e^{-\alpha|t-s|}/(2\alpha).$$

Direct calculation shows that the integral equation for eigenvalues of Z_2^{β} is equivalent to the boundary value problem

$$-f''(t) + \alpha^2 f(t) = \frac{1+\beta^2}{\lambda} f(t); \qquad (f' - \alpha(1+2\beta^{-2})f)(0) = (f' + \alpha f)(1) = 0.$$

By standard method we derive that if $r_1 < r_2 < \ldots$ are the positive roots of transcendental equation

$$F_1(\zeta) := (\zeta^2 - \alpha^2(1+2\beta^{-2}))\frac{\sin(\zeta)}{\zeta} - 2\alpha(1+\beta^{-2})\cos(\zeta) = 0$$

then $\lambda_n(Z_2^{\beta}) = \frac{1+\beta^2}{r_n^2+\alpha^2}$, $n \ge 1$.

Recall that the eigenvalues of the stationary Ornstein–Uhlenbeck process were derived in [22]. By rescaling we obtain $\lambda_n(\sqrt{1+\beta^2}\, U_{(\alpha)}) = \frac{1+\beta^2}{\rho_n^2+\alpha^2}$, $n \ge 1$, where $\rho_1 < \rho_2 < \ldots$ are the positive roots of transcendental equation

$$F_2(\zeta) := (\zeta^2 - \alpha^2)\frac{\sin(\zeta)}{\zeta} - 2\alpha\cos(\zeta) = 0.$$

We claim that $\lambda_n(Z_2^{\beta})$ and $\lambda_n(\sqrt{1+\beta^2}\, U_{(\alpha)})$ are asymptotically close, and therefore, using the Wenbo Li comparison theorem, see [22], we can write

$$\mathbb{P}\{||Z_2^{\beta}||_2 \le \varepsilon\} \sim C_{\text{dist}} \cdot \mathbb{P}\Big\{||U_{(\alpha)}||_2 \le \frac{\varepsilon}{\sqrt{1+\beta^2}}\Big\}, \qquad \varepsilon \to 0, \tag{4}$$

where the distortion constant is given by

$$C_{\text{dist}} = \left(\prod_{n=1}^{\infty} \frac{\lambda_n(\sqrt{1+\beta^2}\, U_{(\alpha)})}{\lambda_n(Z_2^{\beta})}\right)^{\frac{1}{2}} = \left(\prod_{n=1}^{\infty} \frac{r_n^2+\alpha^2}{\rho_n^2+\alpha^2}\right)^{\frac{1}{2}}.$$

To justify (4) we should prove the convergence of the last infinite product. As in [7], we use the complex variable method.

For large N in the disk $|\zeta| < \pi(N-\frac{1}{2})$ there are exactly $2N$ zeros $\pm r_j$, $j = 1, \ldots, N$, of $F_1(\zeta)$, and exactly $2N$ zeros $\pm\rho_j$, $j = 1, \ldots, N$, of $F_2(\zeta)$. By the Jensen theorem, see ([23], Section 3.6.1), we have

$$\ln\left(\frac{|F_1(0)|}{|F_2(0)|} \cdot \prod_{n=1}^{N} \frac{\rho_n^2}{r_n^2}\right) = \frac{1}{2\pi}\int_0^{2\pi} \ln\frac{|F_1(\pi(N-\frac{1}{2})\exp(i\varphi))|}{|F_2(\pi(N-\frac{1}{2})\exp(i\varphi))|}\, d\varphi.$$

It is easy to see that if we take $|\zeta| = \pi(N-\frac{1}{2})$ then

$$\frac{|F_1(\zeta)|}{|F_2(\zeta)|} \rightrightarrows 1, \qquad N \to \infty.$$

Therefore,

$$\prod_{n=1}^{\infty} \frac{r_n^2}{\rho_n^2} = \frac{|F_1(0)|}{|F_2(0)|}. \tag{5}$$

Now we use Hadamard's theorem on canonical product, see ([23], Section 8.24):

$$F_1(\zeta) \equiv F_1(0) \cdot \prod_{n=1}^{\infty} \left(1 - \frac{\zeta^2}{r_n^2}\right); \qquad F_2(\zeta) \equiv F_2(0) \cdot \prod_{n=1}^{\infty} \left(1 - \frac{\zeta^2}{\rho_n^2}\right).$$

In view of (5) this gives

$$C_{\text{dist}}^2 = \prod_{n=1}^{\infty} \frac{r_n^2 + \alpha^2}{\rho_n^2 + \alpha^2} = \frac{|F_1(0)|}{|F_2(0)|} \cdot \prod_{n=1}^{\infty} \left(1 + \frac{\alpha^2}{r_n^2}\right) \Big/ \prod_{n=1}^{\infty} \left(1 + \frac{\alpha^2}{\rho_n^2}\right) = \frac{|F_1(i\alpha)|}{|F_2(i\alpha)|} = 1 + \beta^{-2}.$$

Thus, (4) is proved. Since the small deviation asymptotics of $U_{(\alpha)}$ is known, see ([7], Proposition 2.1) and ([20], Corollary 3), we obtain the following Theorem:

Theorem 2. *The following asymptotic relation holds as $\varepsilon \to 0$:*

$$\mathbb{P}\{||\mathring{U}_{(\alpha)} + \beta U_{(\alpha)}||_2 \le \varepsilon\} \sim \sqrt{\frac{\alpha e^{\alpha}}{\pi}} \cdot \frac{8\varepsilon^2}{|\beta|\sqrt{1+\beta^2}} \cdot \exp\left(-\frac{1+\beta^2}{8}\,\varepsilon^{-2}\right).$$

Finally, we consider the stationary process

$$Z_3^{\beta}(t) = \mathcal{B}_{(\alpha)}(t) + \beta U_{(\alpha)}(t), \qquad t \in [0,1],$$

where $\mathcal{B}_{(\alpha)}$ is the Bogoliubov periodic process ([24–26]) with zero mean and covariance function

$$G_{\mathcal{B}_{(\alpha)}}(s,t) = \frac{1}{2\alpha \sinh(\alpha/2)} \cosh\left(\alpha|t-s| - \frac{\alpha}{2}\right).$$

A portion of tedious calculations gives the boundary value problem for eigenvalues of Z_3^{β}:

$$-f''(t) + \alpha^2 f(t) = \frac{1+\beta^2}{\lambda} f(t); \qquad f'(0) - \mathcal{A}_1 f(0) + \mathcal{A}_2 f(1) = f'(1) + \mathcal{A}_1 f(1) - \mathcal{A}_2 f(0) = 0.$$

Here

$$\mathcal{A}_1 = \alpha\,\frac{(1+\beta^2\gamma)^2 + 1}{(1+\beta^2\gamma)^2 - 1}; \qquad \mathcal{A}_2 = 2\alpha\,\frac{1+\beta^2\gamma}{(1+\beta^2\gamma)^2 - 1}; \qquad \gamma = 1 - e^{-\alpha}.$$

Just as in the previous example we obtain $\lambda_n(Z_3^{\beta}) = \frac{1+\beta^2}{\mathfrak{r}_n^2+\alpha^2}$, $n \ge 1$, where $\mathfrak{r}_1 < \mathfrak{r}_2 < \ldots$ are the positive roots of transcendental equation

$$F_3(\zeta) := (\zeta^2 - \alpha^2)\,\frac{\sin(\zeta)}{\zeta} - 2\mathcal{A}_1 \cos(\zeta) + 2\mathcal{A}_2 = 0.$$

Arguing as before, we derive

$$\mathbb{P}\{||Z_3^{\beta}||_2 \le \varepsilon\} \sim \widetilde{C}_{\text{dist}} \cdot \mathbb{P}\Big\{||U_{(\alpha)}||_2 \le \frac{\varepsilon}{\sqrt{1+\beta^2}}\Big\}, \qquad \varepsilon \to 0,$$

where

$$\widetilde{C}^2_{\text{dist}} = \prod_{n=1}^{\infty} \frac{\tau_n^2+\alpha^2}{\rho_n^2+\alpha^2} = \frac{|F_3(i\alpha)|}{|F_2(i\alpha)|} = \frac{\gamma(1+\beta^2)^2}{\beta^2(2+\beta^2\gamma)},$$

and thus we obtain the following Theorem:

Theorem 3. *The following asymptotic relation holds as* $\varepsilon \to 0$*:*

$$\mathbb{P}\{||\mathcal{B}_{(\alpha)} + \beta U_{(\alpha)}||_2 \le \varepsilon\} \sim \sqrt{\frac{\alpha(e^\alpha-1)}{\pi}} \cdot \frac{8\varepsilon^2}{|\beta|\sqrt{2+\beta^2(1-e^{-\alpha})}} \cdot \exp\left(-\frac{1+\beta^2}{8}\varepsilon^{-2}\right).$$

3. Mixed Processes Related to Polynomials of Covariance Operator

Recall that the covariance operator $\mathbb{G}_X$ related to the Gaussian process X is the integral operator with kernel G_X.

Lemma 1. *Let covariance operators* $\mathbb{G}_X$ *and* $\mathbb{G}_Z$ *are linked by relation* $\mathbb{G}_Z = \mathcal{P}(\mathbb{G}_X)$*, where* $\mathcal{P}$ *is a polynomial*

$$\mathcal{P}(x) := x + a_2x^2 + \cdots + a_{k-1}x^{k-1} + a_kx^k.$$

Then the following asymptotic relation holds as $\varepsilon \to 0$*:*

$$\mathbb{P}\{||Z||_2 \le \varepsilon\} \sim \widehat{C}_{\text{dist}} \cdot \mathbb{P}\{||X||_2 \le \varepsilon\}.$$

Proof. By the Wenbo Li comparison theorem, we should prove that the following infinite product converges:

$$\widehat{C}^2_{\text{dist}} = \prod_{n=1}^{\infty} \frac{\lambda_n(X)}{\lambda_n(Z)}.$$

It is well known that the set of eigenvalues of $\mathcal{P}(\mathbb{G}_X)$ coincides with the set $\mathcal{P}(\{\lambda_n(X)\}_{n\in\mathbb{N}})$. Moreover, since $\mathcal{P}$ increases in a neighborhood of the origin, for sufficiently large n we have just $\lambda_n(Z) = \mathcal{P}(\lambda_n(X))$. Thus,

$$\widehat{C}^2_{\text{dist}} = \prod_{n=1}^{\infty} \frac{\lambda_n(X)}{\mathcal{P}(\lambda_n(X))} = \prod_{n=1}^{\infty}(1+O(\lambda_n(X))).$$

Since X is square integrable, the series $\sum\limits_n \lambda_n(X)$ converges. Therefore, the infinite product also converges, and the lemma follows. □

The first example is the mixed process

$$Z_4^\beta(t) = B(t) + \beta\overline{W}(t), \qquad t \in [0,1],$$

where the integrated centered Wiener process $\overline{W}$ is defined in (2).

The integral equation for the eigenvalues of $\overline{W}$ is equivalent to the boundary value problem [4]

$$y^{(IV)} = \frac{1}{\lambda}y, \qquad y(0) = y(1) = y''(0) = y''(1) = 0. \tag{6}$$

It is easy to see that the operator of the problem (6) is just the square of the operator of the boundary value problem

$$-y'' = \frac{1}{\lambda}y, \qquad y(0) = y(1) = 0, \tag{7}$$

which corresponds to the Brownian bridge. Therefore, we have the relation $\mathbb{G}_{\overline{W}} = \mathbb{G}_B^2$ (surely, this can be checked directly). Thus,

$$\mathbb{G}_{Z_4^\beta} = \mathbb{G}_B + \beta^2\mathbb{G}_B^2.$$

Therefore, we can apply Lemma 1. Since the small ball asymptotics for the Brownian bridge was obtained long ago

$$\mathbb{P}\{||B||_2 \le \varepsilon\} \sim \sqrt{\frac{8}{\pi}} \cdot \exp\Big(-\frac{1}{8}\varepsilon^{-2}\Big),$$

it remains to calculate

$$\widehat{C}^2_{\text{dist}} = \prod_{n=1}^{\infty} \frac{\lambda_n(B)}{\lambda_n(B) + \beta^2\lambda_n^2(B)} = \prod_{n=1}^{\infty} \frac{(\pi n)^2}{(\pi n)^2 + \beta^2}.$$

The application of Hadamard's theorem to the function $F_4(\zeta) = \frac{\sin(\zeta)}{\zeta}$ gives

$$\prod_{n=1}^{\infty} \Big(1 + \frac{\beta^2}{(\pi n)^2}\Big) = \frac{F_4(i\beta)}{F_4(0)} = \frac{\sinh(\beta)}{\beta},$$

and we arrive at the following Theorem:

Theorem 4. *The following asymptotic relation holds as* $\varepsilon \to 0$*:*

$$\mathbb{P}\{||B + \beta\overline{W}||_2 \le \varepsilon\} \sim \sqrt{\frac{8\beta}{\pi\sinh(\beta)}} \cdot \exp\Big(-\frac{1}{8}\varepsilon^{-2}\Big).$$

Now we consider a family of mixed processes ($m \in \mathbb{N}$)

$$\widehat{Z}^{\beta}_{2m}(t) = W(t) + \beta W^{\mathcal{E}}_{2m}(t); \qquad \widehat{Z}^{\beta}_{2m-1}(t) = W(1-t) + \beta W^{\mathcal{E}}_{2m-1}(t), \qquad t \in [0,1],$$

where $W^{\mathcal{E}}_m$ is so-called *Euler integrated* Brownian motion, see [27,28]:

$$W^{\mathcal{E}}_0(t) = W(t); \qquad W^{\mathcal{E}}_{2m-1}(t) = \int_t^1 W^{\mathcal{E}}_{2m-1}(s)\,ds; \qquad W^{\mathcal{E}}_{2m}(t) = \int_0^t W^{\mathcal{E}}_{2m-1}(s)\,ds.$$

It was shown in [28], see also ([6], Proposition 5.1), that the covariance operator of $W^{\mathcal{E}}_m$ can be expressed as

$$\mathbb{G}_{W^{\mathcal{E}}_{2m}} = \mathbb{G}^{2m+1}_{W}; \qquad \mathbb{G}_{W^{\mathcal{E}}_{2m-1}} = \mathbb{G}^{2m}_{\widetilde{W}},$$

where $\widetilde{W}(t) = W(1-t)$. Obviously, the small ball asymptotics for $\widetilde{W}$ and for W coincide.

Thus, we can apply Lemma 1, and it remains to calculate

$$\widehat{C}^2_{\text{dist}} = \prod_{n=1}^{\infty} \frac{\lambda_n(W)}{\lambda_n(W) + \beta^2\lambda_n^{k+1}(W)} = \prod_{n=1}^{\infty} \frac{(\pi(n-\frac{1}{2}))^{2k}}{(\pi(n-\frac{1}{2}))^{2k} + \beta^2}$$

(here $k = 2m$ or $k = 2m-1$).

Application of Hadamard's theorem to the function $\cos(\zeta)$ gives

$$\prod_{n=1}^{\infty} \Big(1 - \frac{\zeta^2}{(\pi(n-\frac{1}{2}))^2}\Big) = \cos(\zeta). \tag{8}$$

Put $z = \exp\big(\frac{i\pi}{2k}\big)$ and multiply relations (8) for $\zeta = \beta^{\frac{1}{k}}z$, $\zeta = \beta^{\frac{1}{k}}z^3, \dots, \zeta = \beta^{\frac{1}{k}}z^{2k-1}$. This gives

$$\widehat{C}^2_{\text{dist}} = \Big(\prod_{j=1}^{k} \cos\big(\beta^{\frac{1}{k}}z^{2j-1}\big)\Big)^{-1}.$$

We take into account that

$$\begin{aligned}\cos\left(\beta^{\frac{1}{k}} z^{2j-1}\right) \cdot \cos\left(\beta^{\frac{1}{k}} z^{2k-2j+1}\right) &= |\cos\left(\beta^{\frac{1}{k}} z^{2j-1}\right)|^2 \\ &= \sinh^2\left(\beta^{\frac{1}{k}} \sin(\tfrac{\pi(2j-1)}{2k})\right) + \cos^2\left(\beta^{\frac{1}{k}} \cos(\tfrac{\pi(2j-1)}{2k})\right)\end{aligned}$$

and obtain the following Theorem:

Theorem 5. *For $m \in \mathbb{N}$, the following asymptotic relations hold as $\varepsilon \to 0$:*

$$\mathbb{P}\{||W + \beta W^{\mathcal{E}}_{2m}||_2 \le \varepsilon\} \sim \frac{4}{\sqrt{\pi}}$$
$$\times \frac{1}{\sqrt{\prod_{j=1}^{m} \left(\sinh^2\left(\beta^{\frac{1}{2m}} \sin(\frac{\pi(2j-1)}{4m})\right) + \cos^2\left(\beta^{\frac{1}{2m}} \cos(\frac{\pi(2j-1)}{4m})\right)\right)}} \cdot \varepsilon \cdot \exp\left(-\frac{1}{8}\varepsilon^{-2}\right);$$

$$\mathbb{P}\{||\widetilde{W} + \beta W^{\mathcal{E}}_{2m-1}||_2 \le \varepsilon\} \sim \frac{4}{\sqrt{\pi \cosh\left(\beta^{\frac{1}{2m-1}}\right)}}$$
$$\times \frac{1}{\sqrt{\prod_{j=1}^{m-1} \left(\sinh^2\left(\beta^{\frac{1}{2m-1}} \sin(\frac{\pi(2j-1)}{4m-2})\right) + \cos^2\left(\beta^{\frac{1}{2m-1}} \cos(\frac{\pi(2j-1)}{4m-2})\right)\right)}} \cdot \varepsilon \cdot \exp\left(-\frac{1}{8}\varepsilon^{-2}\right).$$

4. Discussion

We have initiated the study of the complicated problem of exact small deviations asymptotics in L_2 for mixed Gaussian processes with independent components. After the survey of the problem, we consider the linear combination of two processes whose covariance functions are Green functions for two different boundary value problems to the same differential equation. The simplest example here is given by the standard Wiener process $W(t)$ and the Brownian bridge $B(t)$. Also we provide the exact small deviation asymptotics for more complicated mixtures containing the Ornstein–Uhlenbeck processes.

Next, we deal with pairs of processes whose covariance functions are kernels of integral operators which are powers (or, more general, polynomials) of of the same integral operator. The basic example here is the Brownian bridge and the integrated centered Wiener process

$$\overline{W}(t) = \int_0^t \left(W(s) - \int_0^1 W(u)du \right) ds.$$

Another series of examples is given by the Wiener process and the so-called Euler integrated Wiener process.

It would be interesting to understand the genesis of boundary conditions and integral operators in the more general cases of mixed processes.

Acknowledgments: This work was supported by the grant of RFBR 16-01-00258 and by the grant SPbGU-DFG 6.65.37.2017.

Author Contributions: Both authors contributed equally in the writing of this article.

Conflicts of Interest: The authors declare that there is no conflict of interests.

References

1. Li, W.V.; Shao, Q.M. Gaussian processes: Inequalities, Small Ball Probabilities and Applications. In *Stochastic Processes: Theory and Methods, Handbook of Statistics*; Rao, C.R., Shanbhag, D., Eds.; Elsevier: Amsterdam, The Netherlands, 2001; Volume 19, pp. 533–597.
2. Lifshits, M.A. Asymptotic behavior of small ball probabilities. In *Probability Theory and Mathematical Statistics, Proceedings of the Seventh Vilnius Conference (1998), Vilnius, Lithuania, 12–18 August 1998*; Grigelionis, B., Ed.; VSP/TEV: Rancho Cordova, CA, USA, 1999; pp. 453–468.

3. Lifshits, M.A. Bibliography of Small Deviation Probabilities, on the Small Deviation Website. Available online: http://www.proba.jussieu.fr/pageperso/smalldev/biblio.pdf (accessed on 22 March 2018).
4. Beghin, L.; Nikitin, Y.Y.; Orsingher, E. Exact small ball constants for some Gaussian processes under the L^2-norm. *J. Mathem. Sci.* **2005**, *128*, 2493–2502.
5. Kharinski, P.A.; Nikitin, Y.Y. Sharp small deviation asymptotics in L_2-norm for a class of Gaussian processes. *J. Math. Sci.* **2006**, *133*, 1328–1332.
6. Nazarov, A.I.; Nikitin, Y.Y. Exact small ball behavior of integrated Gaussian processes under L^2-norm and spectral asymptotics of boundary value problems. *Probab. Theory Relat. Fields* **2004**, *129*, 469–494.
7. Nazarov, A.I. On the sharp constant in the small ball asymptotics of some Gaussian processes under L_2-norm. *J. Math. Sci.* **2003**, *117*, 4185–4210.
8. Nazarov, A.I. Exact L_2-small ball asymptotics of Gaussian processes and the spectrum of boundary-value problems. *J. Theor. Prob.* **2009**, *22*, 640–665.
9. Nikitin, Y.Y.; Orsingher, E. Exact small deviation asymptotics for the Slepian and Watson processes in the Hilbert norm. *J. Math. Sci.* **2006**, *137*, 4555–4560.
10. Nikitin, Y.Y.; Pusev, R.S. Exact small deviation asymptotics for some Brownian functionals. *Theor. Probab. Appl.* **2013**, *57*, 60–81.
11. Chigansky, P.; Kleptsyna, M.; Marushkevych, D. Exact spectral asymptotics of fractional processes. *arXiv* **2018**, arXiv:1802.09045. Available online: https://arxiv.org/abs/1802.09045 (accessed on 22 March 2018).
12. Cheridito, P. Mixed fractional Brownian motion. *Bernoulli* **2001**, *7*, 913–934.
13. El-Nouty, C. The fractional mixed fractional Brownian motion. *Stat. Probab. Lett.* **2003**, *65*, 111–120.
14. Cai, C.; Chigansky, P.; Kleptsyna, M. Mixed Gaussian processes: A filtering approach. *Ann. Prob.* **2016**, *44*, 3032–3075.
15. Yor, M. A Gaussian martingale which is the sum of two independent Gaussian non-semimartingales. *Electron. Commun. Probab.* **2015**, *20*, 1–5.
16. Nazarov, A.I.; Nikitin, Y.Y. Logarithmic L2-small ball asymptotics for some fractional Gaussian processes. *Theor. Probab. Appl.* **2005**, *49*, 645–658.
17. Bronski, J.C. Small ball constants and tight eigenvalue asymptotics for fractional Brownian motions. *J. Theor. Probab.* **2003**, *16*, 87–100.
18. Nazarov, A.I. On a set of transformations of Gaussian random functions. *Theor. Probab. Appl.* **2010**, *54*, 203–216.
19. Nazarov, A.I.; Petrova, Y.P. The small ball asymptotics in Hilbertian norm for the Kac–Kiefer–Wolfowitz processes. *Theor. Probab. Appl.* **2016**, *60*, 460–480.
20. Gao, F.; Hannig, J.; Lee, T.-Y.; Torcaso, F. Laplace transforms via Hadamard factorization with applications to Small Ball probabilities. *Electr. J. Probab.* **2003**, *8*, 1–20.
21. Borodin, A.N.; Salminen, P. *Handbook of Brownian Motion: Facts and Formulae*; Birkhäuser: Basel, Switzerland, 1996.
22. Li, W.V. Comparison results for the lower tail of Gaussian seminorms. *J. Theor. Probab.* **1992**, *5*, 1–31.
23. Titchmarsh, E. *Theory of Functions*, 2nd ed.; Oxford University Press: Oxford, UK, 1975.
24. Sankovich, D.P. Gaussian functional integrals and Gibbs equilibrium averages. *Theor. Math. Phys.* **1999**, *119*, 670–675.
25. Sankovich, D.P. Some properties of functional integrals with respect to the Bogoliubov measure. *Theor. Math. Phys.* **2001**, *126*, 121–135.
26. Pusev, R.S. Asymptotics of small deviations of the Bogoliubov processes with respect to a quadratic norm. *Theor. Math. Phys.* **2010**, *165*, 1348–1357.
27. Chang, C.-H.; Ha, C.-W. The Greens functions of some boundary value problems via the Bernoulli and Euler polynomials. *Arch. Math.* **2001**, *76*, 360–365.
28. Gao, F.; Hannig, J.; Lee, T.-Y.; Torcaso, F. Integrated Brownian motions and exact L_2-small balls. *Ann. Probab.* **2003**, *31*, 1320–1337.

mathematics

MDPI

Article

Convergence in Total Variation to a Mixture of Gaussian Laws

Luca Pratelli [1] **and Pietro Rigo** [2,*]

[1] Accademia Navale, Viale Italia 72, 57100 Livorno, Italy; pratel@mail.dm.unipi.it
[2] Dipartimento di Matematica "F. Casorati", Universita' di Pavia, via Ferrata 1, 27100 Pavia, Italy
* Correspondence: pietro.rigo@unipv.it

Received: 29 April 2018; Accepted: 5 June 2018; Published: 11 June 2018

Abstract: It is not unusual that $X_n \overset{dist}{\longrightarrow} VZ$ where X_n, V, Z are real random variables, V is independent of Z and $Z \sim \mathcal{N}(0,1)$. An intriguing feature is that $P(VZ \in A) = E\Big\{\mathcal{N}(0,V^2)(A)\Big\}$ for each Borel set $A \subset \mathbb{R}$, namely, the probability distribution of the limit VZ is a mixture of centered Gaussian laws with (random) variance V^2. In this paper, conditions for $d_{TV}(X_n,\, VZ) \to 0$ are given, where $d_{TV}(X_n,\, VZ)$ is the total variation distance between the probability distributions of X_n and VZ. To estimate the rate of convergence, a few upper bounds for $d_{TV}(X_n,\, VZ)$ are given as well. Special attention is paid to the following two cases: (i) X_n is a linear combination of the squares of Gaussian random variables; and (ii) X_n is related to the weighted quadratic variations of two independent Brownian motions.

Keywords: mixture of Gaussian laws; rate of convergence; total variation distance; Wasserstein distance; weighted quadratic variation

MSC: 60B10; 60F05

1. Introduction

All random elements involved in the sequel are defined on a common probability space $(\Omega, \mathcal{F}, P)$. We let $\mathcal{B}$ denote the Borel σ-field on $\mathbb{R}$ and $\mathcal{N}(a,b)$ the Gaussian law on $\mathcal{B}$ with mean a and variance b, where $a \in \mathbb{R}$, $b \geq 0$, and $\mathcal{N}(a,0) = \delta_a$. Moreover, Z always denotes a real random variable such that:

$$Z \sim \mathcal{N}(0,1).$$

In plenty of frameworks, it happens that:

$$X_n \overset{dist}{\longrightarrow} VZ, \tag{1}$$

where X_n and V are real random variables and V is independent of Z. Condition (1) actually occurs in the CLT, both in its classical form (with $V = 1$) and in its exchangeable and martingale versions (Examples 3 and 4). In addition, condition (1) arises in several recent papers with various distributions for V. See, e.g., [1–8].

An intriguing feature of condition (1) is that the probability distribution of the limit:

$$P(VZ \in A) = \int \mathcal{N}(0,V^2)(A)\, dP, \quad A \in \mathcal{B},$$

is a mixture of centered Gaussian laws with (random) variance V^2. Moreover, condition (1) can be often strengthened into:

$$d_W(X_n, VZ) \to 0, \tag{2}$$

where $d_W(X_n, VZ)$ is the Wasserstein distance between the probability distributions of X_n and VZ. In fact, condition (2) amounts to (1) provided the sequence (X_n) is uniformly integrable; see Section 2.1.

A few (engaging) problems are suggested by conditions (1) and (2). One is:

(*) Give conditions for $d_{TV}(X_n, VZ) \to 0$, where

$$d_{TV}(X_n, VZ) = \sup_{A \in \mathcal{B}} |P(X_n \in A) - P(VZ \in A)|.$$

Under such (or stronger) conditions, estimate the rate of convergence, i.e., find quantitative bounds for $d_{TV}(X_n, VZ)$.

Problem (*) is addressed in this paper. Before turning to results, however, we mention an example.

Example 1. *Let B be a fractional Brownian motion with Hurst parameter H and*

$$X_n = \frac{n^{1+H}}{2} \int_0^1 t^{n-1}(B_1^2 - B_t^2)\, dt.$$

The asymptotics of X_n and other analogous functionals of the B-paths (such as weighted power variations) is investigated in various papers. See, e.g., [5,7–10] and references therein. We note also that:

$$\int_0^1 t^n B_t\, dB_t = \frac{X_n}{n^H} - \frac{H}{2H+n} \quad \textit{for each } H \geq 1/4,$$

where the stochastic integral is meant in Skorohod's sense (it reduces to an Ito integral if $H = 1/2$).

Let $a(H) = 1/2 - |1/2 - H|$ and $V = \sqrt{H\Gamma(2H)}\, B_1 \sim \mathcal{N}(0, H\Gamma(2H))$. In [8], it is shown that, for every $\beta \in (0,1)$, there is a constant k (depending on H and β only) such that:

$$d_{TV}(X_n, VZ) \leq k\, n^{-\beta\, a(H)} \quad \textit{for all } n \geq 1,$$

where Z is a standard normal random variable independent of V. Furthermore, the rate $n^{-\beta\, a(H)}$ is quite close to be optimal; see condition (2) of [8].

In Example 1, problem (*) admits a reasonable solution. In fact, in a sense, Example 1 is our motivating example.

This paper includes two main results.

The first (Theorem 1) is of the general type. Suppose $l_n := \int |t\, \phi_n(t)|\, dt < \infty$, where ϕ_n is the characteristic function of X_n. (In particular, X_n has an absolutely continuous distribution). Then, an upper bound for $d_{TV}(X_n, VZ)$ is provided in terms of l_n and $d_W(X_n, VZ)$. In some cases, this bound allows to prove $d_{TV}(X_n, VZ) \to 0$ and to estimate the convergence rate. In Example 5, for instance, such a bound improves on the existing ones; see Theorem 3.1 of [6] and Remark 3.5 of [7]. However, for the upper bound to work, one needs information on l_n and $d_W(X_n, VZ)$, which is not always available. Thus, it is convenient to have some further tools.

In the second result (Theorem 2), the ideas underlying Example 1 are adapted to weighted quadratic variations; see [5,8,9]. Let B and B' be independent standard Brownian motions and

$$X_n = n^{1/2} \sum_{k=0}^{n-1} f(B_{k/n} - B'_{k/n})\, \{(\Delta B_{k/n})^2 - (\Delta B'_{k/n})^2\},$$

where $f : \mathbb{R} \to \mathbb{R}$ is a suitable function, $\Delta B_{k/n} = B_{(k+1)/n} - B_{k/n}$ and $\Delta B'_{k/n} = B'_{(k+1)/n} - B'_{k/n}$. Under some assumptions on f (weaker than those usually requested in similar problems), it is shown that

$d_{TV}(X_n, VZ) = O(n^{-1/4})$, where $V = 2\sqrt{\int_0^1 f^2(\sqrt{2}B_t)\,dt}$. Furthermore, $d_{TV}(X_n, VZ) = O(n^{-1/2})$ if one also assumes $\inf|f| > 0$. (We recall that, if a_n and b_n are non-negative numbers, the notation $a_n = O(b_n)$ means that there is a constant c such that $a_n \leq c\, b_n$ for all n).

2. Preliminaries

2.1. Distances between Probability Measures

In this subsection, we recall a few known facts on distances between probability measures. We denote by $(S, \mathcal{E})$ a measurable space and by μ and ν two probability measures on $\mathcal{E}$.

The total variation distance between μ and ν is:

$$\|\mu - \nu\| = \sup_{A \in \mathcal{E}} |\mu(A) - \nu(A)|.$$

If X and Y are $(S, \mathcal{E})$-valued random variables, we also write:

$$d_{TV}(X, Y) = \|P(X \in \cdot) - P(Y \in \cdot)\| = \sup_{A \in \mathcal{E}} |P(X \in A) - P(Y \in A)|$$

to denote the total variation distance between the probability distributions of X and Y.

Next, suppose S is a separable metric space, $\mathcal{E}$ the Borel σ-field and

$$\int d(x, x_0)\, \mu(dx) + \int d(x, x_0)\, \nu(dx) < \infty \quad \text{for some } x_0 \in S,$$

where d is the distance on S. The Wasserstein distance between μ and ν is:

$$W(\mu, \nu) = \inf_{X \sim \mu, Y \sim \nu} E[d(X, Y)],$$

where inf is over the pairs (X, Y) of $(S, \mathcal{E})$-valued random variables such that $X \sim \mu$ and $Y \sim \nu$. By a duality theorem, $W(\mu, \nu)$ admits the representation:

$$W(\mu, \nu) = \sup_f \left| \int f\, d\mu - \int f\, d\nu \right|,$$

where sup is over those functions $f : S \to \mathbb{R}$ such that $|f(x) - f(y)| \leq d(x, y)$ for all $x, y \in S$; see, e.g., Section 11.8 of [11]. Again, if X and Y are $(S, \mathcal{E})$-valued random variables, we write:

$$d_W(X, Y) = W\big[P(X \in \cdot),\, P(Y \in \cdot)\big]$$

to mean the Wasserstein distance between the probability distributions of X and Y.

Finally, we make precise the connections between convergence in distribution and convergence according to Wasserstein distance in the case $S = \mathbb{R}$. Let X_n and X be real random variables such that $E|X_n| + E|X| < \infty$ for each n. Then, the following statements are equivalent:

- $\lim_n d_W(X_n, X) = 0$;
- $X_n \xrightarrow{dist} X$ and $E|X_n| \to E|X|$;
- $X_n \xrightarrow{dist} X$ and the sequence (X_n) is uniformly integrable.

2.2. Two Technical Lemmas

The following simple lemma is fundamental for our purposes.

Lemma 1. *If $a_1, a_2 \in \mathbb{R}$, $0 \le b_1 \le b_2$ and $b_2 > 0$, then:*

$$\|\mathcal{N}(a_1,b_1)-\mathcal{N}(a_2,b_2)\| \le 1-\sqrt{\frac{b_1}{b_2}}+\frac{|a_1-a_2|}{\sqrt{2\,\pi\,b_2}}.$$

Lemma 1 is well known; see e.g. Proposition 3.6.1 of [12] and Lemma 3 of [8].
Note also that, if $a_1 = a_2 = a$, Lemma 1 yields:

$$\|\mathcal{N}(a,b_1)-\mathcal{N}(a,b_2)\| \le \frac{|b_1-b_2|}{b_i} \quad \text{for each } i \text{ such that } b_i > 0.$$

The next result, needed in Section 4, is just a consequence of Lemma 1. In such a result, $\mathcal{X}$ and $\mathcal{Y}$ are separable metric spaces, $g_n : \mathcal{X}\times\mathcal{Y}\to\mathbb{R}$ and $g : \mathcal{X}\times\mathcal{Y}\to\mathbb{R}$ Borel functions, and X and Y random variables with values in $\mathcal{X}$ and $\mathcal{Y}$, respectively.

Lemma 2. *Let ν be the probability distribution of Y. If X is independent of Y and*

$$g_n(X,y)\sim\mathcal{N}\big(0,\,\sigma_n^2(y)\big),\quad g(X,y)\sim\mathcal{N}\big(0,\,\sigma^2(y)\big),\quad \sigma^2(y)>0$$

for ν-almost all $y\in\mathcal{Y}$, then:

$$d_{TV}\Big(g_n(X,Y),\,g(X,Y)\Big)\le\min\Big\{\,E\Big(\frac{|\sigma_n(Y)-\sigma(Y)|}{\sigma(Y)}\Big),\;E\Big(\frac{|\sigma_n^2(Y)-\sigma^2(Y)|}{\sigma^2(Y)}\Big)\,\Big\}.$$

Proof. Since X is independent of Y,

$$\begin{aligned}d_{TV}\Big(g_n(X,Y),\,g(X,Y)\Big)&=\sup_{A\in\mathcal{B}}\Big|\int\Big(P\big(g_n(X,y)\in A\big)-P\big(g(X,y)\in A\big)\Big)\,\nu(dy)\Big|\\&\le\int\|P\big(g_n(X,y)\in\cdot\big)-P\big(g(X,y)\in\cdot\big)\|\,\nu(dy).\end{aligned}$$

Thus, since $g_n(X,y)$ and $g(X,y)$ have centered Gaussian laws and $g(X,y)$ has strictly positive variance, for ν-almost all $y\in\mathcal{Y}$, Lemma 1 yields:

$$\begin{aligned}&d_{TV}\Big(g_n(X,Y),\,g(X,Y)\Big)\le\int\frac{|\sigma_n(y)-\sigma(y)|}{\sigma(y)}\,\nu(dy)=E\Big(\frac{|\sigma_n(Y)-\sigma(Y)|}{\sigma(Y)}\Big)\\\text{and}\quad&d_{TV}\Big(g_n(X,Y),\,g(X,Y)\Big)\le\int\frac{|\sigma_n^2(y)-\sigma^2(y)|}{\sigma^2(y)}\,\nu(dy)=E\Big(\frac{|\sigma_n^2(Y)-\sigma^2(Y)|}{\sigma^2(Y)}\Big).\end{aligned}$$

□

3. A General Result

As in Section 1, let X_n, V and Z be real random variables, with $Z\sim\mathcal{N}(0,1)$ and V independent of Z. Since $|V|Z\sim VZ$, it can be assumed $V\ge 0$. We also assume $E|X_n|+E|VZ|<\infty$, so that we can define:

$$d_n=d_W(X_n,\,VZ).$$

In addition, we let:

$$X_n'=X_n+d_n^{1/2}\,U,$$

where U is a standard normal random variable independent of $(X_n, V, Z : n\ge 1)$.

We aim to estimate $d_{TV}(X_n, VZ)$. Under some conditions, however, the latter quantity can be replaced by $d_{TV}(X_n, X_n')$.

Lemma 3. *For each* $\alpha < 1/2$,

$$|d_{TV}(X_n, VZ) - d_{TV}(X_n, X_n')| \leq d_n^{1/2} + d_n^{1/2-\alpha} + P(V < d_n^{\alpha}).$$

In addition, if $E(1/V) < \infty$, *then:*

$$|d_{TV}(X_n, VZ) - d_{TV}(X_n, X_n')| \leq d_n^{1/2}\{1 + E(1/V)\}.$$

Proof. The Lemma is trivially true if $d_n = 0$. Hence, it can be assumed $d_n > 0$. Define $X_n'' = VZ + d_n^{1/2}U$ and note that:

$$|d_{TV}(X_n, VZ) - d_{TV}(X_n, X_n')| \leq d_{TV}(X_n', X_n'') + d_{TV}(X_n'', VZ).$$

For each $A \in \mathcal{B}$,

$$P(X_n' \in A) = \int \mathcal{N}(X_n, d_n)(A)\, dP \quad \text{and} \quad P(X_n'' \in A) = \int \mathcal{N}(VZ, d_n)(A)\, dP.$$

Hence, Lemma 1 yields:

$$\begin{aligned} d_{TV}(X_n', X_n'') &= \sup_{A \in \mathcal{B}} \left| \int \Big(\mathcal{N}(X_n, d_n)(A) - \mathcal{N}(VZ, d_n)(A) \Big)\, dP \right| \\ &\leq \int \|\mathcal{N}(X_n, d_n) - \mathcal{N}(VZ, d_n)\|\, dP \leq \frac{E|X_n - VZ|}{d_n^{1/2}}. \end{aligned}$$

On the other hand, the probability distribution of X_n'' can also be written as:

$$P(X_n'' \in A) = \int \mathcal{N}(0, V^2 + d_n)(A)\, dP.$$

Arguing as above, Lemma 1 implies again:

$$\begin{aligned} d_{TV}(X_n'', VZ) &\leq \int \|\mathcal{N}(0, V^2 + d_n) - \mathcal{N}(0, V^2)\|\, dP \\ &\leq E\Big(1 - \frac{V}{\sqrt{V^2 + d_n}}\Big) \leq E\Big(\frac{d_n^{1/2}}{\sqrt{V^2 + d_n}}\Big) \leq \frac{d_n^{1/2}}{\epsilon} + P(V < \epsilon) \end{aligned}$$

for each $\epsilon > 0$. Letting $\epsilon = d_n^{\alpha}$ with $\alpha < 1/2$, it follows that:

$$|d_{TV}(X_n, VZ) - d_{TV}(X_n, X_n')| \leq \frac{E|X_n - VZ|}{d_n^{1/2}} + d_n^{1/2-\alpha} + P(V < d_n^{\alpha}). \tag{3}$$

Inequality (3) holds true for *every* joint distribution for the pair (X_n, VZ). In particular, inequality (3) holds if such a joint distribution is taken to be one that realizes the Wasserstein distance, namely, one such that $E|X_n - VZ| = d_n$. In this case, one obtains:

$$|d_{TV}(X_n, VZ) - d_{TV}(X_n, X_n')| \leq d_n^{1/2} + d_n^{1/2-\alpha} + P(V < d_n^{\alpha}).$$

Finally, if $E(1/V) < \infty$, it suffices to note that:

$$d_{TV}(X_n'', VZ) \leq E\Big(\frac{d_n^{1/2}}{\sqrt{V^2 + d_n}}\Big) \leq d_n^{1/2} E(1/V).$$

□

For Lemma 3 to be useful, $d_{TV}(X_n, X_n')$ should be kept under control. This can be achieved under various assumptions. One is to ask X_n to admit a Lipschitz density with respect to Lebesgue measure.

Theorem 1. *Let ϕ_n be the characteristic function of X_n and*

$$l_n = \int |t\,\phi_n(t)|\,dt = 2\int_0^\infty t\,|\phi_n(t)|\,dt.$$

Given $\beta \geq 1$, suppose $\sup_n E|X_n|^\beta < \infty$ and $d_n \to 0$. Then, there is a constant k, independent of n, such that:

$$d_{TV}(X_n, X_n') \leq k\left(l_n\, d_n^{1/2}\right)^{\beta/(\beta+1)}.$$

In particular,

$$d_{TV}(X_n, VZ) \leq d_n^{1/2} + d_n^{1/2-\alpha} + P(V < d_n^\alpha) + k\left(l_n\, d_n^{1/2}\right)^{\beta/(\beta+1)}$$

for each $\alpha < 1/2$, and

$$d_{TV}(X_n, VZ) \leq d_n^{1/2}\left\{1 + E(1/V)\right\} + k\left(l_n\, d_n^{1/2}\right)^{\beta/(\beta+1)} \qquad \text{if } E(1/V) < \infty.$$

It is worth noting that, if $\beta = 1$, the condition $\sup_n E|X_n| < \infty$ follows from $d_n \to 0$. On the other hand, $d_n \to 0$ can be weakened into $X_n \overset{dist}{\longrightarrow} VZ$ whenever $\sup_n E|X_n|^\beta < \infty$ for some $\beta > 1$; see Section 2.1.

Proof of Theorem 1. If $l_n = \infty$, the Theorem is trivially true. Thus, it can be assumed $l_n < \infty$.

Since ϕ_n is integrable, the probability distribution of X_n admits a density f_n with respect to Lebesgue measure. In addition,

$$\begin{aligned} |f_n(x) - f_n(y)| &= (1/2\pi)\,\Big|\int (e^{-itx} - e^{-ity})\,\phi_n(t)\,dt\Big| \\ &\leq \frac{|x-y|}{2\pi}\int |t\,\phi_n(t)|\,dt = \frac{l_n\,|x-y|}{2\pi}. \end{aligned}$$

Given $t > 0$, it follows that:

$$\begin{aligned} 2\,d_{TV}(X_n, X_n') &\leq 2\int \|P(X_n \in \cdot) - P(X_n + d_n^{1/2}u \in \cdot)\|\,\mathcal{N}(0,1)(du) \\ &= \int\int |f_n(x) - f_n(x - d_n^{1/2}u)|\,dx\,\mathcal{N}(0,1)(du) \\ &\leq P(|X_n| > t) + P(|X_n'| > t) + \int_{-t}^{t}\int |f_n(x) - f_n(x - d_n^{1/2}u)|\,\mathcal{N}(0,1)(du)\,dx. \end{aligned}$$

Since $\sup_n E|X_n|^\beta < \infty$ and $d_n \to 0$, one obtains:

$$\begin{aligned} P(|X_n| > t) + P(|X_n'| > t) &\leq P(|X_n| > t) + P(|X_n| > t/2) + P(d_n^{1/2}|U| > t/2) \\ &\leq 2\,P(|X_n| > t/2) + \frac{d_n^{\beta/2}E|U|^\beta}{(t/2)^\beta} \\ &\leq \frac{2\,E|X_n|^\beta + d_n^{\beta/2}E|U|^\beta}{(t/2)^\beta} \leq \frac{k^*}{t^\beta} \end{aligned}$$

for some constant k^*. Hence,

$$2\,d_{TV}(X_n,\,X_n') \le \frac{k^*}{t^\beta} + \frac{l_n\,d_n^{1/2}}{2\pi}\int_{-t}^{t}\int |u|\,\mathcal{N}(0,1)(du)\,dx$$
$$\le \frac{k^*}{t^\beta} + \frac{l_n\,d_n^{1/2}}{\pi}\,t \quad \text{for each } t>0.$$

Minimizing over t, one finally obtains:

$$2\,d_{TV}(X_n,\,X_n') \le c(\beta)\,(k^*)^{1/(\beta+1)}\Big(\frac{l_n\,d_n^{1/2}}{\pi}\Big)^{\beta/(\beta+1)},$$

where $c(\beta)$ is a constant that depends on β only. This concludes the proof. □

Theorem 1 provides upper bounds for $d_{TV}(X_n,\,VZ)$ in terms of l_n and d_n. It is connected to Proposition 4.1 of [4], where d_{TV} is replaced by the Kolmogorov distance.

In particular, Theorem 1 implies that $d_{TV}(X_n,\,VZ) \to 0$ provided $V>0$ a.s. and

$$\lim_n d_n^{1/2} l_n = \lim_n d_W(X_n,\,VZ)^{1/2}\int |t\,\phi_n(t)|\,dt = 0.$$

In addition, Theorem 1 allows to estimate the convergence rate. As an extreme example, if $d_n \to 0$, $E(1/V)<\infty$ and $\sup_n\{l_n + E|X_n|^\beta\} < \infty$ for all $\beta \ge 1$, then:

$$d_{TV}(X_n,\,VZ) = O(d_n^\alpha) \quad \text{for every } \alpha < 1/2.$$

We next turn to examples. In each such examples, Z is a standard normal random variable independent of all other random elements.

Example 2. (Classical CLT). *Let $V=1$ and $X_n = (1/\sqrt{n})\sum_{i=1}^n \xi_i$, where $\xi_1,\xi_2,\ldots$ is an i.i.d. sequence of real random variables such that $E(\xi_1)=0$ and $E(\xi_1^2)=1$. In this case, $d_n = O(n^{-1/2})$; see Theorem 2.1 of [13]. Suppose now that $E|\xi_1|^\beta < \infty$ for all $\beta \ge 1$ and ξ_1 has a density f (with respect to Lebesgue measure) such that $\int |f'(x)|\,dx < \infty$. Then, $\sup_n\{l_n + E|X_n|^\beta\} < \infty$ for all $\beta \ge 1$, and Theorem 1 yields:*

$$d_{TV}(X_n, Z) = O(n^{-\alpha}) \quad \textit{for each } \alpha < 1/4.$$

This rate, however, is quite far from optimal. Under the present assumptions on ξ_1, in fact, $d_{TV}(X_n, Z) = O(n^{-1/2})$; see Theorem 1 of [14].

We finally prove $\sup_n\{l_n + E|X_n|^\beta\} < \infty$. It is well known that $E|\xi_1|^\beta < \infty$ for all β implies $\sup_n E|X_n|^\beta < \infty$ for all β. Hence, it suffices to prove $\sup_n l_n < \infty$. Let ϕ be the characteristic function of ξ_1 and $q = \int |f'(x)|\,dx$. An integration by parts yields $|\phi(t)| \le q/|t|$ for each $t \ne 0$. By Lemma 1.4 of [15], one also obtains $|\phi(t)| \le 1 - (1/43)(t/q)^2$ for $|t| < 2q$ (just let $b = 2q$ and $c = 1/2$ in Lemma 1.4 of [15]). Since $\phi_n(t) = \phi(t/\sqrt{n})^n$ for each $t \in \mathbb{R}$,

$$|\phi_n(t)| \le \Big(\frac{q\sqrt{n}}{|t|}\Big)^n \quad \textit{for } |t| \ge q\sqrt{n} \quad \textit{and}$$
$$|\phi_n(t)| \le \Big(1 - \frac{t^2}{43\,q^2 n}\Big)^n \quad \textit{for } |t| < q\sqrt{n}.$$

Using these inequalities, $\sup_n l_n < \infty$ follows from a direct calculation.

As noted above, the rate provided by Theorem 1 in the classical CLT is not optimal. While not exciting, this fact could be expected. Indeed, Theorem 1 is a general result, applying to *arbitrary* X_n, and should not be requested to give optimal bounds in a very special case (such as the classical CLT).

Example 3. (Exchangeable CLT). *Suppose now that (ξ_n) is an exchangeable sequence of real random variables with $E(\xi_1^2) < \infty$. Define*

$$V = \sqrt{E(\xi_1^2 \mid \mathcal{T}) - E(\xi_1 \mid \mathcal{T})^2} \quad \text{and} \quad X_n = \frac{\sum_{i=1}^n \{\xi_i - E(\xi_1 \mid \mathcal{T})\}}{\sqrt{n}},$$

where $\mathcal{T}$ is the tail σ-field of (ξ_n). By de Finetti's theorem,

$$d_{TV}(X_n,\, VZ) \leq E\Big(\|P(X_n \in \cdot \mid \mathcal{T}) - \mathcal{N}(0, V^2)\|\Big).$$

Hence, $d_{TV}(X_n,\, VZ) \to 0$ provided $\|P(X_n \in \cdot \mid \mathcal{T}) - \mathcal{N}(0, V^2)\| \xrightarrow{P} 0$. As to Theorem 1, note that $X_n \xrightarrow{dist} VZ$ (see e.g. Theorem 3.1 of [16]) and

$$\begin{aligned} E(X_n^2) = E\{E(X_n^2 \mid \mathcal{T})\} &= n\, E\Big\{E\Big(\big(\tfrac{\sum_{i=1}^n (\xi_i - E(\xi_1 \mid \mathcal{T}))}{n}\big)^2 \mid \mathcal{T}\Big)\Big\} \\ &= n\, E(V^2/n) = E(V^2) < \infty. \end{aligned}$$

Furthermore, $l_n \leq E\Big\{\int |t|\, |E(e^{itX_n} \mid \mathcal{T})|\, dt\Big\}$. Thus, by Theorem 1, $d_{TV}(X_n,\, VZ) \to 0$ whenever

$$E(\xi_1^2 \mid \mathcal{T}) > E(\xi_1 \mid \mathcal{T})^2 \text{ a.s.} \quad \text{and} \quad \lim_n d_n^{1/2}\, E\Big\{\int |t|\, |E(e^{itX_n} \mid \mathcal{T})|\, dt\Big\} = 0.$$

Example 4. (Martingale CLT). *Let*

$$X_n = \sum_{j=1}^{k_n} \xi_{n,j},$$

where $(\xi_{n,j} : n \geq 1,\, j = 1, \ldots, k_n)$ is an array of real square integrable random variables and $k_n \uparrow \infty$. For each $n \geq 1$, let:

$$\mathcal{F}_{n,0} \subset \mathcal{F}_{n,1} \subset \ldots \subset \mathcal{F}_{n,k_n}$$

be sub-σ-fields of $\mathcal{F}$ with $\mathcal{F}_{n,0} = \{\emptyset, \Omega\}$. A well known version of the CLT (see e.g. Theorem 3.2 of [17]) states that $X_n \xrightarrow{dist} VZ$ provided:

(i) $\xi_{n,j}$ *is $\mathcal{F}_{n,j}$-measurable and $E(\xi_{n,j} \mid \mathcal{F}_{n,j-1}) = 0$ a.s.;*
(ii) $\sum_j \xi_{n,j}^2 \xrightarrow{P} V^2$, $\max_j |\xi_{n,j}| \xrightarrow{P} 0$, $\sup_n E(\max_j \xi_{n,j}^2) < \infty$;
(iii) $\mathcal{F}_{n,j} \subset \mathcal{F}_{n+1,j}$.

Condition (iii) can be replaced by:

(iv) *V is measurable with respect to the σ-field generated by $\mathcal{N} \cup (\cap_{n,j} \mathcal{F}_{n,j})$ where $\mathcal{N} = \{A \in \mathcal{F} : P(A) = 0\}$.*

Note also that, under (i), one obtains $E(X_n^2) = \sum_{j=1}^{k_n} E(\xi_{n,j}^2)$.

Now, in addition to (i)–(ii)–(iii) or (i)–(ii)–(iv), suppose $\sup_n \sum_{j=1}^{k_n} E(\xi_{n,j}^2) < \infty$. Then, Theorem 1 (applied with $\beta = 2$) implies $d_{TV}(X_n,\, VZ) \to 0$ whenever $V > 0$ a.s. and $\lim_n d_n^{1/2}\, l_n = 0$. Moreover,

$$d_{TV}(X_n,\, VZ) = O\Big(\big(l_n\, d_n^{1/2}\big)^{2/3}\Big) \qquad \text{if } E(1/V) + l_n < \infty \text{ for each } n.$$

Our last example is connected to the second order Wiener chaos. We first note a simple fact as a lemma.

Lemma 4. *Let $\xi = (\xi_1, \ldots, \xi_k)$ be a centered Gaussian random vector. Define:*

$$Y = \sum_{j=1}^{k} a_j \{\xi_j^2 - \gamma_j^2\},$$

where $a_j \in \mathbb{R}$ and $\gamma = (\gamma_1, \ldots, \gamma_k)$ is an independent copy of ξ. Then, the characteristic function ψ of Y can be written as:

$$\psi(t) = E\{e^{-t^2 S}\}, \; t \in \mathbb{R}, \text{ where } S = \sum_{i,j} a_i a_j E(\xi_i \xi_j) (\xi_i + \gamma_i) (\xi_j + \gamma_j).$$

Proof. Let $\sigma_{i,j} = E(\xi_i \xi_j)$, $\xi^* = (\xi + \gamma)/\sqrt{2}$ and $\gamma^* = (\xi - \gamma)/\sqrt{2}$. Then,

$$(\xi^*, \gamma^*) \sim (\xi, \gamma), \quad Y = 2 \sum_j a_j \xi_j^* \gamma_j^*, \quad S = 2 \sum_{i,j} a_i a_j \sigma_{i,j} \xi_i^* \xi_j^*.$$

Therefore,

$$\psi(t) = E\{E(e^{itY} \mid \xi^*)\} = E\{e^{-2t^2 \sum_{i,j} a_i a_j \sigma_{i,j} \xi_i^* \xi_j^*}\} = E\{e^{-t^2 S}\}.$$

□

Example 5. (Squares of Gaussian random variables). *For each $n \geq 1$, let $(\xi_{n,1}, \ldots, \xi_{n,k_n})$ be a centered Gaussian random vector and*

$$X_n = \sum_{j=1}^{k_n} a_{n,j} \{\xi_{n,j}^2 - E(\xi_{n,j}^2)\} \qquad \text{where } a_{n,j} \in \mathbb{R}.$$

Take an independent copy $(\gamma_{n,1}, \ldots, \gamma_{n,k_n})$ of $(\xi_{n,1}, \ldots, \xi_{n,k_n})$ and define:

$$Y_n = \sum_{j=1}^{k_n} a_{n,j} \{\xi_{n,j}^2 - \gamma_{n,j}^2\},$$

$$S_n = \sum_{i=1}^{k_n} \sum_{j=1}^{k_n} a_{n,i} a_{n,j} E(\xi_{n,i} \xi_{n,j}) (\xi_{n,i} + \gamma_{n,i}) (\xi_{n,j} + \gamma_{n,j}).$$

Note that S_n is a (random) quadratic form of the covariance matrix $(E(\xi_{n,i} \xi_{n,j}) : 1 \leq i, j \leq k_n)$. Therefore, $S_n \geq 0$.

Since $|\phi_n|^2$ agrees with the characteristic function of Y_n, Lemma 4 yields:

$$|\phi_n(t)|^2 = E\{e^{-t^2 S_n}\}.$$

Being $S_n \geq 0$, it follows that:

$$\begin{aligned} |\phi_n(t)|^2 &= E\Big\{e^{-t^2 S_n} 1_{\{S_n \geq t^{\frac{\epsilon-4}{2}}\}}\Big\} + E\Big\{e^{-t^2 S_n} 1_{\{S_n < t^{\frac{\epsilon-4}{2}}\}}\Big\} \\ &\leq e^{-t^{\epsilon/2}} + P(S_n < t^{\frac{\epsilon-4}{2}}) = e^{-t^{\epsilon/2}} + P(S_n^{-2-\epsilon} > t^{4+\frac{\epsilon(2-\epsilon)}{2}}) \\ &\leq e^{-t^{\epsilon/2}} + E(S_n^{-2-\epsilon})\, t^{-4-\frac{\epsilon(2-\epsilon)}{2}} \quad \text{for all } \epsilon > 0 \text{ and } t > 0. \end{aligned}$$

Hence,

$$\frac{l_n}{2} = \int_0^\infty t\,|\phi_n(t)|\,dt \le 1 + \int_1^\infty t\,|\phi_n(t)|\,dt$$
$$\le 1 + \int_1^\infty t\,e^{\frac{-t^{\epsilon/2}}{2}}\,dt + \sqrt{E\big(S_n^{-2-\epsilon}\big)}\int_1^\infty t^{-1-\frac{\epsilon(2-\epsilon)}{4}}\,dt,$$

so that $\sup_n l_n < \infty$ *whenever* $\sup_n E\big(S_n^{-2-\epsilon}\big) < \infty$ *for some* $\epsilon \in (0,2)$.

To summarize, applying Theorem 1 with $\beta = 2$, *one obtains:*

$$d_{TV}(X_n, VZ) = O(d_n^{1/3}) \tag{4}$$

provided $X_n \xrightarrow{dist} VZ$, *for some V independent of Z, and*

$$E(1/V) + \sup_n\{E\big(S_n^{-2-\epsilon}\big) + E(X_n^2)\} < \infty \quad \textit{for some } \epsilon \in (0,2).$$

The bound (4) requires strong conditions, which may be not easily verifiable in real problems. However, the above result is sometimes helpful, possibly in connection with the martingale CLT of Example 4. As an example, the conditions for (4) are not hard to be checked when $\xi_{n,1},\ldots,\xi_{n,k_n}$ *are independent for fixed n. We also note that, to our knowledge, the bound (4) improves on the existing ones. In fact, letting* $p = 2$ *in Theorem 3.1 of [6] (see also Remark 3.5 of [7]) one only obtains* $d_{TV}(X_n, VZ) = O(d_n^{1/5})$.

4. Weighted Quadratic Variations

Theorem 1 works nicely if one is able to estimate d_n and l_n, which is usually quite hard. Thus, it is convenient to have some further tools. In this section, $d_{TV}(X_n, VZ)$ is upper bounded via Lemma 2. We focus on a special case, but the underlying ideas are easily adapted to more general situations. The results in [8], for instance, arise from a version of such ideas.

For any function $x : [0,1] \to \mathbb{R}$, denote:

$$\Delta x(k/n) = x((k+1)/n) - x(k/n) \quad \text{where } n \ge 1 \text{ and } k = 0,1,\ldots,n-1.$$

Let $q \ge 2$ be an integer, $f : \mathbb{R} \to \mathbb{R}$ a Borel function, and $J = \{J_t : 0 \le t \le 1\}$ a real process. The weighted q-variation of J on $\{0, 1/n, 2/n, \ldots, 1\}$ is:

$$J_n^* = \sum_{k=0}^{n-1} f(J_{k/n})\,(\Delta J_{k/n})^q.$$

As noted in [5], to fix the asymptotic behavior of J_n^* is useful to determine the rate of convergence of some approximation schemes of stochastic differential equations driven by J. Moreover, the study of J_n^* is also motivated by parameter estimation and by the analysis of single-path behaviour of J. See [5,9,18–21] and references therein.

More generally, given an $\mathbb{R}^2$-valued process:

$$(I, J) = \{(I_t, J_t) : 0 \le t \le 1\},$$

one could define:

$$(I, J)_n^* = \sum_{k=0}^{n-1} f(I_{k/n})\,(\Delta J_{k/n})^q.$$

The weight $f(I_{k/n})$ of $(\Delta J_{k/n})^q$ depends now on I. Thus, in a sense, $(I, J)_n^*$ can be regarded as the weighted q-variation of J relative to I.

Here, we focus on:

$$X_n = n^{1/2} \sum_{k=0}^{n-1} f\big(B_{k/n} - B'_{k/n}\big) \left\{ (\Delta B_{k/n})^2 - (\Delta B'_{k/n})^2 \right\}, \tag{5}$$

where B and B' are independent standard Brownian motions. Note that, letting $q = 2$ and $I = B - B'$, one obtains:

$$X_n = n^{1/2} \left\{ (I, B)_n^* - (I, B')_n^* \right\}.$$

Thus, $n^{-1/2} X_n$ can be seen as the difference between the quadratic variations of B and B' relative to $I = B - B'$.

We aim to show that, under mild assumptions on f, the probability distributions of X_n converge in total variation to a certain mixture of Gaussian laws. We also estimate the rate of convergence. The smoothness assumptions on f are weaker than those usually requested in similar problems; see, e.g., [5].

Theorem 2. *Let B and B' be independent standard Brownian motions and Z a standard normal random variable independent of (B, B'). Define X_n by Equation (5) and*

$$V = 2 \sqrt{\int_0^1 f^2\big(\sqrt{2} B_t\big)\, dt}.$$

Suppose $E(1/V^2) < \infty$ and

$$|f(x) - f(y)| \le c\, |x - y|\, e^{|x| + |y|}$$

for some constant c and all $x, y \in \mathbb{R}$. Then, there is a constant k independent of n satisfying:

$$d_{TV}(X_n, VZ) \le k\, n^{-1/4}.$$

Moreover, if $\inf |f| > 0$, one also obtains $d_{TV}(X_n, VZ) \le k\, n^{-1/2}$.

To understand better the spirit of Theorem 2, think of the trivial case $f = 1$. Then, the asymptotic behavior of $X_n = n^{1/2} \sum_{k=0}^{n-1} \{ (\Delta B_{k/n})^2 - (\Delta B'_{k/n})^2 \}$ can be deduced by classical results. In fact, $d_{TV}(X_n, 2Z) = O(n^{-1/2})$ and this rate is optimal; see Theorem 1 of [14]. On the other hand, since $V = 2$, the same conclusion can be drawn from Theorem 2.

We finally prove Theorem 2.

Proof of Theorem 2. First note that $T = (B + B')/\sqrt{2}$ and $Y = (B - B')/\sqrt{2}$ are independent standard Brownian motions and

$$X_n = 2\, n^{1/2} \sum_{k=0}^{n-1} f\big(\sqrt{2}\, Y_{k/n}\big)\, \Delta T_{k/n}\, \Delta Y_{k/n}.$$

Note also that:

$$VZ \sim 2\, T_1 \sqrt{\int_0^1 f^2\big(\sqrt{2} Y_t\big)\, dt}.$$

Thus, in order to apply Lemma 2, it suffices to let $\mathcal{X} = \mathcal{Y} = C[0,1]$, $X = T$, and

$$g_n(x,y) = 2\,n^{1/2} \sum_{k=0}^{n-1} f(\sqrt{2}\,y(k/n))\,\Delta x(k/n)\,\Delta y(k/n),$$

$$g(x,y) = 2\,x(1)\sqrt{\int_0^1 f^2(\sqrt{2}y(t))\,dt}.$$

For fixed $y \in \mathcal{Y}$, $g_n(T,y)$ and $g(T,y)$ are centered Gaussian random variables. Since $E\{(\Delta T_{k/n})^2\} = 1/n$,

$$\sigma_n^2(y) = E\{g_n(T,y)^2\} = 4\sum_{k=0}^{n-1} f^2(\sqrt{2}\,y(k/n))\,(\Delta y(k/n))^2$$
$$\text{and} \quad \sigma^2(y) = E\{g(T,y)^2\} = 4\int_0^1 f^2(\sqrt{2}y(t))\,dt.$$

On noting that $\sigma^2(Y) \sim V^2$, one also obtains:

$$\sigma^2(Y) > 0 \text{ a.s.} \quad \text{and} \quad E\{1/\sigma^2(Y)\} = E(1/V^2) < \infty.$$

Next, define:

$$a_n = (1/4)\,E\{|\sigma_n^2(Y) - \sigma^2(Y)|\}$$
$$= E\Big\{\Big|\sum_{k=0}^{n-1} f^2(\sqrt{2}\,Y_{k/n})\,(\Delta Y_{k/n})^2 - \int_0^1 f^2(\sqrt{2}Y_t)\,dt\Big|\Big\}.$$

By Lemma 2 and the Cauchy–Schwarz inequality,

$$d_{TV}(X_n,\, VZ)^2 = d_{TV}\Big(g_n(T,Y),\, g(T,Y)\Big)^2 \le E\Big(\frac{|\sigma_n(Y) - \sigma(Y)|}{\sigma(Y)}\Big)^2$$
$$\le E\{1/\sigma^2(Y)\}\,E\{(\sigma_n(Y) - \sigma(Y))^2\}$$
$$\le E(1/V^2)\,E\{|\sigma_n^2(Y) - \sigma^2(Y)|\} = 4\,E(1/V^2)\,a_n.$$

If $\inf|f| > 0$, since $\sigma^2(Y) \ge 4\,\inf f^2$, Lemma 2 implies again:

$$d_{TV}(X_n,\, VZ) \le E\Big(\frac{|\sigma_n^2(Y) - \sigma^2(Y)|}{\sigma^2(Y)}\Big) \le \frac{E\Big(|\sigma_n^2(Y) - \sigma^2(Y)|\Big)}{4\,\inf f^2} = \frac{a_n}{\inf f^2}.$$

Thus, to conclude the proof, it suffices to show that $a_n = O(n^{-1/2})$.
Define $c^* = \max(c,\, |f(0)|)$ and note that:

$$|f(s)| \le c^*\,e^{2|s|} \quad \text{and} \quad |f(s)^2 - f(t)^2| \le 2\,c\,c^*|s-t|\,e^{3(|s|+|t|)} \quad \text{for all } s,\, t \in \mathbb{R}.$$

Define also:

$$a_n^{(1)} = E\Big\{\Big|\sum_{k=0}^{n-1} f^2(\sqrt{2}\,Y_{k/n})\,\big((\Delta Y_{k/n})^2 - 1/n\big)\Big|\Big\} \quad \text{and}$$
$$a_n^{(2)} = E\Big\{\Big|(1/n)\sum_{k=0}^{n-1} f^2(\sqrt{2}\,Y_{k/n}) - \int_0^1 f^2(\sqrt{2}Y_t)\,dt\Big|\Big\}.$$

Since $a_n \leq a_n^{(1)} + a_n^{(2)}$, it suffices to see that $a_n^{(i)} = O(n^{-1/2})$ for each i. Since Y has independent increments and $E\{((\Delta Y_{k/n})^2 - 1/n)^2\} = 2/n^2$,

$$
\begin{aligned}
(a_n^{(1)})^2 &= E\Big\{\Big|\sum_{k=0}^{n-1} f^2(\sqrt{2}\,Y_{k/n})\,((\Delta Y_{k/n})^2 - 1/n)\Big|\Big\}^2 \\
&\leq \sum_{k=0}^{n-1} E\Big\{f^4(\sqrt{2}\,Y_{k/n})\,((\Delta Y_{k/n})^2 - 1/n)^2\Big\} \\
&= \sum_{k=0}^{n-1} E\{f^4(\sqrt{2}\,Y_{k/n})\}\,E\{((\Delta Y_{k/n})^2 - 1/n)^2\} \\
&= (2/n^2)\sum_{k=0}^{n-1} E\{f^4(\sqrt{2}\,Y_{k/n})\} \\
&\leq \frac{2\,(c^*)^4\,E\{e^{8\sqrt{2}M}\}}{n} \qquad \text{where } M = \sup_{0\leq t\leq 1} |Y_t|.
\end{aligned}
$$

Similarly,

$$
\begin{aligned}
a_n^{(2)} &= E\Big\{\Big|(1/n)\sum_{k=0}^{n-1} f^2(\sqrt{2}\,Y_{k/n}) - \int_0^1 f^2(\sqrt{2}Y_t)\,dt\Big|\Big\} \\
&\leq \sum_{k=0}^{n-1}\int_{k/n}^{(k+1)/n} E\{|f^2(\sqrt{2}\,Y_{k/n}) - f^2(\sqrt{2}\,Y_t)|\}\,dt \\
&\leq 2\sqrt{2}\,c\,c^*\sum_{k=0}^{n-1}\int_{k/n}^{(k+1)/n} E\{|Y_{k/n} - Y_t|\,e^{6\sqrt{2}M}\}\,dt \\
&\leq 2\sqrt{2}\,c\,c^*\sqrt{E\{e^{12\sqrt{2}M}\}}\sum_{k=0}^{n-1}\int_{k/n}^{(k+1)/n}\sqrt{E\{(Y_{k/n} - Y_t)^2\}}\,dt \\
&\leq 2\sqrt{2}\,c\,c^*\sqrt{E\{e^{12\sqrt{2}M}\}}\,\frac{1}{\sqrt{n}}.
\end{aligned}
$$

Therefore, $a_n^{(i)} = O(n^{-1/2})$ for each i, and this concludes the proof. □

Author Contributions: Each author contributed in exactly the same way to each part of this paper.

Funding: This research was supported by the Italian Ministry of Education, University and Research (MIUR): Dipartimenti di Eccellenza Program (2018-2022) - Dept. of Mathematics "F. Casorati", University of Pavia.

Conflicts of Interest: The authors declare no conflict of interest.

References

1. Azmoodeh, E.; Gasbarra, D. New moments criteria for convergence towards normal product/tetilla laws. *arXiv* **2017**, arXiv:1708.07681.
2. Eichelsbacher, P.; Thäle, C. Malliavin-Stein method for Variance-Gamma approximation on Wiener space. *Electron. J. Probab.* **2015**, *20*, 1–28. [CrossRef]
3. Gaunt, R.E. On Stein's method for products of normal random variables and zero bias couplings. *Bernoulli* **2017**, *23*, 3311–3345. [CrossRef]
4. Gaunt, R.E. Wasserstein and Kolmogorov error bounds for variance-gamma approximation via Stein's method I. *arXiv* **2017**, arXiv:1711.07379.
5. Nourdin, I.; Nualart, D.; Tudor, C.A. Central and non-central limit theorems for weighted power variations of fractional Brownian motion. *Ann. I.H.P.* **2010**, *46*, 1055–1079. [CrossRef]
6. Nourdin, I.; Poly, G. Convergence in total variation on Wiener chaos. *Stoch. Proc. Appl.* **2013**, *123*, 651–674. [CrossRef]

7. Nourdin, I.; Nualart, D.; Peccati, G. Quantitative stable limit theorems on the Wiener space. *Ann. Probab.* **2016**, *44*, 1–41. [CrossRef]
8. Pratelli, L.; Rigo, P. Total Variation Bounds for Gaussian Functionals. 2018, Submitted. Available online: http://www-dimat.unipv.it/rigo/frac.pdf (accessed on 10 April 2018)
9. Nourdin, I.; Peccati, G. Weighted power variations of iterated Brownian motion. *Electron. J. Probab.* **2008**, *13*, 1229–1256. [CrossRef]
10. Peccati, G.; Yor, M. Four limit theorems for quadratic functionals of Brownian motion and Brownian bridge. In *Asymptotic Methods in Stochastics, AMS, Fields Institute Communication Series*; Amer. Math. Soc.: Providence, RI, USA, 2004; pp. 75–87.
11. Dudley R.M. *Real Analysis and Probability*; Cambridge University Press: Cambridge, UK, 2004.
12. Nourdin, I.; Peccati, G. *Normal Approximations with Malliavin Calculus: From Stein's Method to Universality*; Cambridge University Press: Cambridge, UK, 2012.
13. Goldstein, L. L^1 bounds in normal approximation. *Ann. Probab.* **2007**, *35*, 1888–1930. [CrossRef]
14. Sirazhdinov, S.K.H.; Mamatov, M. On convergence in the mean for densities. *Theory Probab. Appl.* **1962**, *7*, 424–428. [CrossRef]
15. Petrov, V.V. *Limit Theorems of Probability Theory: Sequences of Independent Random Variables*; Clarendon Press: Oxford, UK, 1995.
16. Berti, P.; Pratelli, L.; Rigo, P. Limit theorems for a class of identically distributed random variables. *Ann. Probab.* **2004**, *32*, 2029–2052.
17. Hall, P.; Heyde, C.C. *Martingale Limit Theory and Its Applications*; Academic Press: New York, NY, USA, 1980.
18. Barndorff-Nielsen, O.E.; Graversen, S.E.; Shepard, N. Power variation and stochastic volatility: A review and some new results. *J. Appl. Probab.* **2004**, *44*, 133–143. [CrossRef]
19. Gradinaru, M.; Nourdin, I. Milstein's type schemes for fractional SDEs. *Ann. I.H.P.* **2009**, *45*, 1058–1098. [CrossRef]
20. Neuenkirch, A.; Nourdin, I. Exact rate of convergence of some approximation schemes associated to SDEs driven by a fractional Brownian motion. *J. Theor. Probab.* **2007**, *20*, 871–899. [CrossRef]
21. Nourdin, I. A simple theory for the study of SDEs driven by a fractional Brownian motion in dimension one. In *Seminaire de Probabilites*; Springer: Berlin, Germany, 2008; Volume XLI, pp. 181–197.

mathematics

MDPI

Article

Some Notes about Inference for the Lognormal Diffusion Process with Exogenous Factors

Patricia Román-Román †, Juan José Serrano-Pérez † and Francisco Torres-Ruiz *,†

Departamento de Estadística e Investigación Operativa, Facultad de Ciencias, Universidad de Granada, Avenida Fuente Nueva, 18071 Granada, Spain; proman@ugr.es (P.R.-R.); jjserra@ugr.es (J.J.S.-P.)

* Correspondence: fdeasis@ugr.es; Tel.: +34-9582-41000 (ext. 20056)

† These authors contributed equally to this work.

Received: 16 April 2018; Accepted: 15 May 2018; Published: 21 May 2018

Abstract: Different versions of the lognormal diffusion process with exogenous factors have been used in recent years to model and study the behavior of phenomena following a given growth curve. In each case considered, the estimation of the model has been addressed, generally by maximum likelihood (ML), as has been the study of several characteristics associated with the type of curve considered. For this process, a unified version of the ML estimation problem is presented, including how to obtain estimation errors and asymptotic confidence intervals for parametric functions when no explicit expression is available for the estimators of the parameters of the model. The Gompertz-type diffusion process is used here to illustrate the application of the methodology.

Keywords: lognormal diffusion process; exogenous factors; growth curves; maximum likelihood estimation; asymptotic distribution

1. Introduction

The lognormal diffusion process has been widely used as a probabilistic model in several scientific fields in which the variable under consideration exhibits an exponential trend. Originally, the lognormal diffusion process was mainly applied to modeling dynamic variables in the field of economy and finance. Important contributions have been made in this direction by Cox and Ross [1], Markus and Shaked [2], and Merton [3], showing the theoretical and practical importance of the process in that environment. For example, this process is associated with the Black and Scholes model [4] and appears in later extensions as terminal swap-rate models (Hunt and Kennedy [5], Lamberton and Lapeyre [6]).

In 1972, Tintner and Sengupta [7] introduced a modification of the process by including a linear combination of time functions in the infinitesimal mean of the process. The motivation for this was the introduction of external influences on the interest variable (endogenous variable), influences that could contribute to a better explanation of the phenomenon under study. For this reason, these time functions are known as exogenous factors, whose time behavior is assumed to be known or partially known. By using these time functions we can model situations wherein the observed trend shows deviations from the theoretical shape of the trend during certain time intervals, and can therefore use them to help describe the evolution of the process. Furthermore, a suitable choice of the exogenous factors can contribute to the external control of the process for forecasting purposes. Note that the methodology derived from the inclusion of exogenous factors has been applied to several contexts other than the lognormal process (see, for example, Buonocore et al. [8]).

The lognormal diffusion process with exogenous factors has been widely studied in relation to some aspects of inference and first-passage times. It has been applied to the modeling of time variables in several fields (see, for example [9,10]). On occasion, the endogenous variable itself helps identify the exogenous factors. However, there are situations in which external variables to the process that have an influence on the system are not available, or situations in which their functional expressions are unknown. In such cases, Gutiérrez et al. [11] suggested approaching the exogenous factors by means of polynomial functions.

The ability to control the endogenous variable using exogenous factors makes this process particularly useful for forecasting purposes. Some of its main features, such as the mean, mode and quantile functions (that can be expressed as parametric functions of the parameters of the process), can be used for prediction purposes. Therefore, the inference of these functions has been the subject of considerable study, both from the perspective of point estimation and of estimation by confidence intervals. With respect to the former, in [10] a more general study was carried out to obtain maximum likelihood (ML) estimators. In that case, the exact distribution of the estimators was found, and then used to obtain the uniformly minimum variance unbiased (UMVU) estimators. In addition, expressions for the relative efficiency of ML estimators, with respect to UMVU estimators, were obtained. This last study was extended for a class of parametric functions which include the mean and mode functions (together with their conditional versions) as special cases. Concerning estimation by confidence bands, in this paper the authors extended the results obtained by Land [12] on exact confidence intervals for the mean of a lognormal distribution, thus obtaining confidence bands for the mean and mode functions of the lognormal process with exogenous factors and expressing these functions in a more general form.

In most of the works cited, inference has been approached from the ML point of view, considering discrete sampling of the trajectories. To this end, it is essential to have the exact form of the transition density functions from which the likelihood function associated with the sample is constructed. However, alternatives are available for a range of situations. For example, approximating the transition density function using Euler-type schemes derived from the discretization of the stochastic differential equation that models the behavior of the phenomenon under study (sometimes this approach is known as naive ML approach). Other possible alternatives to ML are those derived, for example, from the use of the concept of estimating functions (Bibby et al. [13]) and the generalized method of moments (Hansen [14]). Fuchs in [15] presents a good review of these and other procedures. The Bayesian approach is also present in the study of diffusion processes, as suggested by Tang and Heron in [16].

On the other hand, considering particular choices of the time functions that define the exogenous factors has enabled researchers to define diffusion processes associated to alternative expressions of already-known growth curves. Along these lines, we may cite a Gompertz-type process [17] (applied to the study of rabbit growth), a generalized Von Bertalanffy diffusion process [18] (with an application to the growth of fish species), a logistic-type process [19] (applied to the growth of a microorganism culture), and a Richards-type diffusion process [20]. In [21], a joint analysis of the procedure for obtaining these processes is shown. More recently, Da Luz-Sant'Ana et al. [22] have established, following a similar methodology, a Hubbert diffusion process for studying oil production, while Barrera et al. [23] introduced a process linked to the hyperbolastic type-I curve and applied it in the context of the quantitative polymerase chain reaction (qPCR) technique.

In these last cases, obtaining the ML estimators was a rather laborious task. In fact, the resulting system of equations is exceedingly complex and does not have an explicit solution, and numerical procedures must be employed instead, with the subsequent problem of finding initial solutions (see, for instance [18,19,22]). However, it is impossible to carry out a general study of the system of equations in order to check the conditions of convergence of the chosen numerical method, since it is dependent on sample data. One alternative is then to use stochastic optimization procedures like simulated annealing, variable neighborhood search, and the firefly algorithm [20,23,24]. In any case, the exact distribution of the estimators cannot be obtained. Recently, the asymptotic distribution of the MLestimators and delta method have been used in order to obtain estimation errors, as well as confidence intervals, for the parameters and parametric functions in the context of the Hubbert diffusion model [25].

The main objective of this paper is to provide a unified view of the estimation problem by means of discrete sampling of trajectories, and to cover all the diffusion processes mentioned above. To this end, we will consider the generic expression of the lognormal diffusion process with exogenous factors. In Section 2, a brief summary of the main characteristics of the process is presented. Sections 3 and 4 address the problem of estimation by ML by using discrete sampling. In Section 3, the distribution of the sample is obtained, while in Section 4 the generic form adopted by the system of likelihood equations is derived in terms of the exogenous factor included in the model. Section 5 deals with obtaining the asymptotic distribution of the estimators, after calculating the Fisher information matrix, for which the results of Section 3 are fundamental. Finally, and as an application of the previous developments, Section 6 deals with the particular case of the Gompertz-type process introduced in [17].

2. The Lognormal Diffusion Process With Exogenous Factors

Let $I = [t_0, +\infty)$ be a real interval ($t_0 \geq 0$), $\Theta \subseteq \mathbb{R}^k$ an open set, and $h_{\boldsymbol{\theta}}(t)$ a continuous, bounded and differentiable function on I depending on $\boldsymbol{\theta} \in \Theta$.

The univariate lognormal diffusion process with exogenous factors is a diffusion process $\{X(t); t \in I\}$, taking values on $\mathbb{R}^+$, with infinitesimal moments

$$\begin{aligned} A_1(x,t) &= h_{\boldsymbol{\theta}}(t)x \\ A_2(x) &= \sigma^2 x^2, \qquad \sigma > 0 \end{aligned} \tag{1}$$

and with a lognormal or degenerate initial distribution. This process is the solution to the stochastic differential equation

$$dX(t) = h_{\boldsymbol{\theta}}(t)X(t)dt + \sigma X(t)dW(t), \qquad X(t_0) = X_0,$$

where $W(t)$ is a standard Wiener process independent on $X_0 = X(t_0)$, $t \geq t_0$, being this solution

$$X(t) = X_0 \exp\left(H_{\boldsymbol{\xi}}(t_0,t) + \sigma(W(t) - W(t_0))\right), \qquad t \geq t_0$$

with

$$H_{\boldsymbol{\xi}}(t_0,t) = \int_{t_0}^{t} h_{\boldsymbol{\theta}}(u)du - \frac{\sigma^2}{2}(t - t_0), \qquad \boldsymbol{\xi} = (\boldsymbol{\theta}^T, \sigma^2)^T.$$

An explanation of the main features of the process can be found in [21], where the authors carried out a detailed theoretical analysis. As regards the distribution of the process, if X_0 is distributed according to a lognormal distribution $\Lambda_1\left[\mu_0; \sigma_0^2\right]$, or X_0 is a degenerate variable ($P[X_0 = x_0] = 1$), all the finite-dimensional distributions of the process are lognormal. Concretely, $\forall n \in \mathbb{N}$ and $t_1 < \cdots < t_n$, vector $(X(t_1), \ldots, X(t_n))^T$ has a n-dimensional lognormal distribution $\Lambda_n[\boldsymbol{\varepsilon}, \boldsymbol{\Sigma}]$, where the components of vector $\boldsymbol{\varepsilon}$ and matrix $\boldsymbol{\Sigma}$ are

$$\varepsilon_i = \mu_0 + H_{\boldsymbol{\xi}}(t_0, t_i), \;\; i = 1, \ldots, n$$

and

$$\sigma_{ij} = \sigma_0^2 + \sigma^2(\min(t_i, t_j) - t_0), \;\; i, j = 1, \ldots, n,$$

respectively. The transition probability density function can be obtained from the distribution of $(X(s), X(t))^T$, $s < t$, being

$$f(x,t|y,s) = \frac{1}{x\sqrt{2\pi\sigma^2(t-s)}} \exp\left(-\frac{\left[\ln(x/y) - H_{\boldsymbol{\xi}}(s,t)\right]^2}{2\sigma^2(t-s)}\right), \tag{2}$$

that is, $X(t)|X(s) = y$ follows a lognormal distribution

$$X(t) \mid X(s) = y \rightsquigarrow \Lambda_1 \left(\ln y + H_\xi(s,t), \sigma^2(t-s)\right), \quad s < t.$$

From the previous distributions, one can obtain the characteristics most commonly employed for practical fitting and forecasting purposes. These characteristics can be expressed jointly as

$$G_\xi^{\boldsymbol{\lambda}}(t|y,\tau) = M_\xi(t|y,\tau)^{\lambda_1} \exp\left(\lambda_2 \left(\lambda_3 \sigma_0^2 + \sigma^2(t-\tau)\right)^{\lambda_4}\right), \tag{3}$$

with $\boldsymbol{\lambda} = (\lambda_1, \lambda_2, \lambda_3, \lambda_4)^T$ and where $M_\xi(t|y,\tau) = \exp\left(y + H_\xi(\tau,t)\right)$. Table 1 includes some of these characteristics (the $n-$th moment, and the mode and quantile functions as well as their conditional versions) according to the values of $\boldsymbol{\lambda}$, τ and y.

Table 1. Values used to obtain the n-th moment and the mode and quantile functions from $G_\xi^{\boldsymbol{\lambda}}(t|z,\tau)$. z_α is the α-quantile of a standard normal distribution.

Function	Expression	z	τ	$\boldsymbol{\lambda}$
n-th moment	$E[X(t)^n]$	μ_0	t_0	$(n, n^2/2, 1, 1)^T$
n-th conditional moment	$E[X(t)^n \mid X(s) = y]$	$\ln y$	s	$(n, n^2/2, 0, 1)^T$
mode	$Mode[X(t)]$	μ_0	t_0	$(1, -1, 1, 1)^T$
conditional mode	$Mode[X(t) \mid X(s) = y]$	$\ln y$	s	$(1, -1, 0, 1)^T$
α-quantile	$C_\alpha[X(t)]$	μ_0	t_0	$(1, z_\alpha, 1, 1/2)^T$
α-conditional quantile	$C_\alpha[X(t) \mid X(s) = y]$	$\ln y$	s	$(1, z_\alpha, 0, 1/2)^T$

3. Joint Distribution of *d* Sample-Paths of the Process

Let us consider a discrete sampling of the process, based on d sample paths, at times t_{ij}, $(i = 1, \ldots, d,\ j = 1, \ldots, n_i)$ with $t_{i1} = t_0, i = 1, \ldots, d$. Denote by $\mathbf{X} = \left(\mathbf{X}_1^T | \cdots | \mathbf{X}_d^T\right)^T$ the vector containing the random variables of the sample, where $\mathbf{X}_i^T$ includes the variables of the i-th sample-path, that is $\mathbf{X}_i = (X(t_{i1}), \ldots, X(t_{i,n_i}))^T, i = 1, \ldots, d$.

From Equation (2), and if the distribution of $X(t_1)$ is assumed lognormal $\Lambda_1(\mu_1, \sigma_1^2)$, the probability density function of $\mathbf{X}$ is

$$f_{\mathbf{X}}(\mathbf{x}) = \prod_{i=1}^{d} \frac{\exp\left(-\frac{[\ln x_{i1} - \mu_1]^2}{2\sigma_1^2}\right)}{x_{i1}\sigma_1\sqrt{2\pi}} \prod_{j=1}^{n_i-1} \frac{\exp\left(-\frac{\left[\ln(x_{i,j+1}/x_{ij}) - m_\xi^{i,j,j+1}\right]^2}{2\sigma^2\Delta_i^{j+1,j}}\right)}{x_{ij}\sigma\sqrt{2\pi\Delta_i^{j+1,j}}}$$

where $m_\xi^{i,j+1,j} = H_\xi(t_{ij}, t_{i,j+1})$ and $\Delta_i^{j+1,j} = t_{i,j+1} - t_{ij}$.

Now, we consider vector $\mathbf{V} = \left[\mathbf{V}_0^T | \mathbf{V}_1^T | \cdots | \mathbf{V}_d^T\right]^T = \left[\mathbf{V}_0^T | \mathbf{V}_{(1)}^T\right]^T$, built from $\mathbf{X}$ by means of the following change of variables:

$$\begin{aligned} V_{0i} &= X_{i1},\ i = 1, \ldots, d \\ V_{ij} &= (\Delta_i^{j+1,j})^{-1/2} \ln \frac{X_{i,j+1}}{X_{ij}},\ i = 1, \ldots, d; j = 1, \ldots, n_i - 1. \end{aligned} \tag{4}$$

Taking into account this change of variables, the density of $\mathbf{V}$ becomes

$$f_{\mathbf{V}}(\mathbf{v}) = \frac{\exp\left(-\frac{1}{2\sigma_1^2}(\ln \mathbf{v}_0 - \mu_1 \mathbf{1}_d)^T(\ln \mathbf{v}_0 - \mu_1 \mathbf{1}_d)\right)}{\prod_{i=1}^{d} v_{0i}\left(2\pi\sigma_1^2\right)^{\frac{d}{2}}} \frac{\exp\left(-\frac{1}{2\sigma^2}\left(\mathbf{v}_{(1)} - \gamma^{\xi}\right)^T\left(\mathbf{v}_{(1)} - \gamma^{\xi}\right)\right)}{(2\pi\sigma^2)^{\frac{n}{2}}} \tag{5}$$

with $\ln \mathbf{v}_0 = (\ln v_{01}, \ldots, \ln v_{0d})^T$, $n = \sum_{i=1}^{d}(n_i - 1)$. Here, $\mathbf{1}_d$ represents the d-dimensional vector whose components are all equal to one, while γ^{ξ} is a vector of dimension n with components $\gamma_{ij}^{\xi} = (\Delta_i^{j+1,j})^{-1/2} m_{\xi}^{i,j,j+1}$, $i = 1, \ldots, d; j = 1, \ldots, n_i - 1$.

From Equation (5) it is deduced that:

- $\mathbf{V}_0$ and $\mathbf{V}_{(1)}$ are independents,
- the distribution of $\mathbf{V}_0$ is lognormal $\Lambda_d\left[\mu_1 \mathbf{1}_d; \sigma_1^2 \mathbf{I}_d\right]$,
- $\mathbf{V}_{(1)}$ is distributed as an n-variate normal distribution $N_n\left[\gamma^{\xi}; \sigma^2 \mathbf{I}_n\right]$,

being $\mathbf{I}_d$ and $\mathbf{I}_n$ the identity matrices of order d and n, respectively.

4. Maximum Likelihood Estimation of the Parameters of the Process

Consider a discrete sample of the process in the sense described in the previous section, including the transformation of it given by Equation (4). Denote by $\boldsymbol{\eta} = (\mu_1, \sigma_1^2)^T$ and suppose that $\boldsymbol{\eta}$ and $\boldsymbol{\xi}$ are functionally independent. Then, for a fixed value $\mathbf{v}$ of the sample, the log-likelihood function is

$$L_{\mathbf{v}}(\boldsymbol{\eta}, \boldsymbol{\xi}) = -\frac{(n+d)\ln(2\pi)}{2} - \frac{d \ln \sigma_1^2}{2} - \sum_{i=1}^{d} \ln v_{0i} - \frac{\sum_{i=1}^{d}[\ln v_{0i} - \mu_1]^2}{2\sigma_1^2} - \frac{n \ln \sigma^2}{2} - \frac{Z_1 + \Phi_{\xi} - 2\Gamma_{\xi}}{2\sigma^2} \tag{6}$$

where

$$Z_1 = \sum_{i=1}^{d}\sum_{j=1}^{n_i-1} v_{ij}^2, \qquad \Phi_{\xi} = \sum_{i=1}^{d}\sum_{j=1}^{n_i-1} \frac{\left(m_{\xi}^{i,j+1,j}\right)^2}{\Delta_i^{j+1,j}}, \qquad \Gamma_{\xi} = \sum_{i=1}^{d}\sum_{j=1}^{n_i-1} \frac{v_{ij} m_{\xi}^{i,j+1,j}}{(\Delta_i^{j+1,j})^{1/2}}.$$

Taking into account Equation (6), and since $\boldsymbol{\eta}$ and $\boldsymbol{\xi}$ are functionally independent, the ML estimation of $\boldsymbol{\eta}$ is obtained from the system of equations (Given a function $f : \mathbb{R}^k \to \mathbb{R}$, $\frac{\partial f}{\partial \mathbf{x}^T} = \left(\frac{\partial f}{\partial x_1}, \ldots, \frac{\partial f}{\partial x_k}\right)$. Notation $\frac{\partial f}{\partial \mathbf{x}^T}$ indicates that the result is a row vector).

$$\frac{\partial L_{\mathbf{v}}(\boldsymbol{\eta}, \boldsymbol{\xi})}{\partial \boldsymbol{\eta}^T} = \left(\frac{\partial L_{\mathbf{v}}(\boldsymbol{\eta}, \boldsymbol{\xi})}{\partial \mu_1}, \frac{\partial L_{\mathbf{v}}(\boldsymbol{\eta}, \boldsymbol{\xi})}{\partial \sigma_1^2}\right) = \mathbf{0}$$

resulting in

$$\widehat{\mu}_1 = \frac{1}{d}\sum_{i=1}^{d} \ln v_{0i} \text{ and } \widehat{\sigma}_1^2 = \frac{1}{d}\sum_{i=1}^{d}(\ln v_{0i} - \widehat{\mu}_1)^2.$$

On the other hand, by denoting

$$\Omega_{\xi} = \frac{1}{2}\frac{\partial \Phi_{\xi}}{\partial \boldsymbol{\theta}^T} = \sum_{i=1}^{d}\sum_{j=1}^{n_i-1} \frac{m_{\xi}^{i,j+1,j}}{\Delta_i^{j+1,j}} \frac{\partial m_{\xi}^{i,j+1,j}}{\partial \boldsymbol{\theta}^T}, \qquad \Psi_{\theta} = \frac{1}{2}\frac{\partial \Gamma_{\xi}}{\partial \boldsymbol{\theta}^T} = \sum_{i=1}^{d}\sum_{j=1}^{n_i-1} \frac{v_{ij}}{(\Delta_i^{j+1,j})^{1/2}} \frac{\partial m_{\xi}^{i,j+1,j}}{\partial \boldsymbol{\theta}^T}$$

$$\Upsilon_{\xi} = -\frac{\partial \Phi_{\xi}}{\partial \sigma^2} = \sum_{i=1}^{d} m_{\xi}^{i,n_i,1}, \qquad Z_2 = -2\frac{\partial \Gamma_{\xi}}{\partial \sigma^2} = \sum_{i=1}^{d}\sum_{j=1}^{n_i-1} v_{ij}(\Delta_i^{j+1,j})^{1/2} \tag{7}$$

we have

$$\frac{\partial L_{\mathbf{V}}(\boldsymbol{\eta},\boldsymbol{\xi})}{\partial \boldsymbol{\theta}^T} = \frac{1}{\sigma^2}\left[\Psi_{\boldsymbol{\theta}} - \Omega_{\boldsymbol{\xi}}\right]$$
$$\frac{\partial L_{\mathbf{V}}(\boldsymbol{\eta},\boldsymbol{\xi})}{\partial \sigma^2} = -\frac{n}{2\sigma^2} + \frac{Z_1 + \Phi_{\boldsymbol{\xi}} - 2\Gamma_{\boldsymbol{\xi}}}{2\sigma^4} - \frac{Z_2 - \Upsilon_{\boldsymbol{\xi}}}{2\sigma^2}.$$

Thus, the ML estimation of $\boldsymbol{\xi}$ is obtained as the solution of the following system of $k+1$ equations:

$$\Psi_{\boldsymbol{\theta}} - \Omega_{\boldsymbol{\xi}} = 0 \tag{8}$$
$$Z_1 + \Phi_{\boldsymbol{\xi}} - 2\Gamma_{\boldsymbol{\xi}} - \sigma^2 Z_2 + \sigma^2 \Upsilon_{\boldsymbol{\xi}} = n\sigma^2 \tag{9}$$

In the case where $h_{\boldsymbol{\theta}}$ is a linear function in $\boldsymbol{\theta}$, it is possible to determine an explicit solution for this system of equations (see [10,26]). In other cases, the existence of a closed-form solution can not be guaranteed, and it is therefore necessary to use numerical procedures for its resolution. The fact that these methods require initial solutions has motivated the construction of *ad hoc* procedures which depend on the process derived according to the function $h_{\boldsymbol{\theta}}$ considered (see [18,19,22]). However, it is impossible to carry out a general study of the system of equations in order to check the conditions of convergence of the chosen numerical method, since the system is dependent on sample data and this may lead to unforeseeable behavior. One alternative would be using stochastic optimization procedures like simulated annealing, variable neighborhood search and the firefly algorithm. These algorithms are often more appropriate than classical numerical methods since they impose fewer restrictions on the space of solutions and on the analytical properties of the function to be optimized. Some examples of the application of these procedures in the context of diffusion processes can be seen in [19,21,23,25].

5. Distribution of the ML Estimators of the Parameters and Related Parametric Functions

In this section we will discuss some aspects related to the distribution of the estimators of the parameters of the model, and their repercussions in the corresponding distributions of parametric functions, which can be of interest for several applications.

With regard to the distribution of the estimators of $\boldsymbol{\eta}$, it is immediate to verify that

$$\widehat{\mu_1} \rightsquigarrow N_1[\mu_1; \sigma_1^2/d] \qquad \text{and} \qquad \frac{d\,\widehat{\sigma_1^2}}{\sigma_1^2} \rightsquigarrow \chi^2_{d-1}.$$

If $h_{\boldsymbol{\theta}}$ is linear, it is then possible to calculate exact distributions associated with the estimators of $\boldsymbol{\xi}$, which allows us to establish confidence regions for the parameters as well as UMVU estimators and confidence intervals for linear combinations of $\boldsymbol{\theta}$ and σ^2 (see [10,26]). However, in the non-linear case, the fact that an explicit expression for the estimators of $\boldsymbol{\xi}$ is not always readily available precludes obtaining, in general, exact distributions for them. In that case, asymptotic distributions can be used instead. In fact, on the basis of the properties of the ML estimators, it is known that $\widehat{\boldsymbol{\xi}}$ is asymptotically distributed as a normal distribution with mean $\boldsymbol{\xi}$ and covariance matrix $I(\boldsymbol{\xi})^{-1}$, where $I(\xi)$ is the Fisher's information matrix associated with the full sample (in this case, ignoring the data of the initial distribution).

First we calculate the associated Hessian matrix: (we have adopted the usual expression for the Hessian matrix of $f : \mathbb{R}^k \to \mathbb{R}$ using vectorial notation, that is $\frac{\partial^2 f}{\partial \mathbf{x} \partial \mathbf{x}^T}$).

$$H(\xi) = \frac{\partial^2 L_v(\eta,\xi)}{\partial\xi\partial\xi^T} = \begin{pmatrix} \frac{\partial^2 L_v(\eta,\xi)}{\partial\theta\partial\theta^T} & \left(\frac{\partial^2 L_v(\eta,\xi)}{\partial\sigma^2\partial\theta^T}\right)^T \\ \frac{\partial^2 L_v(\eta,\xi)}{\partial\sigma^2\partial\theta^T} & \frac{\partial^2 L_v(\eta,\xi)}{\partial(\sigma^2)^2} \end{pmatrix}$$

$$= \frac{1}{\sigma^2}\begin{pmatrix} \Pi_\xi - \Xi_\xi & -\frac{1}{\sigma^2}\left[\Psi_\theta^T - \Omega_\xi^T\right] + \frac{1}{2}\left(\frac{\partial \Upsilon_\xi}{\partial\theta^T}\right)^T \\ -\frac{1}{\sigma^2}\left[\Psi_\theta - \Omega_\xi\right] + \frac{1}{2}\frac{\partial \Upsilon_\xi}{\partial\theta^T} & \frac{n}{2\sigma^2} - \frac{Z_1 + \Phi_\xi - 2\Gamma_\xi}{\sigma^4} + \frac{Z_2 - \Upsilon_\xi}{\sigma^2} - \frac{Z_3}{4} \end{pmatrix}$$

where

$$\Pi_\xi = \sum_{i=1}^{d}\sum_{j=1}^{n_i-1} \frac{\partial^2 m_\xi^{i,j+1,j}}{\partial\theta\partial\theta^T}(\Delta_i^{j+1,j})^{-1/2}\left(v_{ij} - (\Delta_i^{j+1,j})^{-1/2} m_\xi^{i,j+1,j}\right)$$

and

$$\Xi_\xi = \sum_{i=1}^{d}\sum_{j=1}^{n_i-1}(\Delta_i^{j+1,j})^{-1}\left(\frac{\partial m_\xi^{i,j+1,j}}{\partial\theta^T}\right)^T \frac{\partial m_\xi^{i,j+1,j}}{\partial\theta^T}, \qquad Z_3 = \sum_{i=1}^{d}\Delta_i^{n_i,1}.$$

Taking into account the distribution of the sample (see Section 3), we have

$$E[\Pi_\xi] = \mathbf{0}, \qquad E[Z_1] = n\sigma^2 + \Phi_\xi, \qquad E[Z_2] = \Upsilon_\xi, \qquad E[\Psi_\theta] = \Omega_\xi, \qquad E[\Gamma_\xi] = \Phi_\xi$$

so, the Fisher's information matrix is given by

$$I(\xi) = -E[H(\xi)] = \frac{1}{\sigma^2}\begin{pmatrix} \Xi_\xi & -\frac{1}{2}\left(\frac{\partial\Upsilon_\xi}{\partial\theta^T}\right)^T \\ -\frac{1}{2}\frac{\partial\Upsilon_\xi}{\partial\theta^T} & \frac{n}{2\sigma^2} + \frac{Z_3}{4} \end{pmatrix},$$

from where it is concluded that $\widehat{\xi} \xrightarrow{D} N_{k+1}\left[\xi; I(\xi)^{-1}\right]$. In addition, and by applying the delta method, for a $q-$parametric function $g(\xi)$ $(q \leq k+1)$ it is verified that

$$g(\widehat{\xi}) \xrightarrow{D} N_q\left[g(\xi); \nabla g(\xi)^T I(\xi)^{-1}\nabla g(\xi)\right]$$

where $\nabla g(\xi)$ represents the vector of partial derivatives of $g(\xi)$ with respect to ξ.

The elements in the diagonal of matrix $I(\xi)^{-1}$ provide asymptotic variances for the estimations of the parameters, while the delta method provides the asymptotic covariance matrix for $g(\widehat{\xi})$ (and consequently the elements of the diagonal are the asymptotic variances for the estimation of each parametric function of $g(\xi)$). For example, if we consider $g(\xi) = G_\xi^\lambda(t|y,\tau)$, that is the general expression for the main characteristics of the process given by Equation (3), then

$$\nabla g(\xi) = g(\xi)\left(\lambda_1\frac{\partial H_\xi(\tau,t)}{\partial\theta^T}, (t-\tau)\left[-\frac{\lambda_1}{2} + \lambda_2\lambda_4\left(\lambda_3\sigma_0^2 + \sigma^2(t-\tau)\right)^{\lambda_4-1}\right]\right).$$

6. Application: The Gompertz-Type Diffusion Process

In this section we focus on the Gompertz-type diffusion process introduced in [17] with the aim of obtaining a continuous stochastic model associated with the Gompertz curve whose limit value depends on the initial value. Concretely

$$f(t) = x_0 \exp\left(-\frac{m}{\beta}\left(e^{-\beta t} - e^{-\beta t_0}\right)\right),\ t \geq t_0 \geq 0,\ m, \beta > 0 \text{ and } x_0 > 0.$$

To this end, the non-homogeneous lognormal diffusion process with infinitesimal moments

$$\begin{aligned} A_1(x,t) &= me^{-\beta t}x \\ A_2(x) &= \sigma^2 x^2 \end{aligned} \tag{10}$$

was considered.

In order to apply the general scheme developed in the preceding sections, we consider the following reparameterization $\boldsymbol{\theta} = (\delta, \alpha)^T = (m/\beta, e^{-\beta})^T$, which leads to expressing the Gompertz curve as

$$f_{\boldsymbol{\theta}}(t) = x_0 \exp\left(-\delta\left(\alpha^t - \alpha^{t_0}\right)\right) \tag{11}$$

whereas the infinitesimal moments (10) are written in the form of Equation (1), with $h_{\boldsymbol{\theta}}(t) = -\delta\alpha^t \ln \alpha$.

Denoting $\varphi^{\alpha}_{i,j+1,j} = \alpha^{t_{i,j+1}} - \alpha^{t_{i,j}}$ and $\omega^{\alpha}_{i,j+1,j} = t_{i,j+1}\alpha^{t_{i,j+1}} - t_{ij}\alpha^{t_{ij}}$, one has $m_{\xi}^{i,j+1,j} = -\delta\varphi^{\alpha}_{i,j+1,j} - \frac{\sigma^2}{2}\Delta_i^{j+1,j}$ and

$$\frac{\partial m_{\xi}^{i,j+1,j}}{\partial \boldsymbol{\theta}^T} = -\left(\varphi^{\alpha}_{i,j+1,j},\ \delta\omega^{\alpha}_{i,j+1,j}\right),$$

so, from Equation (8), and by taking into account of Equation (7), the following system of equations appears

$$X_1^{\alpha} + \delta X_2^{\alpha} + \frac{\sigma^2}{2}X_3^{\alpha} = 0$$

$$X_4^{\alpha} + \delta X_5^{\alpha} + \frac{\sigma^2}{2}X_6^{\alpha} = 0$$

where

$$X_1^{\alpha} = \sum_{i=1}^{d}\sum_{j=1}^{n_i-1}\frac{v_{ij}\varphi^{\alpha}_{i,j+1,j}}{(\Delta_i^{j+1,j})^{1/2}}, \qquad X_2^{\alpha} = \sum_{i=1}^{d}\sum_{j=1}^{n_i-1}\frac{\left(\varphi^{\alpha}_{i,j+1,j}\right)^2}{\Delta_i^{j+1,j}}, \qquad X_3^{\alpha} = \sum_{i=1}^{d}\varphi^{\alpha}_{i,n_i,1}$$

$$X_4^{\alpha} = \sum_{i=1}^{d}\sum_{j=1}^{n_i-1}\frac{v_{ij}\omega^{\alpha}_{i,j+1,j}}{(\Delta_i^{j+1,j})^{1/2}}, \qquad X_5^{\alpha} = \sum_{i=1}^{d}\sum_{j=1}^{n_i-1}\frac{\varphi^{\alpha}_{i,j+1,j}\omega^{\alpha}_{i,j+1,j}}{\Delta_i^{j+1,j}}, \qquad X_6^{\alpha} = \sum_{i=1}^{d}\omega^{\alpha}_{i,n_i,1}.$$

After some algebra, one obtains

$$\delta^{\alpha} = \frac{X_3^{\alpha}X_4^{\alpha} - X_1^{\alpha}X_6^{\alpha}}{X_2^{\alpha}X_6^{\alpha} - X_3^{\alpha}X_5^{\alpha}} \text{ and } \sigma^2_{\alpha} = 2S^{\alpha}, \text{ where } S^{\alpha} = \frac{X_1^{\alpha}X_5^{\alpha} - X_2^{\alpha}X_4^{\alpha}}{X_2^{\alpha}X_6^{\alpha} - X_3^{\alpha}X_5^{\alpha}}.$$

On the other hand, and since

$$\Phi_{\xi} = \delta^2 X_2^{\alpha} + \frac{\sigma^4}{4}Z_3 + \delta\sigma^2 X_3^{\alpha}, \qquad \Gamma_{\xi} = -\delta X_1^{\alpha} - \frac{\sigma^2}{2}Z_2, \qquad \Upsilon_{\xi} = -\delta X_3^{\alpha} - \frac{\sigma^2}{2}Z_3,$$

Equation (9) results in

$$S^{\alpha}\left[2n + S^{\alpha}\right] - \delta^{\alpha}\left[2X_1^{\alpha} + \delta^{\alpha}X_2^{\alpha}\right] - Z_1 = 0 \tag{12}$$

The solution of this equation provides the estimation of α, whereas those of the other parameters are given by $\delta^{\widehat{\alpha}}$ and $\sigma^2_{\widehat{\alpha}}$.

As regards the asymptotic distribution of $\widehat{\xi}$, it is a trivariate normal distribution with mean ξ and covariance matrix given by $I(\xi)^{-1}$, being

$$I(\xi) = \frac{1}{\sigma^2}\begin{pmatrix} X_2^\alpha & \delta\, X_5^\alpha & -X_3^\alpha \\ \delta\, X_5^\alpha & \delta^2\, X_7^\alpha & -\delta\, X_6^\alpha \\ -X_3^\alpha & -\delta\, X_6^\alpha & \frac{n}{2\sigma^2} + \frac{Z_3}{4} \end{pmatrix}$$

with

$$X_7^\alpha = \sum_{i=1}^{d}\sum_{j=1}^{n_i-1} \frac{\left(\omega_{i,j+1,j}^\alpha\right)^2}{\Delta_i^{j+1,j}}.$$

This distribution can be used to obtain the asymptotic standard errors for the estimation of the parameters as well as for some parametric functions of interest (see the last comment of the previous section). In particular, we focus on the inflection time and the corresponding expected value of the process at this instant, conditioned on $X(t_0) = x_0$. Another important parametric function in this context is the upper bound that determines the carrying capacity of the system modeled by the process. Concretely:

- Upper bound, conditioned on $X(t_0) = x_0$, $g_1(\boldsymbol{\theta}) = x_0 \exp\left(\delta\, \alpha^{t_0}\right)$.
- Inflection time, $g_2(\boldsymbol{\theta}) = -\ln\delta / \ln\alpha$.
- Value of the process at the time of inflection, conditioned on $X(t_0) = x_0$, $g_3(\boldsymbol{\theta}) = g_1(\boldsymbol{\theta})/e$.

On the other hand, when using the model for predictive purposes some of the parametric functions of Table 1 can be used. In particular, the conditioned mean function adopts the expression

$$E[X(t)|X(\tau) = y] = g_4(\boldsymbol{\theta}) = y \exp\left(-\delta\left(\alpha^t - \alpha^\tau\right)\right).$$

Note that this curve is of the type of Equation (11). For this reason, this function is useful for forecasting purposes. In this case, it is of interest to provide not only the value of the function at each time instant, but also the standard error of the prediction and a confidence interval determining a range of values that includes, with a given confidence level, the true real value of the forecast.

Application to Real Data

The following example is based on a study developed in [27] on some aspects related to the growth of a population of rabbits. Figure 1 shows the weight (in grams) of 29 rabbits over 30 weeks. The sample paths begin at different initial values, thus showing a sigmoidal behavior, and their bounds are dependent on the initial values. These two aspects suggest that using the Gompertz-type model proposed above would be appropriate.

This data set has been used in various papers to illustrate some aspects of the Gompertz-type process, such as the estimation of the parameters and the study of some time variables that may be of interest in the analysis of growth phenomena of this nature. As regards the estimation of the parameters, in [17] the authors designed an iterative method for solving the likelihood system of equations, while in [24] the maximization of the likelihood function was directly addressed by simulated annealing. In addition, in [28] two time variables of interest for this type of data were analyzed: concretely the inflection time and the time instant in which the process reaches a certain percentage of total growth. Both cases were modeled as first-passage time problems.

In this paper the estimation of the parameters has been carried out from the resolution of Equation (12) by means of the bisection method (see Figure 2) and then by using expressions $\delta^{\widehat{\alpha}}$ and $\sigma^2_{\widehat{\alpha}}$.

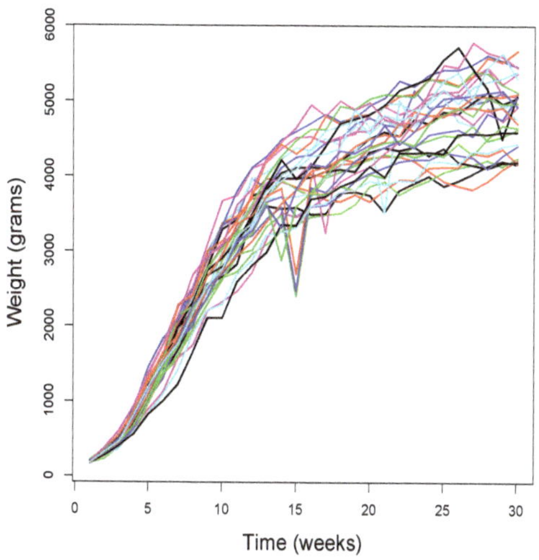

Figure 1. Weight of 29 rabbits over 30 weeks.

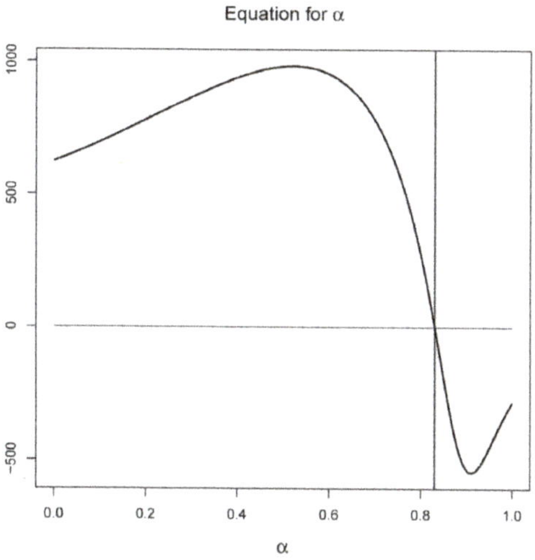

Figure 2. Graph of equation for α.

Table 2 contains the estimated values for the parameters and the inflection time, as well as the asymptotic estimation error and 95% confidence intervals by applying the delta method.

Table 2. Estimated values, standard errors and 95% confidence intervals of the parameters and the inflection time.

Parametric Function	δ	α	σ	$g_2(\theta)$
Estimated value	4.1020	0.8301	0.0708	7.5803
Standard error	0.0556	0.0021	0.0002	0.1053
Confidence interval	(3.9929, 4.1063)	(0.8258, 0.8343)	(0.0704, 0.0713)	(7.3738, 7.7869)

As regards the weight value at the inflection time and the upper bound, remember that these values depend on the one observed at the initial instant. Taking into account the range of observed weight values at the initial instant of observation, several values have been considered within this range. For these values, the expected weight of a rabbit at the moment of inflection has been studied, as well as the possible value of the maximum weight (upper bound). Table 3 contains the estimated values, the asymptotic standard errors, and the 95% confidence intervals.

Function $E[X(t)|X(t_0) = x_0]$ can be used to provide forecasts of the weight of a rabbit that presents an initial weight x_0. Figure 3 shows, for a selection of four of the rabbits used in the study, the estimated mean function together with the 95% asymptotic confidence intervals obtained for each value of this function. Additionally, the observed values are included to check the quality of the adjustment made by the model under consideration. Obviously, this type of representation can also be obtained by considering any value of x_0 in the range of the initial distribution of the weight. Note that the estimated mean functions for each rabbit depend on the initial value, and so do the corresponding confidence intervals for the mean at each time instant. Therefore, the graphs in the figure are different for each rabbit although the estimation of the parameters is unique.

Table 3. Estimated values, standard errors, and 95% confidence intervals of the upper bound and value at the inflection time for several values of the initial weight.

Initial Weight	Upper Bound			Value at Inflection Time		
	$g_3(\hat{\theta})$	St. Error	95% Interval	$g_1(\hat{\theta})$	St. Error	95% Interval
145	1772.836	70.546	(1634.568, 1911.104)	4819.068	191.764	(4443.215, 5194.920)
155	1772.836	75.411	(1625.032, 1920.640)	4819.068	204.990	(4417.295, 5220.841)
165	1883.638	80.276	(1726.298, 2040.978)	5120.260	218.215	(4692.566, 5547.954)
175	2105.243	85.142	(1938.367, 2272.118)	5722.643	231.440	(5269.028, 6176.258)
185	2216.045	90.007	(2039.634, 2392.456)	6023.835	244.665	(5544.299, 6503.371)
195	2216.045	94.872	(2030.098, 2401.992)	6023.835	257.890	(5518.378, 6529.291)
205	2105.243	99.737	(1909.760, 2300.726)	5722.643	271.115	(5191.266, 6254.020)
215	1883.638	104.603	(1678.620, 2088.657)	5120.260	284.341	(4562.961, 5677.558)

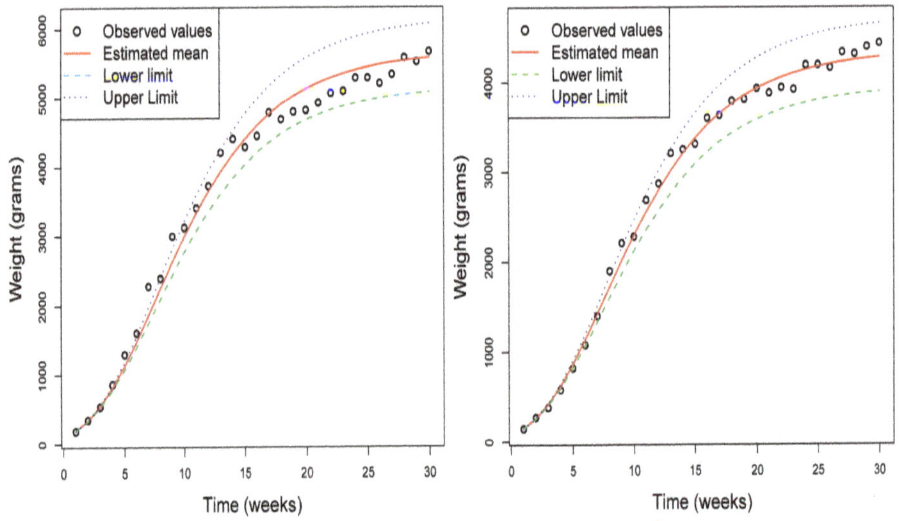

Figure 3. *Cont.*

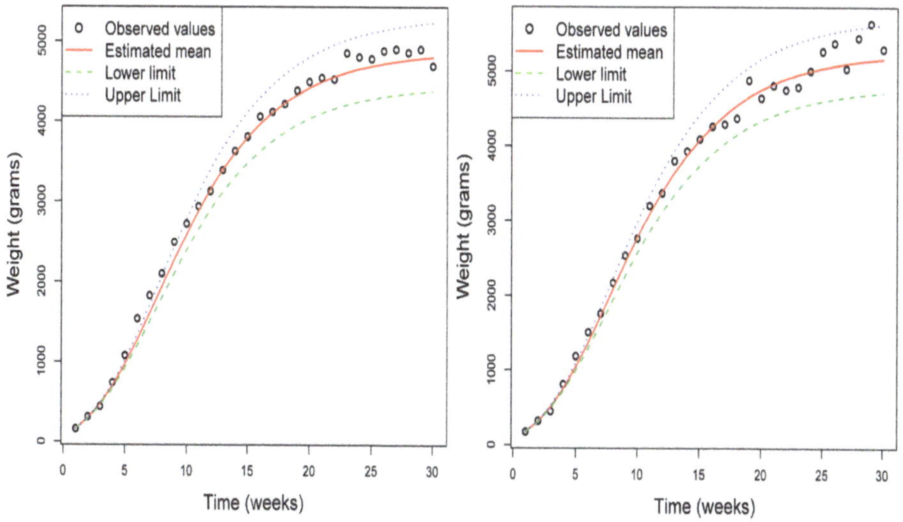

Figure 3. Observed values, estimated mean function, and confidence intervals for a choice of rabbits.

7. Conclusions

The present paper deals with some topics about inference for the non-homogeneous lognormal process (or with exogenous factors). Starting from the general form of the process, we studied the ML estimation of the parameters by using discrete sampling. This general overview enabled us to provide a unified method for several diffusion processes which can be built from particular cases of the non-homogeneous lognormal process for several choices of exogenous factors. In addition, we also looked into the asymptotic distribution of estimators, through which we can calculate the estimation errors and confidence intervals for the estimators of a wide range of parametric functions of interest in many fields. Finally, the process here described is applied to the Gompertz-type diffusion process introduced in [17].

Author Contributions: The three authors have participated equally in the development of this work, either in the theoretical developments or in the applied aspects. The paper was also written and reviewed cooperatively.

Acknowledgments: This work was supported in part by the Ministerio de Economía, Industria y Competitividad, Spain, under Grants MTM2014-58061-P and MTM2017-85568-P.

Conflicts of Interest: The authors declare no conflict of interest.

References

1. Cox, J.C.; Ross, S.A. The evaluation of options for alternative stochastic processes. *J. Financ. Econ.* **1976**, *3*, 145–166. [CrossRef]
2. Marcus, A.; Shaked, I. The relationship between accounting measures and prospective probabilities of insolvency: An application to the banking industry. *Financ. Rev.* **1984**, *19*, 67–83. [CrossRef]
3. Merton, R.C. Option pricing when underlying stock returns are discontinuous. *J. Financ. Econ.* **1976**, *3*, 125–144. [CrossRef]
4. Black, F.; Scholes, M. The pricing of options and corporate liabilities. *J. Political Econ.* **1973**, *81*, 637–654. [CrossRef]
5. Hunt, P.J.; Kennedy, J.G. *Financial Derivatives in Theory and Practice*; Revised Edition; John Wiley and Sons: Chichester, UK, 2004; ISBN 978-0-470-86359-6.
6. Lamberton, D.; Lapeyre, B. *Introduction to Stochastic Calculus Applied to Finance*, 2nd ed.; Chapman and Hall: New York, NY, USA, 2007; ISBN 9781584886266.

7. Tintner, G.; Sengupta, J.K. *Stochastic Economics*; Academic Press: New York, NY, USA, 1972; ISBN 9781483274027.
8. Buonocore, A.; Caputo, L.; Pirozzi, E.; Nobile, A.G. A Non-Autonomous Stochastic Predator-Prey Model. *Math. Biosci. Eng.* **2014**, *11*, 167–188. [CrossRef] [PubMed]
9. D'Onofrio, G.; Lansky, P.; Pirozzi, E. On two diffusion neuronal models with multiplicative noise: The mean first-passage time properties. *Chaos* **2018**, *28*. [CrossRef]
10. Gutiérrez, R.; Román, P.; Romero, D.; Torres, F. Forecasting for the univariate lognormal diffusion process with exogenous factors. *Cybern. Syst.* **2003**, *34*, 709–724. [CrossRef]
11. Gutiérrez, R.; Rico, N.; Román, P.; Romero, D.; Serrano, J.J.; Torres, F. Lognormal diffusion process with polynomial exogenous factors. *Cybern. Syst.* **2006**, *37*, 293–309. [CrossRef]
12. Land, C.E. Hypothesis tests and interval estimates. In *Lognormal Distributions, Theory and Applications*; Crow, E.L., Shimizu, K., Eds.; Marcel Dekker: New York, NY, USA, 1988; pp. 87–112, ISBN 0-8247-7803-0.
13. Bibby, B.; Jacobsen, M.; Sørensen, M. Estimating functions for discretely sampled diffusion type models. In *Handbook of Financial Econometrics*; Aït-Sahalia, Y., Hansen, L., Eds.; North-Holland: Amsterdam, The Netherlands; 2009; pp. 203–268, ISBN 978-0-444-50897-3.
14. Hansen, L. Large sample properties of generalized method of moments estimators. *Econometrica* **1982**, *50*, 1029–1054. [CrossRef]
15. Fuchs, C. *Inference for Diffusion Processes*; Springer: Heidelberg, Germany, 2013; ISBN 978-3-642-25968-5.
16. Tang, S.; Heron, E. Bayesian inference for a stochastic logistic model with switching points. *Ecol. Model.* **2008**, *219*, 153–169. [CrossRef]
17. Gutiérrez, R.; Román, P.; Romero, D.; Serrano, J.J.; Torres, F. A new gompertz-type diffusion process with application to random growth. *Math. Biosci.* **2007**, *208*, 147–165. [CrossRef] [PubMed]
18. Román-Román, P.; Romero, D.; Torres-Ruiz, F. A diffusion process to model generalized von Bertalanffy growth patterns: Fitting to real data. *J. Theor. Biol.* **2010**, *263*, 59–69. [CrossRef] [PubMed]
19. Román-Román, P.; Torres-Ruiz, F. Modelling logistic growth by a new diffusion process: Application to biological system. *BioSystems* **2012**, *110*, 9–21. [CrossRef] [PubMed]
20. Román-Román, P.; Torres-Ruiz, F. A stochastic model related to the Richards-type growth curve. Estimation by means of Simulated Annealing and Variable Neighborhood Search. *App. Math. Comput.* **2015**, *266*, 579–598. [CrossRef]
21. Román-Román, P.; Torres-Ruiz, F. The nonhomogeneous lognormal diffusion process as a general process to model particular types of growth patterns. In *Lecture Notes of Seminario Interdisciplinare di Matematica*; Università degli Studi della Basilicata: Potenza, Italy, 2015; Volume XII, pp. 201–219.
22. Da Luz Sant'Ana, I.; Román-Román, P.; Torres-Ruiz, F. Modeling oil production and its peak by means of a stochastic diffusion process based on the Hubbert curve. *Energy* **2017**, *133*, 455–470. [CrossRef]
23. Barrera, A.; Román-Román, P.; Torres-Ruiz, F. A hyperbolastic type-I diffusion process: Parameter estimation by means of the firefly algorithm. *Biosystems* **2018**, *163*, 11–22. [CrossRef] [PubMed]
24. Román-Román, P.; Romero, D.; Rubio, M.A.; Torres-Ruiz, F. Estimating the parameters of a Gompertz-type diffusion process by means of simulated annealing. *Appl. Math. Comput.* **2012**, *218*, 5121–5131. [CrossRef]
25. Da Luz Sant'Ana, I.; Román-Román, P.; Torres-Ruiz, F. The Hubbert diffusion process: Estimation via simulated annealing and variable neighborhood search procedures. Application to forecasting peak oil production. *Appl. Stoch. Models Bus.* **2018**. [CrossRef]
26. Gutiérrez, R.; Román, P.; Torres, F. Inference on some parametric functions in the univariate lognormal diffusion process with exogenous factors. *Test* **2001**, *10*, 357-373. [CrossRef]
27. Blasco, A.; Piles, M.; Varona, L. A Bayesian analysis of the effect of selection for growth rate on growth curves in rabbits. *Genet. Sel. Evol.* **2003**, *35*, 21–41. [CrossRef] [PubMed]
28. Gutiérrez-Jáimez, R.; Román, P.; Romero, D.; Serrano, J.J.; Torres, F. Some time random variables related to a Gompertz-type diffusion process. *Cybern. Syst.* **2008**, *39*, 467–479. [CrossRef]

mathematics

MDPI

Article

On the Bounds for a Two-Dimensional Birth-Death Process with Catastrophes

Anna Sinitcina [1,†], Yacov Satin [1,†], Alexander Zeifman [2,*,†], Galina Shilova [1,†], Alexander Sipin [1,†], Ksenia Kiseleva [1,†], Tatyana Panfilova [1,†], Anastasia Kryukova [1,†], Irina Gudkova [3,†] and Elena Fokicheva [1,†]

1 Faculty of Applied Mathematics, Computer Technologies and Physics, Vologda State University, 160000 Vologda, Russia; a_korotysheva@mail.ru (An.S.); yacovi@mail.ru (Y.S.); shgn@mail.ru (G.S.); cac1909@mail.ru (Al.S.); ksushakiseleva@mail.ru (K.K.); ptl-70@mail.ru (T.P.); krukovanastya25@mail.ru (A.K.); eafokicheva2007@yandex.ru (E.F.)

2 Faculty of Applied Mathematics, Computer Technologies and Physics, Vologda State University, Institute of Informatics Problems, Federal Research Center "Computer Science and Control" of the Russian Academy of Sciences, Vologda Research Center of the Russian Academy of SciencesSciences, 160000 Vologda, Russia

3 Applied Probability and Informatics Department, Peoples' Friendship University of Russia (RUDN University), Institute of Informatics Problems, Federal Research Center "Computer Science and Control" of the Russian Academy of Sciences, 117198 Moskva, Russia; gudkova_ia@rudn.university

* Correspondence: a_zeifman@mail.ru

† These authors contributed equally to this work.

Received: 19 April 2018; Accepted: 8 May 2018; Published: 11 May 2018

Abstract: The model of a two-dimensional birth-death process with possible catastrophes is studied. The upper bounds on the rate of convergence in some weighted norms and the corresponding perturbation bounds are obtained. In addition, we consider the detailed description of two examples with 1-periodic intensities and various types of death (service) rates. The bounds on the rate of convergence and the behavior of the corresponding mathematical expectations are obtained for each example.

Keywords: continuous-time Markov chains; catastrophes; bounds; birth-death process; rate of convergence

1. Introduction

There is a large number of papers devoted to the research of continuous-time Markov chains and models with possible catastrophes, see for instance [1–21], and the references therein. Such models are widely used in queueing theory and biology, particularly, for simulations in hight-performance computing. In some recent papers, the authors deal with more or less special birth-death processes with additional transitions from and to origin [9–13,18–20]. In [22], a general class of Markovian queueing models with possible catastrophes is analyzed and some bounds on the rate of convergence are obtained. Here we consider a more specific but important model of a two-dimensional birth-death process with possible catastrophes and obtain the upper bounds on the rate of convergence in some weighted norms and the corresponding perturbation bounds.

Ergodicity bounds in l-1 norm (associated with total variation) for such processes can be obtained quite easily due to the possibility of catastrophes, i.e., transitions to zero from any other state. Obtaining the estimates in weighted norms that guarantee the convergence of the corresponding mathematical expectations as well as the construction of the corresponding limiting characteristics are more complex problems.

In addition, we consider in detail two examples with 1-periodic intensities and various types of death (service) rates. The bounds on the rate of convergence and the behavior of the corresponding mathematical expectations are obtained for each example.

Our results seem to be interesting for both queueing theory and biology applications.

Let $\mathbf{X}(t) = (X_1(t), X_2(t))$ be two-dimensional birth-death-catastrophe process (where $X_i(t)$ is the corresponding number of particles of type i, $i = 1,2$) such that in the interval $(t, t+h)$ the following transitions are possible with order h: birth of a particle of type i, death of a particle of type i, and catastrophe (or transition to the zero state $\mathbf{0} = (0,0)$).

Denote by $\lambda_{1,i,j}(t)$, $\lambda_{2,i,j}(t)$, $\mu_{1,i,j}(t)$, $\mu_{2,i,j}(t)$, and by $\xi_{i,j}(t)$ corresponding birth, death, and catastrophe rates for the process. Namely, $\lambda_{1,i,j}(t)$ is the rate of transition from state (i,j) to state $(i+1,j)$ at the moment t, $\lambda_{2,i,j}(t)$ is the rate of transition from state (i,j) to state $(i,j+1)$, $\mu_{1,i,j}(t)$ is the rate of transition from state (i,j) to state $(i-1,j)$, $\mu_{2,i,j}(t)$ is the rate of transition from state (i,j) to state $(i,j-1)$, and finally, $\xi_{i,j}(t)$ is the rate of transition from state (i,j) to state $(0,0)$ at the moment t.

The transition rate diagram associated with the process is presented in Figure 1.

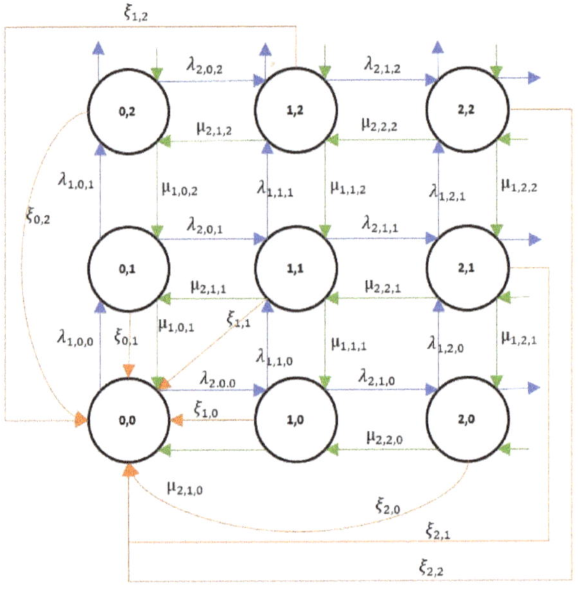

Figure 1. Transition rate diagram.

Suppose that all intensities are nonnegative and locally integrable on $[0;\infty)$ as functions of t. Moreover, we also suppose that the condition of boundedness

$$\lambda_{1,i,j}(t) + \lambda_{2,i,j}(t) + \mu_{1,i,j}(t) + \mu_{2,i,j}(t) + \xi_{i,j}(t) \leq L < \infty, \tag{1}$$

hold for any i,j and almost all $t \geq 0$.

We renumber the states of two-dimensional process $\mathbf{X}(t) = (X_1(t), X_2(t))$ (0,0), (0,1), (1,0), (0,2), (1,1), (2,0), ... by increasing the sum of coordinates, and in the case of the same sum, by increasing the first coordinate. Hence we obtain one-dimensional vector $\mathbf{p}(t) = (p_0(t), p_1(t), \ldots)^T$ of state probabilities in a new numeration, and therefore, we can rewrite the forward Kolmogorov system in the following form:

$$\frac{d\mathbf{p}}{dt} = \mathbf{A}\,(t)\,\mathbf{p}, \quad t \geq 0, \tag{2}$$

where $\mathbf{A}(t) = \left(a_{ij}(t)\right)$ is the corresponding transposed intensity matrix:

$$\mathbf{A}(t) = \begin{pmatrix}
a_{00} & \mu_{1,0,1}+\zeta_{0,1} & \mu_{2,1,0}+\xi_{1,0} & \zeta_{0,2} & \xi_{1,1} & \xi_{2,0} & \zeta_{0,3} & \xi_{1,2} & \xi_{2,1} & \xi_{3,0} & \cdots \\
\lambda_{1,0,0} & a_{11} & 0 & \mu_{1,0,2} & \mu_{2,1,1} & 0 & 0 & 0 & 0 & 0 & \cdots \\
\lambda_{2,0,0} & 0 & a_{22} & 0 & \mu_{1,1,1} & \mu_{2,2,0} & 0 & 0 & 0 & 0 & \cdots \\
0 & \lambda_{1,0,1} & 0 & a_{33} & 0 & 0 & \mu_{1,0,3} & \mu_{2,1,2} & 0 & 0 & \cdots \\
0 & \lambda_{2,0,1} & \lambda_{1,1,0} & 0 & a_{44} & 0 & 0 & \mu_{1,1,2} & \mu_{2,2,1} & 0 & \cdots \\
0 & 0 & \lambda_{2,1,0} & 0 & 0 & a_{55} & 0 & 0 & \mu_{1,2,1} & \mu_{2,3,0} & \cdots \\
0 & 0 & 0 & \lambda_{1,0,2} & 0 & 0 & a_{66} & 0 & 0 & 0 & \cdots \\
\cdots & \cdots & \cdots & \cdots & \cdots & \cdots & \cdots & \cdots & \cdots & &
\end{pmatrix},$$

and $a_{ii}(t) = -\sum_i a_{i,j}(t)$.

Throughout the paper by $\|\cdot\|$ we denote the l_1-norm, i. e., $\|\mathbf{x}\| = \sum |x_i|$, and $\|B\| = \sup_j \sum_i |b_{ij}|$ for $B = (b_{ij})_{i,j=0}^{\infty}$.

Let Ω be a set all stochastic vectors, i.e., l_1 vectors are with nonnegative coordinates and unit norm. Hence the assumption (1) implies the bound $\|\mathbf{A}(t)\| \leq 2L$ for almost all $t \geq 0$. Therefore, the operator function $\mathbf{A}(t)$ from l_1 into itself is bounded for almost all $t \geq 0$ and locally integrable on $[0;\infty)$. Therefore, we can consider the forward Kolmogorov system as a differential equation in the space l_1 with bounded operator.

It is well known, see [23], that the Cauchy problem for such a differential equation has a unique solution for an arbitrary initial condition, and $\mathbf{p}(s) \in \Omega$ implies $\mathbf{p}(t) \in \Omega$ for $t \geq s \geq 0$.

We have

$$\mathbf{p}(t) = U(t,s)\mathbf{p}(s), \tag{3}$$

where $U(t,s)$ is the Cauchy operator of Equation (2).

Note that the vector of state probabilities can be written in 'two-dimensional form' as $\mathbf{p}(t) = (p_{00}(t), p_{01}(t), p_{10}(t), p_{02}(t), p_{11}(t), \dots)^T$.

2. Bounds in l_1 Norm

Consider the first equation in forward Kolmogorov system and rewrite it in the following form:

$$\frac{dp_0}{dt} = -\left(a_{00} + \xi(t)\right)p_0 + \sum_{i\geq 1}\left(a_{0i}(t) - \xi(t)\right)p_i + \xi(t), \tag{4}$$

where $\xi(t) = \inf_{i,j} \zeta_{i,j}(t)$.

Then we have from Equation (2) the following system:

$$\frac{d\mathbf{p}}{dt} = B(t)\mathbf{p} + \mathbf{f}(t), \tag{5}$$

where $\mathbf{f}(t) = (\xi(t), 0, \dots)^T$ and

$$B(t) = \begin{pmatrix}
a_{00}-\xi & \mu_{1,0,1}+\zeta_{0,1}-\xi & \mu_{2,1,0}+\xi_{1,0}-\xi & \zeta_{0,2}-\xi & \xi_{1,1}-\xi & \xi_{2,0}-\xi & \zeta_{0,3}-\xi & \xi_{1,2}-\xi & \xi_{2,1}-\xi & \xi_{3,0}-\xi & \cdots \\
\lambda_{1,0,0} & a_{11} & 0 & \mu_{1,0,2} & \mu_{2,1,1} & 0 & 0 & 0 & 0 & 0 & \cdots \\
\lambda_{2,0,0} & 0 & a_{22} & 0 & \mu_{1,1,1} & \mu_{2,2,0} & 0 & 0 & 0 & 0 & \cdots \\
0 & \lambda_{1,0,1} & 0 & a_{33} & 0 & 0 & \mu_{1,0,3} & \mu_{2,1,2} & 0 & 0 & \cdots \\
0 & \lambda_{2,0,1} & \lambda_{1,1,0} & 0 & a_{44} & 0 & 0 & \mu_{1,1,2} & \mu_{2,2,1} & 0 & \cdots \\
0 & 0 & \lambda_{2,1,0} & 0 & 0 & a_{55} & 0 & 0 & \mu_{1,2,1} & \mu_{2,3,0} & \cdots \\
0 & 0 & 0 & \lambda_{1,0,2} & 0 & 0 & a_{66} & 0 & 0 & 0 & \cdots \\
\cdots & \cdots & \cdots & \cdots & \cdots & \cdots & \cdots & \cdots & \cdots & &
\end{pmatrix}.$$

We have

$$\mathbf{p}(t) = V(t)\mathbf{p}(0) + \int_0^t V(t,\tau)\mathbf{f}(\tau)\,d\tau, \tag{6}$$

where $V(t,\tau)$ is the Cauchy operator of Equation (5).

Then one can estimate the logarithmic norm of $B(t)$ in the space of sequences l_1 (see [24]):

$$\gamma\left(B(t)\right)_1 = \max\left(a_{00}(t) - \xi(t) + \sum_{i\geq 1} a_{i0}(t),\right.$$
$$\left.\sup_{i\geq 1}\left(a_{ii}(t) + a_{0i}(t) - \xi(t) + \sum_{j\neq i, j\geq 1} a_{ji}(t)\right)\right) = -\xi(t). \tag{7}$$

Then for all $0 \leq s \leq t$ we have

$$\|V(t,s)\| \leq e^{-\int\limits_s^t \xi(\tau)\,d\tau}. \tag{8}$$

Therefore, the following statement is correct (see details in [16,18]).

Theorem 1. *Let the intensities of catastrophes be essential, that is*

$$\int_0^\infty \xi(t)\,dt = \infty. \tag{9}$$

Then the process $\mathbf{X}(t)$ *is weakly ergodic and the following bound of the rate of convergence holds:*

$$\|\mathbf{p}^*(t) - \mathbf{p}^{**}(t)\| \leq 2e^{-\int\limits_0^t \xi(\tau)\,d\tau}, \tag{10}$$

for all initial conditions $\mathbf{p}^*(0), \mathbf{p}^{**}(0)$ *and any* $t \geq 0$.

Consider now the "perturbed" process $\bar{\mathbf{X}} = \bar{\mathbf{X}}(t),\ t \geq 0$, adding a dash on top for all corresponding characteristics.

Put $\hat{\mathbf{A}}(t) = \mathbf{A}(t) - \bar{\mathbf{A}}(t)$, and assume that the perturbations are "uniformly small", i.e., for almost all $t \geq 0$ the following inequality is correct

$$\|\hat{\mathbf{A}}(t)\| \leq \varepsilon. \tag{11}$$

Consider the stability bounds of the process $\mathbf{X}(t)$ under these perturbations. In addition, we assume that the process is *exponentially* ergodic, that is, that for some positive M, a and for all s, t, $0 \leq s \leq t$ the following inequality holds

$$e^{-\int_s^t \xi(u)\,du} \leq Me^{-a(t-s)}. \tag{12}$$

Then from Theorem 1:

$$\|\mathbf{p}^*(t) - \mathbf{p}^{**}(t)\| \leq e^{-\int\limits_s^t \xi(\tau)\,d\tau}\|\mathbf{p}^*(s) - \mathbf{p}^{**}(s)\| \leq 2e^{-\int\limits_s^t \xi(\tau)\,d\tau} \leq 2Me^{-a(t-s)}. \tag{13}$$

Here we apply the approach proposed in [25] for a stationary case and generalized for a nonstationary situation in [15,16].

We have

$$\|\mathbf{p}(t) - \bar{\mathbf{p}}(t)\| \leq Me^{-a(t-s)}\|\mathbf{p}(s) - \bar{\mathbf{p}}(s)\| + M\int_s^t \|\hat{A}(u)\|e^{-a(u-s)}du. \tag{14}$$

Therefore we obtain

$$\|\mathbf{p}(t) - \bar{\mathbf{p}}(t)\| \leq \begin{cases} \|\mathbf{p}(s) - \bar{\mathbf{p}}(s)\| + (t-s)\varepsilon, & 0 < t < a^{-1}\log M, \\ a^{-1}(\log M + 1 - Me^{-a(t-s)})\varepsilon + Me^{-a(t-s)}\|\mathbf{p}(s) - \bar{\mathbf{p}}(s)\|, & t \geq a^{-1}\log M. \end{cases} \quad (15)$$

It implies the following statement.

Theorem 2. *If the condition (12) is fulfilled and the perturbations are uniformly small:*

$$\|\hat{A}(t)\| \leq \varepsilon, \quad (16)$$

for almost all $t \geq 0$. Then the following bound holds:

$$\limsup_{t\to\infty} \|\mathbf{p}(t) - \bar{\mathbf{p}}(t)\| \leq \frac{\varepsilon\,(1 + \log M)}{a}, \quad (17)$$

for any initial conditions $\mathbf{p}(0)$, $\bar{\mathbf{p}}(0)$.

Corollary 1. *Let the intensities of the process be 1-periodic and instead of (12) we have the following inequality*

$$\int_0^1 \xi(t)dt \geq \theta > 0. \quad (18)$$

Then (17) is correct for

$$M = e^K, \quad a = \theta, \quad (19)$$

where $K = \sup_{|t-s|\leq 1} \int_s^t \xi(\tau)d\tau < \infty$.

3. Bounds in Weighted Norms

Consider the diagonal matrix $D = diag(d_0, d_1, d_2, d_3, \cdots)$, with entries of the increasing sequence $\{d_n\}$, where $d_0 = 1$, and the corresponding space of sequences $l_{1D} = \{\mathbf{z} = (p_0, p_1, p_2, \ldots)^T\}$ such that $\|\mathbf{z}\|_{1D} = \|D\mathbf{z}\|_1 < \infty$.

Then one can estimate the logarithmic norm of operator $B(t)$ in l_{1D} space.

According to the general approach, we obtain the matrix

$$DB(t)D^{-1} = \begin{pmatrix} a_{00} - \xi & (\mu_{1,0,1} + \xi_{0,1} - \xi)\frac{d_0}{d_1} & (\mu_{2,1,0} + \xi_{1,0} - \xi)\frac{d_0}{d_2} & (\xi_{0,2} - \xi)\frac{d_0}{d_3} & (\xi_{1,1} - \xi)\frac{d_0}{d_4} & (\xi_{2,0} - \xi)\frac{d_0}{d_5} & (\xi_{0,3} - \xi)\frac{d_0}{d_6} & (\xi_{1,2} - \xi)\frac{d_0}{d_7} & \cdots \\ \lambda_{1,0,0}\frac{d_1}{d_0} & a_{11} & 0 & \mu_{1,0,2}\frac{d_1}{d_3} & \mu_{2,1,1}\frac{d_1}{d_4} & 0 & 0 & 0 & \cdots \\ \lambda_{2,0,0}\frac{d_2}{d_0} & 0 & a_{22} & 0 & \mu_{1,1,1}\frac{d_2}{d_4} & \mu_{2,2,0}\frac{d_2}{d_5} & 0 & 0 & \cdots \\ 0 & \lambda_{1,0,1}\frac{d_3}{d_1} & 0 & a_{33} & 0 & 0 & \mu_{1,0,3}\frac{d_3}{d_6} & \mu_{2,1,2}\frac{d_3}{d_7} & \cdots \\ 0 & \lambda_{2,0,1}\frac{d_4}{d_1} & \lambda_{1,1,0}\frac{d_4}{d_2} & 0 & a_{44} & 0 & 0 & \mu_{1,1,2}\frac{d_4}{d_7} & \cdots \\ 0 & 0 & \lambda_{2,1,0}\frac{d_5}{d_2} & 0 & 0 & a_{55} & 0 & 0 & \cdots \\ 0 & 0 & 0 & \lambda_{1,0,2}\frac{d_6}{d_3} & 0 & 0 & a_{66} & 0 & \cdots \\ \cdots & \cdots & \cdots & \cdots & \cdots & \cdots & \cdots & & \end{pmatrix},$$

where $a_{ii}(t) = -\sum_i a_{i,j}(t)$.

Consider now the logarithmic norm

$$\gamma\,(B(t))_{1D} = \gamma\left(DB(t)D^{-1}\right)_1. \quad (20)$$

Let us make the correspondence between the column number of matrix $DB(t)D^{-1}$ and the number of zeros under the main diagonal in this column (till the first nonzero element). Then we obtain the arithmetic progression $\{a_i\}$:

a_1	a_2	a_3	a_4	a_5	a_6	a_7	a_8	a_9	a_{10}	a_{11}	a_{12}	a_{13}	a_{14}	a_{15}	a_{16}	a_{17}	a_{18}	a_{19}	a_{20}	a_{21}	...
1	2	3	4	5	6	7	8	9	10	11	12	13	14	15	16	17	18	19	20	21	...
1	1	2	2	2	3	3	3	3	4	4	4	4	4	5	5	5	5	5	5	6	...

We compose the sequence $\{b_i\}$ of the number of identical entries of the third line: $2, 3, 4, 5, 6, \cdots$. Note that $\sum_{i=1}^{N} b_i$ is equal to the last a_k, corresponding to the number of zeros N in the k-th column.

Then the sum of the first N elements of sequence $\{b_i\}$ is approximately equal to the number of the column $a_n = n$:

$$(2b_1 + (N-1))\,N \approx 2a_n = 2n.$$

Knowing the column number n, one can find the formula for the number of zeros N under the main diagonal in this column till the first nonzero element. We note that the number of zeros over the diagonal till $\mu_{1,i,j}$ is one less.

If N is not an integer, we must take the nearest right to N an integer.

One can see that columns $2, 5, 9, 14, \ldots$ (these columns correspond to sums $\sum_{i=1}^{j} b_i$ and integer N) contain death rates $\mu_{2,i,j}(t)$ only, and columns $3, 6, 10, 15, \ldots$ contain death rates $\mu_{1,i,j}(t)$ only, and all other columns contain the both death intensities.

Consider the following quantities:

for $n = 0$

$$\alpha_n(t) = \lambda_{1,0,0}(t) + \lambda_{2,0,0}(t) + \xi(t) - \lambda_{1,0,0}(t)\tfrac{d_1}{d_0} - \lambda_{2,0,0}(t)\tfrac{d_2}{d_0},$$

for $n = 1$

$$\alpha_n(t) = \lambda_{1,0,1}(t) + \lambda_{2,0,1}(t) + \mu_{1,0,1}(t) + \xi_1(t) - \lambda_{1,0,1}(t)\tfrac{d_3}{d_1} - \lambda_{2,0,1}(t)\tfrac{d_4}{d_1} - (\mu_{1,0,1}(t) + \xi_1(t) - \xi(t))\tfrac{d_0}{d_1},$$

for $n = 2$

$$\alpha_n(t) = \lambda_{1,1,0}(t) + \lambda_{2,1,0}(t) + \mu_{2,1,0}(t) + \xi_2(t) - \lambda_{1,1,0}(t)\tfrac{d_4}{d_2} - \lambda_{2,1,0}(t)\tfrac{d_5}{d_2} - (\mu_{2,1,0}(t) + \xi_2(t) - \xi(t))\tfrac{d_0}{d_2},$$

for integer $\frac{-3+\sqrt{9+8n}}{2}$:

$$\begin{aligned}\alpha_n(t) &= \lambda_{1,s^{-1}(n)}(t) + \lambda_{2,s^{-1}(n)}(t) + \mu_{2,s^{-1}(n)}(t) + \xi_{s(i,j)}(t) - \lambda_{1,s^{-1}(n)}(t)\tfrac{d_{n+N+1}}{d_n} - \lambda_{2,s^{-1}(n)}(t)\tfrac{d_{n+N+2}}{d_n}\\ &-\mu_{2,s^{-1}(n)}(t)\tfrac{d_{n-N-1}}{d_n} - (\xi_{s(i,j)}(t) - \xi(t))\tfrac{d_0}{d_n}, \quad N = \tfrac{-3+\sqrt{9+8n}}{2},\end{aligned}$$

for integer $\frac{-3+\sqrt{9+8(n-1)}}{2}$:

$$\begin{aligned}\alpha_n(t) &= \lambda_{1,s^{-1}(n)}(t) + \lambda_{2,s^{-1}(n)}(t) + \mu_{1,s^{-1}(n)}(t) + \xi_{s(i,j)}(t) - \lambda_{1,s^{-1}(n)}(t)\tfrac{d_{n+N+1}}{d_n} - \lambda_{2,s^{-1}(n)}(t)\tfrac{d_{n+N+2}}{d_n}\\ &-\mu_{1,s^{-1}(n)}(t)\tfrac{d_{n-N}}{d_n} - (\xi_{s(i,j)}(t) - \xi(t))\tfrac{d_0}{d_n}, \quad N = \left\lceil \tfrac{-3+\sqrt{9+8n}}{2} \right\rceil,\end{aligned}$$

in other cases:

$$\begin{aligned}\alpha_n(t) &= \lambda_{1,s^{-1}(n)}(t) + \lambda_{2,s^{-1}(n)}(t) + \mu_{1,s^{-1}(n)}(t) + \mu_{2,s^{-1}(n)}(t) + \xi_{s(i,j)}(t) - \lambda_{1,s^{-1}(n)}(t)\tfrac{d_{n+N+1}}{d_n}\\ &-\lambda_{2,s^{-1}(n)}(t)\tfrac{d_{n+N+2}}{d_n} - \mu_{1,s^{-1}(n)}(t)\tfrac{d_{n-N}}{d_n} - \mu_{2,s^{-1}(n)}(t)\tfrac{d_{n-N-1}}{d_n}\\ &-(\xi_{s(i,j)}(t) - \xi(t))\tfrac{d_0}{d_n}, \quad N = \left\lceil \tfrac{-3+\sqrt{9+8n}}{2} \right\rceil.\end{aligned}$$

Then the following algorithm helps us to correlate the number n and pair (i,j):

(1) $n_1 = n - 1$,

(2) $n_2 = n_1 - 2$,
...
(k) $n_k = n_{k-1} - k$, while $n_k > 0$,
(k+1) $i = n_k, j = k - n_k$.

We obtain

$$\gamma\left(B(t)\right)_{1D} = -\inf_t \alpha_n(t) = -\alpha(t). \tag{21}$$

Therefore, for all $0 \le s \le t$ we have the bound for the corresponding Cauchy operator:

$$\|V(t,s)\|_{1D} \le e^{-\int\limits_s^t \alpha(\tau)\, d\tau}, \tag{22}$$

and the following statement.

Theorem 3. *Let for some sequence $\{d_i\}$ we have the condition*

$$\int_0^\infty \alpha(t)\, dt = \infty. \tag{23}$$

Then the process $\mathbf{X}(t)$ is weakly ergodic and the following bound of the rate of convergence is correct:

$$\|\mathbf{p}^*(t) - \mathbf{p}^{**}(t)\|_{1D} \le e^{-\int\limits_0^t \alpha(\tau)\, d\tau} \|\mathbf{p}^*(0) - \mathbf{p}^{**}(0)\|_{1D}, \tag{24}$$

for any initial conditions $\mathbf{p}^*(0), \mathbf{p}^{**}(0)$ and for all $t \ge 0$.

Mathematical expectations for both processes $X_1(t)$ and $X_2(t)$ can be obtained using formulas:

$$\begin{aligned} E_1(t) &= 1(p_2 + p_4 + p_7 + \ldots) + 2(p_5 + p_8 + p_{12} + \cdots) + \ldots \\ &= 1(p_{10} + p_{11} + p_{12} + \ldots) + 2(p_{20} + p_{21} + p_{22} + \cdots) + \ldots \end{aligned} \tag{25}$$

and

$$\begin{aligned} E_2(t) &= 1(p_1 + p_4 + p_8 + \ldots) + 2(p_3 + p_7 + p_{12} + \cdots) + \ldots \\ &= 1(p_{01} + p_{11} + p_{21} + \ldots) + 2(p_{02} + p_{12} + p_{22} + \cdots) + \ldots. \end{aligned} \tag{26}$$

Let us now introduce a process $N(t) = |\mathbf{X}(t)| = X_1(t) + X_2(t)$, that is the number of all particles at the moment t.

Then one has for the mathematical expectation (the mean) of this process the following equality:

$$\begin{aligned} E_N(t) &= 1(p_{01} + p_{10}) + 2(p_{02} + p_{11} + p_{20}) + 3(p_{03} + p_{12} + p_{21} + p_{30}) + \cdots) + \ldots \\ &= 1(p_1 + p_2) + 2(p_3 + p_4 + p_5) + 3(p_6 + p_7 + p_8 + p_9) + \ldots = E_1(t) + E_2(t). \end{aligned} \tag{27}$$

We note that for $W = \inf_{i \ge 0} \frac{d_i}{i}$ the next inequality holds

$$E_N(t) = 1(p_1 + p_2) + 2(p_3 + p_4 + p_5) + 3(p_6 + p_7 + p_8 + p_9) + \ldots \le \sum_{i \ge 1} i p_i \le \frac{1}{W} \sum_{i \ge 1} d_i p_i = \frac{\|p(t)\|_{1D}}{W}.$$

Denote by $E_N(t,k) = E(|X(t)|/|X(0)| = k)$ the conditional expected number of all particles in the system at instant t, provided that initially (at instant $t = 0$) k particles of both types were present in the system.

Corollary 2. *Let the condition (23) hold and there is a sequence $\{d_i\}$ such that $W > 0$, then the process $N(t)$ has the limiting mean $\phi_N(t) = E_N(t,0)$, and for any j and all $t \geq 0$ the following bound of the rate of convergence is correct:*

$$|E_N(t,j) - E_N(t,0)| \leq \frac{1+d_j}{W} e^{-\int\limits_0^t \alpha(\tau)\,d\tau}. \tag{28}$$

Applying in addition the condition that the perturbations of the intensity matrix are small enough in the corresponding norm, that is $\|\hat{A}(t)\|_{1D} \leq \varepsilon$, one can also obtain $\|\hat{B}(t)\|_{1D} \leq \varepsilon$.

We assume here that the process $\mathbf{X}(t)$ is exponentially ergodic in l_{1D}-norm, that is for some positive M_1, a_1 and for all $s, t, 0 \leq s \leq t$ the following inequality holds:

$$e^{-\int_s^t \alpha(u)\,du} \leq M_1 e^{-a_1(t-s)}. \tag{29}$$

Here we apply the approach from [18].

One can rewrite the original system for the unperturbed process in the form:

$$\frac{d\mathbf{p}}{dt} = \bar{\mathbf{B}}(t)\mathbf{p}(t) + \bar{\mathbf{f}}(t) + \hat{\mathbf{B}}(t)\mathbf{p}(t) + \hat{\mathbf{f}}(t). \tag{30}$$

Then

$$\mathbf{p}(t) = \bar{V}(t,0)\mathbf{p}(0) + \int_0^t \bar{V}(t,\tau)\bar{\mathbf{f}}(\tau)\,d\tau + \int_0^t \bar{V}(t,\tau)\,\hat{B}(\tau)p(\tau) \tag{31}$$

and

$$\bar{\mathbf{p}}(t) = \bar{V}(t,0)\bar{\mathbf{p}}(0) + \int_0^t \bar{V}(t,\tau)\bar{\mathbf{f}}(\tau)\,d\tau. \tag{32}$$

Therefore, in *any* norm for any initial conditions we have the correct bound:

$$\|\mathbf{p}(t) - \bar{\mathbf{p}}(t)\| \leq \int_0^t \|\bar{V}(t,\tau)\| \left(\|\hat{B}(\tau)\|\|\mathbf{p}(\tau)\| + \|\hat{\mathbf{f}}(\tau)\|\right) d\tau. \tag{33}$$

Then we have the following inequality for the logarithmic norm:

$$\gamma(\bar{B}(t))_{1D} \leq \gamma(DB(t)D^{-1})_1 + \|\hat{B}(t)\|_{1D} \leq -\alpha(t) + \varepsilon. \tag{34}$$

On the other hand, one can obtain the estimation using inequality (29):

$$\|\mathbf{p}(t)\|_{1D} \leq \|V(t)\mathbf{p}(0)\|_{1D} + \int\limits_0^t \|V(t,\tau)\mathbf{f}(\tau)\,d\tau\|_{1D} \leq$$
$$\leq M_1 e^{-a_1 t}\|\mathbf{p}(0)\|_{1D} + \frac{LM_1}{a_1} \tag{35}$$

for any initial condition $\mathbf{p}(0)$. Moreover, $\|\hat{\mathbf{f}}(\tau)\|_{1D} \leq \varepsilon$.

Then using bound (33), we have

$$\|\mathbf{p}(t) - \bar{\mathbf{p}}(t)\|_{1D} \leq \int_0^t e^{-\int_\tau^t (\alpha(u)-\varepsilon)du} \left(\varepsilon\left(M_1 e^{-a_1\tau}\|\mathbf{p}(0)\|_{1D} + \frac{LM_1}{a_1}\right) + \varepsilon\right) d\tau \leq$$
$$\leq o(1) + \frac{\varepsilon M_1(1 + LM_1/a_1)}{a_1 - \varepsilon}.$$

Therefore, the following statement is correct.

Theorem 4. *Let inequalities (12) and (29) hold for any initial condition* $\mathbf{p}(0) \in l_{1D}$ *and for all* $t \geq 0$, *then we have*

$$\|\mathbf{p}(t) - \bar{\mathbf{p}}(t)\|_{1D} \leq M_1\varepsilon \left(\frac{LM_1 + a_1}{a_1(a_1 - \varepsilon)} + M_1 t e^{-(a_1-\varepsilon)t} \|\mathbf{p}(0)\|_{1D} \right), \tag{36}$$

and

$$\limsup_{t\to\infty} \|\mathbf{p}(t) - \bar{\mathbf{p}}(t)\|_{1D} \leq \frac{M_1\varepsilon(LM_1 + a_1)}{a_1(a_1 - \varepsilon)}. \tag{37}$$

Corollary 3. *Let in addition the sequence be increasing fast enough, such that* $W > 0$, *then for any* $j, t \geq 0$ *we have*

$$\limsup_{t\to\infty} |E_N(t) - \bar{E}_N(t)| \leq \frac{M_1\varepsilon(LM_1 + a_1)}{W a_1(a_1 - \varepsilon)}. \tag{38}$$

4. Examples

Example 1. *Let* $\lambda_{1,i,j}(t) = \lambda_{2,i,j}(t) = 2 + \sin 2\pi t$, $i, j \geq 0$, $\mu_{2,0,1}(t) = 1 + \cos 2\pi t$, $\mu_{1,1,0}(t) = 2(1 + \cos 2\pi t)$, *and other* $\mu_{1,i,j}(t) = \mu_{2,i,j}(t) = 3(1 + \cos 2\pi t)$, *and let catastrophe intensities be* $\xi_{i,j}(t) = 5$. *Put* $\varepsilon = 10^{-6}$.

Choose $d_n = 1 + \frac{n}{10}$, then $\alpha(t) = \alpha_0(t) - \frac{2}{3} = \frac{56}{10} - \frac{3}{10}\sin 2\pi t$, $a = 5, M = 1, a_1 = 3.8, M_1 = 1.1$, W = 1/10.

We obtain now the following bounds

$$|E_N(t,1) - E_N(t,0)| \leq 3.4 \cdot 10^{-4}, \; t \geq 4, \tag{39}$$

$$\limsup_{t\to\infty} \|\mathbf{p}(t) - \bar{\mathbf{p}}(t)\| \leq 2 \cdot 10^{-4}, \tag{40}$$

$$\limsup_{t\to\infty} |E_N(t,0) - \bar{E}_N(t,0)| \leq 0.023. \tag{41}$$

The values of $\alpha_n(t)$ are shown in Figure 2. The mean for the process $N(t)$ on the interval $t \in [0, 3]$ for different initial conditions are shown in Figures 3–6, and the bounds for the limiting perturbed mean is shown in Figure 7.

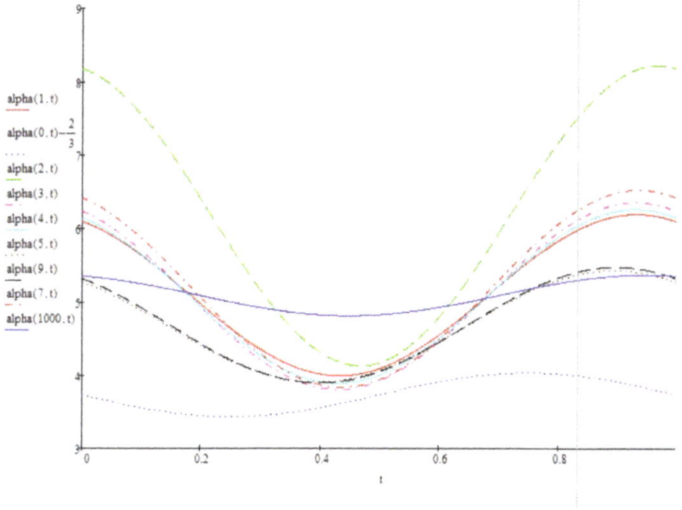

Figure 2. The values of several $\alpha_n(t)$ for Example 1.

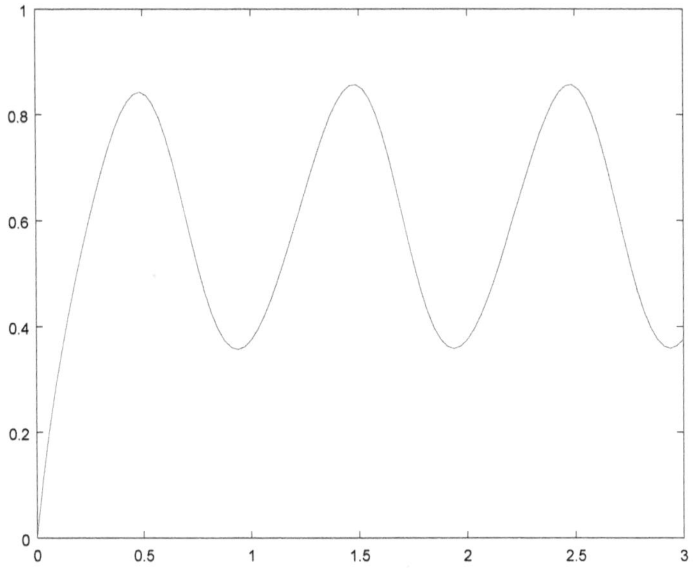

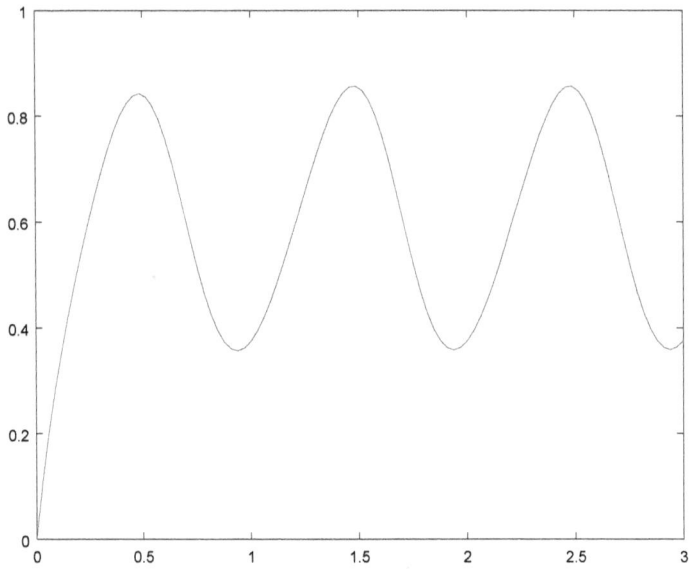

Figure 3. The mean $E_N(t, 0)$ on the interval $t \in [0, 3]$ with initial condition $\mathbf{X}(t) = (0, 0)$ for Example 1.

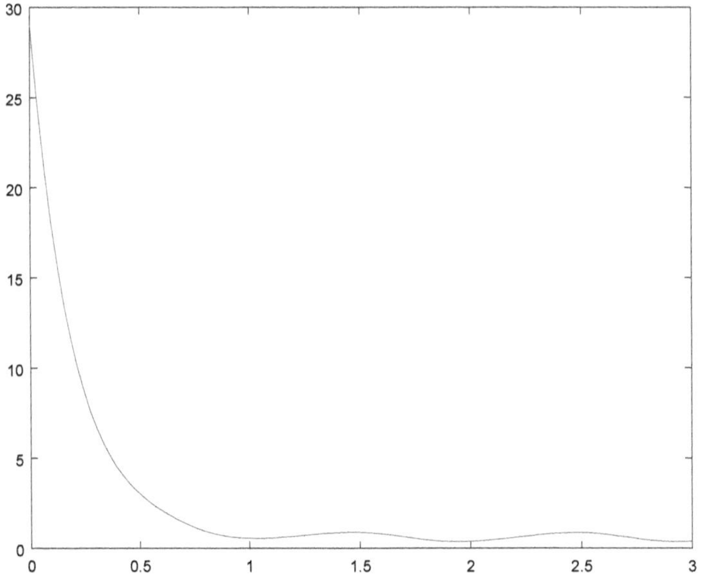

Figure 4. The mean $E_N(t, 29)$ on the interval $t \in [0, 3]$ with initial condition $\mathbf{X}(t) = (0, 29)$ for Example 1.

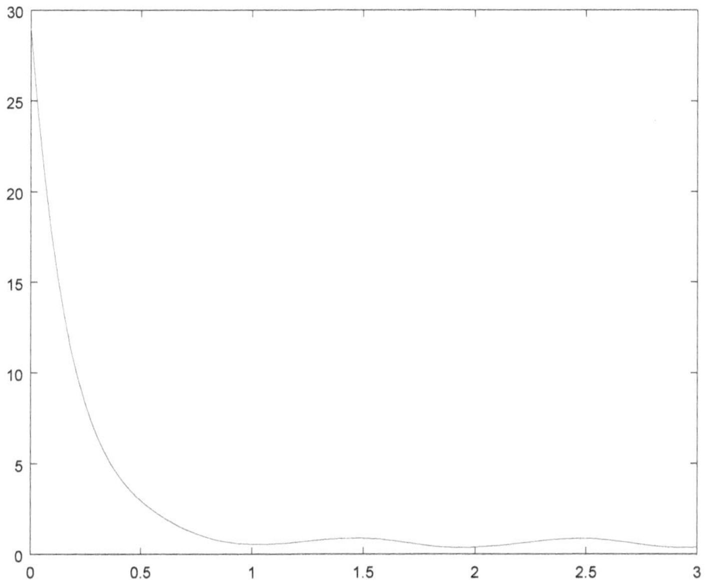

Figure 5. The mean $E_N(t, 29)$ on the interval $t \in [0, 3]$ with initial condition $\mathbf{X}(t) = (14, 15)$ for Example 1.

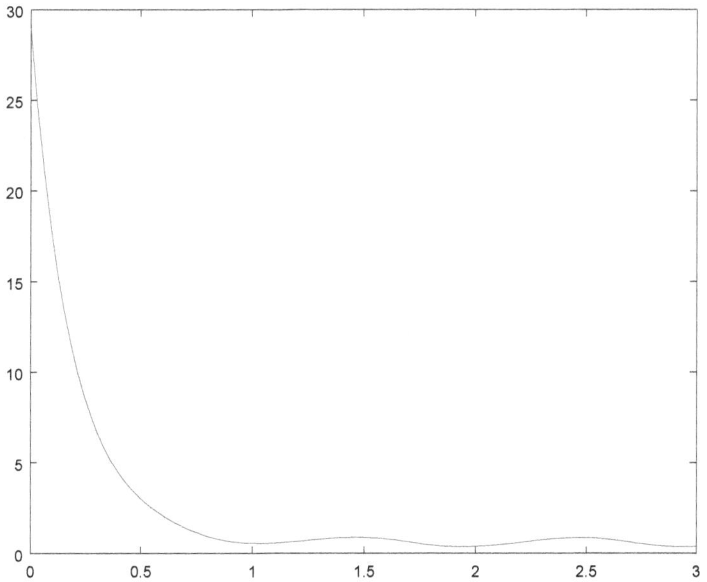

Figure 6. The mean $E_N(t, 29)$ on the interval $t \in [0, 3]$ with initial condition $\mathbf{X}(t) = (29, 0)$ for Example 1.

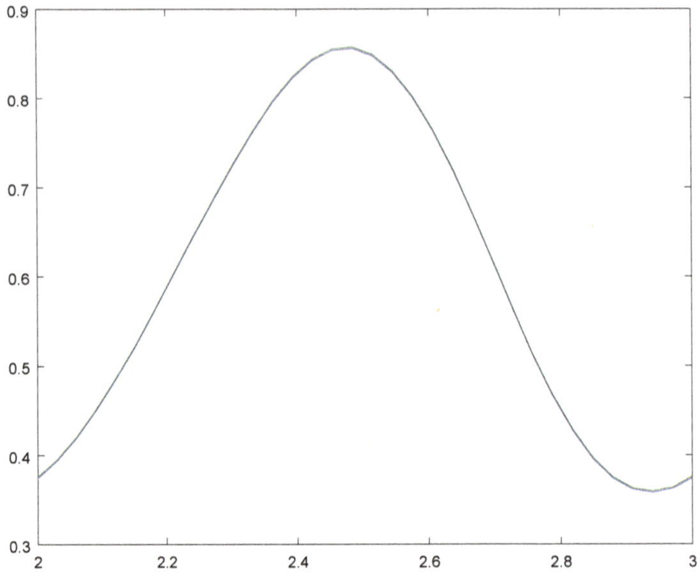

Figure 7. The limiting perturbed mean for $t \in [2,3]$ for Example 1.

Example 2. *Let now* $\lambda_{1,i,j}(t) = \lambda_{2,i,j}(t) = 2 + \cos 2\pi t$, $i, j \geq 0$, $\mu_{1,i,j}(t) = \min(1 + i \cdot j, 3)(1 + \cos 2\pi t)$, $i \geq 1, j \geq 0$, $\mu_{2,i,j}(t) = \min(1 + i \cdot j, 3)(1 + \cos 2\pi t)$, $i \geq 0, j \geq 1$, *and let the catastrophe rates be* $\xi_{i,j}(t) = 5$. *Let* $\varepsilon = 10^{-3}$.

Put $d_n = 1 + \frac{n}{10}$, then $\alpha(t) = \alpha_9(t) - \frac{1}{7} = \frac{548}{133} + \frac{4}{19} \sin 2\pi t - \frac{9}{19} \cos 2\pi t$, $a = 5, M = 1, a_1 = 4.12$, $M_1 = 1.2, W = 1/10$.

Then we obtain

$$|E_N(t,1) - E_N(t,0)| \leq 9.4 \cdot 10^{-4}, \ t \geq 2, \tag{42}$$

$$\limsup_{t \to \infty} \|\mathbf{p}(t) - \bar{\mathbf{p}}(t)\| \leq 2 \cdot 10^{-4}, \tag{43}$$

$$\limsup_{t \to \infty} |E_N(t,0) - \bar{E}_N(t,0)| \leq 0.021. \tag{44}$$

The values of $\alpha_n(t)$ are shown in Figure 8.

The mean for the process $N(t)$ on the interval $t \in [0,3]$ for different initial conditions are shown in Figures 9–12, and the bounds for the limiting perturbed mean is shown in Figure 13.

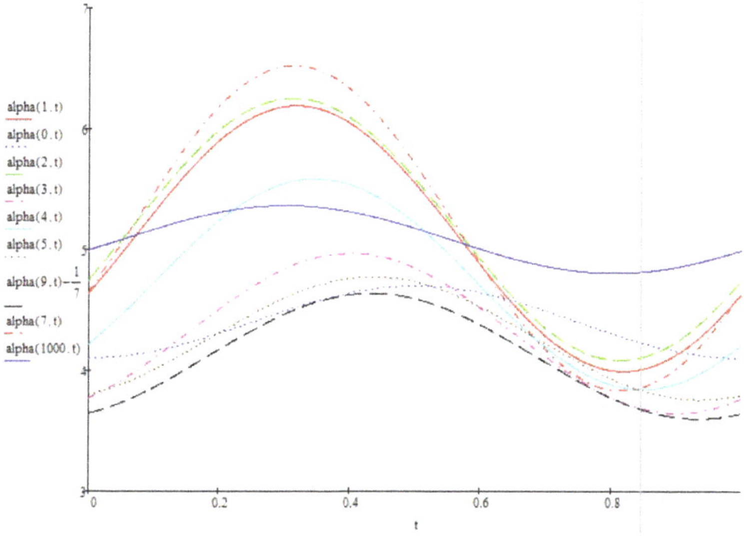

Figure 8. The values of several $\alpha_n(t)$ for Example 2.

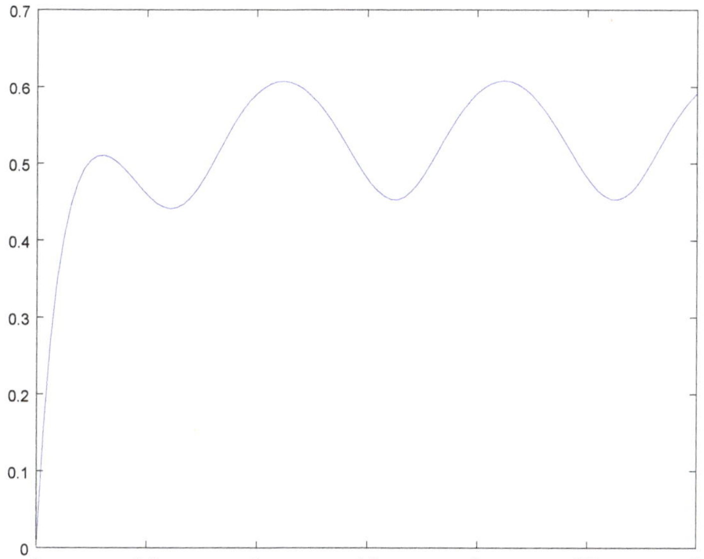

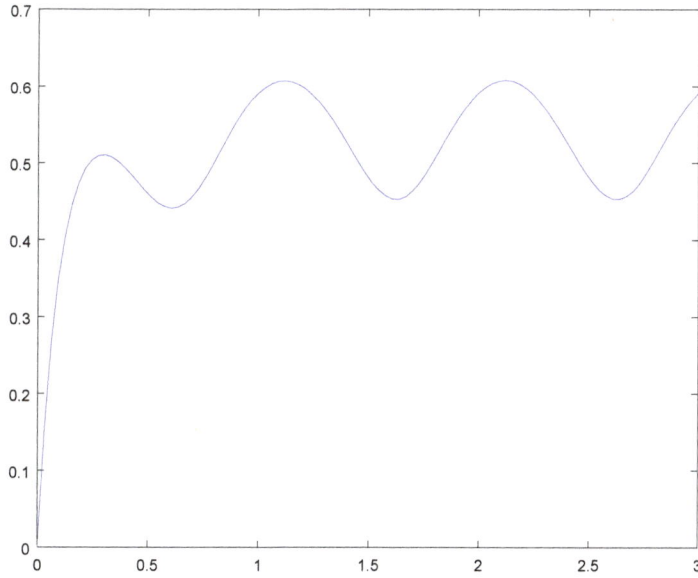

Figure 9. The mean $E_N(t, 0)$ on the interval $t \in [0, 3]$ with initial condition $\mathbf{X}(t) = (0, 0)$ for Example 2.

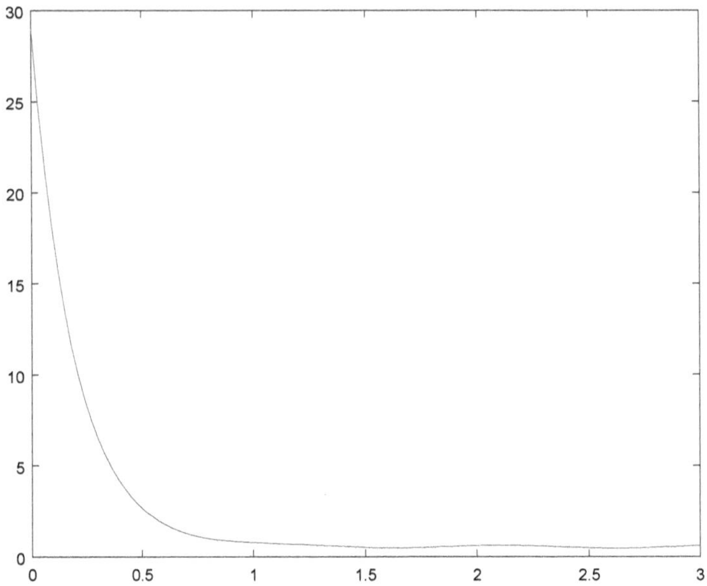

Figure 10. The mean $E_N(t, 29)$ on the interval $t \in [0, 3]$ with initial condition $\mathbf{X}(t) = (0, 29)$ for Example 2.

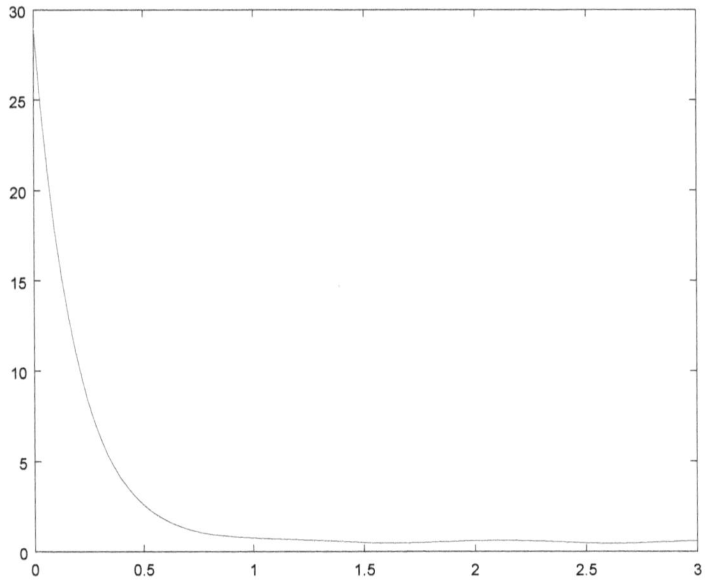

Figure 11. The mean $E_N(t, 29)$ on the interval $t \in [0, 3]$ with initial condition $\mathbf{X}(t) = (14, 15)$ for Example 2.

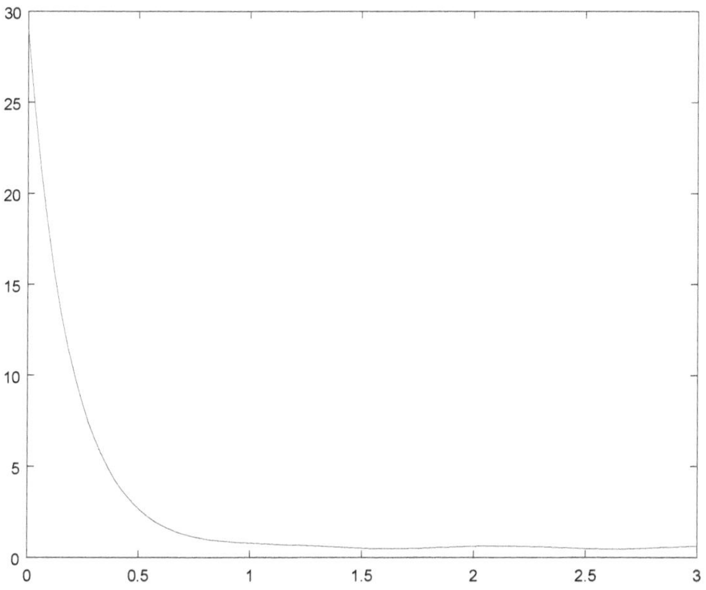

Figure 12. The mean $E_N(t, 29)$ on the interval $t \in [0, 3]$ with initial condition $\mathbf{X}(t) = (0, 29)$ for Example 2.

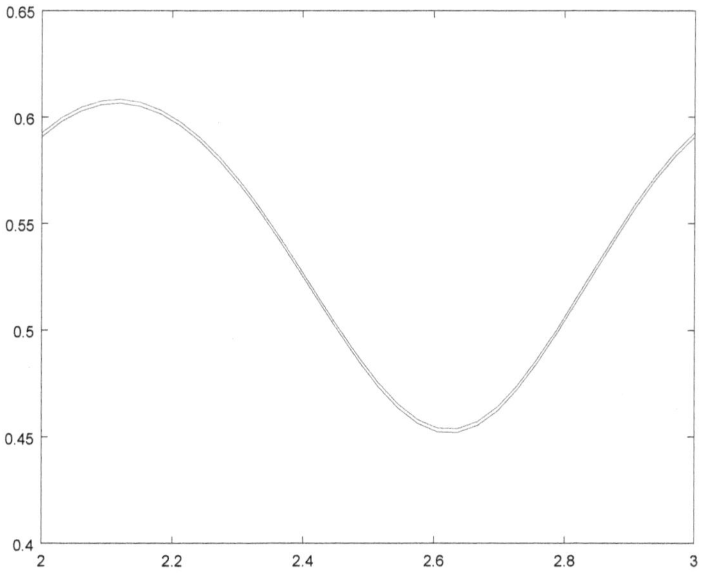

Figure 13. The limiting perturbed mean for $t \in [2, 3]$ for Example 2.

Remark 1. *These graphs give us the additional information on the considered examples. Namely, Figures 2 and 8 show the bounding on the rate of convergence, see Equation (21) for Examples 1, 2 respectively, in Figures 3–6 and 9–12 one can see the mathematical expectation of the number of all particles at the moment t until the stationary behaviour. Finally, the limiting behaviour of the limiting mathematical expectation of the number of all particles for the perturbed process is shown in Figures 7 and 13.*

Author Contributions: Conceptualization, A.Z., Al.S. and G.S.; Methodology, A.Z., K.K., T.P, E.F.; Software, Y.S.; Validation, A.Z., An.S, K.K., A.K. and I.G.; Investigation, An.S., A.Z.; Writing-Original Draft Preparation, An.S, A.Z.; Writing-Review and Editing, K.K., I.G.; Supervision, A.Z.; Project Administration, K.K.

Acknowledgments: The publication was supported by the Ministry of Education and Science of the Russian Federation (project No. 2.882.2017/4.6).

Conflicts of Interest: The authors declare no conflict of interest.

References

1. Di Crescenzo, A.; Giorno, V.; Nobile, A.G.; Ricciardi, L.M. On the M/M/1 queue with catastrophes and its continuous approximation. *Queueing Syst.* **2003**, *43*, 329–347. [CrossRef]
2. Di Crescenzo, A.; Giorno, V.; Nobile, A.G.; Ricciardi, L.M. A note on birth-death processes with catastrophes. *Stat. Probab. Lett.* **2008**, *78*, 2248–2257. [CrossRef]
3. Di Crescenzo, A.; Giorno, V.; Kumar, B.K.; Nobile, A.G. A double-ended queue with catastrophes and repairs, and a jump-diffusion approximation. *Methodol. Comput. Appl. Probab.* **2012**, *14*, 937–954. [CrossRef]
4. Dharmaraja, S.; Di Crescenzo, A.; Giorno, V.; Nobile, A.G. A continuous-time Ehrenfest model with catastrophes and its jump-diffusion approximation. *J. Stat. Phys.* **2015**, *161*, 326–345. [CrossRef]
5. Dudin, A.; Nishimura, S. A BMAP/SM/1 queueing system with Markovian arrival input of disasters. *J. Appl. Probab.* **1999**, *36*, 868–881. [CrossRef]
6. Dudin, A.; Karolik, A. BMAP/SM/1 queue with Markovian input of disasters and non-instantaneous recovery. *Perform. Eval.* **2001**, *45*, 19–32. [CrossRef]
7. Dudin, A.; Semenova, O. A stable algorithm for stationary distribution calculation for a BMAP/SM/1 queueing system with Markovian arrival input of disasters. *J. Appl. Probab.* **2004**, *41*, 547–556. [CrossRef]
8. Do, T.V.; Papp, D.; Chakka, R.; Sztrik, J.; Wang, J. M/M/1 retrial queue with working vacations and negative customer arrivals. *Int. J. Adv. Intell. Paradig.* **2014**, *6*, 52–65. [CrossRef]
9. Chen, A.Y.; Renshaw, E. The $M/M/1$ queue with mass exodus and mass arrives when empty. *J. Appl. Prob.* **1997**, *34*, 192–207. [CrossRef]
10. Chen, A.Y.; Renshaw, E. Markov bulk-arriving queues with state-dependent control at idle time. *Adv. Appl. Prob.* **2004**, *36*, 499–524. [CrossRef]
11. Chen, A.Y.; Pollett, P.; Li, J.P.; Zhang, H.J. Markovian bulk-arrival and bulk-service queues with state-dependent control. *Queueing Syst.* **2010**, *64*, 267–304. [CrossRef]
12. Li, J.; Chen, A. The Decay Parameter and Invariant Measures for Markovian Bulk-Arrival Queues with Control at Idle Time. *Methodol. Comput. Appl. Probab.* **2013**, *15*, 467–484.
13. Zhang, L.; Li, J. The M/M/c queue with mass exodus and mass arrivals when empty. *J. Appl. Probab.* **2015**, *52*, 990–1002. [CrossRef]
14. Zeifman, A.; Korotysheva, A.; Satin, Y.; Shorgin, S. On stability for nonstationary queueing systemswith catastrophes. *Inform. Appl.* **2010**, *4*, 9–15.
15. Zeifman, A.; Korotysheva, A.; Panfilova, T.; Shorgin, S. Stability bounds for some queueing systems with catastrophes. *Inform. Appl.* **2011**, *5*, 27–33.
16. Zeifman, A.; Korotysheva, A. Perturbation Bounds for Mt/Mt/N Queue with Catastrophes. *Stoch. Model.* **2012**, *28*, 49–62. [CrossRef]
17. Zeifman, A.; Satin, Y.; Panfilova, T. Limiting characteristics for finite birth-death-catastrophe processes. *Mater. Biosci.* **2013**, *245*, 96–102. [CrossRef] [PubMed]
18. Zeifman, A.; Satin, Y.; Korotysheva, A.; Korolev, V.; Shorgin, S.; Razumchik, R. Ergodicity and perturbation bounds for inhomogeneous birth and death processes with additional transitions from and to origin. *Int. J. Appl. Math. Comput. Sci.* **2015**, *25*, 787–802. [CrossRef]

19. Zeifman, A.; Satin, Y.; Korotysheva, A.; Korolev, V.; Bening, V. On a class of Markovian queuing systems described by inhomogeneous birth-and-death processes with additional transitions. *Dokl. Math.* **2016**, *94*, 502–505. [CrossRef]
20. Zeifman, A.; Korotysheva, A.; Satin, Y.; Kiseleva, K.; Korolev, V.; Shorgin, S. Ergodicity and uniform in time truncation bounds for inhomogeneous birth and death processes with additional transitions from and to origin. *Stoch. Model.* **2017**, *33*, 598–616. [CrossRef]
21. Giorno, V.; Nobile, A.; Spina, S. On some time non-homogeneous queueing systems with catastrophes. *Appl. Math.* **2014**, *245*, 220–234. [CrossRef]
22. Zeifman, A.; Korotysheva, A.; Satin, Y.; Kiseleva, K.; Korolev, V.; Shorgin, S. Bounds for Markovian Queues with Possible Catastrophes. In Proceedings of the 31th European Conference on Modeling and Simulation, Budapest, Hungary, 23–26 May 2017; pp. 628–634.
23. Daleckii, J.L.; Krein, M.G. Stability of solutions of differential equations in Banach space (No. 43). *Am. Math. Soc.* **2002**, *53*, 34–36.
24. Zeifman, A.; Satin, Y.; Chegodaev, A. On nonstationary queueing systems with catastrophes. *Inform. Appl.* **2009**, *3*, 47–54.
25. Mitrophanov, A.Y. Stability and exponential convergence of continuous-time Markov chains. *J. Appl. Prob.* **2003**, *40*, 970–979. [CrossRef]

mathematics

MDPI

Article

Network Reliability Modeling Based on a Geometric Counting Process

Somayeh Zarezadeh [1], Somayeh Ashrafi [2] and Majid Asadi [2,3,*]

[1] Department of Statistics, Shiraz University, Shiraz 71454, Iran; s.zarezadeh@shirazu.ac.ir
[2] Department of Statistics, University of Isfahan, Isfahan 81744, Iran; s.ashrafi@sci.ui.ac.ir
[3] School of Mathematics, Institute of Research in Fundamental Sciences (IPM), P.O. Box 19395-5746, Tehran, Iran
* Correspondence: m.asadi@sci.ui.ac.ir

Received: 4 August 2018 ; Accepted: 8 October 2018; Published: 11 October 2018

Abstract: In this paper, we investigate the reliability and stochastic properties of an n-component network under the assumption that the components of the network fail according to a counting process called a *geometric counting process* (GCP). The paper has two parts. In the first part, we consider a two-state network (with states *up* and *down*) and we assume that its components are subjected to failure based on a GCP. Some mixture representations for the network reliability are obtained in terms of signature of the network and the reliability function of the arrival times of the GCP. Several aging and stochastic properties of the network are investigated. The reliabilities of two different networks subjected to the same or different GCPs are compared based on the stochastic order between their signature vectors. The residual lifetime of the network is also assessed where the components fail based on a GCP. The second part of the paper is concerned with three-state networks. We consider a network made up of n components which starts operating at time $t = 0$. It is assumed that, at any time $t > 0$, the network can be in one of three states *up*, *partial performance* or *down*. The components of the network are subjected to failure on the basis of a GCP, which leads to change of network states. Under these scenarios, we obtain several stochastic and dependency characteristics of the network lifetime. Some illustrative examples and plots are also provided throughout the article.

Keywords: two-dimensional signature; multi-state network; totally positive of order 2; stochastic order; stochastic process

1. Introduction

In recent years, there has been a great growth in the use of networks (systems), such as communication networks and computer networks, in human life. The networks are a set of nodes that are connected by a set of links to exchange data through the links, where some particular nodes in the network are called terminals. Usually, a network can be modeled mathematically as a graph $G(\mathbb{V}, \mathbb{E}, \mathbb{T})$ in which $\mathbb{V}$ shows the collection of nodes, $\mathbb{E}$ shows the set of links and $\mathbb{T}$ denotes the set of terminals. Depending on the purpose of designing a network, the states of the network can be defined in terms of the connections between the terminals. In the simplest case, the networks have two states: *up* and *down*. However, in some applications, the networks may have several states which are known, in reliability engineering, as the multi-state networks. Multi-state networks have extensive applications in various areas of science and technology. From a mathematical point of view, the states of multi-state networks are usually shown by, $K = 0, 1, \ldots, M$, in which $K = 0$ shows the complete failure of the network and $K = M$ shows the perfect functioning of the network. A large number of research works have been published in literature on the reliability and aging properties of multi-state networks and systems under different scenarios. For the recent works on various applications and reliability properties of networks, we refer to [1–10].

When a network is operating during its mission, its states may change over the time according to the change of the states of its components. From the reliability viewpoint, the change in the states of the components may occur based on a specific stochastic mechanism. In a recent book, Gertsbakh and Shpungin [11] have proposed a new reliability model for a two-state network under the condition that components (with two states) fail according to a renewal process. Motivated by this, Zarezadeh and Asadi [12] and Zarezadeh et al. [13] studied the reliability of networks under the assumption that the components are subject to failure according to a counting process. Under the special case that the process of the components' failure is a nonhomogeneous Poisson process (NHPP), they arrived at some mixture representations for the reliability function of the network lifetime and explored its stochastic and aging properties under different conditions.

The aim of the present paper is to assess the network reliability under the condition that the failure of the components appear according to a recently proposed stochastic process called *geometric counting process* (GCP). We assume that the nodes of the network are absolutely reliable and, throughout the paper, whenever we say that the components of the network fail, we mean that the links of the network fail. Let $\{\xi(t), t \geq 0\}$ be a counting process where $\xi(t)$ denotes the number of events in $[0, t]$. A GCP, introduced in [14], is a subclass of counting process $\{\xi(t), t \geq 0\}$, which satisfies the following necessary conditions (for the sufficiently small $\Delta(t)$)

1. $\xi(0) = 0$,
2. $P(\xi(t + \Delta(t)) - \xi(t) = 1) = \lambda(t)\Delta(t) + o(\Delta(t))$,
3. $P(\xi(t + \Delta(t)) - \xi(t) \geq 2) = o(\Delta(t))$.

To be more precise, a GCP is a counting process $\xi(t)$, with $\xi(0) = 0$ such that, for any interval $(t_1, t_2]$,

$$P(\xi(t_2) - \xi(t_1) = k) = \frac{1}{1 + \Lambda(t_2) - \Lambda(t_1)} \left(\frac{\Lambda(t_2) - \Lambda(t_1)}{1 + \Lambda(t_2) - \Lambda(t_1)}\right)^k, \quad k = 0, 1, \dots, \tag{1}$$

where $\Lambda(t) = E(\xi(t))$ is the mean value function (MVF) of the process. It is usually assumed that $\Lambda(t)$ is a smooth function in the sense that there exists a function $\lambda(t)$ such that $\lambda(t) = d\Lambda(t)/dt$. The function $\lambda(t)$ is called the intensity function of the process. We have to mention here that, as noted by Cha and Finkelstien [14], the NHPP also lies in the class of counting process satisfying (i)–(iii), with an additional property that the increments of the process are independent. The motivation of using the GCP, in comparison with NHPP, is natural in some practical situations as we mention in the following. The GCP model, like the NHPP model, has a simple form and easy to handle mathematical characteristics. In an NHPP model, the increments of the process are independent, while, in the GCP model, the increments of the process have positive dependence. In practice, there are situations in which there is positive dependence of increments in a process that occurs naturally. For instance, assume that the components of a railway network destroyed by an earthquake that occurs according to a counting process. Then, the probability of the next earthquake is often higher if the previous earthquake has happened recently, compared with the situation that it happened earlier (see [14]). Furthermore, the NHPP has a limitation that the mean and the variance of the process are equal, i.e., $E(\xi(t)) = Var(\xi(t))$, while, in GCP, the variance of the process is always greater than the mean, i.e., $Var(\xi(t)) > E(\xi(t))$. This property of the GCP makes it cover many situations that can not be described and covered by the NHPP. For more details on recent mathematical developments and applications of the GCP model, see [14,15].

The reminder of the paper is arranged as follows. In Section 2, we first give the well-known concept of *signature* of a network. Then, we consider a two-state network that consists of n components. We assume that the components of the network fail according to a GCP. We obtain some mixture representations for the reliability of the network based on the signatures. Several aging and stochastic properties of the network are explored. Among others, conditions are investigated under which the monotonicity of the intensity function of the process of component failure implies the monotonicity of the network hazard rate. The reliabilities of the lifetimes of the different networks, subjected to the same or different GCPs, are compared based on the stochastic order between the associated signature vectors. We also study the stochastic properties of the residual lifetime of the network where the components fail based on a GCP. Section 3 is devoted to the reliability assessment of the single-step three-state networks. Recall that a network is said to be single-step if the failure of one component changes the network state at most by one. First, we give the notion of a two-dimensional signature associated with three-state networks. Then, we consider an n-component network and assume that the network has three states *up*, *partial performance* and *down*. We again assume that the components of the network are subjected to failure on the basis of GCP, which results in the change of network states. Under these conditions, we obtain several stochastic and dependency characteristics of the networks based on the two-dimensional signature. Several examples and plots are also provided throughout the article for illustration purposes.

Before giving the main results of the paper, we give the following definitions that are useful throughout the paper. For more details, see [16].

Definition 1. *Let X and Y be two random variables (RVs) with survival functions $\bar{F}_X$ and $\bar{F}_Y$, probability density functions (PDFs) f_X and f_Y, hazard rates h_X and h_Y, and reversed hazard rates r_X and r_Y, respectively:*

- *X is said to be smaller than Y in the usual stochastic order (denoted by $X \leq_{st} Y$) if $\bar{F}_X(x) \leq \bar{F}_Y(x)$ for all x.*
- *X is said to be smaller than Y in the hazard rate order (denoted by $X \leq_{hr} Y$) if $h_X(x) \geq h_Y(x)$ for all x.*
- *X is said to be smaller than Y in the reversed hazard rate order (denoted by $X \leq_{rh} Y$) if $r_X(x) \leq r_Y(x)$ for all x.*
- *X is said to be smaller than Y in the mean residual life order (denoted by $X \leq_{mrl} Y$) if $E(X - x|X > x) \leq E(Y - x|Y > x)$ for all x.*
- *X is said to be smaller than Y in likelihood ratio order (denoted by $X \leq_{lr} Y$) if $f_Y(x)/f_X(x)$ is an increasing function of x.*

It can be shown that, if $X \leq_{lr} Y$, then $X \leq_{hr} Y$ and $X \leq_{rh} Y$. In addition, $X \leq_{hr} Y$ implies $X \leq_{mrl} Y$ and $X \leq_{st} Y$.

Definition 2. *Let $\mathbf{X}$ and $\mathbf{Y}$ be two random vectors with survival functions $\bar{F}_{\mathbf{X}}$ and $\bar{F}_{\mathbf{Y}}$, respectively.*

- *$\mathbf{X}$ is said to be smaller than $\mathbf{Y}$ in the upper orthant order (denoted by $\mathbf{X} \leq_{uo} \mathbf{Y}$) if $\bar{F}_{\mathbf{X}}(\mathbf{x}) \leq \bar{F}_{\mathbf{Y}}(\mathbf{x})$ for all $\mathbf{x} \in R^n$.*
- *$\mathbf{X}$ is said to be smaller than $\mathbf{Y}$ in the usual stochastic order (denoted by $\mathbf{X} \leq_{st} \mathbf{Y}$) if $E(\rho(\mathbf{X})) \leq E(\rho(\mathbf{Y}))$ for every increasing function $\rho(\cdot)$ for which the expectations exist.*

Definition 3.

- *The nonnegative function $g(\mathbf{x})$ is called multivariate totally positive of order 2 (MTP$_2$) if $g(\mathbf{x})g(\mathbf{y}) \leq g(\mathbf{x} \vee \mathbf{y})g(\mathbf{x} \wedge \mathbf{y})$, for all $\mathbf{x}, \mathbf{y} \in R^n$, where $\mathbf{x} \wedge \mathbf{y} = (\min\{x_1, y_1\}, \ldots, \min\{x_n, y_n\})$ and $\mathbf{x} \vee \mathbf{y} = (\max\{x_1, y_1\}, \ldots, \max\{x_n, y_n\})$.*
- *The RVs X and Y are said to be positively quadrant dependent (PQD) if, for every pair of increasing functions $\psi_1(x)$ and $\psi_2(x)$,*
$$Cov(\psi_1(X), \psi_2(Y)) \geq 0.$$
- *The RVs X and Y are said to be associated if for every pair of increasing functions $\psi_1(x, y)$ and $\psi_2(x, y)$, $Cov(\psi_1(X, Y), \psi_2(X, Y)) \geq 0$.*

In a special case when $n = 2$, the MTP$_2$ is known as totally positive of order 2 (TP$_2$).

2. Two-State Networks under GCP of Component Failure

In the reliability engineering literature, several approaches have been employed to assess the reliability of networks and systems. Among various ways that are considered to explore the reliability and aging properties of the networks, an approach is based on the notion of *signature* (or *D-spectrum*). The concept of signature, which depends only on the network design, has proven very useful in the analysis of the networks performance particularly for comparisons between networks with different structures. Consider a network (system) that consists of n components. The *signature* associated with the network is a vector $\mathbf{s} = (s_1, s_2, \ldots, s_n)$, in which the ith element shows the probability that the ith component failure in the network causes the network failure, under the condition that all permutations of order of components failure are equally likely. In other words, the ith element s_i is equal to $s_i = n_i/n!$, $i = 1, \ldots, n$, where n_i is the number of permutations in which the ith component failure changes the network state from up to down. For more details on signatures and their applications in the study of system reliability, see, for example, Refs. [17–20] and references therein. In this section, we give a signature-based mixture representation for the reliability of the network under the condition that the components of the network fail according to a GCP $\{\xi(t), t \geq 0\}$ with MVF $\Lambda(t)$. We have from Equation (1)

$$P(\xi(t) = k) = \frac{1}{1+\Lambda(t)}\Big(\frac{\Lambda(t)}{1+\Lambda(t)}\Big)^k, \quad k = 0, 1, \ldots.$$

Then, the survival function of the kth arrival time of process, ϑ_k, is given as

$$\begin{aligned} \bar{F}_{\vartheta_k}(t) :=& P(\vartheta_k > t) = P(\xi(t) < k) \\ =& 1 - \Big(\frac{\Lambda(t)}{1+\Lambda(t)}\Big)^k, \quad k = 0, 1, \ldots \end{aligned}$$

and the PDF of the kth arrival time ϑ_k is achieved as

$$f_{\vartheta_k}(t) = k\frac{\lambda(t)}{\Lambda(t)(1+\Lambda(t))}\Big(\frac{\Lambda(t)}{1+\Lambda(t)}\Big)^k, \quad k = 0, 1, \ldots.$$

Let T denote the lifetime of a network with n components. The components of network are subjected to failure based on a GCP with MVF $\Lambda(t)$. From the reliability modeling proposed by Zarezadeh and Asadi [12], the reliability of the network lifetime, denoted by $\bar{F}_T$, is represented as

$$\bar{F}_T(t) = \sum_{i=1}^{n} s_i\Big(1 - \Big(\frac{\Lambda(t)}{1+\Lambda(t)}\Big)^i\Big), \quad t > 0, \tag{2}$$

or equivalently as

$$\bar{F}_T(t) = \frac{1}{1+\Lambda(t)}\sum_{i=0}^{n-1} \bar{S}_i\Big(\frac{\Lambda(t)}{1+\Lambda(t)}\Big)^i, \quad t > 0, \tag{3}$$

where $\bar{S}_i = \sum_{k=i+1}^{n} s_k$ is the survival signature of the network. Then, the PDF of T is obtained as

$$f_T(t) = \frac{\lambda(t)}{(1+\Lambda(t))^2}\sum_{i=1}^{n} i s_i\Big(\frac{\Lambda(t)}{1+\Lambda(t)}\Big)^{i-1}, \quad t > 0, \tag{4}$$

where $\lambda(t) = d\Lambda(t)/dt$. In addition, the hazard rate of network lifetime is given as follows:

$$h_T(t) = \frac{\lambda(t)}{\Lambda(t)(1+\Lambda(t))} \cdot \frac{\sum_{i=1}^{n} i s_i \left(\frac{\Lambda(t)}{1+\Lambda(t)}\right)^i}{\sum_{i=1}^{n} s_i \left[1 - \left(\frac{\Lambda(t)}{1+\Lambda(t)}\right)^i\right]}$$

$$= \frac{\lambda(t)}{\Lambda(t)} \frac{\sum_{i=1}^{n} i s_i \left(\frac{\Lambda(t)}{1+\Lambda(t)}\right)^i}{\sum_{i=0}^{n-1} \bar{S}_i \left(\frac{\Lambda(t)}{1+\Lambda(t)}\right)^i}, \quad t > 0.$$

With $h_{\vartheta_k}(t) = f_{\vartheta_k}(t)/\bar{F}_{\vartheta_k}(t)$ as the hazard rate of the kth arrival time of the GCP, it can be seen that the hazard rate of network lifetime can be also written as

$$h_T(t) = \frac{\sum_{k=1}^{n} s_k f_{\vartheta_k}(t)}{\sum_{k=1}^{n} s_k \bar{F}_{\vartheta_k}(t)} = \sum_{k=1}^{n} s_k(t) h_{\vartheta_k}(t), \tag{5}$$

which is a mixture representation with mixing probability vector $\mathbf{s}(t) = (s_1(t), \ldots, s_n(t))$ where, for $k = 1, \ldots, n$,

$$s_k(t) = \frac{s_k \left[1 - \left(\frac{\Lambda(t)}{1+\Lambda(t)}\right)^k\right]}{\sum_{k=1}^{n} s_k \left[1 - \left(\frac{\Lambda(t)}{1+\Lambda(t)}\right)^k\right]} \tag{6}$$

$$= \frac{s_k \left[1 - \left(\frac{\Lambda(t)}{1+\Lambda(t)}\right)^k\right]}{\sum_{i=0}^{n-1} \bar{S}_i \left(1 - \frac{\Lambda(t)}{1+\Lambda(t)}\right) \left(\frac{\Lambda(t)}{1+\Lambda(t)}\right)^i}. \tag{7}$$

One can easily show that $s_k(t)$ is the probability that the lifetime of system is equal to the kth arrival time of the process given that the network lifetime is greater than t.

Let us look at the following example.

Example 1. *Consider a network that consists of six nodes and* 10 *links with the graph depicted in Figure 1.*

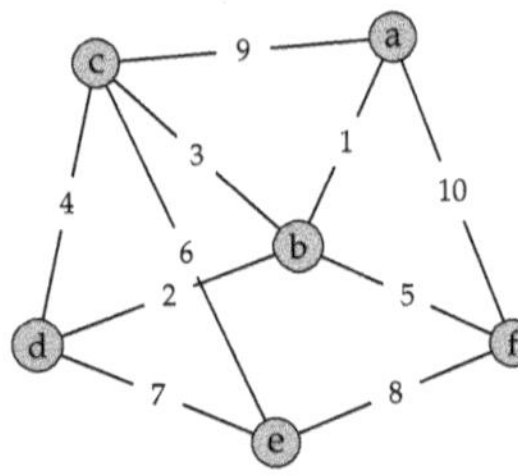

Figure 1. A network with 10 links, and six nodes

The network is assumed to work if there is a connection between some of nodes which we consider them as terminals. We consider two different sets of terminals for the network:

- *First, we consider all nodes as terminals,* $\mathbb{T} = \mathbb{V}$ *(all-terminal connectivity). In this case, we can show that the corresponding signature vector of a network is as follows:*

$$\mathbf{s} = (0, 0, \frac{1}{30}, \frac{9}{70}, \frac{29}{90}, \frac{65}{126}, 0, 0, 0, 0).$$

- *Second, assume that the network is working if there is a connection between nodes c and f. That is, the terminals set is* $\mathbb{T}^\star = \{c, f\}$. *In this case, the signature vector is obtained as*

$$\mathbf{s}^\star = (0, 0, \frac{1}{120}, \frac{37}{840}, \frac{179}{1260}, \frac{379}{1260}, \frac{19}{70}, \frac{1}{6}, \frac{1}{15}, 0).$$

An algorithm for calculating these signatures is available from the authors upon the request.

Assume that, in each case, the network is subjected to failure based on the GCPs with the same MVFs. Denote the lifetimes of the network corresponding to (a) and (b) by T *and* $T^\star$, *respectively. Comparing the corresponding survival signatures of network for two cases shows that* $\bar{S}_i \leq \bar{S}_i^\star$, *for* $i = 0, 1, \ldots, 9$, *and hence based on (3) we have* $T \leq_{st} T^*$ *implying that the network with two-terminal connectivity is more reliable than the network with all-terminal connectivity, as expected intuitively.*

Figure 2a gives the plot of network reliability in the case of all-terminal connectivity, $\mathbb{T} = \mathbb{V}$, *and when* $\Lambda(t) = t^a$, *for different values of a. As seen, the reliability function of network does not order with respect to a for all* $t > 0$. *Of course, this is true in any network when MVF* $\Lambda(t) = t^a$, $a > 0$. *This is so using the fact that*

$$P(\vartheta_k > t; a) = \left[1 - \left(\frac{t^a}{1 + t^a}\right)^k\right]$$

is increasing (decreasing) in a for $0 < t < 1$ $(t > 1)$, *for a general signature vector* $\mathbf{s}$ *the reliability of the network*

$$P(T > t; a) = \sum_{k=1}^{n} s_k P(\vartheta_k > t; a),$$

is also increasing (decreasing) in a for $0 < t < 1$ $(t > 1)$. *Figure 2b represents the hazard rate of network when* $\mathbb{T} = \mathbb{V}$ *for* $a = 0.5, 1, 1.5$. *Figure 3a,b shows the plots of reliability function and hazard rate of the network lifetime when the terminals set is considered as* $\mathbb{T}^\star = \{c, f\}$.

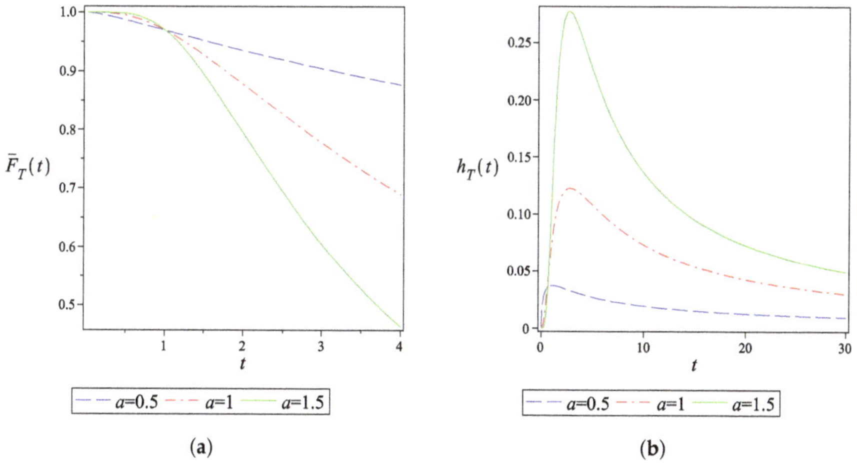

Figure 2. The plots of (**a**) the reliability function; (**b**) the hazard rate of network lifetime when the terminals set is $\mathbb{T} = \mathbb{V}$.

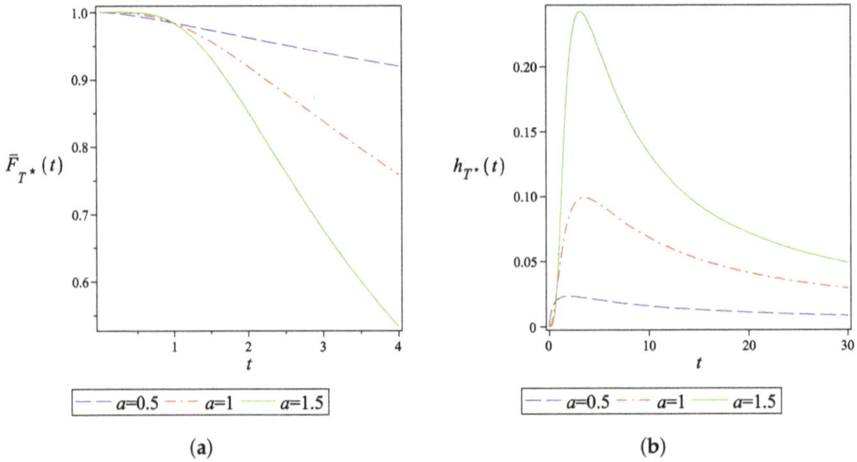

Figure 3. The plots of (**a**) the reliability function; (**b**) the hazard rate of network lifetime when the terminals set is $\mathbb{T}^\star = \{c, f\}$.

It is interesting to compare the network reliability when the failure of components appear according to a GCP and the network reliability when the failure of components occur based on an NHPP. In the sequel, we show that, if the network has a series structure, then the reliability of the network in the GCP model dominates the reliability of the network in an NHPP model. Consider a two-state series network with the property that the first and the last components are considered to be terminals. We assume that the network fails if the linkage between the two terminals are disconnected. This occurs at the time of the first component failure. Let T_{NP} and T_{GP} denote the lifetimes of the network when the component failure appears according to NHPP and GCP with the same MVF $\Lambda(t)$, respectively. If $\vartheta_{1,NP}$ and $\vartheta_{1,GP}$ denote the arrival times of the first component failure based on NHPP and GCP, respectively, then, from inequality $e^x > (1+x)$, $x > 0$, we can write

$$P(\vartheta_{1,NP} > t) = e^{-\Lambda(t)} < \frac{1}{1+\Lambda(t)} = P(\vartheta_{1,GP} > t).$$

Hence, based on the fact that, for a series network $\mathbf{s} = (1, 0, \ldots, 0)$, relation (2) implies that $T_{NP} \leq_{st} T_{GP}$.

The following example reveals that the above result, proved for the series network, is not necessarily true for any network.

Example 2. *Consider the network described in Example 1. Figure 4 shows that the reliability functions of the network for part (a). As the plots show the reliability functions are not ordered in NHPP and GCP models with the same MVFs $\Lambda(t) = t$. The reliability of the network in the NHPP model is higher than the GCP model for the early times of operating of the network. However, when the time goes ahead, the network reliability in NHPP declines rapidly and stays below the reliability of the GCP model.*

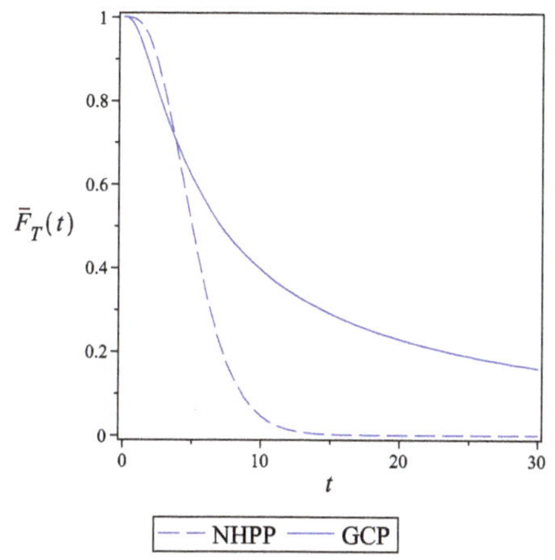

Figure 4. The reliability of network in Example 2.

The next theorem explores the monotonicity relation of the intensity function of the process and the hazard rate of the network.

Theorem 1. *Let the components of a network fail based on a GCP with increasing intensity function $\lambda(t)$. Then, the hazard rate of network is increasing if and only if $\psi(u)$ is increasing in u where*

$$\psi(u) = \frac{\sum_{k=0}^{n-1}(k+1)s_{k+1}u^k}{\sum_{k=0}^{n-1}\bar{S}_k u^k}. \tag{8}$$

Proof. From (3) and (4), the hazard rate of network can be written as

$$h_T(t) = \frac{f_T(t)}{\bar{F}_T(t)} = \lambda(t)\psi(\mu(t)), \tag{9}$$

where $\psi(\cdot)$ is defined in (8) and $\mu(t) = \Lambda(t)/(1+\Lambda(t))$ is increasing in t. If $\psi(t)$ is increasing, from (9), it can be easily seen that $h_T(t)$ is increasing. This completes the 'if' part of the theorem. To prove 'only if' part of the theorem, let $h_T(t)$ be increasing and $\psi(t)$ be decreasing in the interval (a,b). For MVF $\Lambda(t) = ct$, $c > 0$, and hence $\lambda(t) = c$ as an increasing function, we conclude that $h_T(t)$ is decreasing on interval (a,b), which contradicts with the assumption that the hazard rate of network is increasing for all t. □

Theorem 2. *Let T and $T^\star$ denote the lifetimes of two networks with signature vectors $\mathbf{s} = (s_1, \ldots, s_n)$ and $\mathbf{s}^\star = (s_1^\star, \ldots, s_n^\star)$, respectively. Suppose that the components of networks fail based on GCPs with MVFs $\Lambda(t)$ and $\Lambda^\star(t)$, respectively. If $\mathbf{s} \leq_{st} \mathbf{s}^\star$ and $\Lambda(t) \geq \Lambda^\star(t)$ for all $t \geq 0$, then $T \leq_{st} T^\star$.*

Proof. Let ϑ_k and $\vartheta_k^\star$ denote the kth arrival times of the two processes, $k = 1, \ldots, n$. Since $P(\vartheta_k \leq \vartheta_{k+1}) = 1$, then, from Theorem 1.A.1 of [16], we can write $\vartheta_k \leq_{st} \vartheta_{k+1}$. Hence,

$$P(T > t) = \sum_{k=1}^{n} s_k \bar{F}_{\vartheta_k}(t) \leq \sum_{k=1}^{n} s_k^\star \bar{F}_{\vartheta_k}(t). \tag{10}$$

In addition, the condition $\Lambda(t) \geq \Lambda^\star(t)$ for all $t \geq 0$ implies that $\vartheta_k \leq_{st} \vartheta_k^\star$, $k = 1, ..., n$. Then, we get

$$\sum_{k=1}^{n} s_k^\star \bar{F}_{\vartheta_k}(t) \leq \sum_{k=1}^{n} s_k^\star \bar{F}_{\vartheta_k^\star}(t) = P(T^\star > t). \tag{11}$$

Hence, the result follows from (10) and (11). □

Theorem 3. *For two networks as described in Theorem 2, assume that the failure of components of both networks appear according to the same GCPs:*

1. *If* $\mathbf{s} \leq_{st} \mathbf{s}^\star$, *then* $T \leq_{st} T^\star$,
2. *If* $\mathbf{s} \leq_{hr} \mathbf{s}^\star$, *then* $T \leq_{hr} T^\star$,
3. *If* $\mathbf{s} \leq_{rh} \mathbf{s}^\star$, *then* $T \leq_{rh} T^\star$,
4. *If* $\mathbf{s} \leq_{lr} \mathbf{s}^\star$, *then* $T \leq_{lr} T^\star$.

Proof. It can be easily shown that $\vartheta_k \leq_{lr} \vartheta_{k+1}$ $k = 1, 2, \ldots$. Since lr-ordering implies hr-, rh- and st-ordering, parts (i), (ii), (iii), and (iv) are proved, by using (2), from Theorems 1.A.6, 1.B.14, 1.B.52, and 1.C.17 of [16], respectively. □

From part (ii) of Theorem 3, since hr-ordering implies mrl-ordering, we conclude that if $\mathbf{s} \leq_{hr} \mathbf{s}^\star$ then $T \leq_{mrl} T^\star$. However, the following example shows that the assumption $\mathbf{s} \leq_{hr} \mathbf{s}^\star$ can not be replaced with $\mathbf{s} \leq_{mrl} \mathbf{s}^\star$ to have $T \leq_{mrl} T^\star$.

Example 3. *Consider two networks with signature vectors* $\mathbf{s} = (0, 2/3, 1/3)$ *and* $\mathbf{s}^\star = (1/3, 0, 2/3)$. *It is easy to see that* $\mathbf{s} \leq_{mrl} \mathbf{s}^\star$. *However,* $\mathbf{s} \nleq_{st} \mathbf{s}^\star$ *and hence* $\mathbf{s} \nleq_{hr} \mathbf{s}^\star$. *Assume that the components of both networks fail based on the same GCPs with MVF* $\Lambda(t) = t^2$. *Then, a straightforward calculation gives* $E(T) = 2.5525 > E(T^\star) = 2.4871$, *which, in turn, implies that* $T \nleq_{mrl} T^\star$.

2.1. Residual Lifetime of a Working Network

Let T denote the lifetime of a network whose components are subjected to failure based on a GCP with MVF $\Lambda(t)$. If the network is up at time t, then the residual lifetime of the network is presented by the conditional RV $(T - t | T > t)$ with conditional reliability function given as

$$\begin{aligned} P(T - t > x | T > t) &= \frac{1}{P(T > t)} \sum_{k=1}^{n} s_k P(\vartheta_k > t + x) \\ &= \sum_{k=1}^{n} s_k(t) P(\vartheta_k - t > x | \vartheta_k > t), \end{aligned}$$

where $s_k(t)$ is the kth element of vector $\mathbf{s}_k(t)$ as defined in (6). This shows that the reliability function of the residual lifetime of the network is a mixture of the reliability functions of residual lifetimes of the first n arrival times of GCP, where the mixing probability vector is $\mathbf{s}(t) = (s_1(t), \ldots, s_n(t))$. As we have already mentioned, $s_k(t)$ is in fact the probability that the kth component failure causes the failure of the network, given that the lifetime of the network is more than t; that is,

$$s_k(t) = P(T = \vartheta_k | T > t), \quad k = 1, 2, ..., n.$$

In what follows, we call the vector $\mathbf{s}(t)$ as the conditional signature of the network. In the sequel, we give some stochastic properties of the conditional signature of network under the condition that the components of the network fail based on GCP model.

Theorem 4. *Consider a network with signature vector* $\mathbf{s} = (s_1, s_2, \ldots, s_n)$. *With* $M = \max\{i | s_i > 0\}$, $\lim_{t\to 0} \mathbf{s}(t) = \mathbf{s}$ *and* $\lim_{t\to\infty} \mathbf{s}(t) = \tilde{\mathbf{s}}$ *where* $\tilde{\mathbf{s}} = (\tilde{s}_1, \ldots, \tilde{s}_M)$ *and* $\tilde{s}_i = \frac{i s_i}{\sum_{k=1}^{n} k s_k}$, $i = 1, \ldots, n$.

Proof. From (6), we can write

$$s_i(t) = \frac{s_i\left[1-\left(\frac{\Lambda(t)}{1+\Lambda(t)}\right)^i\right]}{\sum_{k=1}^{M} s_k\left[1-\left(\frac{\Lambda(t)}{1+\Lambda(t)}\right)^k\right]}, \qquad i=1,\ldots,n.$$

Then, it is easily seen that, for any i, $\lim_{t\to 0} s_i(t) = s_i$ and hence

$$\lim_{t\to 0} \mathbf{s}(t) = \mathbf{s}.$$

On the other hand, for $i = 1,\ldots,n$,

$$\begin{aligned}\lim_{t\to\infty} s_i(t) &= \lim_{u\to 1} \frac{s_i(1-u^i)}{\sum_{k=1}^{n} s_k(1-u^k)},\\ &= \lim_{u\to 1} \frac{is_iu^{i-1}}{\sum_{k=1}^{n} ks_ku^{k-1}} = \frac{is_i}{\sum_{k=1}^{n} ks_k},\end{aligned}$$

and consequently the result follows. □

Example 4. *For the network in Example 1, with $\Lambda(t) = t^a$, we have*

$$h_{\vartheta_i}(t) = \frac{akt^{ak}}{(1+t^a)\left[(1+t^a)^k - t^{ak}\right]}, \qquad i=1,2,\ldots.$$

Hence, it is easily seen that $\lim_{t\to\infty} h_{\vartheta_i}(t) = 0$. Thus, based on (5) and Theorem 4, we have $\lim_{t\to\infty} h_T(t) = 0$ for any network structure.

Theorem 5. *Consider a network whose components fail based on two different GCPs with MVFs $\Lambda(t)$ and $\Lambda^\star(t)$, respectively. Denote by $\mathbf{s}(t)$ and $\mathbf{s}^\star(t)$ the corresponding conditional signatures of the two networks. Then,*

- *$\mathbf{s}(t)$ and $\mathbf{s}^\star(t)$ are increasing in the sense of st-ordering with respect to t;*
- *$\Lambda(t) \le \Lambda^\star(t)$ implies that $\mathbf{s}(t) \le_{st} \mathbf{s}^\star(t)$ for all $t \ge 0$.*

Proof. With $\delta(i,u) = 1-u^i$, from (6), for each MVF $\Lambda(t)$, $s_i(t)$ can be written as

$$s_i(t) = \frac{s_i\delta(i,\mu(\Lambda(t)))}{\sum_{i=1}^{n} s_i\delta(i,\mu(\Lambda(t))}, \qquad i=1,\ldots,n, \tag{12}$$

where $\mu(v) = v/(1+v)$. Then, we have

$$\sum_{i=k}^{n} s_i(t) = \beta_k(\mu(\Lambda(t))),$$

in which

$$\beta_k(u) = \frac{\sum_{i=k}^{n} s_i[1-u^i]}{\sum_{i=1}^{n} s_i[1-u^i]}.$$

For $i_1 \leq i_2$,

$$\begin{aligned}\frac{\delta(i_2,u)}{\delta(i_1,u)} =& \frac{1-u^{i_2}}{1-u^{i_1}} = \frac{\sum_{j=0}^{i_2-1} u^j}{\sum_{j=0}^{i_1-1} u^j} \\ =& \frac{\sum_{j=0}^{i_1-1} u^j + \sum_{j=i_1}^{i_2-1} u^j}{\sum_{j=0}^{i_1-1} u^j} \\ =& 1 + \sum_{j'=i_1}^{i_2-1} \frac{1}{\sum_{j=0}^{i_1-1} u^{j-j'}} \end{aligned} \tag{13}$$

is increasing in u and hence $\delta(i,u)$ is TP_2 in i and u. Using this fact and Lemma 2.4 of [13]:

- Since $\mu(u)$ is increasing in u, then, for an arbitrary MVF $\Lambda(\cdot)$, we have

$$\sum_{i=k}^{n} s_i(t) \leq \sum_{i=k}^{n} s_i(t'), \qquad t \leq t'.$$

- Since $\mu(u)$ is increasing in u, and $\Lambda(t) \leq \Lambda^\star(t)$ for all $t \geq 0$, then

$$\sum_{i=k}^{n} s_i(t) \leq \sum_{i=k}^{n} s_i^\star(t), \qquad t \geq 0.$$

Therefore, the proof of the theorem is complete. □

The following lemma from [13] is useful to get some stochastic properties of conditional signature expressed in (6). Before expressing the lemma, we recall that a non-negative function $f(x)$, $x \geq 0$, is said to be upside-down bathtub-shaped if it is increasing on $[0,a]$, is constant on $[a,b]$ and is decreasing on $[b,\infty)$ where $0 \leq a \leq b \leq \infty$.

Lemma 1. *Let $\alpha(.)$ and $\beta(.)$ be non-negative discrete functions and γ be positive and real-valued. Define*

$$\tau(u) = \frac{\gamma \sum_{i=l'}^{k'} \alpha(i) u^i}{\sum_{i=l}^{k} \alpha(i)\beta(i) u^i}, \qquad u > 0, \tag{14}$$

where $l \leq l'$, $l' < k'$ and $k' \leq k$. Assume that $\beta(.)$ is a non-constant decreasing (increasing) function on $\{l', l'+1, ..., k'\}$. Then, for $l = l'$ and $k' < k (k = k'$ and $l' > l)$, we have

1. *$\tau(u)$ is upside-down bathtub-shaped with a single change-point;*
2. *$\tau(u)$ is bounded above by $\gamma/\beta(k')$ $(\gamma/\beta(l'))$.*

Now, we have the following theorem.

Theorem 6. *For a network with signature vector $\mathbf{s} = (s_1, \ldots, s_n)$,*

1. *$s_m(t)$ is decreasing in t and $s_M(t)$ is increasing in t where $m = \min\{i|s_i > 0\}$ and $M = \max\{i|s_i > 0\}$;*
2. *$s_j(t)$, $m \leq j \leq M$, is upside-down bathtub-shaped with a single change-point;*
3. *$s_j(t)$, $m \leq j \leq M$, is bounded above by $s_j/\bar{S}_{j-1}$;*
4. *The maximum value of $s_j(t)$, $m \leq j \leq M$, does not depend on the MVF $\Lambda(t)$.*

Proof. Assume that $\mu(t) = t/(1+t)$. From (12), we can write

$$s_m(t) = s_m\Big(\sum_{i=m}^{M} s_i \frac{\delta(i,\mu(\Lambda(t)))}{\delta(m,\mu(\Lambda(t)))}\Big)^{-1} \qquad s_M(t) = s_M\Big(\sum_{i=m}^{M} s_i \frac{\delta(i,\mu(\Lambda(t)))}{\delta(M,\mu(\Lambda(t)))}\Big)^{-1}. \tag{15}$$

As seen in (13), $\delta(i_2, x)/\delta(i_1, x)$, for $i_2 \geq i_1$, is increasing in x. Since $\mu(\Lambda(t))$ is increasing in t, then it can be concluded that $\delta(i_2, \mu(\Lambda(t)))/\delta(i_1, \mu(\Lambda(t)))$ is also increasing in t, for $i_2 \geq i_1$. Based on this fact and (15), we observe that $s_m(t)$ is a decreasing function of t and $s_M(t)$ is an increasing function of t. This completes the proof of part (a).

From relation (7), we have

$$s_j(t) = \frac{s_j \sum_{i=0}^{j-1} \left(\mu(\Lambda(t))\right)^i}{\sum_{i=0}^{n-1} \bar{S}_i \left(\mu(\Lambda(t))\right)^i}.$$

With $\gamma = s_j$, $\alpha(i) = 1$, $\beta(i) = \bar{S}_i$, $l = l' = 0$, $k' = j - 1$ and $k = n - 1$ in (14), define $\omega_j(u) := \tau(u)$. Then, we can write

$$s_j(t) = \omega_j(\mu(\Lambda(t))).$$

Since $\mu(\Lambda(t))$ is increasing in t, parts (b) and (c) follow from parts (i) and (ii) of Lemma 1.

Part (d) can be proved from the fact that

$$\max_{t>0} s_j(t) = \max_{t>0} \omega_j(\mu(\Lambda(t))) = \max_{t>0} \omega_j(t).$$

□

The following theorem compares the performance of two used networks based on their conditional signatures.

Theorem 7. *Let $\mathbf{s}(t)$ and $\mathbf{s}^\star(t)$ be the conditional signatures of two networks with lifetimes T and $T^\star$, respectively. Suppose that the component failure in both networks appear based on the same GCPs.*

1. *If $\mathbf{s}(t) \leq_{st} \mathbf{s}^\star(t)$, then $(T - t|T > t) \leq_{st} (T^\star - t|T^\star > t)$;*
2. *If $\mathbf{s}(t) \leq_{hr} \mathbf{s}^\star(t)$, then $(T - t|T > t) \leq_{hr} (T^\star - t|T^\star > t)$;*
3. *If $\mathbf{s}(t) \leq_{rh} \mathbf{s}^\star(t)$, then $(T - t|T > t) \leq_{rh} (T^\star - t|T^\star > t)$;*
4. *If $\mathbf{s}(t) \leq_{lr} \mathbf{s}^\star(t)$, then $(T - t|T > t) \leq_{lr} (T^\star - t|T^\star > t)$.*

Proof. It can be easily seen that $\vartheta_k \leq_{lr} \vartheta_{k+1}$, $k = 1, \ldots, n-1$. Then, from Theorem 1.C.6 of [16], we have, for $k = 1, \ldots, n-1$ and $t \geq 0$,

$$(\vartheta_k - t|\vartheta_k > t) \leq_{lr} (\vartheta_{k+1} - t|\vartheta_{k+1} > t).$$

Hence, these residual lifetimes are also hr-, rh- and st-ordered. Since $\mathbf{s}(t) \leq_{st} \mathbf{s}^\star(t)$, from Theorem 1.A.6 of [16], we have, for all $x > 0$,

$$\begin{aligned} P(T - t > x|T > t) &= \sum_{k=1}^{n} s_k(t) P(\vartheta_k - t > x|\vartheta_k > t) \\ &\leq \sum_{k=1}^{n} s_k^\star(t) P(\vartheta_k - t > x|\vartheta_k > t) = P(T^\star - t > x|T^\star > t). \end{aligned}$$

This establishes part (a). The proof of parts (b), (c) and (d) are obtained similarly by using Theorems 1.B.14, 1.B.52, and 1.C.17 of [16], respectively. □

3. Three-State Networks under GCP of Component Failure

In this section, we study the reliability of the lifetimes of the networks with three states under the condition that the components fail according to a GCP with MVF $\Lambda(t) = \lambda t$. In order to develop the results, we need the notion of two-dimensional signature that has been defined for single-step three-state networks by Gertsbakh and Shpungin [11]. Throughout this section, we are dealing with a single-step three-state network consisting of n binary components where we assume that the network has three states: *up* (denoted by $K = 2$), *partial performance* (denoted by $K = 1$) and *down* (denoted by $K = 0$). Suppose that the network starts to operate at time $t = 0$ where it is in state $K = 2$. Denote by $\mathcal{T}_1$ the time that the network remains in state $K = 2$ and by $\mathcal{T}_2$ the network lifetime i.e., the entrance time into state $K = 0$. Let I (J) be the number of failed components when the network enters into state $K = 1$ ($K = 0$). Gertsbakh and Shpunging [11] introduced the notion of two-dimensional signature as

$$s_{i,j} = P(I = i, J = j) = \frac{n_{i,j}}{n}, \quad 1 \le i < j \le n, \tag{16}$$

where $n_{i,j}$ represents the number of permutations in which the ith and the jth components failure change the network states from $K = 2$ to $K = 1$ and from $K = 1$ to $K = 0$, respectively. We denote by matrix $\mathcal{S}$ the two-dimensional signature with elements defined in (16). In the following, we first obtain the joint reliability function of $(\mathcal{T}_1, \mathcal{T}_2)$. Under the assumption that all orders of components failure are equally probable, we have

$$\begin{aligned} P(\mathcal{T}_1 > t_1, \mathcal{T}_2 > t_2) &= \sum_{i=0}^{n} \sum_{j=i+1}^{n} P(I = i, J = j) P(\xi(t_1) < i, \xi(t_2) < j | I = i, J = j) \\ &= \sum_{i=0}^{n} \sum_{j=i+1}^{n} s_{i,j} P(\xi(t_1) < i, \xi(t_2) < j), \end{aligned}$$

in which the second equality follows from the fact that the event $\{I = i, J = j\}$ depends only on the network structure and does not depend on the mechanism of the components failure. In addition, it can be shown, by changing the order of summations, that

$$P(\mathcal{T}_1 > t_1, \mathcal{T}_2 > t) = \sum_{i=0}^{n} \sum_{j=i}^{n} \bar{S}_{i,j} P(\xi(t_1) = i, \xi(t_2) = j), \tag{17}$$

where $\bar{S}_{i,j} = \sum_{k=i+1}^{n-1} \sum_{j=\max\{k,j\}+1}^{n} s_{k,l}$.

Suppose that the component failures occur at random times $\vartheta_1, \ldots, \vartheta_n$ that are corresponding to the first n arrival times of the GCP $\{\xi(t), t \ge 0\}$. Using the fact that the event $(\xi(t) = i)$ occurs if and only if $(\vartheta_i \le t < \vartheta_{i+1})$, it can be shown that

$$P(\mathcal{T}_1 > t_1, \mathcal{T}_2 > t_2) = \sum_{i=0}^{n} \sum_{j=i+1}^{n} s_{i,j} P(\vartheta_i > t_1, \vartheta_j > t). \tag{18}$$

Assuming that the MVF of the GCP is $\Lambda(t) = \lambda t$, Di Crescenzo and Pellerey [15] obtained the PDF of $(\vartheta_1, \ldots, \vartheta_n)$ as

$$f_{\vartheta_1,\ldots,\vartheta_n}(t_1, \ldots, t_n) = \frac{n! \lambda^n}{(1 + \lambda t_n)^{n+1}}, \qquad 0 < t_1 < \cdots < t_n. \tag{19}$$

Using (19), the joint PDF of ϑ_i and ϑ_j is achieved as

$$f_{\vartheta_i,\vartheta_j}(t_i, t_j) = \frac{j! \lambda^j}{(1 + \lambda t_j)^{j+1}} \frac{t_i^{i-1} (t_j - t_i)^{j-i-1}}{(i-1)!(j-i-1)!}, \qquad 0 < t_i < t_j. \tag{20}$$

In the following, we present an example of a three-state network whose components fail according to a GCP with MVF $\Lambda(t) = \lambda t$.

Example 5. *Consider a network with a graph as depicted in Figure 5. The network has 14 links and eight nodes in which the dark nodes are considered to be terminals. Suppose that the nodes are absolutely reliable and the links are subjected to failure.*

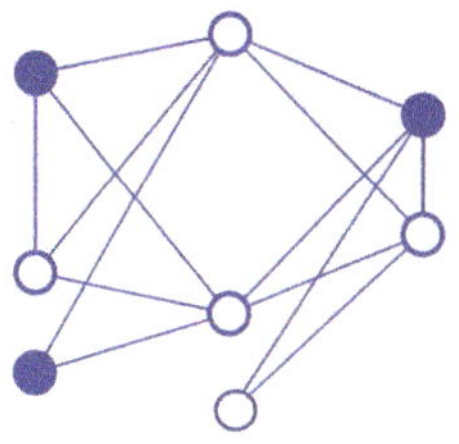

Figure 5. The network with eight nodes, and 14 links.

Assume that the network is in up state if all terminals are connected, in partial performance state if two terminals are connected and in down state if all terminals are disconnected. Let the network components fail according to a GCP with intensity function $\lambda(t) = 1$ and all orders of links failure are equally likely. Figure 6 presents the plot of joint reliability function of the network lifetimes $(\mathcal{T}_1, \mathcal{T}_2)$. The elements $s_{i;j}$ of the two-dimensional signature are calculated using an algorithm by the authors, which can be provided to the readers upon the request.

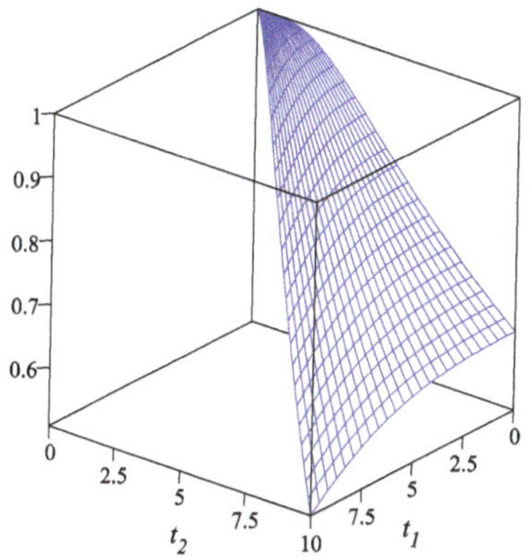

Figure 6. The joint reliability function of $(\mathcal{T}_1, \mathcal{T}_2)$ in Example 5.

In the following theorem, we compare the state lifetimes of two three-state networks. In order to do this, we need the following Lemma.

Lemma 2. *Assume that $\{\xi_1(t), t \geq 0\}$ and $\{\xi_2(t), t \geq 0\}$ are two GCPs with intensity functions $\lambda_1(t) = \lambda_1 > 0$, and $\lambda_2(t) = \lambda_2 > 0$, respectively. Let $\vartheta_{1,1}, \vartheta_{1,2}, \dots$ and $\vartheta_{2,1}, \vartheta_{2,2}, \dots$ denote the arrival times corresponding to the two processes, respectively. If $\lambda_1 \geq \lambda_2$, then $(\vartheta_{1,1}, \dots, \vartheta_{1,n}) \leq_{st} (\vartheta_{2,1}, \dots, \vartheta_{2,n})$ for every $n \geq 1$.*

Proof. Using relation (19), it can be seen that $f_{\vartheta_{i,1},\dots,\vartheta_{i,n}}$ is MTP$_2$, which implies that $\vartheta_{i,1}, \dots, \vartheta_{i,n}$ are associated, $i = 1, 2$. In addition, we have

$$\frac{f_{\vartheta_{2,1},\dots,\vartheta_{2,n}}(t_1, \dots, t_n)}{f_{\vartheta_{1,1},\dots,\vartheta_{1,n}}(t_1, \dots, t_n)} = (\frac{\lambda_2}{\lambda_1})^n (\frac{1 + \lambda_1 t_n}{1 + \lambda_2 t_n})^{n+1}, \quad 0 < t_1 < \dots < t_n,$$

which is increasing in $(t_1, \dots, t_n)$. Therefore, the required result is concluded from Theorem 6.B.8 of [16]. □

Theorem 8. *Consider two three-state networks that each consist of n components having signature matrices $\mathcal{S}_1$ and $\mathcal{S}_2$, respectively. Let the components of ith network fail according to GCP $\{\xi_i(t), t \geq 0\}$, $i = 1,2$ with intensity function $\Lambda_i(t) = \lambda_i$. Let $\vartheta_{i,1}, \vartheta_{i,2}, \ldots$ denote the arrival times corresponding to $\{\xi_i(t), t \geq 0\}$. Suppose that $(\mathcal{T}_1^{(1)}, \mathcal{T}_2^{(1)})$ and $(\mathcal{T}_1^{(2)}, \mathcal{T}_2^{(2)})$ are the corresponding state lifetimes of the two networks, respectively:*

- *If $\lambda_1 \geq \lambda_2$ and $\mathcal{S}_1 \leq_{uo} \mathcal{S}_2$, then $(\mathcal{T}_1^{(1)}, \mathcal{T}_2^{(1)}) \leq_{uo} (\mathcal{T}_1^{(2)}, \mathcal{T}_2^{(2)})$,*
- *If $\lambda_1 \geq \lambda_2$ and $\mathcal{S}_1 \leq_{st} \mathcal{S}_2$, then $(\mathcal{T}_1^{(1)}, \mathcal{T}_2^{(1)}) \leq_{st} (\mathcal{T}_1^{(2)}, \mathcal{T}_2^{(2)})$.*

Proof. From Lemma 2, if $\lambda_1 \geq \lambda_2$, then $(\vartheta_{1,1}, \ldots, \vartheta_{1,n}) \leq_{st} (\vartheta_{2,1}, \ldots, \vartheta_{2,n})$, which implies $(\vartheta_{1,i}, \vartheta_{1,j}) \leq_{st(uo)} (\vartheta_{2,i}, \vartheta_{2,j})$ for all $1 \leq i < j \leq n$.

- Using representation (17), we have

$$\begin{aligned}
P(\mathcal{T}_1^{(1)} > t_1, \mathcal{T}_2^{(1)} > t_2) &= \sum_{i=0}^{n}\sum_{j=i}^{n} \bar{S}_{1,i,j} P(\xi_1(t_1) = i, \xi_1(t_2) = j) \\
&\leq \sum_{i=0}^{n}\sum_{j=i}^{n} \bar{S}_{2,i,j} P(\xi_1(t_1) = i, \xi_1(t_2) = j) \\
&= \sum_{i=0}^{n}\sum_{j=i+1}^{n} s_{2,i,j} P(\vartheta_{1,i} > t_1, \vartheta_{1,j} > t_2) \\
&\leq \sum_{i=0}^{n}\sum_{j=i+1}^{n} s_{2,i,j} P(\vartheta_{2,i} > t_1, \vartheta_{2,j} > t_2) \\
&= P(\mathcal{T}_1^{(2)} > t_1, \mathcal{T}_2^{(2)} > t_2),
\end{aligned}$$

  where the first inequality follows from the assumption that $\mathcal{S}_1 \leq_{uo} \mathcal{S}_2$ and the second inequality follows from $(\vartheta_{1,i}, \vartheta_{1,j}) \leq_{uo} (\vartheta_{2,i}, \vartheta_{2,j})$ for $1 \leq i < j \leq n$.
- Using the fact that $(\vartheta_{1,i}, \vartheta_{1,j}) \leq_{st} (\vartheta_{2,i}, \vartheta_{2,j})$ for all $i < j$ and the assumption that $\mathcal{S}_1 \leq_{st} \mathcal{S}_2$, the required result is concluded from Theorem 3.3 of [21].

□

In the sequel, we investigate the dependency between $\mathcal{T}_1$ and $\mathcal{T}_2$ based on the dependency between RVs I and J. In fact, we show that, if I and J are PQD (associated), then $\mathcal{T}_1$ and $\mathcal{T}_2$ are also PQD (associated). Before that, let

$$s_i^{(1)} = P(I = i), \quad s_j^{(2)} = P(J = j),$$

and $\bar{S}_i^{(k)} = \sum_{l=i+1}^{n} s_l^{(k)}$, $k = 1,2$.

Theorem 9. *Let $\mathcal{T}_1$ be the lifetime of a three-state network in state $K = 2$ and $\mathcal{T}_2$ be the lifetime of the network. Let the components failure of the network appear according to the GCP $\{\xi(t), t \geq 0\}$ with arrival times $\vartheta_1, \ldots, \vartheta_n$.*

- *If I and J are PQD, then $\mathcal{T}_1$ and $\mathcal{T}_2$ are PQD.*
- *If I and J are associated, then $\mathcal{T}_1$ and $\mathcal{T}_2$ are associated.*

Proof. From representation (20), one can show that $f_{\vartheta_i, \vartheta_j}(t_i, t_j)$ is TP_2, which implies that ϑ_i and ϑ_j are associated and PQD.

(a) Let $\phi(\cdot)$ and $\psi(\cdot)$ be two increasing functions. From representation (18), we have

$$\begin{aligned} E(\phi(\mathcal{T}_1)\psi(\mathcal{T}_2)) &= \sum_{i=1}^{n-1}\sum_{j=i+1}^{n} s_{i,j}E(\phi(\vartheta_i)\psi(\vartheta_j)) \\ &\geq \sum_{i=1}^{n-1}\sum_{j=i+1}^{n} s_{i,j}E(\phi(\vartheta_i))E(\psi(\vartheta_j)) \\ &\geq \sum_{i=1}^{n} s_i^{(1)}E(\phi(\vartheta_i))\sum_{j=1}^{n} s_j^{(2)}E(\psi(\vartheta_j)) \\ &= E(\phi(\mathcal{T}_1))E(\psi(\mathcal{T}_2)), \end{aligned}$$

where the first inequality follows from the fact that ϑ_i and ϑ_j are PQD and the second inequality follows from the assumption that I and J are PQD.

(b) Proof of part (b) is the same as the proof of part (a) using the fact that, for every two-variate increasing functions $\phi'(\cdot,\cdot)$ and $\psi'(\cdot,\cdot)$,

$$E(\phi'(\mathcal{T}_1,\mathcal{T}_2)\psi'(\mathcal{T}_1,\mathcal{T}_2)) = \sum_{i=1}^{n-1}\sum_{j=i+1}^{n} s_{i,j}E(\phi'(\vartheta_i,\vartheta_j)\psi'(\vartheta_i,\vartheta_j)).$$

□

The results of the theorem are interesting in the sense that the PQD (associated) property of I and J, which is non-aging and depends only on the structure of the network, is transferred to the PQD (associated) property of $\mathcal{T}_1$ and $\mathcal{T}_2$, which is the aging characteristic of the network.

Example 6. *Consider again the network presented in Example 5. It can be seen that, for every $i, j = 1, \ldots, 14$, $\bar{S}_{i,j} \geq \bar{S}_i^{(1)}\bar{S}_j^{(2)}$. This implies that I and J are PQD. Hence, if the components fail according to a GCP, then $\mathcal{T}_1$ and $\mathcal{T}_2$ are PQD.*

4. Conclusions

In this article, we studied the reliability, aging and stochastic characteristics of an n-component network whose components were subjected to failure according to a geometric counting process (GCP). We first considered the case that the network has two states (up and down). Some mixture representations of the network reliability were obtained in terms of signature of the network and the reliability functions of the arrival times of the GCP. We studied the conditions under which the hazard rate of the network is increasing in the case that intensity function of the process is increasing. Stochastic comparisons were made between the lifetimes of different networks, subjected to GCPs, based on the stochastic comparisons between their signatures. The residual lifetime of the network was also explored. In the second part of the paper, we considered the networks with three states: up, partial performance, and down. The components of the network were assumed to fail based on a GCP with mean value function $\Lambda(t) = \lambda t$, which leads to the change of the network states. Under these circumstances, we arrived at several stochastic and dependency properties of the networks with the same and different structures. The results of Section 3 were obtained under the special case that the MVF of the GCP is $\Lambda(t) = \lambda t$. The developments of the paper were mainly dependent on the notions of signature and two-dimensional signature. The results of the paper were illustrated by several examples.

Author Contributions: The collaboration of the authors in conceptualization, methodology and formal analysis of the paper are in the order of names appeared in the article.

Funding: Asadi's research work was carried out in the IPM Isfahan branch and was in part supported by a grant from IPM (No. 96620411).

Acknowledgments: The authors would like to thank the editor and two anonymous reviewers for their comments and suggestions that led to improving the exposition of the article.

Conflicts of Interest: The authors declare no conflict of interest.

References

1. Ashrafi, S.; Asadi, M. On the stochastic and dependence properties of the three-state systems. *Metrika* **2015**, *78*, 261–281. [CrossRef]
2. Ashrafi, S.; Asadi, M. Dynamic reliability modeling of three-state networks. *J. Appl. Prob.* **2014**, *51*, 999–1020. [CrossRef]
3. Ashrafi, S.; Zarezadeh, S. A Shock-Based Model for the Reliability of Three-State Networks. *IEEE Trans. Reliab.* **2018**, *67*, 274–284. [CrossRef]
4. Eryilmaz, S. Mean residual and mean past lifetime of multi-state systems with identical components. *IEEE Trans. Reliab.* **2010**, *59*, 644–649. [CrossRef]
5. Eryilmaz, S.; Xie, M. Dynamic modeling of general three-state k-out-of-n:G systems: Permanent based computational results. *J. Comput. Appl. Math.* **2014**, *272*, 97–106. [CrossRef]
6. Huang, J.; Zuo, M.J.; Wu, Y. Generalized multi-state k-out-of-n:G systems. *IEEE Trans. Reliab.* **2000**, *49*, 105–111. [CrossRef]
7. Lisnianski, A.; Levitin, G. *Multi-State System Reliability: Assessment, Optimizationand Applications;* World Scientific Publishing: Singapore, 2003.
8. Tian, Z.; Yam, R.C.M.; Zuo, M.J.; Huang, H.Z. Reliability bounds for multi-state k-out-of-n systems. *IEEE Trans. Reliab.* **2008**, *57*, 53–58. [CrossRef]
9. Zhao, X.; Cui, L. Reliability evaluation of generalized multi-state k-out-of-n systems based on FMCI approach. *Int. J. Syst. Sci.* **2010**, *41*, 1437–1443. [CrossRef]
10. Zuo, M.J.; Tian, Z. Performance evaluation for generalized multi-state k-out-of-n systems. *IEEE Trans. Reliab.* **2006**, *55*, 319–327. [CrossRef]
11. Gertsbakh, I.; Shpungin, Y. *Network Reliability and Resilience;* Springer: Berlin, Germany, 2011.
12. Zarezadeh, S.; Asadi, M. Network reliability modeling under stochastic process of component failures. *IEEE Trans. Reliab.* **2013**, *62*, 917–929. [CrossRef]
13. Zarezadeh, S.; Asadi, M.; Balakrishnan, N. Dynamic network reliability modeling under nonhomogeneous Poisson processes. *Eur. J. Oper. Res.* **2014**, *232*, 561–571. [CrossRef]
14. Cha, J.H.; Finkelstein, M. A note on the class of geometric counting processes. *Prob. Engin. Inform. Sci.* **2013**, *27*, 177–185. [CrossRef]
15. Di Crescenzo, A.; Pellerey, F. Some results and applications of geometric counting processes. *Methodol. Comput. Appl. Probab.* **2018**.10.1007/s11009-018-9649-9. [CrossRef]
16. Shaked M.; Shanthikumar, J.G. *Stochastic Orders*; Springer: Berlin, Germany, 2007.
17. Lindqvist, B.H.; Samaniego, F.J.; Huseby, A.B. On the equivalence of systems of different sizes, with applications to system comparisons. *Adv. Appl. Prob.* **2016**, *48*, 332–348. [CrossRef]
18. Navarro, J.; Samaniego, F.J.; Balakrishnan, N. Signature-based representations for the reliability of systems with heterogeneous components. *J. Appl. Prob.* **2011**, *48*, 856–867. [CrossRef]
19. Samaniego, F.J. *System Signatures and their Applications in Engineering Reliability;* Springer: Berlin, Germany, 2007.
20. Samaniego, F.J.; Navarro, J. On comparing coherent systems with heterogeneous components. *Adv. Appl. Prob.* **2016**, *48*, 88–111. [CrossRef]
21. Belzunce, F.; Mercader, J.A.; Ruiz, J.M.; Spizzichino, F. Stochastic comparisons of multivariate mixture models. *J. Multivariate Anal.* **2009**, *100*, 1657–1669. [CrossRef]

MDPI
St. Alban-Anlage 66
4052 Basel
Switzerland
Tel. +41 61 683 77 34
Fax +41 61 302 89 18
www.mdpi.com

Mathematics Editorial Office
E-mail: mathematics@mdpi.com
www.mdpi.com/journal/mathematics

www.ingramcontent.com/pod-product-compliance
Lightning Source LLC
Chambersburg PA
CBHW051721210326
41597CB00032B/5558
* 9 7 8 3 0 3 9 2 1 7 2 8 1 *